Hierarchical Protection for Smart Grids

Hierarchical Protection for Smart Grids

Jing Ma and Zengping Wang

State Key Laboratory of Alternate Electrical Power Systems with Renewable Energy Sources
North China Electric Power University
Beijing, China

This edition first published 2018 by John Wiley & Sons Singapore Pte. Ltd under exclusive licence granted by Science Press for all media and languages (excluding simplified and traditional Chinese) throughout the world (excluding Mainland China), and with non-exclusive license for electronic versions in Mainland China.

Registered Offices
John Wiley & Sons, Inc., 111 River Street, Hoboken, NJ 07030, USA
John Wiley & Sons Singapore Pte. Ltd, 1 Fusionopolis Walk, #07-01 Solaris South Tower, Singapore 138628

Editorial Office
1 Fusionopolis Walk, #07-01 Solaris South Tower, Singapore 138628

For details of our global editorial offices, customer services, and more information about Wiley products visit us at www.wiley.com.

Wiley also publishes its books in a variety of electronic formats and by print-on-demand. Some content that appears in standard print versions of this book may not be available in other formats.

Library of Congress Cataloging-in-Publication Data

Names: Ma, Jing (Electrical engineer), author. | Wang, Zengping, author.
Title: Hierarchical protection for smart grids / by Jing Ma and Zengping Wang.
Description: Hoboken, NJ : John Wiley & Sons, 2018. | Includes
bibliographical references and index. |
Identifiers: LCCN 2017016757 (print) | LCCN 2017036536 (ebook) |
ISBN 9781119304838 (pdf) | ISBN 9781119304821 (epub) | ISBN 9781119304807 (cloth)
Subjects: LCSH: Smart power grids. | Electric power distribution–Security measures.
Classification: LCC TK3105 (ebook) | LCC TK3105 .M3 2017 (print) | DDC 621.31/7–dc23
LC record available at https://lccn.loc.gov/2017016757

Cover design by Wiley
Cover image: © Spectral-Design/Gettyimages

Set in 10/12pt Warnock by SPi Global, Pondicherry, India
Printed in Singapore by C.O.S. Printers Pte Ltd

10 9 8 7 6 5 4 3 2 1

Contents

About the Author

Professor Jing Ma has been working in this area since 2003. His research mainly concentrates on power system analysis/control/protection/operation/modelling/simulation and smart grids. For more than 12 years he has carried out systematic research and practice on power system hierarchical protection, especially the approaches of local and substation area protection and studies on wide area protection. He was the first to introduce the two-terminal network and voltage drop equation to the areas of local protection, and also the first to apply limited overlapping multiple differential regions for the protection of a substation area. He has invented a variety of wide area protection strategies using electrical information and operating signals to establish a wide area protection system with high accuracy and efficiency. A series of papers has been published in authoritative journals such as *IEEE Transactions on Power Systems* and *IEEE Transactions on Power Delivery*. The work has been widely acknowledged and cited by international peers, and some of his research results have been used in many practical engineering projects to accelerate the application and spread of wide area control technology. In recent years, he has undertaken many major projects in China, such as guiding a project of the National Natural Science Foundation of China to study wide area backup protection. He set up an advanced real-time dynamic simulation laboratory for fault transient analysis of power systems, and pioneered the design and realization of the corresponding protection techniques. He has also been responsible for several projects for governments and enterprises on the study of the hierarchical protection in smart grids and was also a major member of the National Basic Research Program of China (973 Program) on the study of wide area protection and control for complicated power systems. He cooperated with China Electric Power Research Institute in guiding the study of integrated protection systems of the substation area and wide area. He has taught courses on power system protective relaying for years, and much of the material in this book has been taught to students and other professionals.

Foreword

We were pleased to have Dr Jing Ma visit us at Virginia Tech as a visiting researcher. Virginia Tech has made very significant contributions to the field of phasor measurements and wide area measurement systems and their application to practical power system problems. Dr Jing Ma was an active participant in our research in this area, and worked very well with our research team.

I am glad to see that he is now publishing a book on *Hierarchical Protection for Smart Grids*. He is very well qualified to write such a book, dealing with topics on the protection of power systems. After reviewing traditional protection topics, this book goes into protection of renewable energy systems, substation area protection, and wide area protection principles. Many new ideas are presented in these later chapters, and I am sure they will be carefully studied by serious students and researchers.

It is significant that Dr Jing Ma is working closely with power system engineers and utility companies of China. This is one of the key characteristics of a successful engineering professor: to get the opinion of practising power system engineers on the direction and results of his research.

I am well familiar with the very active research programme at the North China Electric Power University, and this work is a testament to the vibrant research traditions of this university. I expect to see other research results from Dr Jing Ma and his team in the coming years.

Following some of the major power blackouts in recent years, the efficacy of PMUs and WAMS to help identify causes and provide countermeasures for dealing with widespread disturbances on power grids has been well recognized. These measurements have provided a very accurate situational awareness of the current state of the grids, as well as their vulnerabilities. The promising applications arising out of this early work are improved monitoring, protection and control of the power grids. Within a span of about 20 years after the invention of PMUs, many research teams around the world have been developing applications of this technology, and their work will surely lead to improved performance of electric power grids in the coming years.

This book, *Hierarchical Protection for Smart Grids* presents a comprehensive view of synchronized phasor measurement technology and its applications. It combines academic rigour with pragmatic considerations in dealing with the emerging discipline of smart grid technologies. I am pleased to see the work presented in this book, and I am sure it will be a valuable reference for students, researchers and practising power system engineers of the future.

Prof. Arun G. Phadke
University Distinguished Research Professor
Department of Electrical and Computer Engineering
Virginia Tech
Blacksburg, Virginia
USA

Preface

As the construction of smart grids is being vigorously promoted, hierarchical protection has been proposed and has quickly become the focus of research. This book is the very first to conduct a comprehensive discussion on smart grid hierarchical protection and a detailed analysis of specific protection schemes.

With the integration of large-scale renewable energy and the development of AC/DC hybrid EHV/UHV interconnected power grids, it is difficult for stage protection to utilize only local information to adapt to the changeable network structure and operating mode. Meanwhile, the integration of distributed generation causes the distribution network to change from a single-source radiation network to a double-source or multi-source network, where the distribution of power flow and the size and direction of fault current change fundamentally. Thus, the original protection schemes based on a fixed setting value have major limitations in striking an effective balance between the sensitivity and selectivity of relay protection. In many blackouts in China and elsewhere, the improper operation of protection is usually one of the main causes of fault occurrence and expansion, which can eventually contribute to the collapse of a power grid.

Without the limitations of local information, hierarchical relay protection could solve the above problems in traditional protection from the global perspective of a power system. In recent years, many colleges, universities and power companies have actively participated in the exploration of hierarchical protection. A lot of progress has been made in theoretical research, together with some local engineering demonstrations, which have laid the foundation for the construction of hierarchical protection and overcoming the difficulties in traditional relay protection. On the basis of summarizing the existing research findings and learning from the experience and lessons of traditional relay protection, with research achievements of the author as the main body, this book conducts a forward-looking discussion of the key technical problems of hierarchical protection construction in particular breadth and depth – including the constitution mode of hierarchical protection, local area protection, substation area protection and wide area protection – trying to point to an evolutionary direction for the construction of hierarchical protection.

This book strives to make the explanation of basic theories understandable and the derivation of formulas rigorous and complete. On this basis, through large numbers of case studies, rigorous verification of the hierarchical protection schemes introduced in the book which fit engineering practice is conducted.

It should be noted that, since hierarchical protection is still in the ascendant, the illustrations in the book may not be the final solution. For problems that have not yet achieved a unified understanding, the author proposes distinctive options in this book, in the sincere hope that readers will be inspired to make more excellent research achievements. Due to limited space, problems that cannot be discussed in detail are included in the references for in-depth study by readers.

This book is applicable to graduate students in universities, scientific and technical personnel in research institutes and professional personnel with a degree of theoretical knowledge and practical experience, for scientific research on hierarchical protection and relevant technical innovation. Due to the limited knowledge of the author, mistakes are inevitable in the book, and any criticism and correction from readers will be welcomed.

I'd like to express sincere thanks for the great support from the National Natural Science Foundation of China (No. 51277193,50907021), National Basic Research Program of China (973 program) (No. 2012CB215200), the Chinese University Scientific Fund Project (No. 2014ZZD02), Henry Fok Education Fund (No. 141057), Beijing Metropolis Beijing Nova Program (Z141101001814012), the Excellent Talents in Beijing City (2013B009005000001) and the Fund of Fok Ying Tung Education Foundation (141057) in the process of the creation of this book.

Jing Ma
Beijing, China
2016

Introduction

In this book, the latest research results on local area protection, station area protection and wide area protection in the smart grid are introduced systematically.

This book is divided into six chapters. The first mainly introduces the basic theories of relay protection and the constitution mode of hierarchical protection. The second chapter introduces local area conventional protection such as transformer and line protection. Chapter 3 seeks to introduce the fault characteristics of renewable power generation and local area renewable energy protection. The following chapter introduces topology analysis and fault tolerance identification. Chapter 5 introduces substation area protection based on logical and electrical variables, while the final chapter introduces wide area protection based on logical and electrical variables.

Rich and novel in content, this book covers almost every aspect of smart grid hierarchical protection research. It could provide a useful reference for teachers and students of electrical engineering in colleges and universities as well as scientific and technical researchers.

1

Basic Theories of Power System Relay Protection

1.1 Introduction

As the first defence line to ensure the security of a power grid, relay protection is very important for the fast isolation of faults and the effective control of fault expansion [1,2]. However, in recent years, with the continuous integration of large-scale renewable energy sources, the structure of modern power grids has become more and more complex, and more and more problems in traditional relay protection have been exposed, such as difficulties in backup protection setting and cooperation, unexpected changes of grid structure or operating conditions which may cause protection to malfunction or refuse to operate, and cause major load transfer which easily leads to cascading, tripping and even blackout, etc.

To solve these problems, the smart grid hierarchical relay protection system was proposed. Hierarchical relay protection is based on the smart grid [3] and information sharing technology, and is composed of bay level protection, substation area protection and wide area protection. Bay level protection, which is also called local area protection, aims to realize primary protection for components in the substation through independent and decentralized configuration. Substation area protection aims to realize backup protection for components in the substation by centralizing the information of components to the substation host computer. Wide area protection aims to realize local backup protection and remote backup protection between substations through the interaction of information between relevant substation protection units.

In this chapter, first the basic theories of power system relay protection are introduced, the functions and basic requirements of relay protection are summarized, and the basic principles of relay protection are illustrated. Then, the composition mode of hierarchical relay protection is analysed in detail. The cooperation between local area protection, substation area protection and wide area protection and the protection range of each are discussed, laying the foundation for subsequent chapters in this book.

1.2 Function of Relay Protection

A power system is an energy transmission network composed of various electrical devices corresponding to electric power production, transformation, transmission, distribution and use, which are connected according to certain technical and

Hierarchical Protection for Smart Grids, First Edition. Jing Ma and Zengping Wang.

economic requirements. Generally, the devices through which the electric power flows are called the primary equipment of the power system: for example, the generator, transformer, circuit breaker, bus, transmission line, compensation capacitor, shunt capacitors, shunt reactors, motor and the other power consumption equipment. The devices for the monitoring, measuring, control and protection of the operating state of the primary equipment are called the secondary equipment of the power system. Through voltage and current transformers, the high voltage and large current signals of the primary equipment are converted in proportion to low voltage and small current signals for the secondary equipment [4].

The operating state of a power system can usually be described by the operating parameters. The main operating parameters include active power, reactive power, voltage, current, frequency and the angular difference between emf phasors. According to different operating conditions, the operating state of a power system can be divided into normal state, abnormal state and fault state.

When a power system is in normal operation, the primary equipment and main operating parameters are all within the allowed deviation ranges, and the power system can operate continuously to provide electric power. However, when a disturbance occurs in a power system, the balance of the main operating parameters will be broken, and the power system operating state will change.

After a power system has been disturbed, then, according to the degree of disturbance, two circumstances may result. One is that the power system transits from the original stable state to a new stable state, the deviation of operating parameters from normal values remaining within the allowed ranges – for example, an increase or decrease of load, or the regulation of the prime mover – and the system could continue in normal operation. The other is that when a fault occurs in a power system, the operation of the system will change dramatically, resulting in local failure of the power system, electrical equipment and normal power supply to electricity users, even global failure.

If there is a fault state, and no special measures are taken, it is difficult to restore the system to normal operation, which could have a major impact on industrial and agricultural production, national defence, construction or the lives of ordinary people. Many types of fault can occur in a power system, including short circuit, phase disconnection and successive occurrence of multiple faults. The most common and most dangerous faults are various forms of short circuit, including three-phase short circuit, phase-to-phase short circuit, two-phase grounding fault, single-phase grounding fault, and motor and transformer winding turn-to-turn short circuit. In addition, there may be disconnection of one phase or two phases and complex faults such as some of the above faults occurring in succession.

Since the devices in a power system are connected to each other, a fault on one device will soon affect the other parts of the system. Thus, the time to clear the faulty device must be very short, sometimes even as short as tens of milliseconds, i.e. a small number of cycles. In such a short period of time, it is impossible for the operating staff to identify the fault and clear the faulty device. Automatic devices are needed to do that, i.e. relay protection devices.

A relay protection device is an automatic device installed on the components of the whole power system, which can respond to various faults or abnormal operating states of electrical components in the designated area quickly and accurately, and operate within the preset time limit to issue tripping signals to the circuit breaker. The term 'relay protection' generally refers to the relay protection technology or relay protection system composed of various relay protection devices.

The basic tasks of relay protection are [5–8]:

1) Clear the faulty components from the power system automatically, quickly and selectively, and ensure that the non-faulty parts remain in normal operation.
2) Respond to the abnormal state of the electrical devices and issue signals to inform the duty personnel, or automatically make adjustments, even issuing tripping commands.

When a power system is in normal operation, relay protection does not operate, it simply monitors the operating state of the power system and the components. Once a fault or abnormal operating state is detected, relay protection will quickly operate to isolate the fault and issue a warning to ensure the safety of the power system. Relay protection plays an important role in ensuring the safe operation of the system and its power quality, and preventing the expansion and occurrence of faults.

1.3 Basic Requirements of Relay Protection

Technically, relay protection which operates to trip switches should meet four basic requirements, i.e. reliability, selectivity, speed and sensitivity. These four basic requirements are the important criteria to analyse, evaluate and study relay protection.

1.3.1 Reliability

The reliability of relay protection refers to the capability of the relay protection to operate reliably when a fault occurs within the protection range, without any refusal to operate, and not malfunctioning in any case where protection should not operate.

Reliability is the basic requirement of relay protection. It depends on the design, manufacture and operational maintenance levels. To ensure reliability, protection schemes with performance that meets the requirements and with simple principles should be used. Reliable hardware and software with anti-jamming capability should be used to form the protection device. In addition, there should be essential automatic detection, locking and warning measures, with convenient setting, debugging and operational maintenance.

An important index for evaluating the reliability of relay protection is the correct operational rate of relay protection, which is calculated as follows:

$$R_{\mathrm{c}} = \frac{k_{correct}}{k_{total}} \times 100\% \tag{1.1}$$

where R_c is the correct operational rate of relay protection, $k_{correct}$ is the correct operational times of relay protection, k_{total} is the total operational times of relay protection, which includes the correct operational times, malfunctioning times and refusing-to-operate times.

1.3.2 Selectivity

Selectivity means that a particular fault should be cleared by the protection of the faulty device itself, and only when the protection or circuit breaker of the faulty device refuses to operate will protection of the adjacent device or the breaker failure protection be allowed to clear the fault. Thus, fault clearance can be limited to the minimum range, and the safe operation of the non-faulty part of the system is guaranteed.

To ensure selectivity, apart from using a time delay to make the backup protection and primary protection of a line cooperate correctly with each other, the correct cooperation between the backup protection of adjacent components also needs to be considered. On the one hand, the sensitivity of backup protection of the higher-level component should be lower than that of the lower-level component. On the other hand, the operational time of backup protection of the higher-level component should be longer than that of the lower-level component.

1.3.3 Speed

Speed refers to clearing the fault as quickly as possible, in order to improve the stability of the system, reduce the damage to equipment, limit the range affected by the fault and improve the effectiveness of power restoration.

The main reasons for the requirement for fast fault clearance are as follows:

1) The heating power and electrodynamic force that affect the degree of equipment damage are both proportional to the fault clearance time. The shorter the fault clearance time, the more beneficial it is to the reduction of equipment damage.
2) The longer the short circuit point arc is ignited, the more likely the fault is to expand. A single-phase grounding short circuit fault may develop into a phase-to-phase short circuit fault, or even to a three-phase short circuit fault, which is more damaging to system stability; an instantaneous short circuit fault that could be restored may develop into a permanent short circuit fault that cannot be restored.
3) It is beneficial to improving the power restoration effect of automatic devices such as auto-reclosing and standby power auto-switching, as well as the self-start and restoration of a motor.

The requirement on speed should be determined according to system stability, wiring and specific conditions of the protected device. Improving the operating speed must be on the premise that reliability is satisfied. When the requirement on operating speed is met, slightly slowing down the operating speed means more time to obtain electrical information, which is beneficial to improving the reliability of relay protection.

The fault clearance time is the sum of protection operating time and circuit breaker operating time. The operating time of fast protection is usually 0.06~0.12 s; the shortest could be 0.01~0.04 s. The circuit breaker operating time is usually 0.06~0.15 s; the shortest could be 0.02~0.06 s.

1.3.4 Sensitivity

Sensitivity of relay protection refers to the capability of responding to a fault or abnormal operating state in the protected range. A protection device that meets the requirement on sensitivity can respond correctly to any short circuit fault within the preset protection range under any system operating conditions, no matter what type of fault, where the fault is located, whether via fault resistance or not. This is usually measured by the sensitivity coefficient.

1) For protection that operates in response to an increase of an electrical variable, the sensitivity coefficient is:

$$\textit{sensitivity coefficient} = \frac{I_{\mathrm{f\,min}}}{I_{\mathrm{set}}} \times 100\% \tag{1.2}$$

   where I_{set} is the setting value of the relay protection, $I_{\mathrm{f\,min}}$ is the minimum short circuit current calculated value when a metallic short circuit occurs within the protection zone.
2) For protection that operates in response to a decrease of an electrical variable, the sensitivity coefficient is:

$$\textit{sensitivity coefficient} = \frac{I_{\mathrm{set}}}{I_{\mathrm{f\,max}}} \times 100\% \tag{1.3}$$

   where I_{set} is the setting value of the relay protection, $I_{\mathrm{f\,max}}$ is the maximum calculated short circuit current value when a metallic short circuit occurs within the protection zone.

The four requirements form the basis of the analysis, evaluation and research on relay protection. The emphasis on each of the four requirements should be 'modest' and balanced, and meet power system safe operational standards, otherwise adverse effects will result. Usually, selectivity and speed are contradictory; sensitivity and reliability are contradictory; anti refusing-to-operate and anti malfunction are contradictory. Therefore, the four requirements should be in dialectical unity in the configuration of protection, according to the actual operating conditions of a power system and the function of the protected equipment. The scientific research, design, manufacture and operation of relay protection are mostly centred on how to deal with the dialectical unity between the four requirements. It is of vital importance that protection devices are configured with the same principle on components at different locations in the power system and that they cooperate; it should be possible to maximize the way of configuring the corresponding relay protection for the same power component installed at different locations in a power system to preserve the operational efficiency of the protected power system. These issues fully demonstrate the scientific nature of relay protection theoretical research and the technical nature of relay protection engineering practice.

1.4 Basic Principles of Relay Protection

1.4.1 Over-Current Protection

When three-phase and phase-to-phase short circuit faults occur at any point in a power system, the approximate formula for calculating the power frequency periodic component of the short circuit current on the line between the fault point and the power source is:

$$I_k = \frac{E_\varphi}{Z_\Sigma} = K_\varphi \frac{E_\varphi}{Z_S + Z_k} \tag{1.4}$$

where E_φ is the phase emf of the system equivalent power source. Z_k is the impedance between the fault point and the relaying point. Z_S is the impedance between the relaying point and the system equivalent power source. K_φ is the short circuit type coefficient; for three-phase short circuit $K_\varphi = 1$, for phase-to-phase short circuit $K_\varphi = \frac{\sqrt{3}}{2}$. As there is a change in (a) the startup mode of a power system, (b) the topology of the network between the relaying point and the power source and (c) the load level, E_φ and Z_S will both change. As the distance between the fault point and relaying point and the short circuit type differ, the values of Z_k and K_φ will also differ, and so will the short circuit current.

Within the protection range, the amplitude of the short circuit current is always bigger than that of the load current. Over-current protection distinguishes between the normal operating state and the short circuit state according to the amplitude of the current at the relaying point. The principle is simple and reliable, and is easy to implement. The operating equation of over-current protection is:

$$I > I_{set} \textit{ fault occurs within the protection range, tripping} \tag{1.5}$$

where I is the current at the relaying point and I_{set} is the setting value of over-current protection. The amplitude of current at the relaying point is closely related to Z_S, E_φ and K_φ, and varies with the distance between the fault point and the equivalent power source. The longer the distance, the smaller the current.

1.4.2 Directional Current Protection

Over-current protection only uses the increase of current amplitude after phase-to-phase short circuit to distinguish between faulty and normal operating states. It is difficult to apply this principle to multi-source networks. For the double-source system shown in Figure 1.1, since there are power sources on both sides, circuit breakers and protection devices are installed at both ends of line AB.

Suppose the capacities of the two power sources are different, and the fault current at relay B in the case of a k1 fault is smaller than the fault current at relay B in the case of a k2 fault. In order to protect line AB, the setting value of over-current

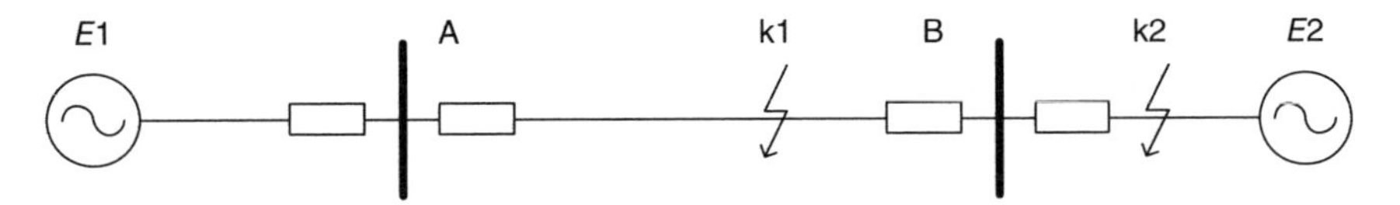

Figure 1.1 Double-source system.

protection at relay B must be smaller than the fault current at relay B in the case of a k2 fault, thus when a fault occurs at k2, over-current protection at relay B will malfunction. This problem could be solved by adding directional protection at relay B.

Directional protection can identify the direction of short circuit power flow, and it operates only when the power flows from bus to line (forward direction). Combined with over-current protection, directional current protection uses not only the amplitude of the current, but also the direction of power flow, and thus it can clear the fault quickly and selectively.

Directional current protection identifies the direction of the fault by measuring the angular difference between current and voltage. Since the line is inductive, when a fault occurs in the forward direction, the short circuit current at the relay will lag behind the bus voltage by a phase angle φ_k (i.e. the impedance angle of line from bus to fault point), and $0° < \varphi_k \leq 90°$. The output of power direction component varies with the angular difference between the input voltage and current. For the power direction component to be the most sensitive to the most common short circuit faults, the maximum sensitivity angle should be $\varphi_{\text{sen}} = \varphi_k$. And to ensure that the directional component can operate reliably when the fault resistance causes line impedance angle φ_k to vary $0°{\sim}90°$, the operating angle should be in a certain range, and is usually equal to φ_k. The operating equation of directional current protection can be expressed as:

$$\varphi_{\text{sen}} + 90° > \arg\frac{\dot{U}_{\text{m}}}{\dot{I}_{\text{m}}} > \varphi_{\text{sen}} - 90° \tag{1.6}$$

where $\dot{U}_{\text{m}}$ is the measured voltage of protection, and $\dot{I}_{\text{m}}$ is the measured current of protection.

The direction component should operate reliably when various kinds of faults occur in the forward direction, and should not operate when a fault occurs in the reverse direction. Rather, it should be sensitive to faults in the forward direction [9].

1.4.3 Distance Protection

The protection range and sensitivity of current protection are greatly affected by variations in system operating mode. In order to meet the requirements of a complex network, relay protection principles with better performance must be applied, among which is distance protection. Distance protection has a relatively stable protection range and can identify the direction of the short circuit point, so it is widely applied in power systems [10,11].

Current protection only uses the single feature of current increase in a short circuit fault, while distance protection uses the dual features of voltage decrease and current increase in a short circuit fault. Calculating the ratio of measured voltage to measured current reflects the distance from the fault point to the relaying point, and then distance protection will operate if the calculated value is smaller than the setting value.

In distance protection, the measured impedance Z_m is the ratio of measured voltage $\dot{U}_m$ to measured current $\dot{I}_m$ at the relaying point, i.e.

$$Z_m = \frac{\dot{U}_m}{\dot{I}_m} \tag{1.7}$$

where Z_m is a vector, which could be expressed in the form of polar coordinates, or in the form of rectangular coordinates on the complex plane.

When the power system is in normal operation, $\dot{U}_m$ is approximately the rated voltage, $\dot{I}_m$ is the load current and Z_m is the load impedance. The magnitude of the load impedance is relatively large, and the impedance angle is the power factor angle, which is relatively small; thus the impedance is mostly resistive. When a fault occurs in a power system, $\dot{U}_m$ decreases, $\dot{I}_m$ increases and Z_m becomes the line impedance between the fault point and the relaying point, the magnitude of which is relatively small, and the impedance angle is the line impedance angle (relatively large). According to the variation of magnitude and phase angle of Z_m, distance protection could identify whether a fault has occurred or not.

In an ideal case, when metallic short circuit faults occur at different locations, the measured impedance is the complex variable on the line AB in Figure 1.2. Note, however, that the secondary side measured impedance is affected by the errors of current and voltage transformers and transmission line impedance angle, as well as the fault resistance. Usually, the protection range of an impedance component is expanded to the form of a circle, as shown in Figure 1.2, the circular area being referred to as the operational characteristics. When the measured impedance falls in the range of the operational characteristics, distance protection will operate. In Figure 1.2, Z_{set} is the diameter of the circle, which represents the setting impedance of distance protection, and φ_{set} is the maximum sensitivity angle.

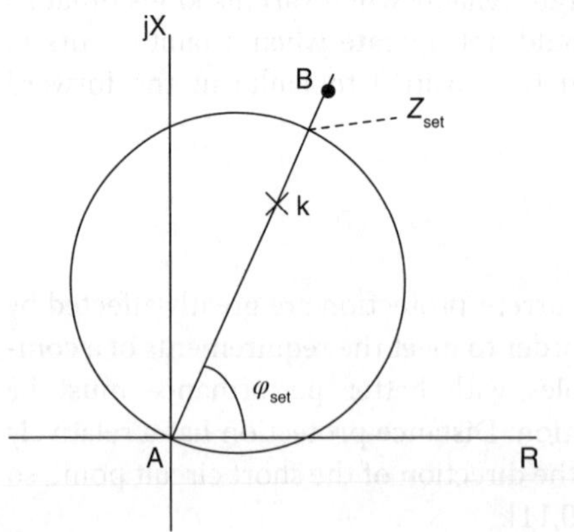

Figure 1.2 Operational characteristics of distance protection.

Although distance protection is not affected by the system operating mode, for distance protection in a high-voltage system, the reliability will be affected by a variety of unfavourable factors, including system oscillation, overload and fault resistance. These factors will cause distance protection to malfunction or refuse to operate, posing a hidden threat to the safe operation of a large-scale complex power grid. Thus, corresponding countermeasures need to be studied to eliminate or reduce the impact of these factors.

The main protection principles in traditional protection have been introduced above. In a single-source network, over-current protection could achieve a good protection effect. In a loop network or double-source network, directional protection should be added to complement over-current protection. In high-voltage and extra-high-voltage (EHV) power grids, distance protection is less effective, so a system operating mode should be applied. However, although the performance of distance protection is good, when oscillation or overload occurs in the system, or when there is fault resistance, using distance protection also runs the risk of incorrect operation.

1.5 Hierarchical Relay Protection

Due to constraints such as the microcomputer protection hardware device, CPU processing ability and specialization, traditional relay protection has used the decentralized and independent configuration mode, and has achieved successful operation in practice over quite a long period of time. However, with the development of power systems, the structure and operating mode of power grids has become more and more complex, and more and more problems with traditional relay protection have been discovered.

1) The complicated cooperation between protection components, and long operating time delay have adverse effects on system stability.
2) Problems include: difficulty of setting up, lack of adaptability to variation of system operating mode and protection mismatch and lack of sensitivity.
3) They are incapable of distinguishing between internal faults and overload caused by power flow transfer, which easily leads to cascading tripping and blackout, greatly endangering the safety of the power grid.

With the development of smart grids and information sharing technology, relay protection can get more information from more extensive sources, which brings new opportunities for solving the problems of traditional relay protection. In recent years, scholars have put forward a novel hierarchical relay protection system which is composed of substation bay layer local protection (hereinafter referred to as local area protection), substation layer protection (hereinafter referred to as substation area protection) and wide area layer protection (hereinafter referred to as wide area protection). Thus, a vertically hierarchical relay protection system may be realized, and through the combination of time and the coordination of the operational information between areas, the performance of relay protection could be improved. Different levels of the hierarchical relay protection system are illustrated as follows.

1.5.1 Local Area Protection

Local area protection is oriented to a single protected object. By retaining conventional primary protection and simplifying the backup protection, with the aim of isolating the fault component quickly and accurately, local area protection can realize fast and reliable protection of components using the information about the protected object itself to make independent decisions. Local area protection is not dependent on external communication channels, thus, even if the communication channel is damaged, it can still complete its protection function. Local area protection mainly includes line protection, main transformer protection and bus protection.

There are two main reasons for retaining local area protection. On the one hand, local area protection is the implementation layer for hierarchical protection. In view of the reliability and speed requirements of the main protection, substation area protection or wide area protection are difficult to use to replace the main protection or integrated main protection. Thus, currently, substation area protection and wide area protection are mainly for backup protection. On the other hand, substation area protection and wide area protection are both dependent on the reliability of the communication system. When communication fails, the backup functions of both will fail. Therefore, local area protection must be retained as the last 'life defence line' of a power system to provide the last backup measure for fault clearance in the case of communication system failure.

Apart from completing its own protection function, local area protection also communicates with the substation area protection to provide organizational information, including protection information, measurement and control information, fault information and electric power information, which are generated by relevant functional equipment in the bay layer, as the data source for device function application and information transmission in the substation control layer. The information is transmitted in the substation in the form of messages, according to the data interface model defined by the IEC 61850 standard [12–15]. The substation is divided into the substation control layer, bay layer and process layer. The information exchange between bay layer and process layer is mainly through the process layer network, including the sampling value information and switch status information. The information exchange between bay layer and substation control layer is mainly through the substation control layer network, including protection operation messages, operation of device and warning and operation and control commands.

In summary, local area protection retains the conventional main protection function, simplifies the backup protection function and improves the performance of conventional backup protection with complicated setting and coordination. Meanwhile, as a node of substation area protection, local area protection is integrated into the substation area protection system and cooperates with substation area protection to realize substation level protection. Local area protection is the cornerstone of hierarchical protection, providing data support for and executing the decisions of substation area protection.

With transmission of power from west to east China, the construction of an EHV power grid and the integration of large-scale renewable energy, local area protection using only the information of a single component will experience many problems. For conventional local area protection, the excitation inrush current of parallel transformer no-load input can cause transformer differential protection to malfunction; the strong zero-sequence mutual inductance coupling between double-circuit lines on the same tower, and complicated cross-line faults will directly affect the operational characteristics and performance of distance protection or directional protection. For local area renewable energy protection, such as the new energy station protection of centralized power generation, if the transient regulation and weak feed characteristics of a new energy current converter are not considered, the collector line protection may malfunction and the sensitivity of outgoing transmission line protection may be reduced.

This book will discuss in detail the problems in conventional local area protection and the new local area energy protection and will introduce corresponding solutions in order to improve the function of local area protection and lay a solid foundation for the construction of hierarchical protection.

1.5.2 Substation Area Protection

Substation area protection is mainly responsible for the protection of components inside the station and backup protection of the tie line between stations. According to the multi-source electrical variable and logical variable information shared by the generic object oriented substation event (GOOSE) network and the sampled value (SV) network [16], the substation host computer could accurately identify substation faults and quickly clear them.

The biggest advantage of substation area protection over traditional protection is that it utilizes multi-source information to make comprehensive decisions and judgements. The comprehensive information from the station, and from adjacent stations, includes direct information such as voltage, current, circuit breaker and switch state and intermediate information or indirect information such as the operational result. By using the redundant direct information and indirect information, the coordination between different protection devices can be optimized, and the contradiction between the selectivity and speed of protection can be alleviated.

Substation area protection could also simplify the setting and cooperation of backup protection. In modern power grids where the network topology is ever more complex, for local backup protection and remote backup protection based on local information, due to limited information acquisition, the protection ranges overlap with one another, and cooperation is very complicated with a large amount of setting value calculation. When the communication is normal, substation area protection can correctly isolate the fault after collecting the information in the station and making a simple identification according to the substation network topology. Therefore, substation area protection could not only effectively improve the performance of protection, but also significantly simplify the setting and coordination of backup protection.

Currently, substation area protection mainly has the following two forms:

1) Distributed substation area protection – This protection algorithm is mainly aimed at distributed buses. By decentralizing traditional centralized bus protection into some bus protection units, each protection unit collecting the

current in each loop, converting it into digital form which is then uploaded to the network, and acquiring the current information of all the other loops from the network, distributed bus protection could identify whether the fault is at the bus according to the calculation results based on the principle of bus differential protection. If the fault is identified to be at a particular bus, then only the circuit breaker in the loop where the protection unit is will be tripped. If one protection unit malfunctions, then only the corresponding loop is wrongly tripped. The other loops connected to the bus will not all be cleared. Thus, compared with traditional centralized bus protection, distributed bus protection is more reliable. However, the design of a distributed substation protection system is complicated, the hardware cost is high and the economic efficiency is poor.

2) Centralized substation area protection – In substation centralized protection, all information is centralized to a computer system for centralized processing. Substation area protection collects all the information about electrical variables and state variables through the process layer network, and the central decision-making point of the protection system analyses and calculates the information within the substation protection range, thus realizing comprehensive judgement and decision-making of protection. The system structure is convenient for data transmission and sharing, and could realize protection of multiple bay layers in the station. Compared with distributed substation area protection, substation centralized protection could acquire more information, locate the fault from the whole station level, simplify the cooperation of protection operating time and also improve the selectivity and reliability of protection, which is beneficial for intelligent protection decision-making.

1.5.3 Wide Area Protection

Wide area relay protection is mainly responsible for protection of the tie line between substations and remote backup protection of substation components. According to the node current, voltage and circuit breaker state shared through the data network, the wide area host computer calculates the basic information of the fault and realizes the function of wide area protection. The advantage of wide area protection over traditional protection based on single-end variables is mainly reflected in that, wide area protection uses not only the information of the protected device, but also the information of other relevant devices, the network topology information and model parameter information. Thus, the function of the existing protection can be optimized, and the selectivity of protection can be improved. Wide area protection uses wide area information to identify the fault directly, without considering the cooperation between different protection devices, which simplifies the setting of the protection, and the speed is improved. When the line information is missing, wide area protection can make up for the missing information by analysing and utilizing the wide area information, thus the reliability of protection can be improved.

According to the information collection range of the regional protection system, the information transmission mode, and the position and function of information processing and decision-making unit, the basic structure of a wide area backup protection system can be divided into two forms: the substation centralized form (without host) and the wide area centralized form (with host).

1) Substation centralized form (without host) – The substation centralized structure (without host) refers to setting a wide area host computer and substation host computer in each substation, i.e. all substations in the region are equivalent to one another in their position. The first layer of the substation centralized structure is similar to the structure with a host, with the intelligent electronic device (IED) realizing the digital acquisition and execution in each substation. In the second layer or substation middle layer, the substation protection function is realized through the substation host computer in each substation. In the third layer or wide area power grid protection layer, the protection function is realized on the basis of information exchange between the station and its adjacent station through the wide area host computer in each substation. Under special circumstances, if communication with the adjacent station is interrupted, then the wide area host computer in this substation could exchange information with the substation host computer of the second adjacent station, in order to realize the function of wide area protection.

   For a substation centralized structure, each substation is a wide area host computer and is responsible for the substation area protection of this station and wide area protection of this station and adjacent stations, as the decision-making centre. However, for this structure, due to the information exchange between adjacent substations, the requirement on the reliability of the communication system is relatively high.
2) Wide area centralized form (with host) – The wide area centralized structure (with host) refers to dividing power grid protection into different areas, with a number of substations and the outgoing lines as one protection area. For each protection area, a wide area host station is set as the decision centre. Wide area protection is realized by the wide area host computer (in the host station), and substation host computers are set in the other stations to realize substation area protection. This protection structure contains three levels: data acquisition and execution completed by intelligent electronic devices (IEDs) in each substation function; the middle layer substation area protection completed by the substation host computer in the protection area; wide area protection completed by the wide area host computer in the host station of the protection area.

   For a wide area centralized structure, a substation is selected as the wide area host station in each protection area, as the decision centre of the wide area protection. The advantage is that the communication between the decision centre and substations is relatively easy, and the amount of communication needed is relatively small, thus the communication network is simple and reliable.

1.5.4 Constitution Mode of Hierarchical Relay Protection

According to Sections 1.5.2 and 1.5.3, the structures of substation centralized protection and regional centralized wide area protection have obvious advantages, which could better meet the application requirements of hierarchical protection.

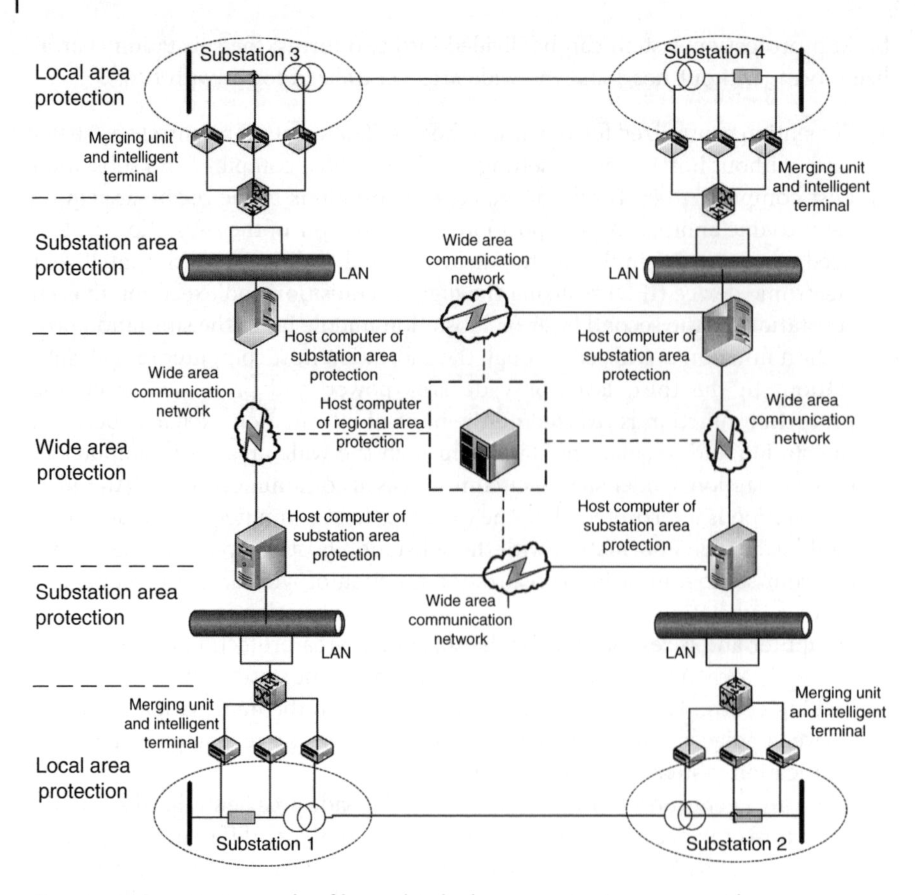

Figure 1.3 Constitution mode of hierarchical relay protection in a smart grid.

The constitution mode of hierarchical relay protection used in this book is based on these two structures. As shown in Figure 1.3, local area protection is responsible for the protection of the bay layer component; substation area protection is responsible for local backup protection of substation components, realized by centralizing the internal information through the substation host computer set in each substation; wide area protection is responsible for local and remote backup protection of inter-station tie lines, and remote backup protection of substation components under special circumstances, through information exchange between the wide area host computer and substation host computers.

In the constitution mode, the protection of substation components and the protection of the inter-station tie line are carried out separately. The number of substation components is relatively small and the space taken up is limited, thus information exchange and data synchronization are easy. For the substation host computer to make a centralized decision, the performance of protection could be improved by using multi-source information, and the configuration of substation protection could be simplified, so construction investment could be saved. In regional protection, the information used is the uploaded data relating to fault identification, preprocessed by the substation host computer, thus the workload of the regional host computer is reduced, unnecessary data interaction is avoided and the pressure on the communication network is relieved.

1.6 Summary

Relay protection is an indispensable part of maintaining the safe and stable operation of a power system. The basic task of relay protection is to identify the fault and quickly clear it, and to ensure that the non-faulty part can continue in normal operation. Relay protection with good performance should meet the requirements of reliability, selectivity, speed and sensitivity.

In the complex modern power grid, protection principles such as current protection, directional protection and distance protection could hardly meet the four requirements of protection. The drawbacks of traditional protection have become increasingly evident, and hierarchical relay protection is proposed as a solution to the above problems. In the constitution mode of hierarchical relay protection, local area protection based on new protection principles serves as the main protection, which could realize reliable protection of the bay layer protection unit; substation area protection based on the communication technology of an intelligent substation could integrate all the information in the station, and optimize the performance of backup protection of substation components; wide area protection could centralize redundant information from a wider range and realize backup protection of the tie line between substations.

Hierarchical relay protection is the focus of this book. The subsequent chapters are all based on the constitution mode of hierarchical relay protection.

References

1 Anderson, P. (2009) *Power System Protection,*Wiley-IEEE Press, New Jersey.
2 Anthony, F. S. (2009) *Protective relay principles,* CRC Press, Florida.
3 Momoh, J. (2012) *Smart Grid: Fundamentals of Design and Analysis, Smart Grid Architectural Designs,* John Wiley & Sons, Ltd, Chichester.
4 State Grid Electric Power Dispatching and Communication Center (2012) *Relay protection training materials of State Grid Corporation of China,* China Electric Power Press, Beijing.
5 Anderson, P. M. (1973) *Analysis of faulted power systems,* Iowa State University Press, Iowa.
6 Anderson, P. M. (1998) *Power System Protection,* IEEE Press, New York.
7 Mason, C. R. (1956) *The Art and Science of Protective Relaying,* John Wiley & Sons, Ltd, Chichester.
8 Horowitz, S. H. (1980) *Protective Relaying for Power Systems,* IEEE Press, New York.
9 Horowitz, S. H. and Phadke, A.G. (2008) *Power System Relaying,* Third edition, John Wiley & Sons, Ltd, Chichester.
10 Domin, T.J. (2008) *Protective relaying-principles and applications,* Third edition, CRC Press, Florida.

11 Stevenson, Jr. W. D. (1980) *Elements of Power System Analysis*, McGraw-Hill, New York.
12 IEEE 1588-Precision clock synchronization protocol for networked measurement and control systems.
13 IEC 61850-9-2LE implementation guideline for digital interface to instrument transformers using IEC 61850-9-2, UCA International Users Group, http://www.ucainternational.org.
14 IEC 61850-5, Communication networks for power utility automation, Second edition, http://www.iec.ch.
15 Rabiner, L. R. and Gold, B. (1975) *Theory and Application of Digital Signal Processing*, McGraw-Hill, New York.
16 IEC 61850, Communication networks and systems in substations, First edition, http://www.iec.ch.

2

Local Area Conventional Protection

2.1 Introduction

With the development of hybrid AC–DC EHV/UHV interconnected power grids and new, large-scale energy sources connected to the grid, the local protection of independent and decentralized configuration, aimed at protecting equipment and serving as the first line of defence to ensure the security of a power grid, is facing an unprecedented level of test severity. Higher demand is putting stress on the reliability, rapidity and sensitivity of local area conventional protection. Strengthening local protection is still the most fundamental means of reducing the impact of faults on the power system. Meanwhile, with the rapid development of electronic, computer and communication technologies, conventional local area protection is provided with a wider development space. In view of the problems of conventional transformer protection and transmission line protection, with the help of voltage and current information of newer transformers, the existing protection scheme is being improved and new protection principles are being put forward to improve the performance of conventional local area protection and help it to develop in the direction of information, integration and intelligence.

With regard to transformer protection, the demand for high reliability and rapid response are on the increase, with larger capacity transformers constantly being brought into operation. There are already widespread problems in UHV/EHV transformer differential protection, such as lack of sensitivity with no-load switching at faults and malfunction in case of TA saturation due to an external fault [1,2]. In view of the above problems, waveform singular characteristics, instantaneous power and leak inductance are used to construct new main protection schemes that are widely different from differential protection.

For transmission line protection, distance protection is widely used on UHV/EHV lines due to its characteristic of being less affected by the system operating mode and the grid structure. However, there are disadvantages in distance protection:

1) Its performance is prone to being affected by transition resistance and it may malfunction or refuse to act in case of a non-metallic fault.
2) It is difficult to do accurate fault location and quick fault isolation when used for high-voltage, double-circuit transmission lines on a single tower.

Hierarchical Protection for Smart Grids, First Edition. Jing Ma and Zengping Wang.

3) Special conditions such as line overload and false phase selection may contribute to the malfunction of distance protection, and may even cause blackouts.

Taking into account these problems of distance protection, the corresponding solution is put forward in this chapter.

2.2 Transformer Protection

An improved scheme for transformer differential protection and new protection schemes different from differential protection are introduced in this section. In Sections 2.2.1–2.2.4, new methods are put forward based on mathematical morphology [3], and we look at the variation feature of fundamental current amplitude and normalized grille curves [4,5] for discrimination between inrush currents and internal faults with the help of current information, and on this basis, the TA saturation identification method is introduced. In Sections 2.2.5 and 2.2.6 we utilize voltage and current information comprehensively and construct two new transformer protection schemes based on equivalent instantaneous leakage inductance [6] and generalized instantaneous power [7]. These schemes can efficiently solve the problems of lack of sensitivity when no-load switching at faults and malfunction in the case of TA saturation due to external fault.

2.2.1 Adaptive Scheme of Discrimination between Internal Faults and Inrush Currents of Transformer Using Mathematical Morphology

2.2.1.1 Mathematical Morphology

A. Fundamental Concepts and Basic Operations of Mathematical Morphology

Mathematical morphology (MM) [8,9] is known as an image processing technique, where the key points of an image are described by transformations called dilations and erosions. Dilation is the expansion of a particular shape into another, bigger shape, while erosion is shrinking one shape into another shape. Let $f(x)$ shown in Figure 2.1(a) and $g(x)$ shown in Figure 2.1(b) denote a one-dimensional signal and a structure element (SE) respectively, whose domains of definition are D_f, D_g; $D_f = \{1,\ldots,M\}$, $D_g = \{1,\ldots,N\}$, and $M > N$. Dilation and erosion of $f(x)$ by $g(x)$ can be computed from the direct formulas:

$$(f \oplus g)(x) = \max\{f(x-y) + g(y) | (x-y) \in D_f; y \in D_g\} \tag{2.1}$$

$$(f \Theta g)(x) = \min\{f(x+y) - g(y) | (x+y) \in D_f; y \in D_g\} \tag{2.2}$$

where $\oplus$ and Θ denote morphological dilation shown in Figure 2.1(c) and erosion shown in Figure 2.1(d), respectively.

Usually, dilation and erosion are not mutually inversed. They can be combined through cascade connection to form new transforms. If dilation is next to erosion, such a cascade transform is an opening transform. The opposite is a closing transform. The transform can be computed using the following formulas, respectively.

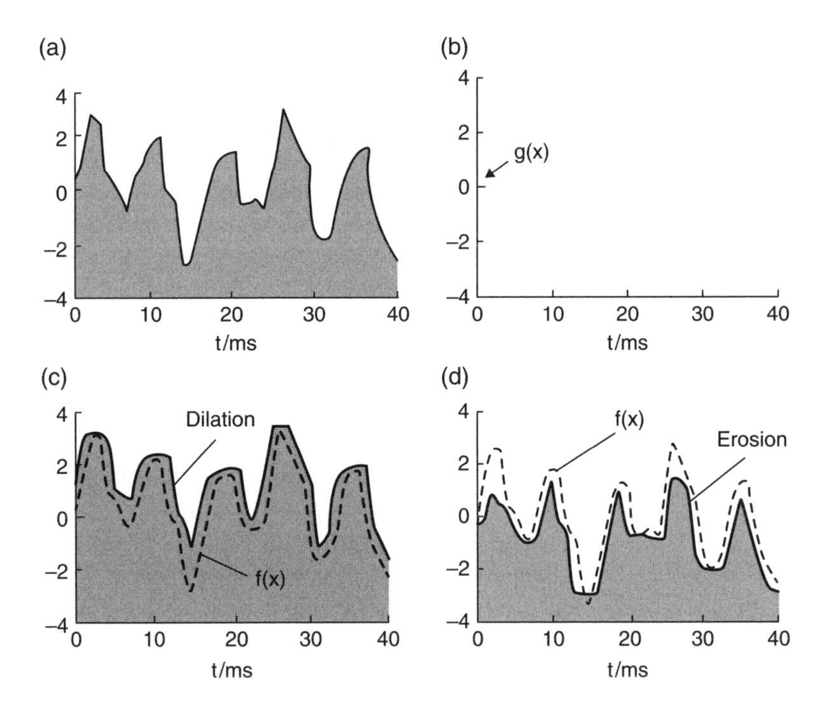

Figure 2.1 Morphological approach to signal using a flat structuring element. (a) Original signal. (b) Structure element. (c) Dilation transformation. (d) Erosion transformation.

$$f \circ g = f \Theta g \oplus g \tag{2.3}$$

$$f \bullet g = f \oplus g \Theta g \tag{2.4}$$

Due to the expansibility of the opening transform, it can be used to remove the peaks in the signal. Due to the inverse expansibility of the closing transform, it can be used to fill the valleys in the signal. In order to reject both positive noise and negative noise and to extract transient signals simultaneously, the differential operation between the morphological opening and closing transforms is proposed.

B. Novel Morphological Gradient

The basic morphological gradient (MG) is defined as the arithmetic difference between the dilated and eroded function $f(x)$ by the SE $g(x)$ of the considered grid. The definition of MG is given by

$$G_{grad} = (f \oplus g)(x) - (f \Theta g)(x) \tag{2.5}$$

There is a distinct difference in the meaning of the MG from gradient in physics. Frequently, the MG is used for edge detection in image and signal processing.

The results of the opening transform and dilation preserve sudden negative changes, while the results of the closing transform and erosion preserve positive sudden changes, so a novel morphological gradient (NMG) can be used to depress the steady components and hence enhance sudden changes.

$$G_{nmg} = (f \circ g \oplus g)(x) - (f \bullet g \ominus g)(x) \tag{2.6}$$

The SE acts as a filtering window, in which the data are smoothed to have a similar morphological structure to the SE. The effectiveness and accuracy of the extraction depend not only on the combination mode of different transforms, but also on the shape and width of the SE. An SE with a simple geometrical shape, such as a circle or triangle, is preferable. The shape of the SE should be selected according to the shape of the processed series. In order to extract the ascending and descending edges of sudden changes, a symmetrical triangular-shaped SE is preferable, defined as: {0, ..., v, ..., 0}. Only if the width of the SE is shorter than that of sudden changes can all sudden changes be extracted. To obtain the width of the narrowest sudden change, the definition of Open is introduced and is described as follows:

$$\text{Open}\,(i) = |n-j| \quad \text{when} \quad |\,\text{Peak}(j) - x(n)| < \delta \tag{2.7}$$

where Open (i) is the ith Open value, $i = 1, 2, \ldots, P_n - 1$, P_n is the sum of local maximum points, Peak(j) is the ith local maximum at the jth point, $x(n)$ is the value at the nth point, $n = 1, 2, \ldots, N$, and δ is a very small predefined threshold.

Based on the above definition of Open, the width of the narrowest sudden change in the data, L_{p} is the minimum of the Open values. That is

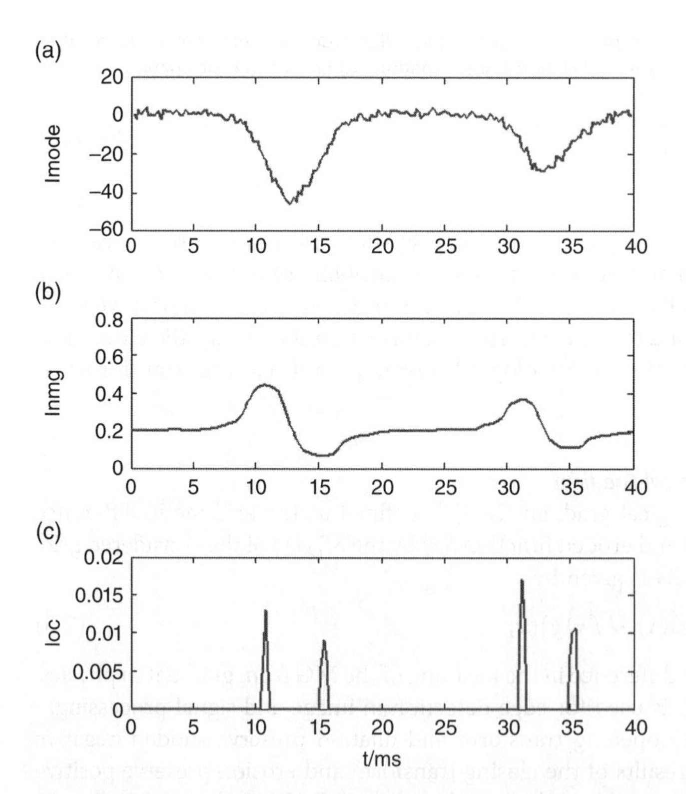

Figure 2.2 Processing results of the Electrical Power Dynamic Laboratory testing data by the mathematical morphology method.

$$L_{\mathrm{p}} = \min\left[\mathrm{Open}(i)\right], \quad i = 1,2,\ldots,P_n - 1 \tag{2.8}$$

To verify the effectiveness of the NMG with the Open value, a good example obtained from the Electrical Power Dynamic Laboratory (EPDL) is illustrated in Figure 2.2. Various kinds of white noise are added. The signal-to-noise ratio (SNR) is about 35 dB. The noisy data are shown in Figure 2.2(a). The formula for SNR is defined as

$$D_{SNR} = 20\log(P_S/P_N) \tag{2.9}$$

where P_S is the variance of the original data, P_N is the variance of noise.

Figure 2.2(b) shows that the NMG has immunity from random noise, since the time-domain variation periods, and thus their Open values, differ greatly from sudden changes. The testing data are then processed by the differential operation between the morphological opening and closing transforms, and the transient signals are eventually extracted, as shown in Figure 2.2(c). This is easy to apply on-line with clear physical meaning and little computational cost.

2.2.1.2 Principle and Scheme Design

A. Basic Principle

Inrush can be generated when a transformer is switched onto the transmission line or an external line fault is cleared. Due to the difference of the magnetic permeability in the iron core of the transformer between inrush currents and internal fault currents, the main magnetic flux varies alternately from non-saturation to saturation in one cycle under inrush currents, which makes their waveforms distorted. However, the transformer operates in the linear section of the magnetizing curve for internal fault conditions. Therefore, 'sudden change' characteristics will not be observed in their waveforms besides the fault point.

Since MM possesses great capability to characterize and recognize unique features in the waveforms – a series of sudden changes in the waveforms – the proposed principle to distinguish between the magnetizing inrush and internal fault is thus feasible.

B. Scheme Design

Here we consider differential current waveforms between phases as the object to analyse. Taking A–B phase, for instance, differential currents between the primary and secondary currents of all phases are calculated and then applied to form three mode signals (I_{abxj}, I_{bcxj}, I_{caxj}) by the following formulas:

$$\begin{cases} I_{\mathrm{a}} = I_{\mathrm{a1}} - I_{\mathrm{a2}} \\ I_{\mathrm{b}} = I_{\mathrm{b1}} - I_{\mathrm{b2}} \\ I_{\mathrm{abxj}} = I_{\mathrm{a}} - I_{\mathrm{b}} \end{cases} \tag{2.10}$$

where I_{a1}, I_{b1} are primary currents; I_{a2}, I_{b2} are secondary currents; I_{abxj} is the differential current between phases A and B.

The NMG of each current mode is calculated using Equation (2.6) and then is extracted by use of the differential operation between the morphological opening and closing transforms, through which some random noise may still exist in the series after being processed and this can be used as the detector of the starting point, illustrated as follows.

Because of various types of transformer and various fault conditions, in order to improve feasibility and practicability of the scheme, a floating threshold I_{fd} is proposed to evaluate whether or not it is the starting point of protection. I_{fd} is defined as

$$I_{fd} > K_k I_{ranmax} \tag{2.11}$$

where I_{ranmax} is the maximum value of random noise under normal conditions and is detected using a sliding data window in a half cycle (10 ms), whose value is much smaller than the value of fault-generated or inrush-generated transient signals. K_k is the proportion factor.

Then, N, the number of fault- or inrush-generated transient signals, is added from the starting point in one and a half cycles if $I_{ts} > k' I_{fd}$, where I_{ts} is the value of fault- or inrush-generated transient signals; k' is the attenuation factor.

If N of each mode is more than or equal to three, it indicates that there is a current inrush, and the relay is inhibited from tripping. By contrast, however, the relay gives the correct response if there is an internal fault. A transformer protection scheme block diagram is shown in Figure 2.3.

2.2.1.3 Testing Results and Analysis

A. Electrical Power Dynamic Laboratory Testing System

In order to verify the feasibility of the proposed method, the author obtained a large amount of actual data through the Electrical Power Dynamic Laboratory (EPDL) simulation. Figure 2.4 shows the connection scheme of the EPDL testing system. The transformer is a Yd11 connection transformer consisting of three single-phase units. The parameters of each single phase are as follows: $S_{\text{rated}} = 10\text{kVA}$,

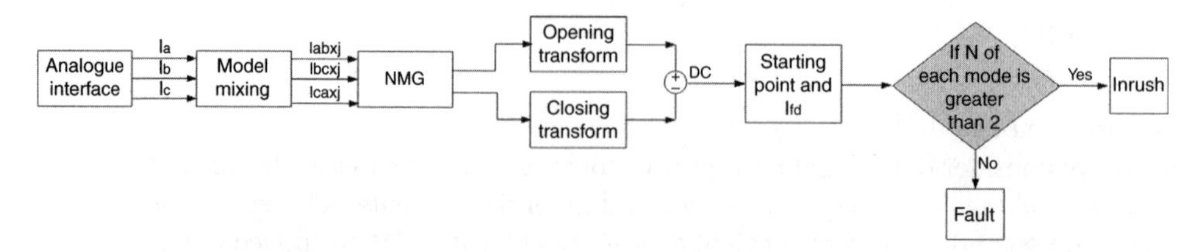

Figure 2.3 Transformer protection scheme block diagram.

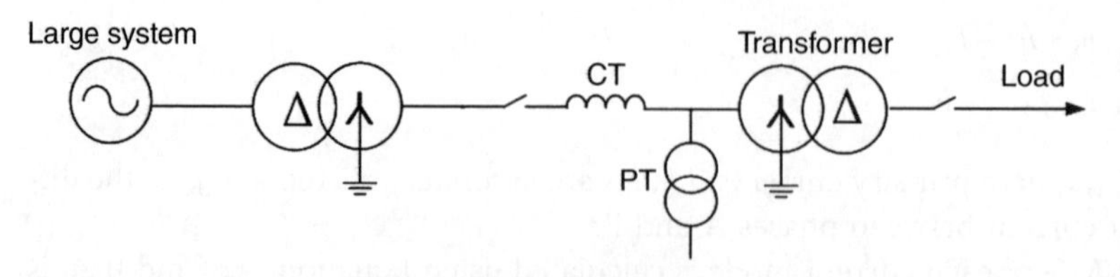

Figure 2.4 Connection scheme of EPDL testing system.

Rated voltage ratio: $U_{1N}/U_{2N} = 1000\,\text{V}/380\,\text{V}$, $I_{\text{no-load}} = 1.45\%$, $U_{\text{shortcircuit}} = 9.0 \sim 15.0\%$, loss of open circuit is 1%, loss of short circuit is 0.35%. The sampling frequency used is 5 kHz.

B. Responses to Different Inrush Conditions

Figures 2.5(a) and 2.6(a) show the current modes obtained from the typical asymmetrical and symmetrical inrush conditions, respectively. Figures 2.5(b) and 2.6(b) show the corresponding signals from the NMG outputs, respectively. As shown in the figures, the variations in inrush wave shapes have resulted in variations in the NMG outputs.

Figures 2.5(c) and 2.6(c) show the transient signals extracted by the differential operation between morphological opening and closing transforms, respectively. DC components are depressed by the differential operation, while peaks and valleys of sudden changes can be extracted effectively. According to the scheme introduced above, N is equal to four in the case of asymmetrical inrush, and N is equal to five in the case of symmetrical inrush. Therefore, the relay will be inhibited from issuing a trip signal in both cases. Compared with this method, Fourier transform-based schemes have difficulties in filtering DC components if some disturbing signals are superimposed.

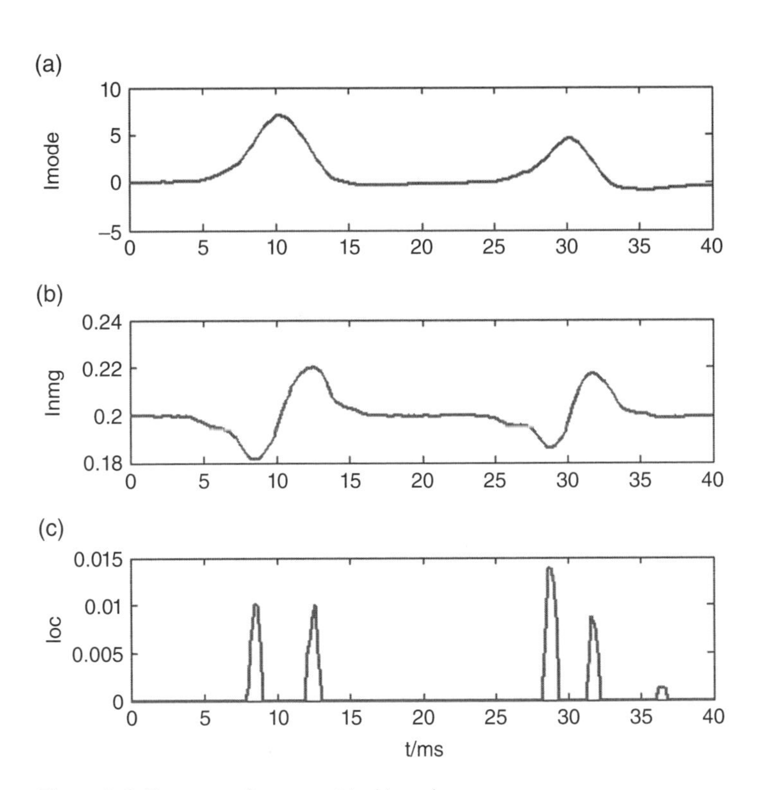

Figure 2.5 The case of asymmetrical inrush.

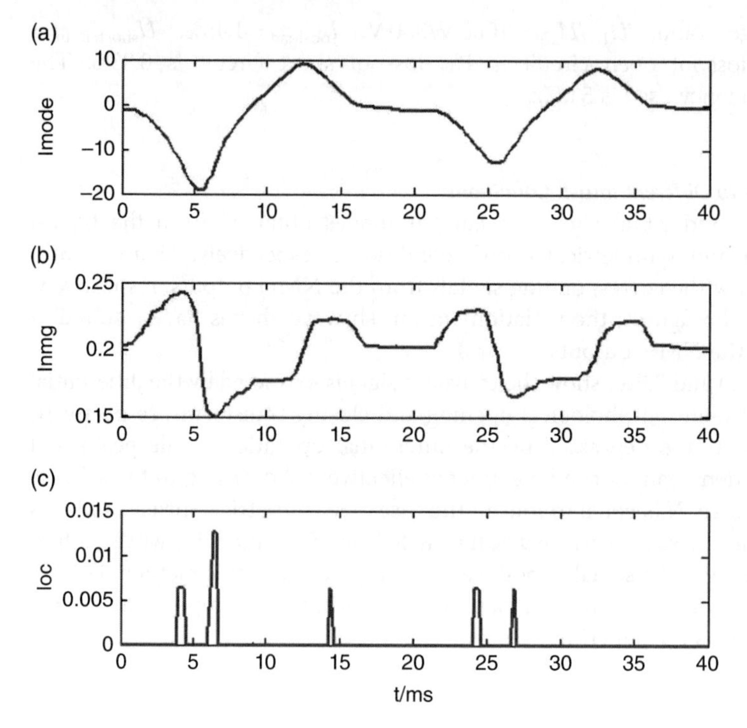

Figure 2.6 The case of symmetrical inrush.

C. Responses to Internal Fault Conditions

Figure 2.7 shows the corresponding responses to a 9% turn-turn fault on phase A. The NMG output in Figure 2.7(b) illustrates that sudden changes of internal faults are much less severe compared with inrush currents. With the transient signals shown in Figure 2.7(c) fast decaying, internal fault currents can be discriminated from inrush currents since N, the number of fault-generated transient signals, is only equal to one. Though there is still a little random noise, the value is too small to be considered as I_{ts}. As a result, the relay operates.

D. Responses to Simultaneous Fault and Inrush Conditions

Figure 2.8 shows the corresponding responses of the inrush to a light internal fault. As shown in the figure, where the variation trends are similar to Figure 2.7, the high-frequency transient signals decay instantaneously right after the fault point. N is also equal to one in this case, which leads to the final decision that this is an internal fault, as shown in Figure 2.8(c).

For inrush due to an internal fault, the distortion degree attenuates rapidly, while the waveforms of magnetizing inrush currents are more severely distorted due to enlargement of the dead angle.

E. Responses to Current Transformer Saturation Conditions

If the conditions are severe enough, it is possible that the distortion may be even worse, and saturation can start to occur even sooner. Severe saturation can cause

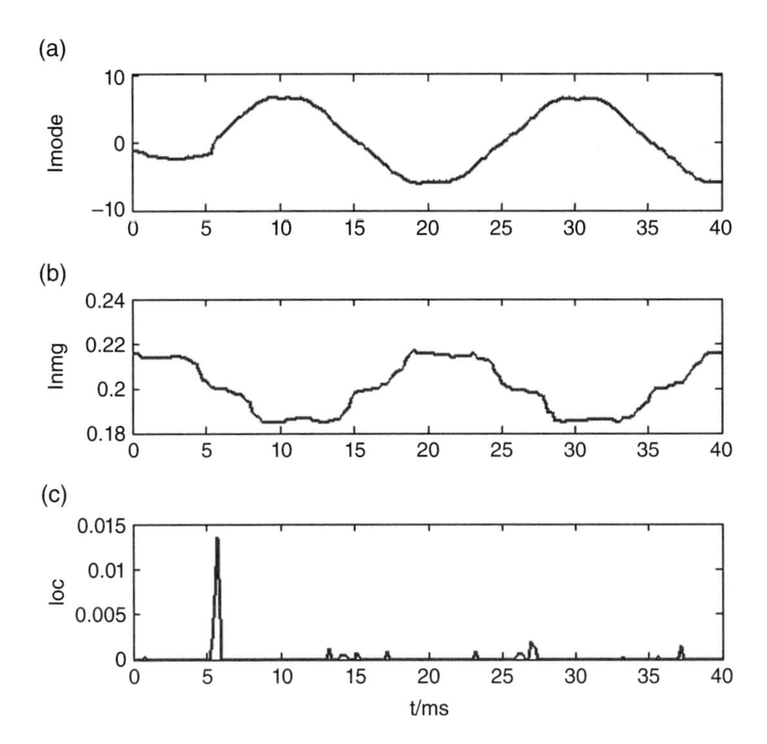

Figure 2.7 The case of an internal fault current.

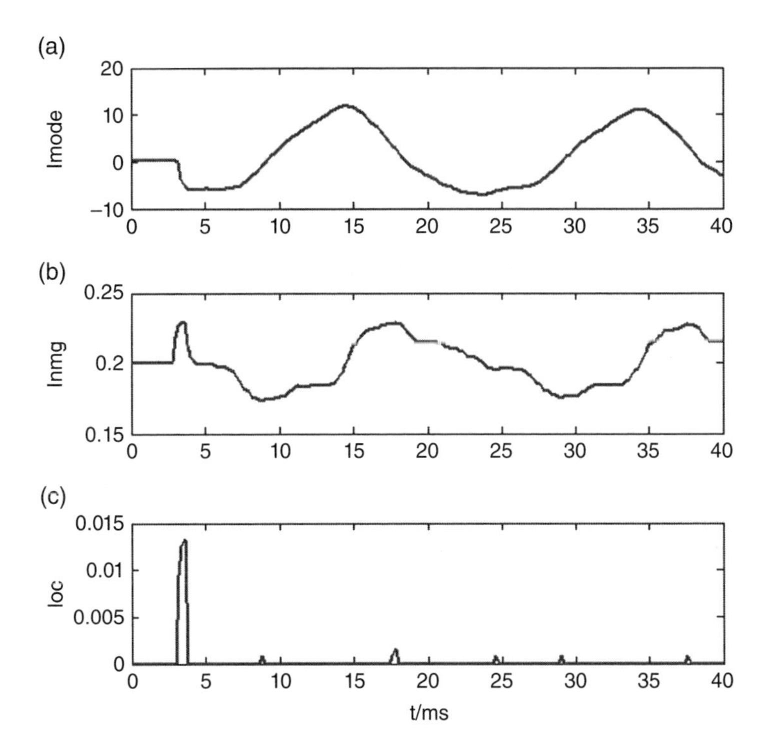

Figure 2.8 The case of inrush due to a light internal fault current.

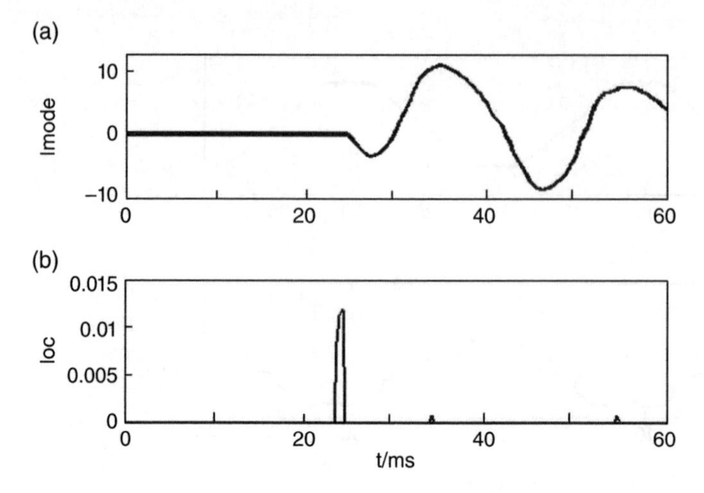

Figure 2.9 The case of simultaneous internal fault current and CT saturation.

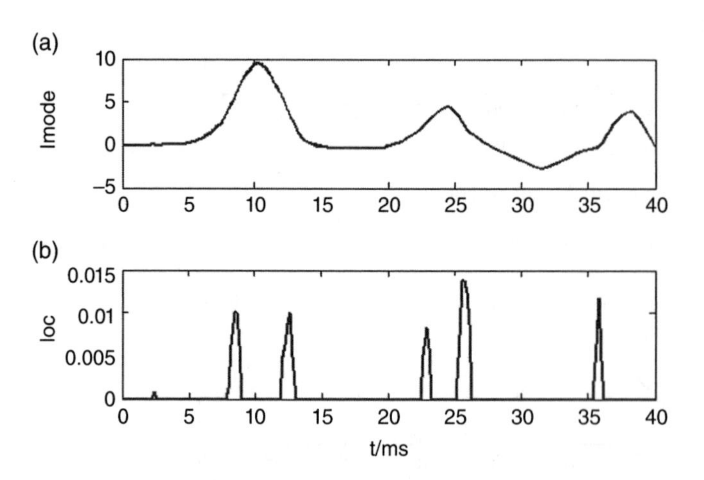

Figure 2.10 The case of simultaneous asymmetrical inrush current and CT saturation.

problems in transformer differential relays. Figure 2.9(a) shows the responses obtained due to the simultaneous occurrence of an internal fault and CT (located at the low-voltage side) saturation. As shown in Figure 2.9(b), the variations in inrush waveforms due to current transformer (CT) saturation do not have any effect on the relay responses, since N is also equal to one and so the relay operates in this case. Compared with the restraining algorithm based on second harmonics, the ratio of second harmonics after CT saturation is 26.7% and this exceeds the predefined threshold of 15%, which makes the relay give an incorrect response.

Furthermore, due to the slowly decaying offset components of inrush currents, CTs can and do saturate during inrush with no internal fault. Due to simultaneous asymmetrical inrush current and CT saturation, the waveform of an asymmetrical inrush current is distorted more and more severely, while the dead angle is extinct after about one power frequency cycle, as shown in Figure 2.10(a). However, in this case, the output of N is equal to five, as shown in Figure 2.10(b), so the relay will be inhibited from issuing a trip signal. Compared with the restraining algorithm based on second harmonics, the ratios of second harmonics before and after CT saturation are 26.9% and 32.1%, respectively, which makes the relay give a correct response. In addition, compared with the waveform correlation scheme, the waveform coefficients are 0.17 before CT saturation and 0.43 after CT saturation. The relay can also be inhibited from tripping. However, the redundancy of this scheme is small.

F. Responses to Internal Faults with External Shunt Capacitance

Since the second harmonic components in fault currents are increased, together with the capacitance in the power system, the proposed technique is used to avoid needless relay blocking when a transformer has an internal failure. Figure 2.11 shows a waveform of differential current in the case of a short circuit fault on phase C of a 670 MVA transformer, the parameters of which are shown in Table 2.1. A 20.26kA power source, a 257 km overhead 500 kV transmission line

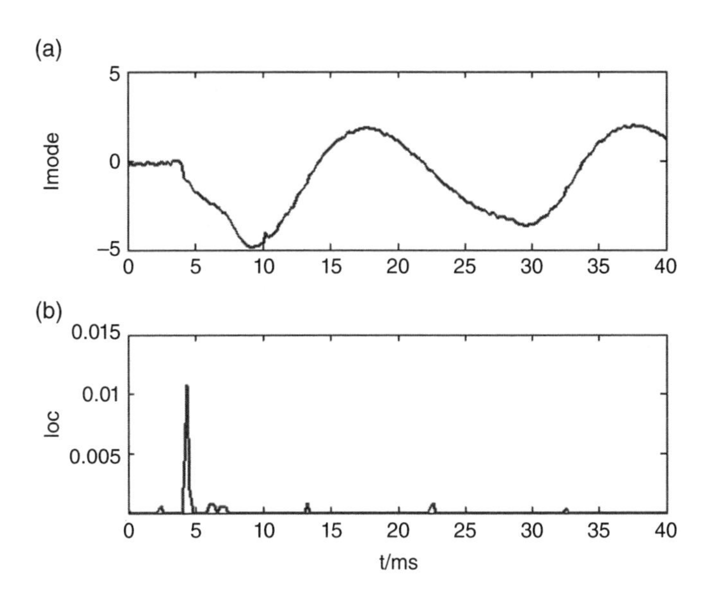

Figure 2.11 The case of an internal fault current with external shunt capacitance.

Table 2.1 Parameters of the test transformer.

Transformer type	3-phase, 2-winding
Rated apparent power	670 MVA
Rated voltage ratio	19/500 kV
Connection style	YNd11
Short circuit reactance	13.0%
Rated frequency	50 Hz

and the transformer mentioned above were simulated by EPDL. This included nearly 33.9% second harmonic and locked conventional relays. However, in this case, N is equal to one, which indicates that this is an internal fault and the relay operates correctly.

The results of EPDL indicate that the proposed technique can distinguish magnetizing inrush from internal faults in power transformers and avoid the symmetrical inrush current and CT saturation. It can also deal with the sampled data containing various kinds of noise and DC components and is stable during internal faults with external shunt capacitance in a long EHV transmission line.

G. Test Results and Analysis

The performance of the proposed technique was evaluated for different types of internal faults and magnetizing inrush currents. Dynamic simulation results and N for three model signals under various states are shown in Table 2.2. In the column of results, '0' indicates that the relay will be inhibited from issuing a trip signal, while '1' indicates that this is an internal fault and the relay operates.

For convenience, the faults and inrush currents in Table 2.2 are numbered to indicate operating conditions. Cases 0–6 are different magnetizing inrush conditions. The N values of three model signals are all more than or equal to four.

Case 7 is a 3% turn-turn fault in phase A. Case 8 is a 5% turn-turn fault in phase B. Case 9 is a 5% turn-turn fault in phase C. Case 10 is a turn-to-earth fault in phase A. Case 11 is a turn-to-earth fault in phase B. Case 12 is an A–B phase fault. Case 13 is a B–C phase fault. The N values for three model signals are all less than or equal to one.

Table 2.2 EPDL simulation cases.

Case	Description	
1–6	Inrush currents only	
7	3% turn-turn fault in phase A	Internal fault only
8	5% turn-turn fault in phase B	
9	5% turn-turn fault in phase C	
10	turn to earth fault in phase A	
11	turn to earth fault in phase B	
12	A–B phase fault	
13	B–C phase fault	
14	3% turn-turn fault in phase A	Simultaneous fault and inrush
15	4% turn-turn fault in phase B	
16	4% turn-turn fault in phase C	
17	turn to earth fault in phase A	
18	turn to earth fault in phase B	
19	A–B phase fault	
20	B–C phase fault	

Table 2.3 Dynamic simulation results of three model signals under various states.

Case	N of I_{abxj}	N of I_{bcxj}	N of I_{caxj}	Results
1	4	4	5	0
2	5	8	4	0
3	6	5	4	0
4	4	7	4	0
5	8	4	6	0
6	6	4	8	0
7	1	0	1	1
8	1	1	1	1
9	0	1	1	1
10	1	1	1	1
11	1	1	0	1
12	1	1	1	1
13	1	1	1	1
14	1	4	2	1
15	1	1	3	1
16	4	1	1	1
17	1	2	1	1
18	1	1	1	1
19	1	1	2	1
20	1	2	1	1

Case 14 is the inrush to a 3% turn-turn fault in phase A. Case 15 is the inrush to a 4% turn-turn fault in phase B. Case 16 is the inrush to a 4% turn-turn fault in phase C. Case 17 is the inrush to a turn-to-earth fault in phase A. Case 18 is the inrush to a turn-to-earth fault in phase B. Case 19 is the inrush to an A–B phase fault. Case 20 is the inrush to a B–C phase fault. Mainly due to the loose coupling between phases in the condition of inrush to light internal faults, partial model signals in cases 14–16 cannot meet the requirement that N is less than or equal to two, but taking all three model signals for comprehensive consideration, the relay operates correctly as a result. As shown in Table 2.3, inrush currents can be distinguished from internal fault currents by this scheme.

H. Comparison between the MM and Wavelet Transform (WT) Methods

A comparison between the proposed MM technique and the WT method has been undertaken. A 2B-spline wavelet is employed to process the input signal. The same alternate trend is also denoted in the 2B-spline wavelet transform. However, compared with the filtered series shown in Figure 2.5(b), the trend in Figure 2.12(b) is less stationary and smooth and some small concave variations

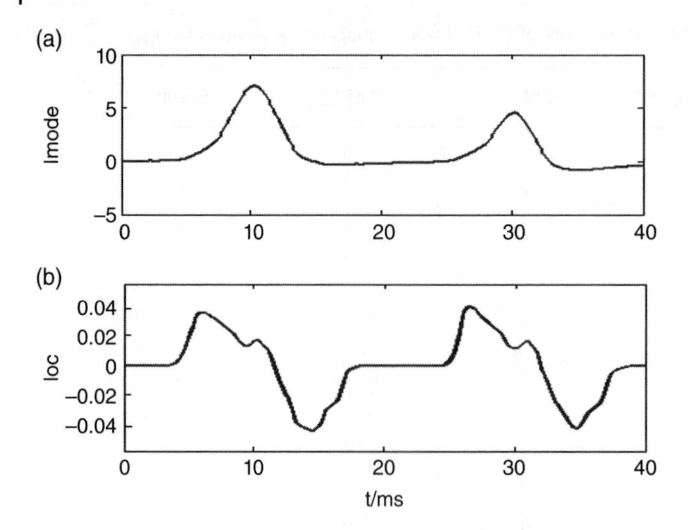

Figure 2.12 Waveform of the asymmetrical inrush current and its 2B-spline wavelet transform.

still exist in the series, which may generate false transient signals and cause error number N. As a result, the relay will give an incorrect response.

2.2.2 Algorithm to Discriminate Internal Fault Current and Inrush Current Utilizing the Variation Feature of Fundamental Current Amplitude

2.2.2.1 Basic Principles

A. The Two-Instantaneous-Value-Product Algorithm

Suppose

$$\begin{cases} i_{t1} = I_{\rm m}\sin\omega t_1 \\ i_{t2} = I_{\rm m}\sin\omega t_2 = I_{\rm m}\sin\omega(t_1 + T_{\rm s}) \end{cases} \tag{2.12}$$

where t_1 is the current sampling point and t_2 is the next sampling point. $T_{\rm s}$ is the time interval between two sampling points, i.e. $T_{\rm s} = t_2 - t_1$.

Applying the two-instantaneous-value-product algorithm [10], the fundamental current amplitude is

$$I_{\rm m} = \sqrt{\frac{i_{t1}^2 + i_{t2}^2 - 2i_{t1}i_{t2}\cos\omega T_{\rm s}}{\sin^2\omega T_{\rm s}}} \tag{2.13}$$

B. Characteristic of the Inrush Current Fundamental Amplitude

When the transformer is no-load closed, the inrush current waveform is greatly distorted (in a concave manner) where the magnetic core enters and exits from saturation, as shown in Figure 2.13 [11].

A total of 56 cases were carried out in this situation. The inrush current waveform is a function of the different core residual magnetization and the switching instant, so the inrush current waveforms are different from each other.

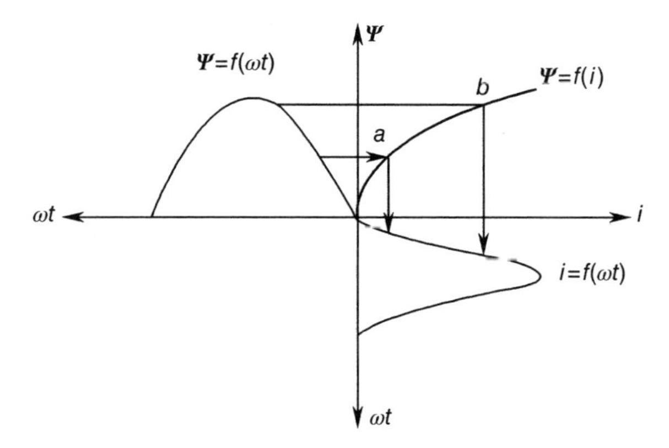

Figure 2.13 Waveform of inrush current.

Figure 2.14 shows the differential current waveforms of the transformer no-load closed under different conditions in the dynamic testing system and the corresponding fundamental current amplitude variation curves. The asymmetric inrush current on the lower half plane, and the asymmetric and symmetrical inrush currents on the upper half plane are shown in Figures 2.14(a), (c) and (e), respectively.

Choose any three adjacent sampling points i_k, i_{k+1} and i_{k+2} in the magnetic core saturation area. Applying Equation (2.13), the fundamental current amplitudes I_{m1} and I_{m2}, which are based on i_k and i_{k+1}; i_{k+1} and i_{k+2} respectively, can be obtained. The quadratic difference between I_{m1} and I_{m2} is

$$\begin{aligned} I_{m2}^2 - I_{m1}^2 &= \left[\left(i_{k+2}^2 + i_{k+1}^2 - 2i_{k+2}i_{k+1}\cos\omega T_s\right) - \frac{i_{k+1}^2 + i_k^2 - 2i_{k+1}i_k\cos\omega T_s}{\sin^2\omega T_s}\right] \\ &= \left(i_{k+2} - i_k\right)\frac{\left[i_{k+2} + i_k - 2i_{k+1}\cos\omega T_s\right]}{\sin^2\omega T_s} \end{aligned} \tag{2.14}$$

When the initial current is the asymmetric inrush current on the lower half plane, as shown in Figure 2.14(a), according to the concave characteristic of the inrush current, we have

$$\frac{i_{k+2} - 2i_{k+1}\cos\omega T_s + i_k}{T_s^2} < \frac{i_{k+2} - 2i_{k+1} + i_k}{T_s^2} < 0 \tag{2.15}$$

In the meantime, the descending degree of the inrush current $i_{k+2} - i_k < 0$, and the ascending degree $i_{k+2} - i_k > 0$. According to Equation (2.14), when the magnetic core enters the saturation area (the descending section of the waveform), $I_{m2} > I_{m1}$ and the fundamental current amplitude keeps increasing monotonically; while when the magnetic core exits from saturation (the ascending section of the waveform), $I_{m2} < I_{m1}$ and the fundamental current amplitude keeps decreasing monotonically.

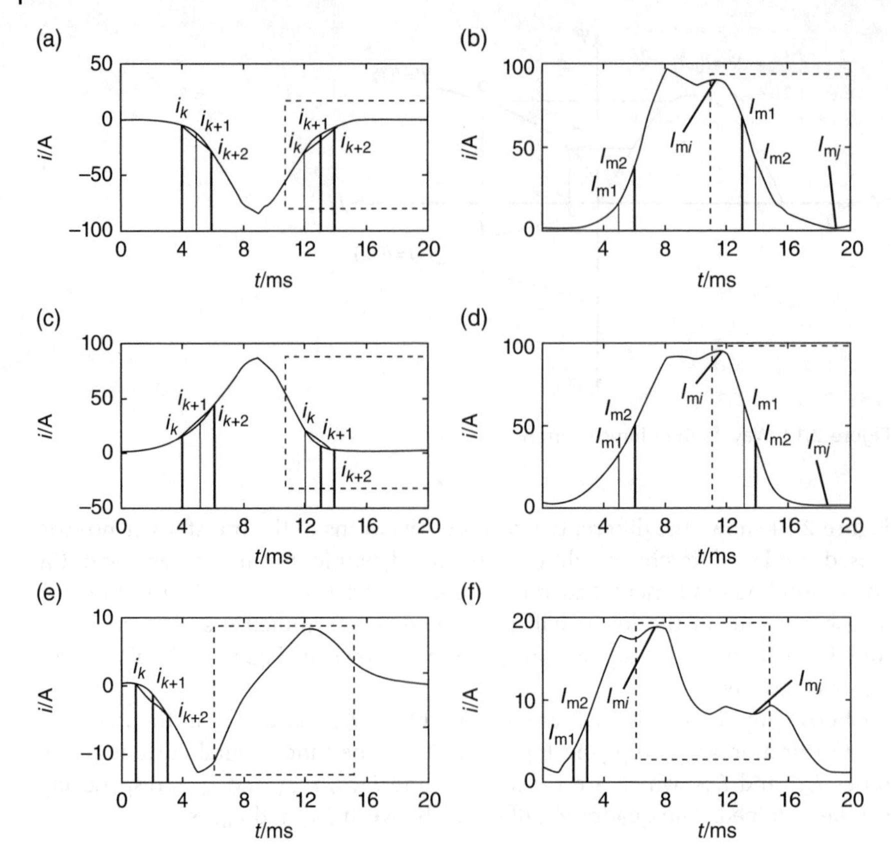

Figure 2.14 Inrush currents and their fundamental current amplitudes.

When the initial current is the asymmetric inrush current on the upper half plane, as shown in Figure 2.14(c), in the vicinity of the saturation area we have

$$\frac{i_{k+2} - 2i_{k+1}\cos\omega T_s + i_k}{T_s^2} > \frac{i_{k+2} - 2i_{k+1} + i_k}{T_s^2} > 0 \tag{2.16}$$

In the meantime, the ascending degree of the inrush current $i_{k+2} - i_k > 0$, and the descending degree $i_{k+2} - i_k < 0$. According to Equation (2.14), when the magnetic core enters the saturation area (the ascending section of the waveform), $I_{m2} > I_{m1}$ and the fundamental current amplitude keeps increasing monotonically. When the magnetic core exits from saturation (the descending section of the waveform), $I_{m2} < I_{m1}$ and the fundamental current amplitude keeps decreasing monotonically.

It should be noted that after the magnetic core enters the deep saturation area (i.e. the linear part of the ferromagnetic curve), the inrush current waveform will be sinusoidal, which means the fundamental current amplitude will keep steady thereafter.

The test results of the fundamental current amplitude under the conditions given in Figures 2.14(a) and (c) are shown in Figures 2.14(b) and (d), respectively.

Clearly, the variation trends of the fundamental current amplitude are consistent with the theoretical analysis above.

The waveforms of symmetrical inrush currents and the corresponding fundamental current amplitude curves are shown in Figures 2.14(e) and (f), respectively. Similarly, it can be concluded that the fundamental amplitude tends to increase when the magnetic core enters saturation and decrease when the magnetic core exits from saturation.

C. Characteristic of the Fundamental Current Amplitude When No-Load Closing at an Internal Fault

When taking into account the transformer winding resistance, the hysteresis loss and the eddy current loss, the loop voltage equation of the transformer, based on the equivalent instantaneous inductance, can be set as

$$u_1 = r_k i_\mathrm{d} + L_k \frac{\mathrm{d}i_\mathrm{d}}{\mathrm{d}t} \tag{2.17}$$

where i_d is the current difference between the primary side and the secondary side; r_k is the equivalent resistance; L_k is the equivalent instantaneous inductance; u_1 is the terminal voltage of the primary side.

Consider that transformers are usually highly inductive, especially large-capacity power transformers, whose equivalent resistance is very small. Thus, the first term in Equation (2.17) is much smaller than the second term. Neglecting the first term, the differential expression of i_d can be obtained as

$$\frac{\mathrm{d}i_\mathrm{d}}{\mathrm{d}t} \approx \frac{u_1}{L_k} \tag{2.18}$$

According to Expression (2.18), the waveform characteristic of i_d is decided by the ratio of the primary side terminal voltage to the equivalent instantaneous inductance. The primary side terminal voltage, u_1 mutates only when an internal fault occurs (whether the transformer is loaded or not), and u_1 stays steady when the transformer operates in its normal state or in the post-fault steady state. Therefore, on the basis of a whole current cycle, it is the instantaneous inductance L_k that really counts. When an internal fault occurs in the transformer, the magnetic core is no longer saturated, so that the excitation inductance and leakage inductance are constant. Thus, the instantaneous inductance L_k is also constant. Therefore, the waveform of the current difference i_d remains sinusoidal except for the fault point. The fundamental current amplitude calculated using the two-instantaneous-value-product algorithm is supposed to remain the same in this area.

2.2.2.2 Protection Scheme Design

According to the analysis above, the difference between an inrush current and fault current lies in the descending degree of the fundamental current amplitude during the magnetic core's exit from saturation. Based on this conclusion, a new identification criterion is proposed here, as follows.

1) Identify the current bias direction – Determine the maximum and minimum values of the current i_{max} and i_{min} within half a cycle of the data window after

the transformer is no-load closed. If $|i_{max}| - |i_{min}| \geq 0$, then the current is forward biased; otherwise, if $|i_{max}| - |i_{min}| < 0$, then the current is reverse biased.

2) Identify the exit-saturation area of the transformer magnetic core – Starting from half a cycle after the transformer's no-load closing, summarize the sampling points of each cycle to form curve S_k. For forward-biased current, the sampling sequence corresponding to the minimum positive value on the descending section of curve S_k is chosen to be the exit-saturation area (see the dashed line frame in Figure 2.14(c)). For reverse-biased current, the sampling sequence corresponding to the maximum negative value on the ascending section of curve S_k is chosen to be the exit-saturation area (see the dashed line frames in Figures 2.14(a) and (e)). After the exit-saturation area is decided, take the maximum and minimum values of the fundamental current amplitude in this area as I_{mi} and I_{mj}, as shown in Figures 2.14(b), (d) and (f).
3) Calculate the changing rate of the fundamental current amplitude k according to Equation (2.19).

$$k = \left| \frac{I_{mi} - I_{mj}}{I_{mj}} \right| \tag{2.19}$$

If $k \geq k_{set}$ (k_{set} is the setting value), then it can be determined to be an inrush current, thus the differential protection should be locked; otherwise, if $k - k_{set}$ and the starting requirement is also met, then an internal fault is determined and relay protection should operate to trip.

2.2.2.3 Simulation Verification

To verify the effectiveness of the proposed method, experimental tests have been carried out at the Electrical Power Dynamic Laboratory (EPDL).

A. The Dynamic Testing System

The experimental transformer is a three-phase, two-winding transformer bank with YNd11 connection, which is fed by a low impedance source, as shown in Figure 2.15. The parameters of the two-winding transformers are given in Table 2.4. Three identical current transformers (CTs) are connected in Δ on the primary side, and another three identical CTs are connected in Y on the secondary side of the power transformer.

The experiments provide different switching and clearing instants for inrush currents, as well as different faults and different numbers of turns for internal faults. A total of 216 cases were tested, divided into five main categories: 56 cases

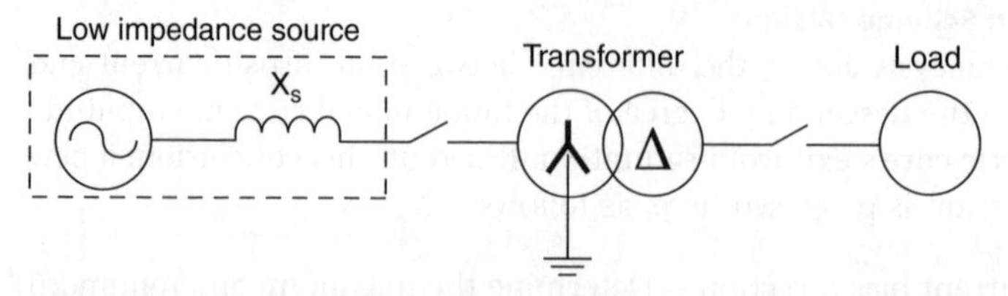

Figure 2.15 Connection scheme of the dynamic analogy testing system.

Table 2.4 Parameters of each single-phase unit of transformer used in the test.

Rated power	10 kVA
Rated voltage ratio	1000/380 V
Rated frequency	50 Hz
No load current	1.45%
No load loss	1%
Short circuit voltage	9.0 ~ 15.0%
Short circuit loss	0.35%

for switching the transformer with no load, 52 cases for simultaneous internal fault and inrush conditions, 54 cases for faulty conditions only, and 54 cases for external faults with the CT in saturation conditions, to test the algorithm.

B. Test Result Analysis

The switching in the transformer bank with no load often causes the inrush current of non-fault phases, which has been verified by a total of 52 cases with simultaneous inrush currents and internal faults. An example taken from these cases is given in Figure 2.16, where the no-load transformer closes at a phase A 3.6% (turn ratio) turn-to-turn fault. The waveforms of the fault phase differential current are

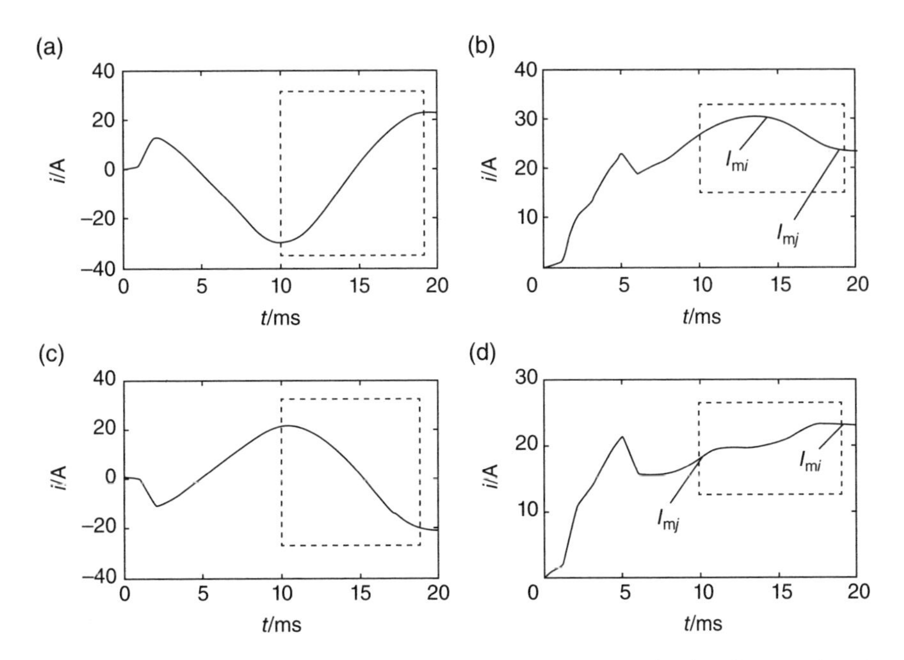

Figure 2.16 Current waveforms and their fundamental current amplitudes when switching in a transformer with an internal turn-turn fault.

shown in Figures 2.16(a) and (c), respectively. Their fundamental current amplitudes are shown in Figures 2.16(b) and (d), respectively. From Figures 2.16(a) and (c), it can be seen that, in the first half cycle, the differential current manifests obvious inrush current features, while in the second half cycle, as the magnetic core exits from saturation, the differential current displays the characteristic of a fault current. According to the identification criterion proposed above, the fundamental current amplitudes in the exit-saturation area corresponding to the differential currents in Figures 2.16(a) and (c) are calculated and shown in Figures 2.16(b) and (d), respectively (the dashed line frame). Applying Equation (2.19), the changing rates of the fundamental current amplitudes are obtained, i.e. $k_b = 0.27$, $k_d = 0.25$.

Data from a total of 54 cases were used to test the algorithm under fault conditions. An example taken from these cases is given in Figure 2.17, where a phase A 3.6% turn-to-turn fault occurs in the loaded transformer. The waveforms of the fault phase differential current are shown in Figures 2.17(a) and (c), respectively. Their fundamental current amplitudes are shown in Figures 2.17(b) and (d), respectively. From Figures 2.17(b) and (d), it can be seen that, for either forward-biased or reverse-biased current, the fundamental current amplitude remains constant in the exit-saturation area. Applying the identification criterion, the changing rates of the fundamental current amplitudes in Figures 2.17(b) and (d) can be calculated, i.e. $k_b = 0.14$, $k_d = 0.13$.

The experiments provide different switching and clearing instants for inrush currents, as well as different faults and different numbers of turns for internal faults. The changing rates of the fundamental current amplitude k gained under different operational states of the transformer are shown in Table 2.5. In the case

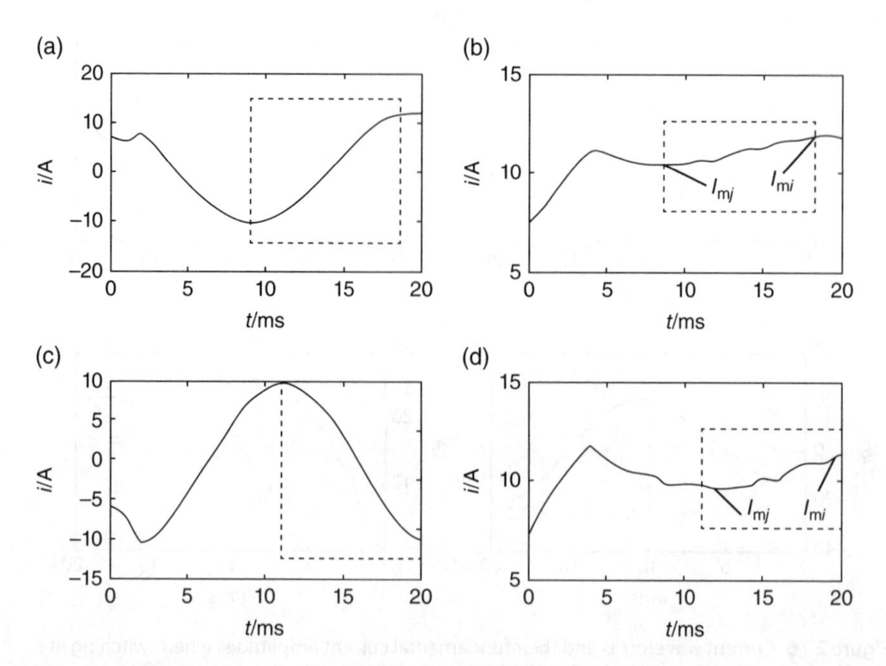

Figure 2.17 Current waveforms and their fundamental current amplitudes when an internal turn-to-turn fault occurs.

Table 2.5 Calculation results of k under different operating states of a two-winding transformer.

Operating state				k	Testing number
	No-load closing			1.435 ~ 87.84	1
Faults at Y-side	Closing at faults	Turn ratio	A3.6%	0.23 ~ 0.32	2
			A6.1%	0.05 ~ 0.12	3
			A9.0%	0.04 ~ 0.11	4
			B18%	0.03 ~ 0.05	5
			C18%	0.03 ~ 0.06	6
		Phase to ground	A	0.02 ~ 0.04	7
			B	0.03 ~ 0.05	8
	Faults occurring when transformer loaded	Turn to turn	A3.6%	0.11 ~ 0.14	9
			A6.1%	0.05 ~ 0.09	10
			A9%	0.04 ~ 0.06	11
			B18%	0.02 ~ 0.03	12
			C18%	0.03 ~ 0.05	13
		Phase to ground	A	0.01 ~ 0.03	14
			B	0.01 ~ 0.02	15
	No-load closing at Δ-side		A4.5%	0.18 ~ 0.29	16
	Faults at Δ-side		A4.5%	0.10 ~ 0.19	17

of inrush currents, the changing rate of the fundamental current amplitude for each phase is calculated, while in the case of fault currents (including faults at no-load closing), only the fault phase is analysed. It can be seen from Table 2.5 that the minimum value of k for inrush currents is 1.435, while the maximum value of k for internal faults is 0.32. The changing rates being more than four times different, by properly setting the threshold value, the distinction between inrush currents and fault currents can be reliably made.

With the same test data, the calculation results using the second harmonic restraint principle and the waveform comparison principle are shown in Table 2.6. For the waveform comparison principle, the waveform coefficient of the differential current is displayed. In the case of no-load closing, the waveform coefficients of all three phases are given, while in the case of no-load closing at faults, only the maximum coefficient of the three phases is given. For the second harmonic restraint principle, the percentage of the second harmonic component in the fundamental component of the differential current is displayed. In the case of no-load closing, the second harmonic percentages of all three phases are given, while in the case of no-load closing at faults, only the minimum percentage of the three phases is given.

For the waveform comparison principle, when the phase constraint method is used with the threshold value set to 0.5, some testing data (the $*$-marked ones)

Table 2.6 Dynamic simulation results of wave comparison and second harmonic restraint algorithm.

Operating state				Waveform comparison principle	Second harmonic restraint principle (%)
No-load closing				0.014 ~ 0.451	15.30 ~ 72.69
Faults at Y-side	Closing at faults	Turn to turn	A3.6%	0.286 ~ 0.374*	29.57 ~ 38.65*
			A6.1%	0.428 ~ 1.107*	8.58 ~ 15.79*
			A9%	0.349 ~ 0.887*	8.17 ~ 13.56
			B18%	0.641 ~ 1.080	5.54 ~ 9.52
			C18%	0.655 ~ 1.143	3.81 ~ 10.34
		Phase to round	A	0.609 ~ 1.058	2.16 ~ 3.95
			B	0.627 ~ 1.106	2.55 ~ 3.36
No-load closing at Δ-side			A4.5%	0.860 ~ 1.115	9.68 ~ 12.97

Note: The *-marked data indicate the cases that cannot be clearly identified with a waveform comparison or the second harmonic restraint principle.

cannot be clearly identified, as shown in Table 2.6. For example, in the cases of no-load closing at internal turn-to-turn faults (including turn-to-turn faults on the Y-side with turn ratio 3.6%, 6.1% and 9%), since none of the three phase current waveform coefficients exceeds 0.5, relay protection will operate after a long time delay. For the second harmonic restraint principle, if the threshold value is set to 15%, operation of the protective relay will also be delayed when the transformer closes with no load at certain internal turn-to-turn faults.

Comparing Tables 2.5 and 2.6, it is obvious that the proposed new principle has better performance than the second harmonic restraint principle and the waveform comparison principle in identifying an inrush current. This is mainly due to the application of the two-instantaneous-value-product algorithm in the new principle, which facilitates its localization capability of dynamically portraying the spatial distribution of the fundamental current amplitude.

C. TA Saturation Caused by External Faults

The differential current is designed to restrain under normal load flows and for external faults. However, if the TAs saturate, external short circuits can, in fact, result in very large differential currents, which has been proved in a total of 54 cases. Therefore, the way TA saturation influences the protection criterion when there is an external fault is dealt with here. In the transient period after the external fault occurs, the short circuit current contains a big DC component. Since TA transforms DC components and produces a large error, a rapid TA saturation will result, which then causes the induced electromotive force on the secondary winding to drop to zero immediately, and the current on the secondary winding

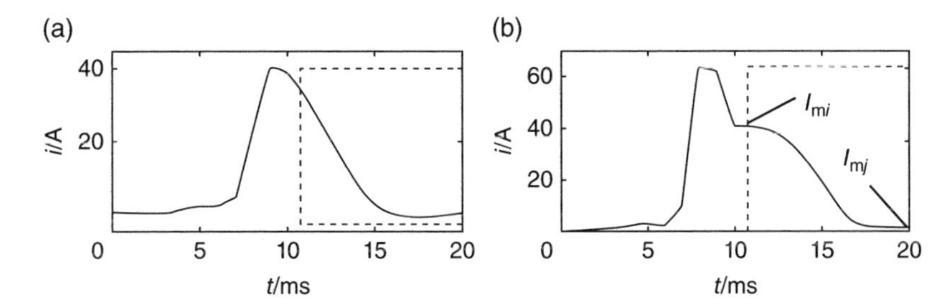

Figure 2.18 Current waveform and its fundamental current amplitudes under TA saturation at an out-zone fault.

remains zero in the same period. With the current on the primary winding decreasing, the TA will exit from saturation, thus the induced electromotive force and current on the secondary winding will increase accordingly.

It can be concluded that, as the TA enters and exits from saturation, the differential current on the secondary winding will experience severe distortion (concave in form) in the waveform. In accordance with the distortion parts, the fundamental current amplitude will also experience severe distortion, as shown in Figure 2.18. The changing rate of the fundamental current amplitude in this case is calculated to be $k = 35.72$. Analysis of TA saturation under different types of external faults shows that the protection principle based on the changing rate of the fundamental current amplitude is reliable in locking the differential protection.

2.2.3 Identifying Transformer Inrush Current Based on a Normalized Grille Curve (NGC)

2.2.3.1 Normalized Grille Curve

A. *Introduction to the Grille Curve*

Here, N_d is defined as the number of needed square grids with the sampling time d as the side length, covering the differential current for a window of half cycle $[t_k - T/2, t_k]$, as shown in Figures 2.19 and 2.20.

The number of needed square grids between two sampling points is calculated by division between their vertical distance and the sampling time d. A grid with solid lines indicates that it is a full square grid, whereas a grid with dashed lines indicates that it is a fractional square grid. Furthermore, N_d, the number of needed square grids for a window of a half cycle, is calculated by adding all the square grids (full square grids and fractional square grids) in a half cycle period. Suppose that the signal has $n + 1$ sampling points at $(t_{i-n}, t_{i-n+1}, \ldots, t_i)$ within the data window $[t_k - T/2, t_k]$, then

$$N_d(i) = \frac{1}{d}\sum_{j=i-n}^{j=i-1} \left|i(t_j) - i(t_{j+1})\right| \tag{2.20}$$

where d is the sampling time.

Two cases are given to illustrate conceptually the advantages of the grille curve method.

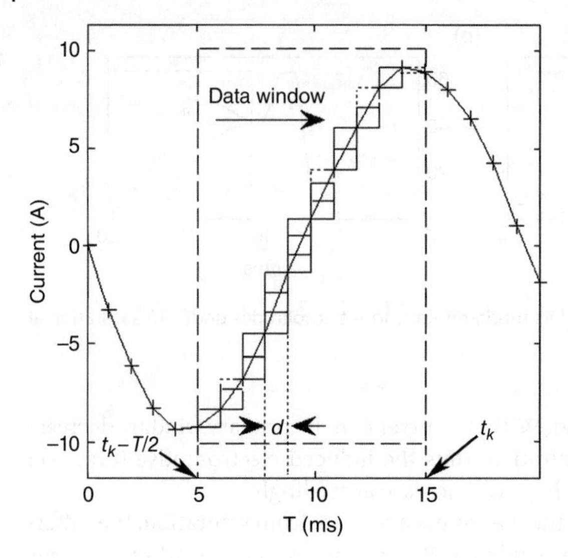

Figure 2.19 The square grids needed to cover the differential current at a certain interval when an internal fault occurs.

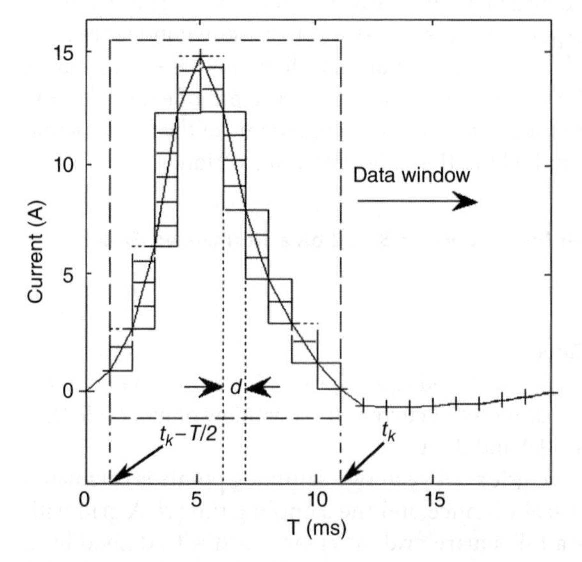

Figure 2.20 The square grids needed to cover the differential current at a certain interval when the transformer is energized.

Figure 2.19 shows the internal fault case of a power transformer. The iron core is not saturated and the magnetizing current is very small, which results in the approximate sine waveform due to its operating in the linear region of the magnetizing characteristic. If t is a half cycle, the N_d curve, which is also called the grille curve, as shown in Figure 2.21, is kept almost constant during any t interval.

In the case of inrush current, the iron core will alternate between saturation and non-saturation, which causes distortions and discontinuities (much different from the normal sine waveform), as shown in Figure 2.20. In this instance, the grille curve, as shown in Figure 2.22, keeps on changing markedly. Therefore, the variation of the grille curve can be used to discriminate between an inrush current and an internal fault current.

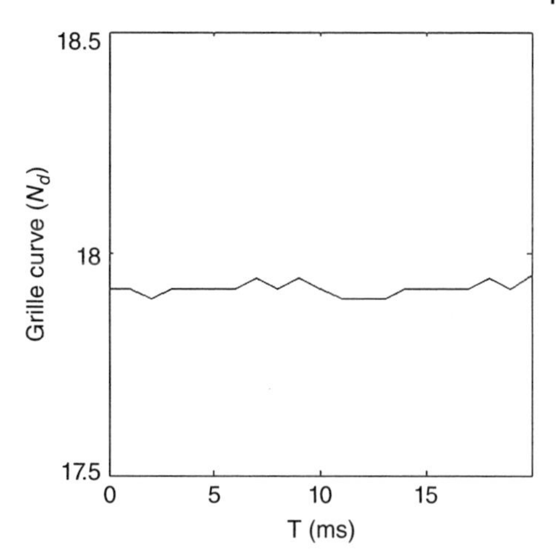

Figure 2.21 The grille curve (N_d) when an internal fault occurs.

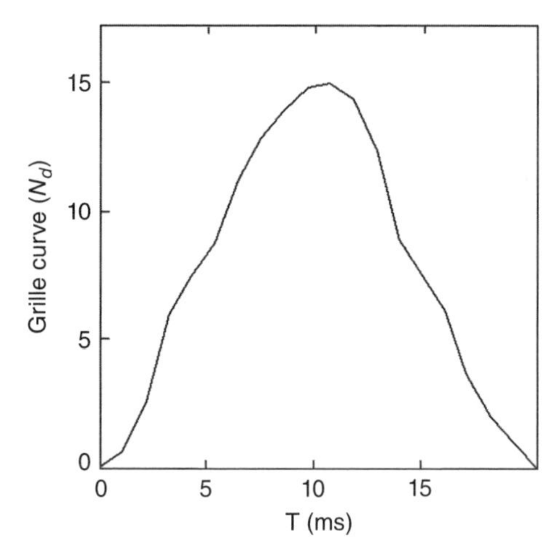

Figure 2.22 The grille curve (N_d) when the transformer is energized.

B. Calculation of Normalized Grille Curves

The normalization of the grille curve is given by

$$B_d(i) = N_d(i)/\max(N_d) \tag{2.21}$$

where $i = 1, \ldots, M$. M is the number of values of N_d. If $M = 1$, the maximum value of N_d is N_d (1). If $M = 2$, max(N_d) is the maximum value of N_d (1) and N_d (2). If N_d (1), …, N_d (M) are obtained, max(N_d) is the maximum value among N_d (1), …, N_d (M). $B_d(i)$ is the ith value in the normalized grille curve. The maximum value of B_d is 1, and the range of B_d is between 0 and 1.

2.2.3.2 NGC-Based Criteria to Identify the Inrush

Two criteria are respectively proposed in the time domain and the frequency domain. The method in the time domain directly detects the variation of the NGC, but that in the frequency domain indirectly reflects the variation by using the ratio between the fundamental frequency component and the DC component.

A. Time Domain Method

The variation of the NGC is directly calculated by using the RMS amplitude given by

$$E_d = \frac{1}{M}\sum_{i=1}^{M} B_d(i) \tag{2.22}$$

$$g = \frac{1}{E_d}\sqrt{\frac{1}{M}\sum_{i=1}^{M}(E_d - B_d(i))^2} \tag{2.23}$$

g is employed to distinguish the inrush current from the internal fault according to the following criterion:

If g exceeds a threshold, the relay judges that there is an inrush current and rejects the tripping. Otherwise, the relay judges that an internal fault has occurred if g is less than the threshold. The threshold should be set to avoid needless operation by the measurement error and the calculation error.

B. Frequency Domain Method

The fundamental frequency component of the NGC is almost zero and the DC component of the NGC is noticeable during an internal fault. However, both the fundamental frequency and DC components of the NGC reveal different characteristics during an inrush current, for the reason that the NGC not only markedly varies but it also has a periodical interval between two minimum values. Therefore, the ratio p between the fundamental frequency and DC components can be used as the criterion for identifying the inrush as follows. If p is larger than a threshold, we make the decision that there is an inrush current in the transformer and block the relay tripping of the differential protection. Otherwise, we make the decision that there is an internal fault and let the relay trip. The ratio p can be calculated by using Fourier analysis.

C. Analysis of NGCs in the Time and Frequency Domains

The calculated NGCs of the two cases in Figures 2.21 and 2.22 were analysed in the time and frequency domains, respectively. For the case of an internal fault, both g and p are negligible and are less than 0.05, as shown in Figures 2.23(a) and (b), respectively. However, for the case of inrush current, both g and p are noticeable, as shown in Figures 2.24(a) and (b), respectively. g is more than 0.65 and p is more than 0.45.

2.2.3.3 Experimental System

To verify the effectiveness of the proposed method, experimental tests were carried out at the Electrical Power Dynamic Laboratory (EPDL). The experimental transformer is a three-phase, two-winding transformer bank with YNd11 connection, which is fed by a low impedance source, as shown in Figure 2.25. The parameters of the two-winding transformers are given in Table 2.7. Three identical current transformers (CTs) are connected in Δ on the primary side, and another three identical CTs are connected in Y on the secondary side of the power transformer.

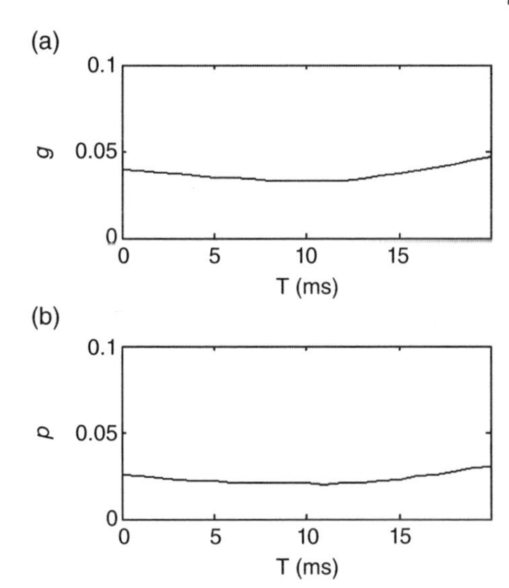

Figure 2.23 The g and p when an internal fault occurs. (a) Analysis of the calculated NGCs in the time domain. (b) Analysis of the calculated NGCs in the frequency domain.

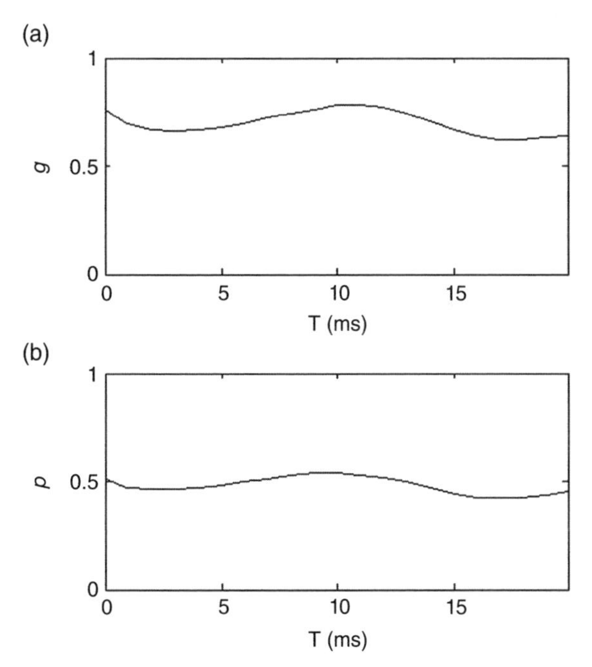

Figure 2.24 The g and p when the transformer is energized. (a) Analysis of the calculated NGCs in the time domain. (b) Analysis of the calculated NGCs in the frequency domain.

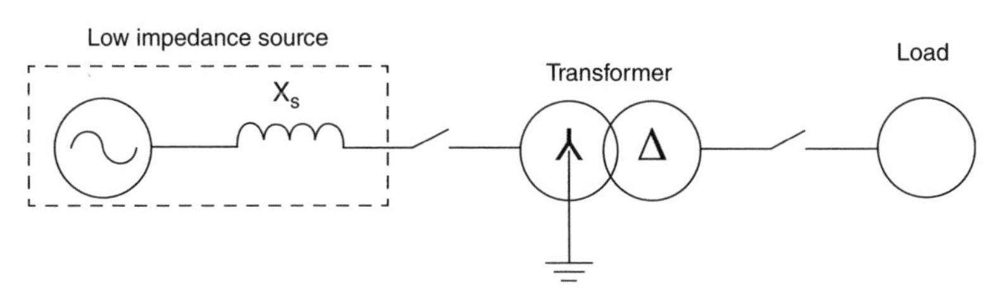

Figure 2.25 Experimental system.

Table 2.7 Parameters of each single-phase unit of transformer used in the test.

Rated power	10 kVA
Rated voltage ratio	1000/380 V
Rated frequency	50 Hz
No load current	1.45%
No load loss	1%
Short circuit voltage	9.0 ~ 15.0%
Short circuit loss	0.35%

The experiments provide different switching and clearing instants for inrush currents, as well as different faults and different numbers of turns for internal faults. A total of 268 cases were tested and divided into five main categories: 56 cases for switching the transformer with no load, 54 cases for faulty conditions only, 52 cases for simultaneous internal fault and inrush conditions, 54 cases for external faults with the CT in saturation, and 52 cases for internal faults with the CT in saturation, to test the algorithm.

2.2.3.4 Testing Results and Analysis

Figures 2.26–2.30 show some examples of the experimental test results: the differential currents and the waveforms of the calculated NGC along with the resulting analysis. In addition, the threshold of g in the time domain is set to 0.25 and the threshold of p in the frequency domain is set to 0.15.

A. Responses to Different Inrush Conditions

A total of 56 cases were carried out in this situation. The inrush current waveform is a function of the different core residual magnetization and the switching instant, so the inrush current waveforms are different from each other. However, in each case, the calculated NGCs alternate between 0 and 1 with a period of one cycle and the maximum value of the NGCs is 1. An example taken from these cases is given in Figure 2.26, where the differential currents of three phases present the inrush currents. The NGCs of the three phases along with their respective results in the time and frequency domains are shown in Figures 2.26(b), (c) and (d), respectively. We find that the calculated NGCs present distorted oscillatory waveforms, which is the key feature of the inrush current. Their analysis results in the frequency domain also show us the noticeable ratio between the fundamental frequency and DC components in the calculated NGCs. These results indicate the severe variation of NGCs and the alternate saturation and exit-saturation of the transformer core during the inrush current period.

B. Responses to Internal Fault Conditions Only

Data from a total of 54 cases were used to calculate the NGCs respectively. Their results were analysed both in the time domain and in the frequency domain.

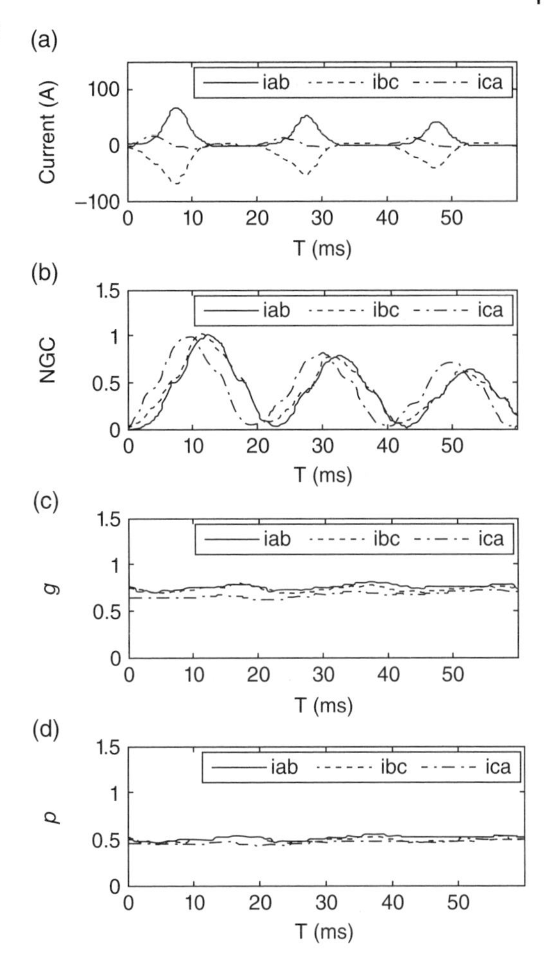

Figure 2.26 Differential currents and experimental results when the transformer is energized. (a) Differential currents. (b) Calculated NGCs. (c) Analysis of the calculated NGCs in the time domain. (d) Analysis of the calculated NGCs in the frequency domain.

An example is shown in Figure 2.27, where a 6.1% turn-to-turn internal fault occurs in phase A.

In Figure 2.27(a), we find that the differential currents i_{ab} and i_{ca} are larger than the nominal magnetizing current and contain a slowly decaying DC component, whereas i_{bc} is within the range of nominal values. Therefore, we need to calculate the NGCs of i_{ab} and i_{ca}, and further analyse them in the time and frequency domains. The calculated NGCs along with their analysis in the time and frequency domains are shown in Figures 2.27(b), (c) and (d), respectively. In the time domain, due to the slowly decaying DC component, the positive and negative half cycles of i_{ab} and i_{ca} are not symmetrical while the DC component is significant. Accordingly, the variation of each NGC calculated by Equation (2.23) presents higher amplitude in the first 30 ms compared with that in the next 30 ms. But even in the first 30 ms, the amplitude of variation is still less than the threshold, as shown in Figure 2.27(c). In the frequency domain, the calculated NGCs of both phases contain negligible ratios between the fundamental frequency and DC components, as shown in Figure 2.27(d), which proves that i_{ab} and i_{ca} are both internal

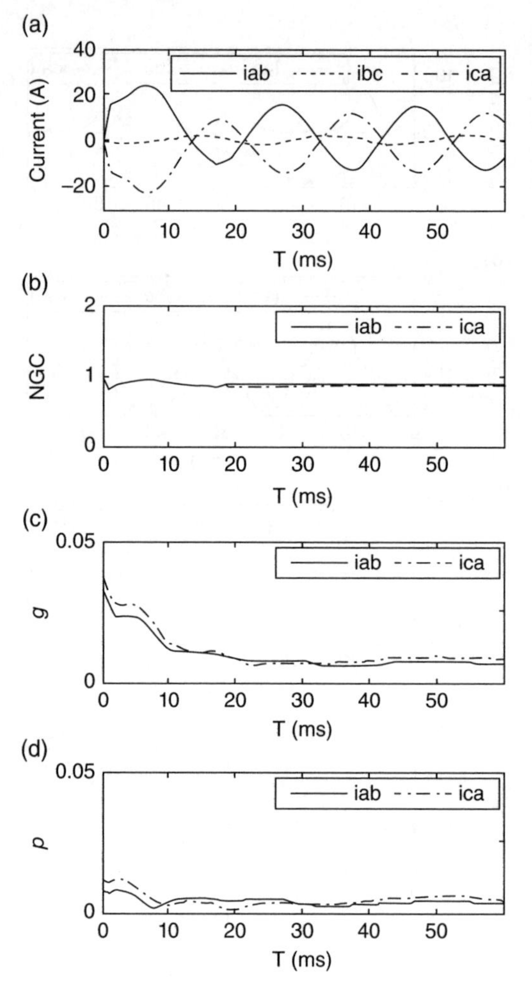

Figure 2.27 Differential currents and experimental results when a 6.1% turn-to-turn internal fault occurs. (a) Differential currents. (b) Calculated NGCs. (c) Analysis of the calculated NGCs in the time domain. (d) Analysis of the calculated NGCs in the frequency domain.

fault currents. In a total of 54 cases, it was found that the method was able to detect the internal fault current even with the decaying DC component.

C. *Responses to Simultaneous Fault and Inrush Conditions*

The switching in the transformer bank with no load often causes the inrush current of non-fault phases, which has been verified by a total of 52 cases with simultaneous inrush currents and internal faults. Figure 2.28, as an example, shows this situation, which is obtained by switching in the transformer bank with no load and a 6.1% turn-to-turn internal fault in phase A. The differential currents of three phases are all distorted severely, as shown in Figure 2.28(a). After analysis in the frequency domain, it can be found that the magnitudes of the second harmonic in fault phases A and C are greater than those of some magnetizing inrush currents. Consequently, the commonly employed conventional differential protection technique based on the second harmonic will have difficulty in distinguishing between an internal fault and an inrush current.

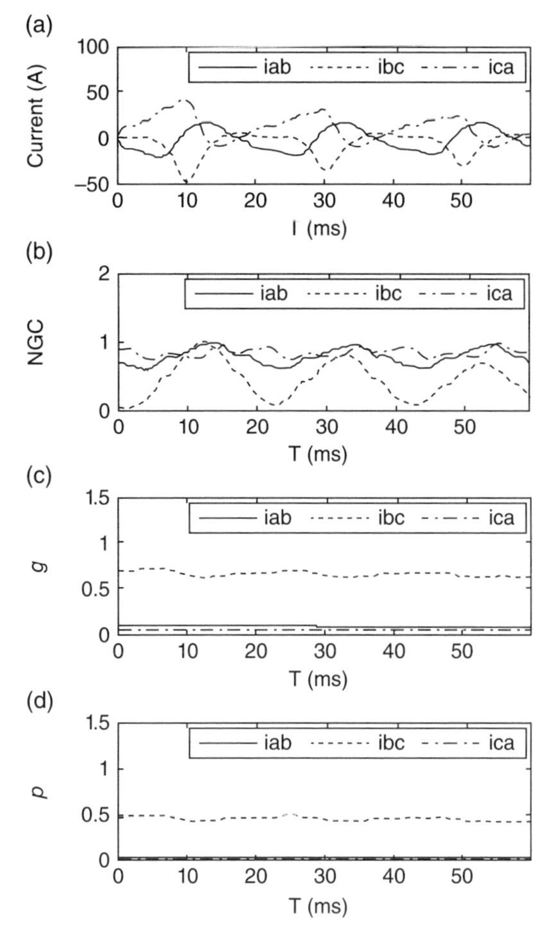

Figure 2.28 Differential currents and experimental results when the transformer is switched with no load and a 6.1% turn-to-turn internal fault occurs. (a) Differential currents. (b) Calculated NGCs. (c) Analysis of the calculated NGCs in the time domain. (d) Analysis of the calculated NGCs in the frequency domain.

Now, the NGCs of three phases are calculated using Equation (2.21). Figures 2.28(b), (c) and (d) show the respective calculated NGCs along with their analysis results in the time and frequency domains. In Figure 2.28(b), the NGC of i_{bc} shows severe variation, whereas the NGCs of i_{ab} and i_{ca} are almost kept constant and only have a little variation resulting from harmonics in the internal fault currents. According to the proposed criterion in the time domain, it is apparent that both i_{ab} and i_{ca} are internal fault currents, and i_{bc} is an inrush current. The same result can be obtained from the frequency domain analysis. The NGCs of i_{ab} and i_{ca} contain negligible ratios between the fundamental frequency and DC components, as shown in Figure 2.28 (d). However, the NGC of i_{bc} is characterized by its noticeable ratio between two components. These results are in accordance with the practical state of the transformer. In the total of 52 cases, the identical results verify that the proposed technique can be used to discriminate internal faults from inrush currents when simultaneous inrush currents and faults occur in the transformer.

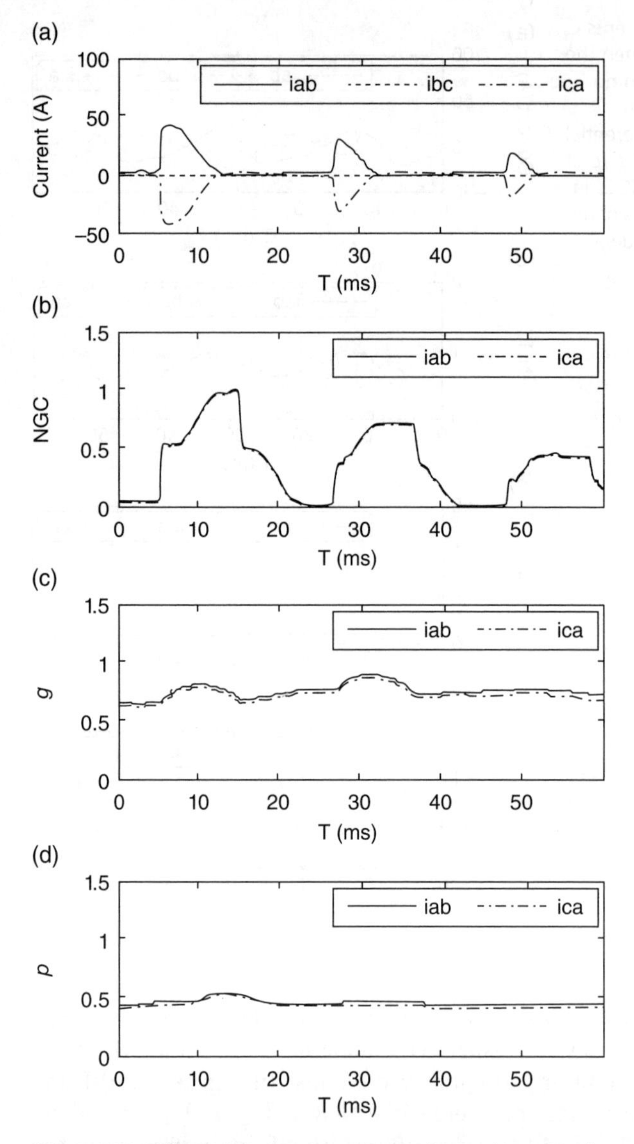

Figure 2.29 Differential currents and experimental results when an external fault occurs with CT saturation. (a) Differential currents. (b) Calculated NGCs. (c) Analysis of the calculated NGCs in the time domain. (d) Analysis of the calculated NGCs in the frequency domain.

D. Responses to External Faults under CT Saturation Conditions

The differential current is designed to restrain under normal load flows and for external faults. However, if the CTs saturate, external short circuits can, in fact, result in very large differential currents, which has been proved by a total of 54 cases. It is thus crucial to be able to ascertain CT saturation effects on the measured currents under external faults.

As an example, Figure 2.29 shows the significant differential currents for an external fault with CT saturation of phase A. In Figure 2.29(a), i_{bc} is within the

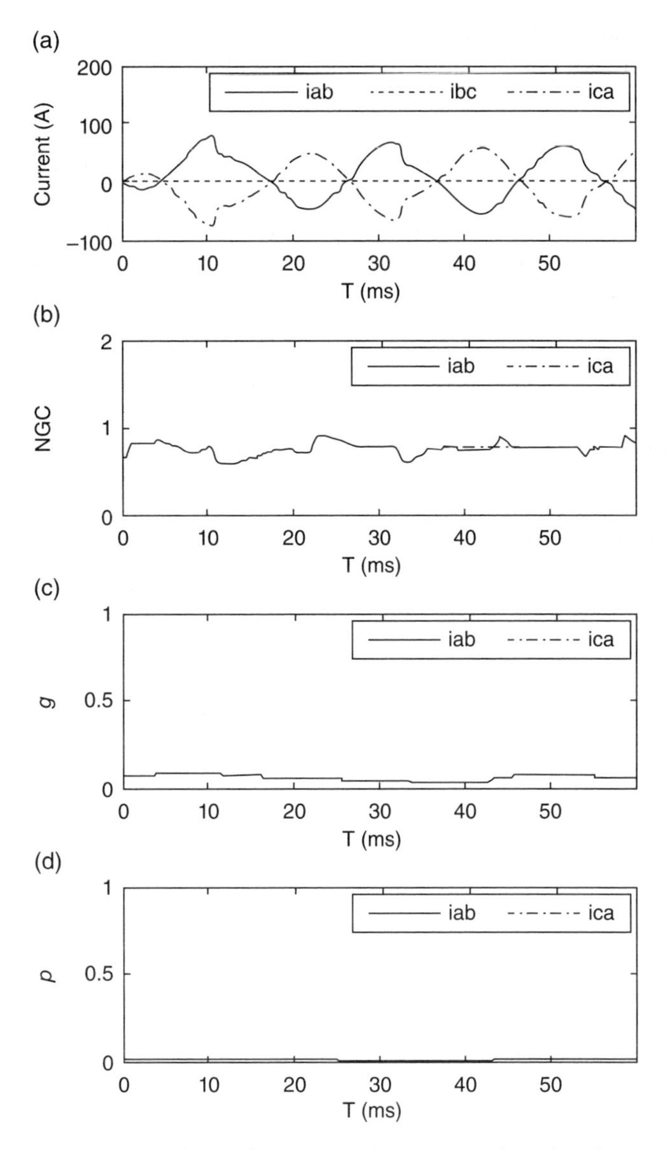

Figure 2.30 Differential currents and experimental results when an internal fault occurs with CT saturation. (a) Differential currents. (b) Calculated NGCs. (c) Analysis of the calculated NGCs in the time domain. (d) Analysis of the calculated NGCs in the frequency domain.

range of nominal magnetizing current, but i_{ab} and i_{ca} are larger than the nominal value. Therefore, the NGCs of i_{ab} and i_{ca} need to be calculated and analysed in the time and frequency domains. The NGCs of i_{ab} and i_{ca} are characterized by distorted oscillatory waveforms, as shown in Figure 2.29(c). In Figure 2.29(d), the noticeable ratios between the fundamental frequency and DC components in the frequency domain are presented. Compared with the internal fault conditions, it is apparent that the method is capable of recognizing it as a non-fault condition. It is found from all 54 cases that this method has strong anti-saturation capability for external faults of transformers.

E. Responses to Internal Faults under CT Saturation Conditions

The internal fault current with CT saturation may include a distorted oscillatory waveform for the first few cycles. If conditions are severe enough, it is possible that the distortion may be even worse, and the saturation can start to occur even sooner. An example taken from 52 cases is shown in Figure 2.30, where an internal fault occurs with CT saturation of phase A. The differential currents i_{ab} and i_{ca} are larger than the nominal magnetizing current in Figure 2.30(a). Therefore, we need to calculate the NGCs of i_{ab} and i_{ca}, and further analyse them in the time and frequency domains. The variations in waveforms of i_{ab} and i_{ca} only have a small effect on their NGCs, as shown in Figure 2.30(b). In this case, the internal fault can be identified easily since the g and p are both negligible and much less than the threshold, as shown in Figures 2.30(c) and (d), respectively. In the total of 52 cases, the protection can issue the trip command accurately, which verifies that the method can be used to identify internal faults under CT saturation conditions.

2.2.4 Adaptive Method to Identify CT Saturation Using Grille Fractal

2.2.4.1 Analysis of the Behaviour of CT Transient Saturation

CT saturation is divided into steady-state saturation and transient saturation: steady-state saturation is the saturation of a CT iron core when the primary current is in the steady state. For a low-voltage grid, it is not difficult for general protection devices to adapt to CT steady-state saturation, but there will always be a transient process after an actual line fault, during which the DC component is greater than the AC component in the excitation circuit. So the CT transient error increases, the performance of which is quite different from the steady-state case. At this point, the CT saturation is called transient saturation. The influence, when a CT is transient saturated, on transfer characteristics is discussed as follows.

Suppose that an external fault occurs when the CT is not saturated, and the primary current of the CT is

$$i_1 = I_{1m}(A - \cos\omega t) \tag{2.24}$$

where I_{1m} is the maximal value of short-circuit current of the primary side; $A = e^{-t/T}$ is the attenuation DC component; T is the time constant of the primary system. The excitation current is very small when the CT is not saturated, namely $i_\mu = 0$. During the initial stage of the fault, the primary current i_1 is transmitted to the secondary side without any distortion. The transient current of the secondary side is

$$i_2 = I_{2m}(A - \cos\omega t) \tag{2.25}$$

where I_{2m} is the maximal value of the short-circuit current of the secondary side.

From the equivalent circuit we can find that

$$W_2 \frac{d\phi'}{dt} = i_2 r_2 \tag{2.26}$$

$$\Delta\phi' = \frac{1}{W_2}\int_{t_1}^{t_2} i_2 r_2 dt \tag{2.27}$$

where W_2 is the number of turns of the secondary side; $\Delta\phi'$ is the increment of iron core flux ϕ' within one cycle; r_2 is the secondary load of the instrument transformer with only resistance accounted for, and reactance excluded. If Equation (2.25) is substituted into Equation (2.27), we obtain

$$\Delta\phi' = \frac{2I_{2m}r_2}{\omega W_2}\left[A(\pi - \arccos A) + \sqrt{1-A^2}\right] \tag{2.28}$$

Continuously after n cycles, the iron core flux reaches the saturation value of CT, and that value is

$$\phi'_r + n\Delta\phi' = \phi'_s \tag{2.29}$$

where ϕ'_r is the iron core remanence of the CT. We can show that the number of cycles when the CT begins saturation is

$$n = \frac{\omega W_2(\phi'_s - \phi'_r)}{2I_{2m}r_2\left[A(\pi - \arccos A) + \sqrt{1-A^2}\right]} \tag{2.30}$$

From Equation (2.28), we can see that although the CT will not be saturated immediately, there is a region of linear transmission, the size of which, however, is constrained by many factors: (1) the size and the degree of deviation of the primary-side short-circuit current; (2) the time constant of the primary-side system; (3) the magnitude of the load of the secondary side; (4) the magnitude and direction of the remanence of the CT and the saturation value and so on. When an external fault occurs, and the magnitude and direction of the remanence are close to those of the saturation value, the CT is likely to be saturated in a very short time, which brings some difficulties for the time difference method.

From the preceding analysis, we can see that for the two types of CT saturation when inrush current and external fault occur, the iron core will periodically enter and exit the saturation region. There will be singular points when the iron core enters and exits the saturation region. The symmetry of the inrush current is no exception. When a fault occurs inside transformers, due to the iron core being in the linear region, there will be singular points only during the fault time. Although the number of high-frequency details of the fault current increases with CT saturation, in fault time the high-frequency details are less obvious than singular points. So the fault inside the transformer is recognized from the singular characteristics of the current waveform.

2.2.4.2 The Basic Principle and the Algorithm of the Grille Fractal

The fractal is a type of self-similar graphic and structure that has no characteristic length but a particular form, the typical nature of which is local similarity shown in small scale. A singular signal has self-similarity, and the theory of fractals has been successfully applied in singular signal detection in many fields in power systems. Because the sampling frequency is not required for the grille fractal algorithm, and there is no need to divide the band, it is suitable for real-time signal processing, and is favoured by more and more researchers. The definition of the grille is referenced to the grille fractal, and a brief and fast criterion of signal singularity detection is proposed. The description of the theory is as follows:

For signal X, suppose that there are $n + 1$(n is even) sampling points ($x_1, x_2, \ldots, x_{n+1}$) in a period $[t_k - \Delta t, t_k]$. Making $\delta = \Delta t/n$, then

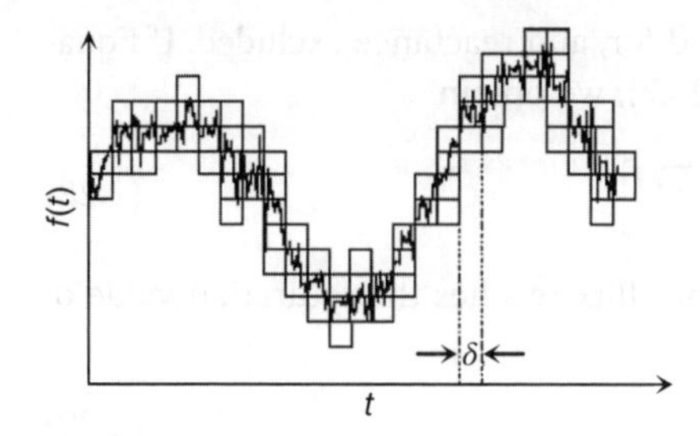

Figure 2.31 Grille definition.

$$N_\delta = \frac{1}{\delta}\sum_{j=1}^{n}\left|x_j - x_{j+1}\right| \tag{2.31}$$

where δ is the time difference of two sampling points; N_δ is the required number of grid lattice square grids, δ is the side length, overlaying the signal in the period of $[t_k - \Delta t,\, t_k]$, as shown in Figure 2.31.

The differential current of the transformer during normal operation approximates to a pure sinusoidal signal. Suppose that Δt is half a cycle; the N_δ of any signal is the same for any period Δt. If a fault occurs outside the transformer, due to the DC components in the short-circuit current during transient processing, the error in the DC component transmission of the CT is so large that the CT will be saturated and the induced electromotive force of the secondary side will go down to zero immediately. During that period, i_2 is also zero. Along with the instantaneous decrease in the value of the current and with the CT exiting saturation, the induced electromotive force of the secondary side increases and i_2 begins to increase again. It is shown that, on entering and exiting saturation of the CT, the waveform of i_2 will be seriously distorted and N_δ will change with the distortion points. Similarly, due to the highly non-linear characteristics of the iron core, there must be a large number of high-frequency details in the inrush current around the time that the CT enters and exits saturation. Performed in the grille curve, the trend of increasing and then decreasing will be caught by both of them.

When a fault occurs inside transformers, a high-frequency singular signal in the fault current will be generated at the fault point. N_δ will change exclusively until the waveform is restored to a sinusoidal waveform. In the grille curve, there will be a trend of increasing and then becoming steady.

From the analysis above, it can be shown that internal and external faults of transformers can be effectively recognized when the CT is saturated using the grille curve variation, which will not be influenced by inrush current. In practice, the data of a transformer computer protection device sampled from the primary side of the power system is easily corrupted by interference and noise. In a reaction to the grille curve, N_δ may fluctuate, which leads to an increase in the number of extremal points of the curves. If an internal fault occurs and the CT is saturated, there will be a trend of increasing and then decreasing for the grille curve, which may lead to protection malfunction if only the grille fractal is analysed. So a self-adaptive generalized morphological filter is proposed. The singular signal is sampled and the noise and interference restrained.

2.2.4.3 Self-Adaptive Generalized Morphological Filter

In recent years, the filtering method based on mathematical morphology has been applied increasingly due to its clear physical meaning, efficiency and practicality. The algorithm is mainly concentrated on morphological opening and closing operation, morphological closing and opening operation and their average combination. The structural elements of different sizes B_1 and B_2 are selected to remove noise and interference signals effectively. Suppose that the width of the post-structural element B_2 is twice that of the pre-structural element B_1. Based on morphological opening and closing operation – a kind of generalized morphological opening–closing and closing–opening filter – the definitions are, respectively:

$$y_1(n) = \mathrm{GOC}(x(n)) = (x \circ B_1 \bullet B_2)(n) \tag{2.32}$$

$$y_2(n) = \mathrm{GCO}(x(n)) = (x \bullet B_1 \circ B_2)(n) \tag{2.33}$$

Because there is still a statistical bias phenomenon in the two kinds of filter, it is difficult to obtain the best filtering effect used alone. Therefore, the average combination of the two kinds of generalized filter is adopted. The filtered output signal $y(n)$ is

$$y(n) = 0.5*(y_1(n) + y_2(n)) \tag{2.34}$$

The shape and size of the structural element is another important factor to determine the effect of morphological filtering. Taking into account the characteristics of the computation and grille curve variation, flat structural elements, which are the most suitable for smoothing, are selected for the analysis. In addition, the opening operation can guarantee the width of the structural element being greater than the width of the maximum of the noise. In order to maximize noise suppression, the width of the structural element B_1 must be greater than the maximal opening.

2.2.4.4 The Design of Protection Program and the Verification of Results

A. Dynamic Simulation System

According to the above analysis, the protection criterion is formulated to recognize faults inside transformers when the CT is saturated. The extreme phenomenon, where the CT of one side is deeply saturated and that of the other side transmits correctly, is considered. At that time, the differential value of the secondary current of the CT is very large, which is likely to cause differential protection malfunction. The practicality and feasibility of the criterion is tested by a dynamic simulation system. The wiring of the dynamic simulation system and the parameters of the electrical components are shown in Figure 2.32.

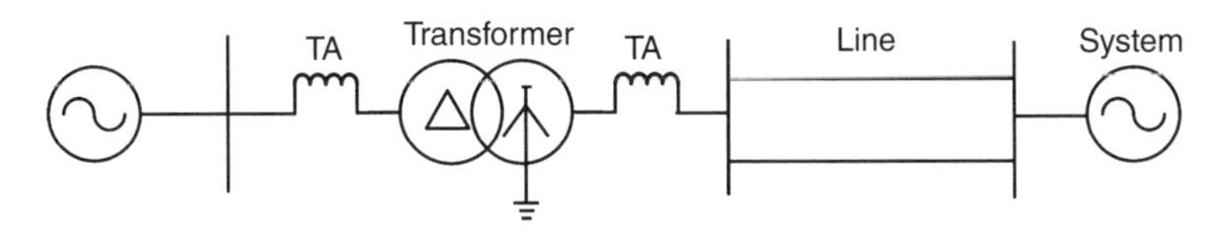

Figure 2.32 Electrical Power Dynamic Laboratory test model and parameters.

In Figure 2.32, the transformer is a two-winding, three-phase transformer, and the wiring of the transformer is YNd11, variable ratio 19 kV/550 kV, rated volume $S_{\text{rated}} = 670$ MVA, short-circuit reactance $X_{\text{short-circuit}} = 13\%$; the parameters of the power source are $P_{\text{rated}} = 600$ MW, $U_{\text{rated}} = 19$ kV, $I_{\text{rated}} = 20.26$kA; the input capacity of the system is $System = 11{,}000$ MVA; the length of line is 257 km, parameter $Z_1 = 0.01808 + \text{j}0.27747\ \Omega/\text{km}, C_1 = 0.012917\ \mu\text{F/km}, Z_0 = 0.23084 + \text{j}0.9728\ \Omega/\text{km}, C_0 = 0.0081161\ \mu\text{F/km}$.

B. Analysis of External Fault and CT Saturation

The waveform of differential current in the case of an external fault and CT saturation is shown in Figure 2.33(a). Compared with the grille variation curve, it is shown that the fluctuation of differential current is small and N_δ stays at a low level before the fault. After the fault, N_δ varies with the increase and decrease of the distortion degree of the current waveform, and the singular points of the waveform correspond to the maximal value or minimal value of N_δ. It is clear that the grille curve retains bumps or pits formed by disturbance or noise signals, which are easily confused with the extreme points, to which the singular points correspond. Because flat structural elements are applied in morphological filters, bumps and pits are smoothed and small pieces of smooth region near extreme points can be caught, as shown in Figure 2.33(c). Even though the noise or interference signal is not completely covered, transformer protection can be blocked by the waveform characteristics of the minimal value of the smooth domain – maximal value of the smooth domain – minimal value of the smooth domain.

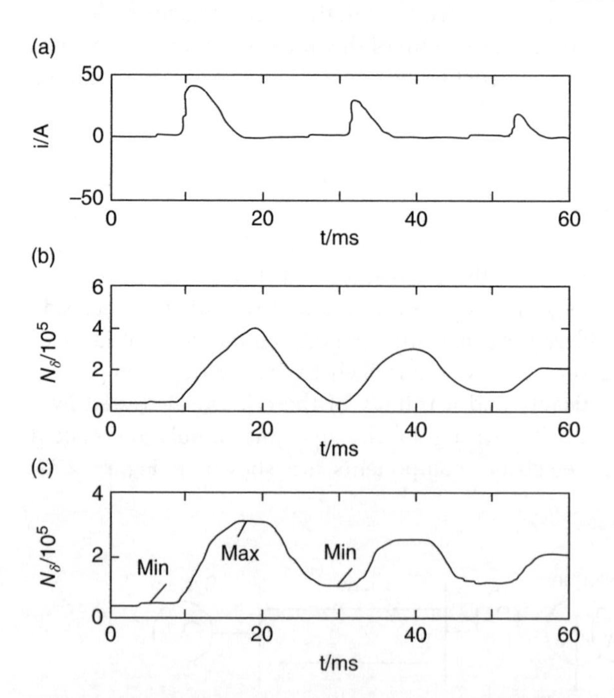

Figure 2.33 The differential current waveform and its results under CT saturation at an out-zone fault. (a) Differential current. (b) Grille variation curve. (c) Processed grille variation curve.

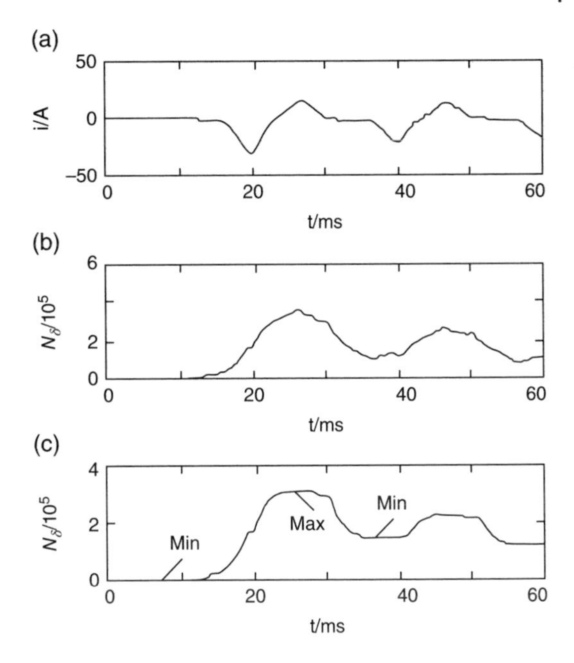

Figure 2.34 The symmetrical inrush current and its results. (a) Differential current. (b) Grille variation curve. (c) Processed grille variation curve.

C. Analysis on Inrush Current

Symmetrical inrush current in the case of transformer no-load closing is shown in Figure 2.34(a). As the waveform is no longer biased to the timeline side, the differential relay of pattern BCH-1 blocked by DC current components will malfunction. From the grille variation curve shown in Figure 2.34(b), it can be seen that the singular points formed by the transformer iron core entering and exiting saturation correspond to the extreme points of N_δ. Smoothed by the filter, the waveform shown in Figure 2.34(c) also shows the characteristic of the minimal value of the smooth domain – maximal value of the smooth domain – minimal value of the smooth domain. The differential protection restraint can be achieved by this characteristic.

D. Analysis of Internal Fault and CT Saturation

The waveform in the case of internal fault and CT saturation is shown in Figure 2.35(a). From Figure 2.35(b), it can be seen that N_δ varies exclusively with the distortion of the differential current, and then the grille curve flattens with the reduced degree of waveform distortion. There are still small fluctuations on the curve due to CT saturation and the influences of interference. The variation trend of rising before becoming steady is not clear. The noise and disturbance signals are effectively restrained by generalized self-adaptive filters, and the singular characteristics near fault points will not be fuzzied. As shown in Figure 2.35(c), only the waveform characteristic of the minimal value of the smooth domain – maximal value of the smooth domain is retained after processing. Compared with Figures 2.33(c) and 2.34(c), the waveform in Figure 2.35(c) is lacking the variation trend from minimal value domain to maximal value domain. So it is pivotal for

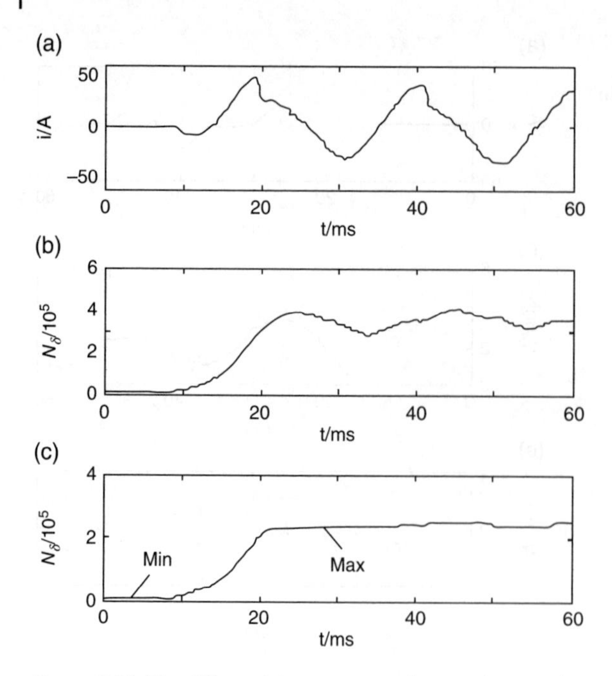

Figure 2.35 The differential current waveform and its results under CT saturation in the case of internal fault. (a) Differential current. (b) Grille variation curve. (c) Processed grille variation curve.

recognizing the maximal value domain for reliable protection action. After the start of transformer differential protection, the N_δ of the differential current at time t_k and N_δ at a time of 1 ms before t_k should be measured in real time. If they are equal and five points taken sequentially are also equal, it is determined that the points after time $(t_k - 1)$ ms fall in the maximal smooth domain.

After the maximal value smooth domain is determined, the average of the grille variation can be calculated from Equation (2.35) from $(t_k - 1)$ ms, $M = 10$ points.

$$B_e = \left| \frac{1}{M} \sum_{i=0}^{M-1} \frac{N_\delta^2(i)}{N_\delta^2(i+S)} - 1 \right| \tag{2.35}$$

where: $N_\delta(0)$ is the number of the grille at time $(t_k - 1)$ ms; S is the number of sampling points during half a cycle.

The results of three typical cases, respectively measured in 15 dynamic simulation tests, are listed in Table 2.8. The transformer internal short-circuit faults can be recognized by setting a threshold value of 0.5.

E. Analysis of a Special Situation

When considering an external fault and a CT not in immediate saturation, it should be distinguished by the conversion of the fault zone. Protection is blocked by the ratio restraint characteristic when an external fault occurs and the CT is not saturated. When the CT is saturated after several cycles, protection is blocked by

Table 2.8 Calculation results of B_e in various kinds of states.

	CT saturation		Symmetrical inrush current
	Internal fault	External fault	
Half cycle before $N_\delta/10^5$	5.7997 ~ 6.6973	2.7816 ~ 3.4374	2.7655 ~ 3.2906
Half cycle after $N_\delta/10^5$	5.7369 ~ 6.6492	0.7988 ~ 1.2859	1.1852 ~ 1.6547
$B_e/10^5$	0.019 ~ 0.034	4.5936 ~ 12.1857	2.7481 ~ 5.3794

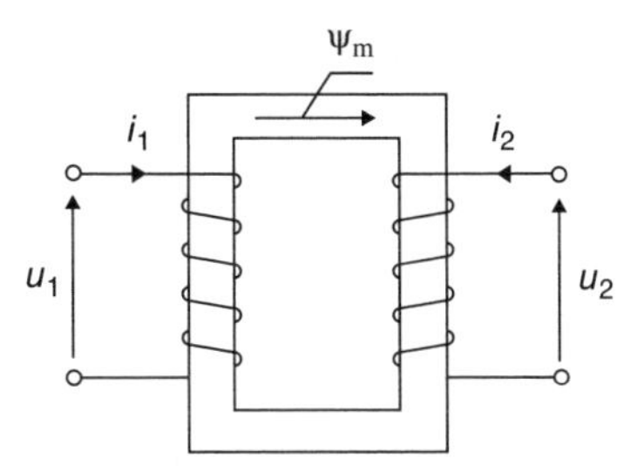

Figure 2.36 A two-winding, single-phase transformer.

the above methods; a transitional fault occurs and the CT is saturated after several cycles, which belongs to a complex of faults. The external fault current, superimposed on the internal fault current waveform, attenuating after several cycles, shows the characteristics of an internal fault, namely there will not be an obvious singular signal near the differential current zero-crossing point, so the protection can act reliably.

In summary, the use of grid fractal theory and adaptive generalized morphological filtering technology to identify CT saturation has the following characteristics:

1) A signal singular feature is not fuzzied by adaptive generalized morphological filtering technology; the noise and interference signals can be effectively filtered; the DC component is restrained very well.
2) Using the relative size of the grille curve, the maximal value and the minimal value of the smooth domain, transformer faults and CT saturation can be correctly recognized by setting a reasonable threshold.
3) The disadvantages of a DC speed saturation relay are overcome, and are not affected by inrush current. The characteristics are obvious, and easy to achieve. The effectiveness and feasibility of this algorithm are verified by dynamic simulation data.

2.2.5 Algorithm for Discrimination Between Inrush Currents and Internal Faults Based on Equivalent Instantaneous Leakage Inductance

2.2.5.1 Basic Principle

A. Physical Theory

Consider a two-winding, single-phase transformer, as shown in Figure 2.36. The primary and secondary voltages can be expressed as:

$$u_1 = i_1 r_1 + L_1 \frac{di_1}{dt} + \frac{d\psi_m}{dt} \tag{2.36}$$

$$u_2 = i_2 r_2 + L_2 \frac{di_2}{dt} + \frac{d\psi_m}{dt} \tag{2.37}$$

where u_1 and u_2 are the voltages of the primary and secondary windings, respectively. i_1 and i_2 are the currents of the primary and secondary windings, respectively. r_1 and r_2 are the resistances of the primary and secondary windings, respectively. L_1 and L_2 are the leakage inductances of the primary and secondary windings, respectively. ψ_m is the mutual flux linkage. The equations consider that the transformation ratio is one.

The mutual flux linkage of the primary and secondary windings is equal and can be eliminated by using Equations (2.36) and (2.37) as follows:

$$u_{12} = L_1 \frac{di_1}{dt} - L_2 \frac{di_2}{dt} \tag{2.38}$$

with

$$u_{12} = u_1 - u_2 - i_1 r_1 + i_2 r_2 \tag{2.39}$$

Under the assumption that the parameters of the transformer L_1 and L_2 are given, it is then possible to calculate the right-hand side of Equation (2.38). The computed values of the right-hand side and the actual values of the left-hand side are equal during magnetizing inrush and normal operations. However, these values are not equal during internal faults. Therefore, this equation is an inherent feature of internal faults, which can be used to discriminate inrush currents from internal faults.

B. *Equivalent Instantaneous Leakage Inductance*

In fact, it is very difficult to calculate the right-hand side of Equation (2.38) because it depends on the leakage inductances which are determined by the size, configuration and location of the transformer windings. Furthermore, because the physical dimensions vary, internal faults make it barely possible for even an approximation of the leakage inductances to be obtained from the transformer design data. This problem is solvable when the EILI definition is employed as a solution.

The trapezoid principle is adopted in Equation (2.38) to transform the continuous differential equation into a discrete difference equation. The digital expressions at kT and $(k+1)\,T$ instants are given by

$$u_{12}(k) = L_{1k} \frac{i_1(k+1) - i_1(k-1)}{2T} - L_{2k} \frac{i_2(k+1) - i_2(k-1)}{2T} \tag{2.40}$$

$$u_{12}(k+1) = L_{1k} \frac{i_1(k+2) - i_1(k)}{2T} - L_{2k} \frac{i_2(k+2) - i_2(k)}{2T} \tag{2.41}$$

where T is the sampling cycle.

Each of L_{1k} and L_{2k} is defined as the EILI, which will be constant when there is an inrush current and during normal operation, but it will no longer be constant when there is an internal fault. Therefore, the EILI, which is equivalent to leakage inductances in the discrimination between internal faults and inrush current, exactly presents the inherent status of the transformer.

Consider that the parameters of the transformer, r_1 and r_2, are known. The differential currents and voltages at kT and $(k+1)\,T$ instants are utilized and the calculated EILIs of primary and secondary windings at kT instant are written as:

$$L_{1k} = 2T[u_{12}(k)i_2(k+2) - u_{12}(k)i_2(k) - u_{12}(k+1)i_2(k+1) + u_{12}(k+1)i_2(k-1)] \\ *[(i_1(k+1) - i_1(k-1))*(i_2(k+2) - i_2(k)) - (i_1(k+2) - i_1(k))*(i_2(k+1) - i_2(k-1))]^{-1} \tag{2.42}$$

$$L_{2k} = 2T[u_{12}(k)i_1(k+2) - u_{12}(k)i_1(k) - u_{12}(k+1)i_1(k+1) + u_{12}(k+1)i_1(k-1)] \\ *[(i_2(k+2) - i_2(k))*(i_1(k+1) - i_1(k-1)) - (i_2(k+1) - i_2(k-1))*(i_1(k+2) - i_1(k))]^{-1} \tag{2.43}$$

C. Two-Winding, Three-Phase Δ/Y Transformer

Figure 2.37 shows the connections of the primary and secondary windings of a Δ/Y transformer. The following equations express the Δ and Y connected windings as functions of the mutual flux linkages and the currents in the windings:

$$u_a = i_a r + L_a \frac{di_a}{dt} + \frac{d\psi_{ma}}{dt} \tag{2.44}$$

$$u_b = i_b r + L_b \frac{di_b}{dt} + \frac{d\psi_{mb}}{dt} \tag{2.45}$$

$$u_c = i_c r + L_c \frac{di_c}{dt} + \frac{d\psi_{mc}}{dt} \tag{2.46}$$

$$u_A = i_A R + L_A \frac{di_A}{dt} + \frac{d\psi_{ma}}{dt} \tag{2.47}$$

$$u_B = i_B R + L_B \frac{di_B}{dt} + \frac{d\psi_{mb}}{dt} \tag{2.48}$$

$$u_C = i_C R + L_C \frac{di_C}{dt} + \frac{d\psi_{mc}}{dt} \tag{2.49}$$

where the parameters of the secondary side have been converted to the primary side by the transformer ratio. u_a, u_b and u_c are the voltages of primary windings a, b and c, respectively. i_a, i_b and i_c are the currents of the primary windings. L_a, L_b and L_c are the leakage inductances of the primary windings. r is the resistance of the

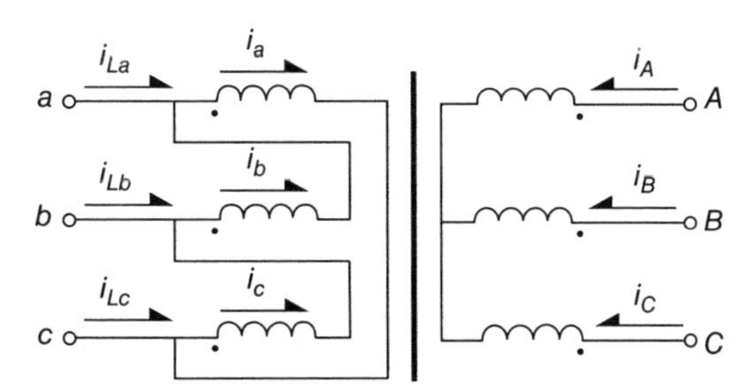

Figure 2.37 A two-winding, three-phase Δ/Y transformer.

primary windings. u_A, u_B and u_C are the voltages of secondary windings A, B and C, respectively. i_A, i_B and i_C are the currents in the secondary windings. L_A, L_B and L_C are the leakage inductances of the secondary windings. R is the resistance of the secondary windings. ψ_{ma}, ψ_{mb} and ψ_{mc} are the mutual flux linkages.

i_a, i_b and i_c are the currents of delta-connected windings, which it is not possible to measure in many situations. In order not to position the CTs within the delta loop to get the exact phase current, the line currents as functions of the currents in the delta-connected windings are expressed as follows:

$$i_{La} = i_a - i_b \tag{2.50}$$

$$i_{Lb} = i_b - i_c \tag{2.51}$$

$$i_{Lc} = i_c - i_a \tag{2.52}$$

Consider the leakage inductances to be constant and equal in the normal operational state and during the inrush current period: $L_a = L_b = L_c = L_1$, $L_A = L_B = L_C = L_2$. The equations of the primary and secondary sides can be written as

$$u_a - u_b = i_{La}r + L_1\frac{di_{La}}{dt} + \frac{d(\psi_{ma} - \psi_{mb})}{dt} \tag{2.53}$$

$$u_b - u_c = i_{Lb}r + L_1\frac{di_{Lb}}{dt} + \frac{d(\psi_{mb} - \psi_{mc})}{dt} \tag{2.54}$$

$$u_c - u_a = i_{Lc}r + L_1\frac{di_{Lc}}{dt} + \frac{d(\psi_{mc} - \psi_{ma})}{dt} \tag{2.55}$$

$$u_A - u_B = (i_A - i_B)R + L_2\frac{d(i_A - i_B)}{dt} + \frac{d(\psi_{ma} - \psi_{mb})}{dt} \tag{2.56}$$

$$u_B - u_C = (i_B - i_C)R + L_2\frac{d(i_B - i_C)}{dt} + \frac{d(\psi_{mb} - \psi_{mc})}{dt} \tag{2.57}$$

$$u_C - u_A = (i_C - i_A)R + L_2\frac{d(i_C - i_A)}{dt} + \frac{d(\psi_{mc} - \psi_{ma})}{dt} \tag{2.58}$$

The flux linkages mutual to the primary and secondary windings of each phase are equal and can be eliminated by using Equations (2.53)–(2.58) as follows:

$$u_{AaBb} = L_1\frac{di_{La}}{dt} - L_2\frac{d(i_A - i_B)}{dt} \tag{2.59}$$

$$u_{BbCc} = L_1\frac{di_{Lb}}{dt} - L_2\frac{d(i_B - i_C)}{dt} \tag{2.60}$$

$$u_{CcAa} = L_1\frac{di_{Lc}}{dt} - L_2\frac{d(i_C - i_A)}{dt} \tag{2.61}$$

with

$$u_{AaBb} = u_a - u_b - u_A + u_B - i_{La}r + (i_A - i_B)R \tag{2.62}$$

$$u_{BbCc} = u_b - u_c - u_B + u_C - i_{Lb}r + (i_B - i_C)R \tag{2.63}$$

$$u_{CcAa} = u_c - u_a - u_C + u_A - i_{Lc}r + (i_C - i_A)R \tag{2.64}$$

The trapezoid principle is adopted in Equation (2.59) to transform the continuous differential equation into a discrete difference equation. The digital expressions at kT and $(k + 1)\,T$ instants are given by:

$$u_{\mathrm{AaBb}}(k) = L_{1k}\frac{i_{La}(k+1)-i_{La}(k-1)}{2T} - L_{2k}\frac{(i_{A}(k+1)-i_{B}(k+1))-(i_{A}(k-1)-i_{B}(k-1))}{2T} \tag{2.65}$$

$$u_{\mathrm{AaBb}}(k+1) = L_{1k}\frac{i_{La}(k+2)-i_{La}(k)}{2T} - L_{2k}\frac{(i_{A}(k+2)-i_{B}(k+2))-(i_{A}(k)-i_{B}(k))}{2T} \tag{2.66}$$

The EILIs of L_{1k} and L_{2k} at kT instant can be calculated in real time by utilizing Equations (2.65) and (2.66). A similar procedure provides the other two groups of L_{1k} and L_{2k} by using Equations (2.60) and (2.61), respectively.

D. Three-Winding, Three-Phase Y_0/Y/Δ Transformer

Consider a three-winding, three-phase transformer, Y_0/Y/Δ connection, whose primary windings are A1, B1 and C1, the secondary windings are A2, B2 and C2, and the tertiary windings are A3, B3 and C3. The following equations express the voltages of the windings as functions of the mutual flux linkages and the currents of the windings,

$$u_{a1} = i_{a1}r_1 + L_1\frac{di_{a1}}{dt} + m_{21}\frac{di_{a2}}{dt} + m_{31}\frac{di_{a3}}{dt} + \frac{d\psi_{ma}}{dt} \tag{2.67}$$

$$u_{b1} = i_{b1}r_1 + L_1\frac{di_{b1}}{dt} + m_{21}\frac{di_{b2}}{dt} + m_{31}\frac{di_{b3}}{dt} + \frac{d\psi_{mb}}{dt} \tag{2.68}$$

$$u_{c1} = i_{c1}r_1 + L_1\frac{di_{c1}}{dt} + m_{21}\frac{di_{c2}}{dt} + m_{31}\frac{di_{c3}}{dt} + \frac{d\psi_{mc}}{dt} \tag{2.69}$$

$$u_{a2} = i_{a2}r_2 + L_2\frac{di_{a2}}{dt} + m_{12}\frac{di_{a1}}{dt} + m_{32}\frac{di_{a3}}{dt} + \frac{d\psi_{ma}}{dt} \tag{2.70}$$

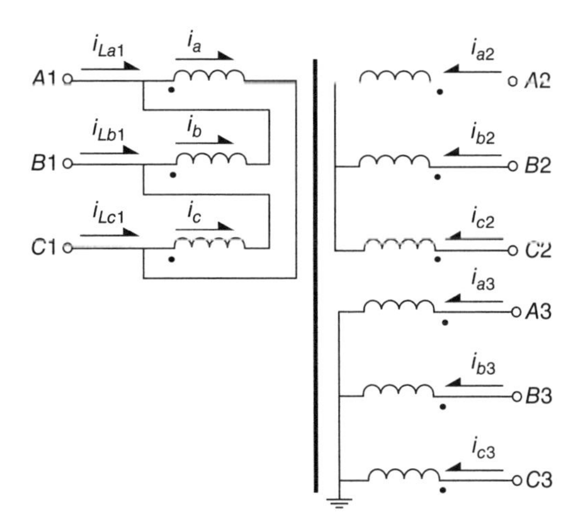

Figure 2.38 A three-winding, three-phase Y_0/Y/Δ transformer.

$$u_{b2} = i_{b2}r_2 + L_2\frac{di_{b2}}{dt} + m_{12}\frac{di_{b1}}{dt} + m_{32}\frac{di_{b3}}{dt} + \frac{d\psi_{mb}}{dt} \quad (2.71)$$

$$u_{c2} = i_{c2}r_2 + L_2\frac{di_{c2}}{dt} + m_{12}\frac{di_{c1}}{dt} + m_{32}\frac{di_{c3}}{dt} + \frac{d\psi_{mc}}{dt} \quad (2.72)$$

$$u_{a3} = i_{a3}r_3 + L_3\frac{di_{a3}}{dt} + m_{13}\frac{di_{a1}}{dt} + m_{23}\frac{di_{a2}}{dt} + \frac{d\psi_{ma}}{dt} \quad (2.73)$$

$$u_{b3} = i_{b3}r_3 + L_3\frac{di_{b3}}{dt} + m_{13}\frac{di_{b1}}{dt} + m_{23}\frac{di_{b2}}{dt} + \frac{d\psi_{mb}}{dt} \quad (2.74)$$

$$u_{c3} = i_{c3}r_3 + L_3\frac{di_{c3}}{dt} + m_{13}\frac{di_{c1}}{dt} + m_{23}\frac{di_{c2}}{dt} + \frac{d\psi_{mc}}{dt} \quad (2.75)$$

where the parameters of the secondary and the tertiary sides have been converted to the primary side by the transformer ratio. u_{a1}, u_{b1} and u_{c1} are the voltages of the primary windings A1, B1 and C1, respectively. i_{a1}, i_{b1} and i_{c1} are the currents of the primary windings. i_{La1}, i_{Lb1} and i_{Lc1} are the line currents of the primary windings. r_1 is the resistance of the primary windings. L_1 is the self-leakage inductance of the primary windings. u_{a2}, u_{b2} and u_{c2} are the voltages of the secondary windings A2, B2 and C2, respectively. i_{a2}, i_{b2} and i_{c2} are the currents of the secondary windings. r_2 is the resistance of the secondary windings. L_2 is the self-leakage inductance of the secondary windings. u_{a3}, u_{b3} and u_{c3} are the voltages of the tertiary windings A3, B3 and C3, respectively. i_{a3}, i_{b3} and i_{c3} are the currents of the tertiary windings. r_3 is the resistance of the tertiary windings. L_3 is the self-leakage inductance of the tertiary windings. m_{12} and m_{21} are the mutual leakage inductances between the primary and secondary windings. m_{31} and m_{13} are the mutual leakage inductances between the primary and tertiary windings. m_{32} and m_{23} are the mutual leakage inductances between the secondary and tertiary windings. ψ_{ma}, ψ_{mb} and ψ_{mc} are the mutual flux linkages. Consider the mutual leakage inductances to be constant and equal during the normal operational conditions, inrush currents and external faults: $m_{12} = m_{21}$, $m_{13} = m_{31}$, $m_{23} = m_{32}$.

The procedure described in the two-winding, three-phase transformer provided Equation (2.59) from Equations (2.53) and (2.56). Processing Equations (2.67)–(2.68) and Equations (2.70)–(2.71) in a similar manner provides Equation (2.76). A similar procedure provides Equations (2.77)–(2.78):

$$u_{ab12} = -(L_1 - m_{12})\frac{di_{La1}}{dt} + (L_2 - m_{21})\frac{d(i_{a2} - i_{b2})}{dt} + (m_{32} - m_{31})\frac{d(i_{a3} - i_{b3})}{dt} \quad (2.76)$$

$$u_{bc12} = -(L_1 - m_{12})\frac{di_{Lb1}}{dt} + (L_2 - m_{21})\frac{d(i_{b2} - i_{c2})}{dt} + (m_{32} - m_{31})\frac{d(i_{b3} - i_{c3})}{dt} \quad (2.77)$$

$$u_{ca12} = -(L_1 - m_{12})\frac{di_{Lc1}}{dt} + (L_2 - m_{21})\frac{d(i_{c2} - i_{a2})}{dt} + (m_{32} - m_{31})\frac{d(i_{c3} - i_{a3})}{dt} \quad (2.78)$$

with

$$u_{ab12} = u_{b1} - u_{a1} + u_{a2} - u_{b2} + i_{La1}r - (i_{a2} - i_{b2})r_2 \quad (2.79)$$

$$u_{bc12} = u_{c1} - u_{b1} + u_{b2} - u_{c2} + i_{Lb1}r_1 - (i_{b2} - i_{c2})r_2 \tag{2.80}$$

$$u_{ca12} = u_{a1} - u_{c1} + u_{c2} - u_{a2} + i_{Lc1}r_1 - (i_{c2} - i_{a2})r_2 \tag{2.81}$$

The parameters of the self and mutual leakage inductances cannot be obtained even from the no-load test and the steady-state short circuit test, which is another obstacle for the algorithm presented. However, application of the EILI is a reasonable method for solving this problem. The trapezoid principle is adopted in Equation (2.76) to transform the continuous differential equation into three discrete difference equations at $(k-1)T$, kT and $(k+1)\,T$ instants, respectively.

$$\begin{aligned} u_{ab12}(k-1) = &-(L_{1k} - m_{12k})\frac{i_{La1}(k-2) - i_{La1}(k)}{2T} + (L_{2k} - m_{21k})\frac{i_{ab2}(k-2) - i_{ab2}(k)}{2T} \\ &+ (m_{32k} - m_{31k})\frac{i_{ab3}(k-2) - i_{ab3}(k)}{2T} \end{aligned} \tag{2.82}$$

$$\begin{aligned} u_{ab12}(k) = &-(L_{1k} - m_{12k})\frac{i_{La1}(k+1) - i_{La1}(k-1)}{2T} + (L_{2k} - m_{21k})\frac{i_{ab2}(k+1) - i_{ab2}(k-1)}{2T} \\ &+ (m_{32k} - m_{31k})\frac{i_{ab3}(k+1) - i_{ab3}(k-1)}{2T} \end{aligned} \tag{2.83}$$

$$\begin{aligned} u_{ab12}(k+1) = &-(L_{1k} - m_{12k})\frac{i_{La1}(k+2) - i_{La1}(k)}{2T} + (L_{2k} - m_{21k})\frac{i_{ab2}(k+2) - i_{ab2}(k)}{2T} \\ &+ (m_{32k} - m_{31k})\frac{i_{ab3}(k+2) - i_{ab3}(k)}{2T} \end{aligned} \tag{2.84}$$

The EILIs of $L_{1k} - m_{12k}$, $L_{2k} - m_{21k}$ and $m_{32k} - m_{31k}$ at the kT instant can be calculated in real time by utilizing Equations (2.82)–(2.84). A similar procedure is valid for Equations (2.77)–(2.78) to provide the other two groups of $L_{1k} - m_{12k}$, $L_{2k} - m_{21k}$ and $m_{32k} - m_{31k}$, respectively.

Similarly, utilizing Equations (2.70)–(2.72) and Equations (2.72)–(2.75), we can calculate three groups of $L_{2k} - m_{23k}$, $L_{3k} - m_{32k}$ and $m_{12k} - m_{13k}$. Meanwhile, utilizing Equations (2.77)–(2.79) and Equations (2.73)–(2.75), we can calculate three groups of $L_{3k} - m_{31k}$, $L_{1k} - m_{13k}$ and $m_{21k} - m_{23k}$.

Subtracting $L_{2k} - m_{21k}$ from $L_{1k} - m_{12k}$ provides

$$(L_{1k} - m_{12k}) - (L_{2k} - m_{21k}) = L_{1k} - L_{2k} \tag{2.85}$$

A similar procedure can be followed to obtain $L_{2k} - L_{3k}$ and $L_{1k} - L_{3k}$ as follows:

$$(L_{2k} - m_{23k}) - (L_{3k} - m_{32k}) = L_{2k} - L_{3k} \tag{2.86}$$

$$(L_{1k} - m_{13k}) - (L_{3k} - m_{31k}) = L_{1k} - L_{3k} \tag{2.87}$$

2.2.5.2 EILI-based Criterion

Two criteria are respectively proposed for dealing with the two-winding transformer and the three-winding transformer. Both of them cooperate with the differential relay to perform the protection task.

A. Criterion of Two-Winding Transformer

At instant kT, the difference of the EILIs among the three groups of primary windings can be expressed as

$$\Delta L_{1k} = \sqrt{\frac{1}{L_{\min 1}}\left((L_{1ka}-L_{1kb})^2 + (L_{1kb}-L_{1kc})^2 + (L_{1kc}-L_{1ka})^2\right)} \tag{2.88}$$

$$L_{\min 1} = \min(L_{1ka}, L_{1kb}, L_{1kc}) \tag{2.89}$$

where L_{1ka}, L_{1kb} and L_{1kc} are the EILIs of primary windings calculated by Equations (2.59), (2.60) and (2.61), respectively.

A similar procedure applies to the EILIs of the secondary windings, as follows:

$$\Delta L_{2k} = \sqrt{\frac{1}{L_{\min 2}}\left((L_{2ka}-L_{2kb})^2 + (L_{2kb}-L_{2kc})^2 + (L_{2kc}-L_{2ka})^2\right)} \tag{2.90}$$

$$L_{\min 2} = \min(L_{2ka}, L_{2kb}, L_{2kc}) \tag{2.91}$$

ΔL_{1k} and ΔL_{2k} are used to distinguish an inrush current from an internal fault, according to the following criterion: if ΔL_{1k} or ΔL_{2k} exceeds a threshold, the relay judges that an internal fault has occurred and lets the relay trip. Otherwise, the relay judges that there is an inrush current and rejects the tripping if both ΔL_{1k} and ΔL_{2k} are less than the threshold. In theory, the threshold is close to zero.

B. Criterion of Three-Winding Transformer

The difference of three groups of the $L_{1k} - L_{2k}$ can be described as

$$\Delta L'_{1k} = \sqrt{\frac{1}{L'_{\min 1}}\left((L_{1ka}-L_{2ka})^2 + (L_{1kb}-L_{2kb})^2 + (L_{1kc}-L_{2kc})^2\right)} \tag{2.92}$$

$$L'_{\min 1} = \min(L_{1ka}, L_{1kb}, L_{1kc}, L_{2ka}, L_{2kb}, L_{2kc}) \tag{2.93}$$

where $L_{1ka} - L_{2ka}$, $L_{1kb} - L_{2kb}$ and $L_{1kc} - L_{2kc}$ are the three groups of EILIs calculated by Equation (2.85).

Two similar procedures provide the difference of three groups of the $L_{2k} - L_{3k}$ and the difference of three groups of the $L_{1k} - L_{3k}$ as follows:

$$\Delta L'_{2k} = \sqrt{\frac{1}{L'_{\min 2}}\left((L_{2ka}-L_{3ka})^2 + (L_{2kb}-L_{3kb})^2 + (L_{2kc}-L_{3kc})^2\right)} \tag{2.94}$$

$$\Delta L'_{3k} = \sqrt{\frac{1}{L'_{\min 3}}\left((L_{1ka}-L_{3ka})^2 + (L_{1kb}-L_{3kb})^2 + (L_{1kc}-L_{3kc})^2\right)} \tag{2.95}$$

$$L'_{\min 2} = \min(L_{2ka}, L_{2kb}, L_{2kc}, L_{3ka}, L_{3kb}, L_{3kc}) \tag{2.96}$$

$$L'_{\min 3} = \min(L_{1ka}, L_{1kb}, L_{1kc}, L_{3ka}, L_{3kb}, L_{3kc}) \tag{2.97}$$

where $L_{2ka} - L_{3ka}$, $L_{2kb} - L_{3kb}$ and $L_{2kc} - L_{3kc}$ are the three groups of EILIs calculated by Equation (2.86). $L_{1ka} - L_{3ka}$, $L_{1kb} - L_{3kb}$ and $L_{1kc} - L_{3kc}$ are the three groups of EILIs calculated by Equation (2.87).

If the amplitude of $\Delta L'_{1k}$, $\Delta L'_{2k}$ and $\Delta L'_{3k}$ are all less than a threshold, we make a decision of the detection of an inrush current in the transformer and block the relay tripping of the differential protection. Otherwise, we make a decision of the detection of an internal fault and let the relay trip. In theory, the threshold is close to zero.

2.2.5.3 Experimental Results and Analysis

To verify the effectiveness of the proposed method, experimental tests have been carried out at the Electric Power Research Institute (EPRI). The experimental system is a single-machine model with a two-winding, three-phase Yd11 connected transformer bank, as shown in Figure 2.39. The system includes two parallel lines. The system parameters are given in Table 2.9. Current transformers (CTs) with Y/Y connection are used as transducers to measure the line currents of the transformer bank.

The experiments provide samples of line currents and terminal voltages in each phase when the transformer is energized or when a fault occurs or when both occur simultaneously. A total of 147 cases have been divided into four main categories: 27 cases for switching the transformer with no load, 27 cases for clearing an external line fault, 49 cases for simultaneous internal fault and inrush conditions

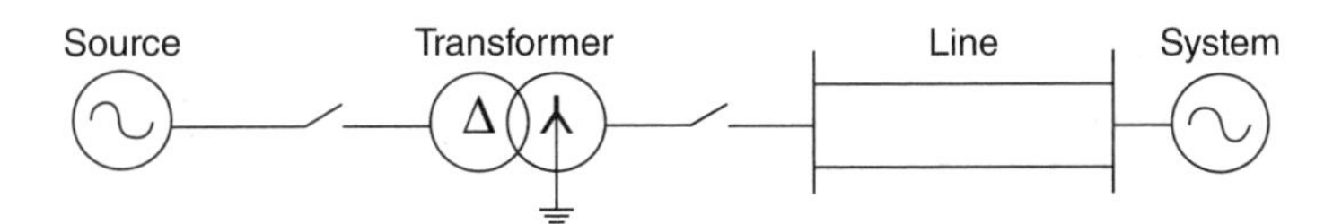

Figure 2.39 Experimental system.

Table 2.9 Parameters of the test model.

Source	
Rated power	600 MW
Rated voltage	19 kV
Rated current	20.26 kA
Transformer	
Rated capacity	670 MVA
Rated voltage ratio	19 kV/550 kV
Short circuit voltage	13.0%
Line	
Length	257 km
Voltage class	500 kV
System	
System capacity	11000 MVA
Rated frequency	50 Hz

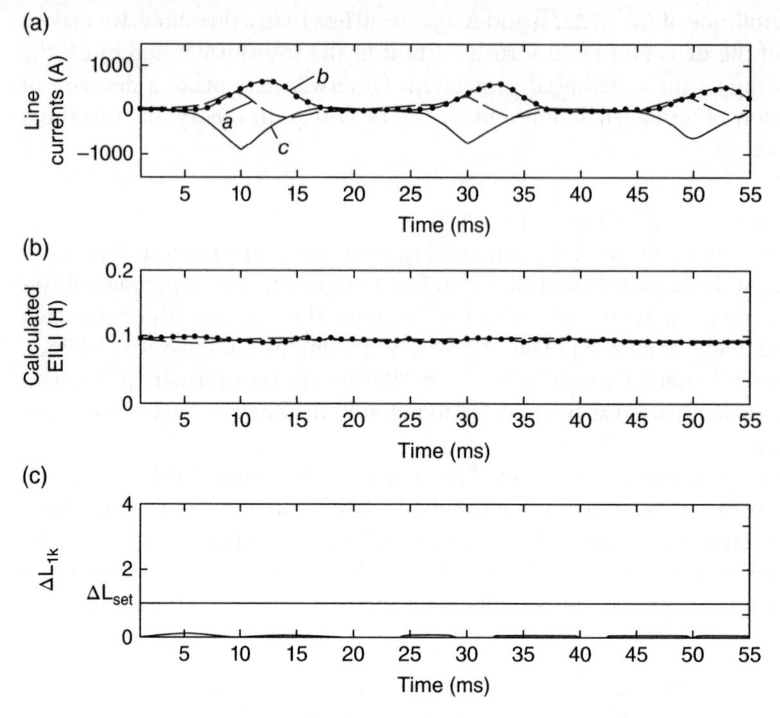

Figure 2.40 Experimental results when the transformer is energized. (a) Line currents. (b) Calculated EILIs. (c) Analysis of the calculated EILIs.

and 44 cases for faulty conditions only; these were used to test the various features of the algorithm. Different switching and clearing instants for the inrush current, as well as different faults and short circuit turn ratios for the internal fault were considered in the tests. The measured values are used as an input to the developed algorithm to identify its response.

Figures 2.40–2.44 show some examples of the experimental test results: the line currents and the waveforms of the calculated EILIs, along with the resulting analysis.

A. Responses to Different Inrush Conditions Only

The magnetizing inrush current is often generated when a transformer is energized or an external line fault is cleared. Data from a total of 54 cases were tested in both situations: 27 cases for switching the transformer with no load and 27 cases for clearing the external line fault. Energization at the primary side of the transformer is given in Figure 2.40, where the line currents, the EILIs of the primary windings calculated by using Equations (2.59)–(2.61), and the analysis results are shown in Figures 2.40(a), (b) and (c), respectively. We can see that the calculated EILIs of the three phases are almost kept constant and have very little variation resulting from the measurement and calculation errors, as shown in Figure 2.40(b). The variation of the ΔL_{1k} based on Equations (2.88) and (2.89) is close to zero and much less than the threshold ΔL_{set}, as shown in Figure 2.40(c).

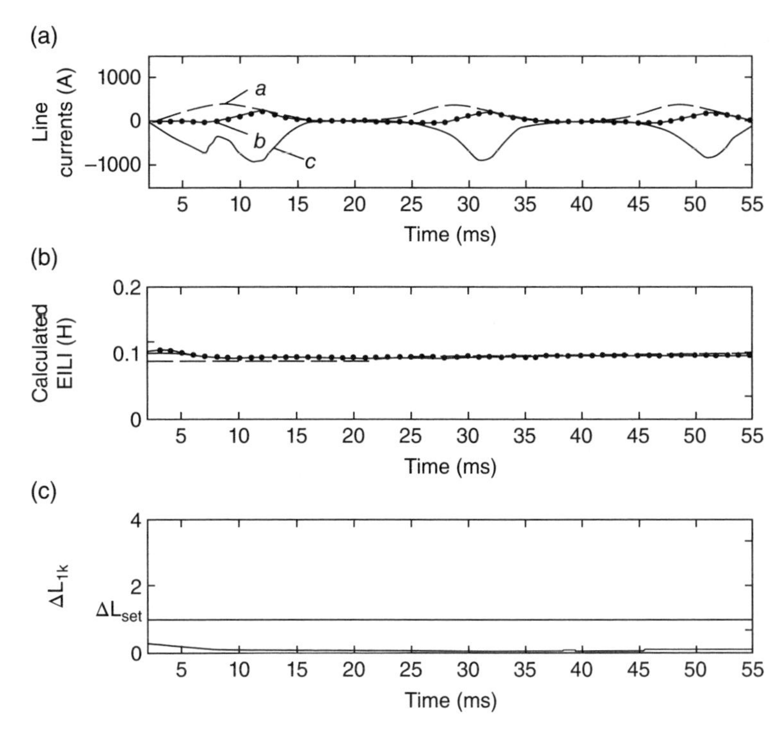

Figure 2.41 Experimental results when an external line fault is cleared. (a) Line currents. (b) Calculated EILIs. (c) Analysis of the calculated EILIs.

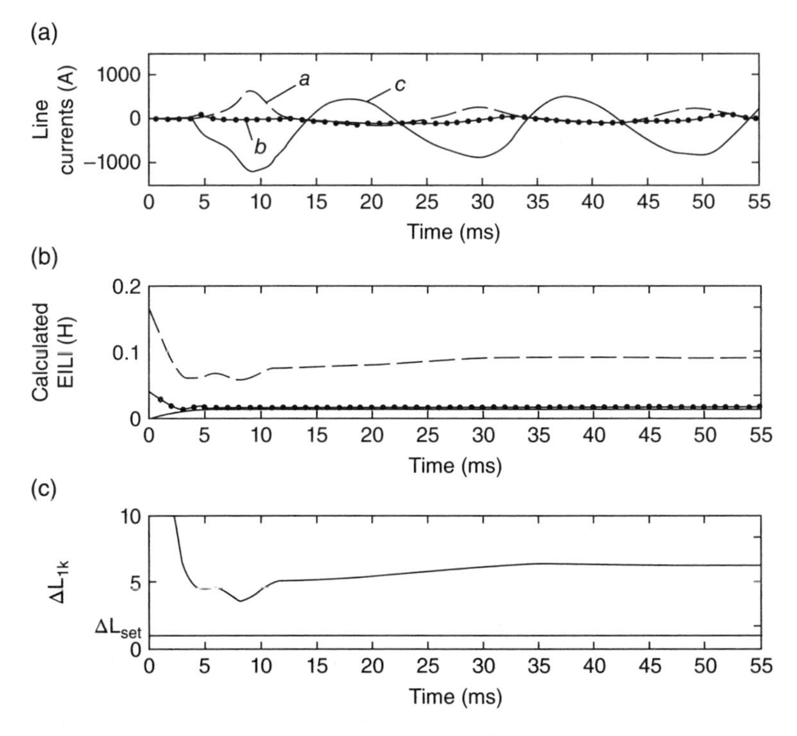

Figure 2.42 Experimental results when the transformer is energized with a 2% turn-to-turn internal fault. (a) Line currents. (b) Calculated EILIs. (c) Analysis of the calculated EILIs.

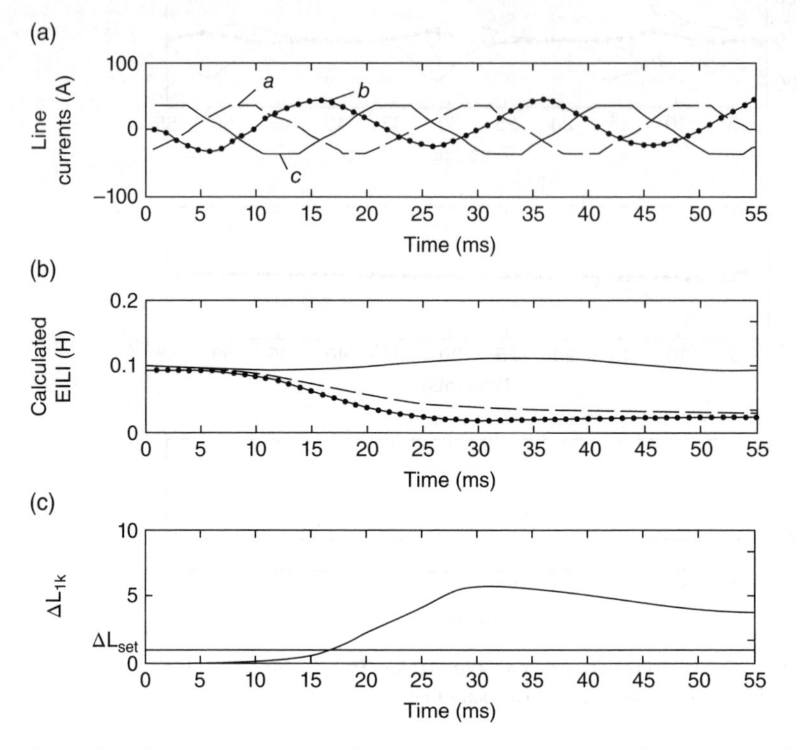

Figure 2.43 Experimental results when a 3% turn-to-turn internal fault occurs at the secondary side. (a) Line currents at the primary side. (b) Calculated EILIs of the primary windings. (c) Analysis of the calculated EILIs.

The same result can be obtained from the case of clearing an external line fault, as shown in Figure 2.41, where the EILIs of three phases only have slight difference at the fault-clearing instant. According to the criterion for a two-winding transformer, the protection will be blocked, although the EILIs of the secondary windings are not available because the line currents of the secondary side are zero in these situations. In the total of 54 cases, the same feature of the inrush current is presented in the calculated EILIs.

B. Responses to Simultaneous Fault and Inrush Conditions

The switching in the transformer bank with no load often causes the inrush current of non-fault phases, which has been verified by a total of 49 cases with simultaneous inrush currents and internal fault currents. As an example, Figure 2.42(a) shows this situation, which is obtained by switching in the transformer bank with no load and a 2% turn-to-turn internal fault in phase C. The line current of phase A presents the nominal magnetizing current, but the line current of phase B is larger than the nominal magnetizing current. Therefore, the EILIs are calculated by using Equations (2.59)–(2.61). Figures 2.42(b) and (c) show the calculated EILIs and the analysis result, respectively. The difference is very noticeable in the

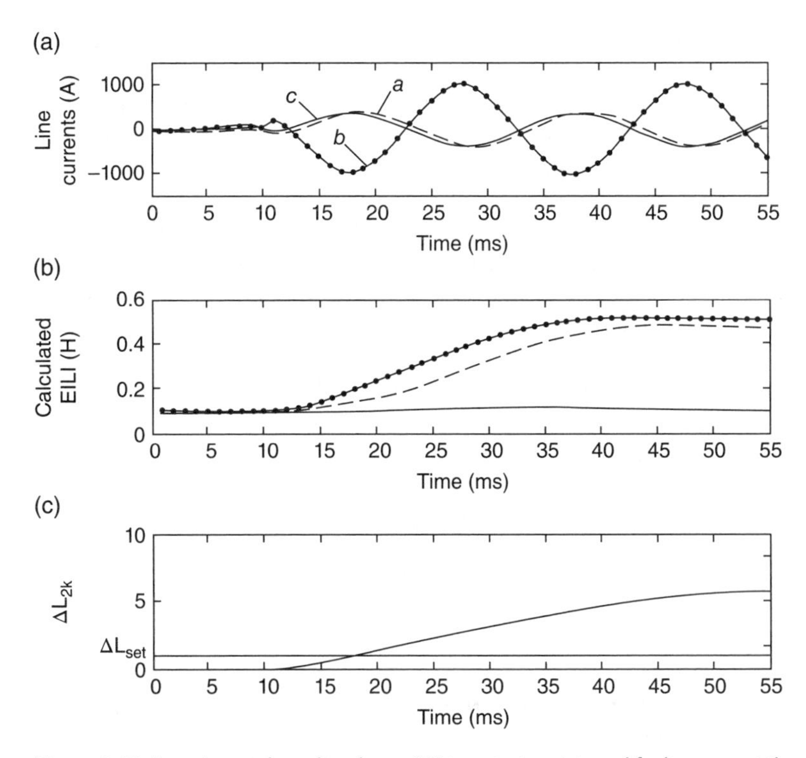

Figure 2.44 Experimental results when a 3% turn-to-turn internal fault occurs at the secondary side. (a) Line currents at the secondary side. (b) Calculated EILIs of the secondary windings. (c) Analysis of the calculated EILIs.

calculated EILIs of the three phases. According to the proposed criterion, it is obvious that the protection will operate rapidly and correctly. These results are in accordance with the practical state of the transformer bank.

In the total of 49 cases, the identical results verify that the proposed techniques can be used to identify internal faults when a simultaneous inrush current and fault occur in the transformer bank. Moreover, it is difficult to discriminate low-level (less than 9%) turn-to-turn internal faults from inrush currents when the algorithm proposed by M. S. Sachdev is applied, which provides less accurate results than the method to calculate EILIs.

C. Responses to Internal Fault Conditions Only

Data from a total of 44 cases were used to calculate the EILIs based on Equations (2.59)–(2.61). In all of the 44 cases, the calculated EILIs are not constant, which results from the variation of the physical dimensions and the deformation of the windings during the internal fault. As shown in Figures 2.43(a) and 2.44(a), a 3% turn-to-turn internal fault occurs in phase B at the secondary side during steady operation. The calculated EILI waveforms of the primary and secondary windings are shown in Figures 2.43(b) and 2.44(b), respectively, which present different variations right after the fault occurrence. Their analysis results are shown in Figures 2.43(c) and 2.44(c), respectively, which show that not only ΔL_{1k} but also

ΔL_{2k} exceeds the threshold ΔL_{set} soon after the fault occurs. The operating time is only about 8 ms.

In addition, the EILIs during the other internal faults (such as grounding internal faults and phase-to-phase internal faults) do not remain constant, which will be effectively used to distinguish internal faults and inrush currents without the parameters of the leakage inductances of the primary and secondary windings. Meanwhile, we do not see any problems in applying the proposed technique for the impact of CTs during internal faults. However, since quite a high false line current may emerge if the CT at one side saturates in depth, we are presently working on the effectiveness of the proposed technique when the CT is in heavy saturation.

In order to validate the proposed method more thoroughly, further practical studies will be continued in planned future work. First, the proposed method will be tested on several three-winding, three-phase transformer banks. Also, different transformers will be used in a large number of practical tests, in which the EILI during different types of internal faults and inrush currents will be analysed to verify the proposed method. These results will be useful for the application of the proposed method in practical transformer products.

2.2.6 A Two-Terminal, Network-Based Method for Discrimination between Internal Faults and Inrush Currents

This is a new method, based on generalized instantaneous power for discrimination between internal faults and inrush currents.

2.2.6.1 Basic Principle

A. Two-Terminal Network of a Single-Phase Transformer

A two-winding, single-phase transformer is shown in Figure 2.45. The primary and secondary voltages can be expressed as:

$$u_1 = i_1 r_1 + L_1 \frac{di_1}{dt} + \frac{d\psi_m}{dt} \tag{2.98}$$

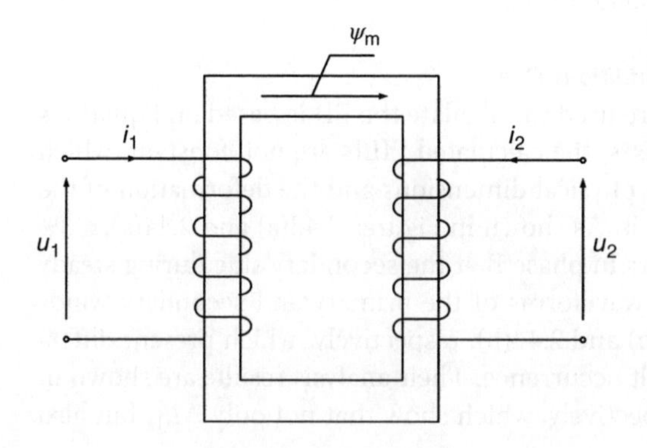

Figure 2.45 A two-winding, single-phase transformer.

$$u_2 = i_2 r_2 + L_2 \frac{\mathrm{d}i_2}{\mathrm{d}t} + \frac{\mathrm{d}\psi_\mathrm{m}}{\mathrm{d}t} \tag{2.99}$$

where the parameters of the secondary side have been converted to the primary side by the transformer ratio. u_1 and u_2 are the voltages of primary and secondary windings, respectively. i_1 and i_2 are the currents in the primary and secondary windings, respectively. r_1 and r_2 are the resistances of the primary and secondary windings, respectively. L_1 and L_2 are the leakage inductances of the primary and secondary windings, respectively. ψ_m is the mutual flux linkage.

The mutual flux linkage of the primary and secondary windings is equal and can be eliminated by using Equations (2.98) and (2.99), as follows:

$$u_\mathrm{d} = i_\mathrm{d} r_1 + L_1 \frac{\mathrm{d}i_\mathrm{d}}{\mathrm{d}t} \tag{2.100}$$

with

$$u_\mathrm{d} = u_1 - u_2 + i_2 r_k + \frac{x_k}{\omega} \frac{\mathrm{d}i_2}{\mathrm{d}t} \tag{2.101}$$

$$i_\mathrm{d} = i_1 + i_2 \tag{2.102}$$

$$r_k = r_1 + r_2 \tag{2.103}$$

$$x_k = \omega(L_1 + L_2) \tag{2.104}$$

where i_d and u_d are the differential current and virtual differential voltage between the primary and secondary windings, respectively. r_k and x_k are the winding resistance and the short-circuit reactance, respectively.

Using Equation (2.100), we can obtain a two-terminal network containing just the winding resistance and the leakage inductance, as shown in Figure 2.46.

Here, P_f and P_c are defined as the active powers flowing into and consumed by the two-terminal network, respectively. They can be expressed as:

$$P_\mathrm{f} = \frac{1}{T} \int_0^T (u_\mathrm{d}(t) i_\mathrm{d}(t)) \mathrm{d}t \tag{2.105}$$

$$P_\mathrm{c} = \frac{1}{T} \int_0^T \left(i_\mathrm{d}^2(t) r_1\right) \mathrm{d}t \tag{2.106}$$

where T is one cycle data window.

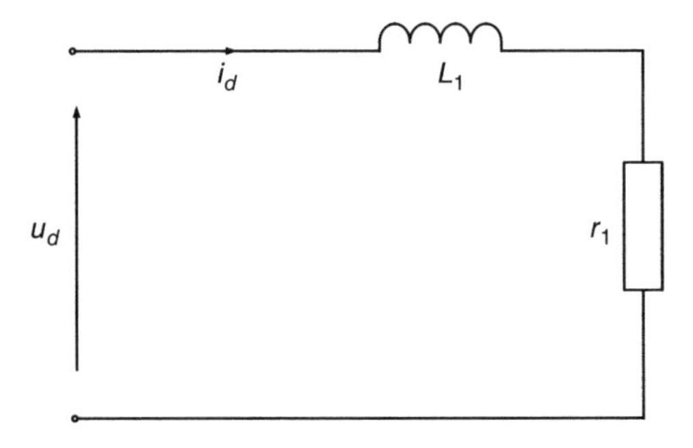

Figure 2.46 A two-terminal network.

In the magnetizing inrush and normal operational cases of the power transformer, P_f is very close to P_c. However, when an internal fault occurs, P_f and P_c are both affected and P_f is not close to P_c anymore, owing to the arcing discharge.

$$P = |P_f - P_c| \tag{2.107}$$

P is defined as the absolute difference of active power (ADOAP) between P_f and P_c. If ADOAP is less than a threshold, the relay determines that there is an inrush current and rejects the tripping. Otherwise, the relay determines that an internal fault has occurred. The threshold should be set to avoid needless operation by measurement error or calculation error. In theory, the threshold is close to zero.

Owing to the elimination of the mutual flux linkage, the technique does not require data on the B–H curve or knowledge of iron losses. Also, from Equations (2.105)–(2.107), we can find that ADOAP does not make use of leakage inductances of primary and secondary windings to distinguish an internal fault from an inrush current.

B. The ADOAPs of a Two-Winding, Three-Phase Transformer

Figure 2.47 shows the connections of the primary and secondary windings of a Δ/Y transformer.

The following equations express the Δ and Y connected windings as functions of the mutual flux linkages and the currents of the windings,

$$u_a = i_a r + L_a \frac{di_a}{dt} + \frac{d\psi_{ma}}{dt} \tag{2.108}$$

$$u_b = i_b r + L_b \frac{di_b}{dt} + \frac{d\psi_{mb}}{dt} \tag{2.109}$$

$$u_c = i_c r + L_c \frac{di_c}{dt} + \frac{d\psi_{mc}}{dt} \tag{2.110}$$

$$u_A = i_A R + L_A \frac{di_A}{dt} + \frac{d\psi_{ma}}{dt} \tag{2.111}$$

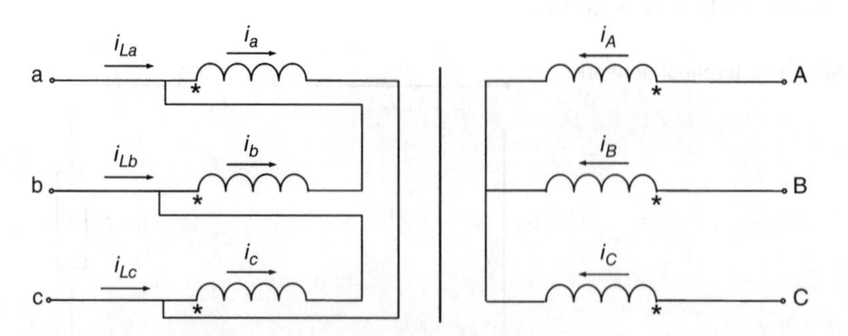

Figure 2.47 A two-winding, three-phase Δ/Y transformer.

$$u_B = i_B R + L_B \frac{di_B}{dt} + \frac{d\psi_{mb}}{dt} \quad (2.112)$$

$$u_C = i_C R + L_C \frac{di_C}{dt} + \frac{d\psi_{mc}}{dt} \quad (2.113)$$

where the parameters of the secondary side have been converted to the primary side by the transformer ratio. u_a, u_b and u_c are the voltages of the primary windings a, b and c, respectively. i_a, i_b and i_c are the currents of the primary windings. L_a, L_b and L_c are the leakage inductances of the primary windings. r is the resistance of the primary windings. u_A, u_B and u_C are the voltages of the secondary windings. i_A, i_B and i_C are the currents of the secondary windings. L_A, L_B and L_C are the leakage inductances of the secondary windings. R is the resistance of the secondary windings. ψ_{ma}, ψ_{mb} and ψ_{mc} are the mutual flux linkages.

The line currents in the Δ-connected windings are obtained as follows

$$i_{La} = i_a - i_b \quad (2.114)$$

$$i_{Lb} = i_b - i_c \quad (2.115)$$

$$i_{Lc} = i_c - i_a \quad (2.116)$$

Consider the leakage inductances to be constant and equal in the normal operational state and during the inrush current period. $L_a = L_b = L_c = L_1$, $L_A = L_B = L_C = L_2$. The equations of the primary and secondary sides can be written as

$$u_a - u_b = i_{La} r + L_1 \frac{di_{La}}{dt} + \frac{d(\psi_{ma} - \psi_{mb})}{dt} \quad (2.117)$$

$$u_b - u_c = i_{Lb} r + L_1 \frac{di_{Lb}}{dt} + \frac{d(\psi_{mb} - \psi_{mc})}{dt} \quad (2.118)$$

$$u_c - u_a = i_{Lc} r + L_1 \frac{di_{Lc}}{dt} + \frac{d(\psi_{mb} - \psi_{ma})}{dt} \quad (2.119)$$

$$u_A - u_B = (i_A - i_B)R + L_2 \frac{d(i_A - i_B)}{dt} + \frac{d(\psi_{ma} - \psi_{mb})}{dt} \quad (2.120)$$

$$u_B - u_C = (i_B - i_C)R + L_2 \frac{d(i_B - i_C)}{dt} + \frac{d(\psi_{mb} - \psi_{mc})}{dt} \quad (2.121)$$

$$u_C - u_A = (i_C - i_A)R + L_2 \frac{d(i_C - i_A)}{dt} + \frac{d(\psi_{mc} - \psi_{ma})}{dt} \quad (2.122)$$

The flux linkages mutual to the primary and secondary windings of each phase are equal and can be eliminated by using Equations (2.115)–(2.120), as follows:

$$\begin{cases} u_{dA} = i_{dA} r_2 + L_2 \dfrac{di_{dA}}{dt} \\ u_{dB} = i_{dB} r_2 + L_2 \dfrac{di_{dB}}{dt} \\ u_{dC} = i_{dC} r_2 + L_2 \dfrac{di_{dC}}{dt} \end{cases} \quad (2.123)$$

with

$$\begin{cases} u_{\mathrm{dA}} = u_{\mathrm{a}} - u_{\mathrm{b}} - u_{\mathrm{A}} + u_{\mathrm{B}} - i_{L\mathrm{a}} r_k - \dfrac{x_k}{\omega} \dfrac{\mathrm{d}i_{L\mathrm{a}}}{\mathrm{d}t} \\ u_{\mathrm{dB}} = u_{\mathrm{b}} - u_{\mathrm{c}} - u_{\mathrm{B}} + u_{\mathrm{C}} - i_{L\mathrm{b}} r_k - \dfrac{x_k}{\omega} \dfrac{\mathrm{d}i_{L\mathrm{b}}}{\mathrm{d}t} \\ u_{\mathrm{dC}} = u_{\mathrm{c}} - u_{\mathrm{a}} - u_{\mathrm{C}} + u_{\mathrm{A}} - i_{L\mathrm{c}} r_k - \dfrac{x_k}{\omega} \dfrac{\mathrm{d}i_{L\mathrm{c}}}{\mathrm{d}t} \end{cases} \tag{2.124}$$

$$\begin{cases} i_{\mathrm{dA}} = -\left(i_{L\mathrm{a}} + i_{\mathrm{A}} - i_{\mathrm{B}}\right) \\ i_{\mathrm{dB}} = -\left(i_{L\mathrm{b}} + i_{\mathrm{B}} - i_{\mathrm{C}}\right) \\ i_{\mathrm{dC}} = -\left(i_{L\mathrm{c}} + i_{\mathrm{C}} - i_{\mathrm{A}}\right) \end{cases} \tag{2.125}$$

$$\begin{cases} r_k = r_1 + r_2 \\ x_k = \omega\left(L_1 + L_2\right) \end{cases} \tag{2.126}$$

Using Equation (2.124), we can obtain three groups of two-terminal networks containing winding resistances and leakage inductances. Then a procedure similar to the single-phase transformer provides P_1, P_2 and P_3, which are all defined as the ADOAPs.

$$\begin{cases} P_1 = \dfrac{1}{T} \left| \displaystyle\int_0^T \left(u_{\mathrm{dA}}(t) i_{\mathrm{dA}}(t) - i_{\mathrm{dA}}^2(t) r_2\right) \mathrm{d}t \right| \\ P_2 = \dfrac{1}{T} \left| \displaystyle\int_0^T \left(u_{\mathrm{dB}}(t) i_{\mathrm{dB}}(t) - i_{\mathrm{dB}}^2(t) r_2\right) \mathrm{d}t \right| \\ P_3 = \dfrac{1}{T} \left| \displaystyle\int_0^T \left(u_{\mathrm{dC}}(t) i_{\mathrm{dC}}(t) - i_{\mathrm{dC}}^2(t) r_2\right) \mathrm{d}t \right| \end{cases} \tag{2.127}$$

If the ADOAPs of three phases are all less than the threshold, the relay determines that there is an inrush current and rejects the tripping. Otherwise, the relay determines that there is an internal fault.

C. *The ADOAPs of a Three-Winding, Three-Phase Transformer*

A three-winding, three-phase transformer with $\Delta/\mathrm{Y}/\mathrm{Y}_0$ connection is shown in Figure 2.48.

The following equations express the voltages of the windings as functions of the mutual flux linkages and the currents of the windings:

$$u_{\mathrm{a1}} = i_{\mathrm{a1}} r_1 + L_1 \frac{\mathrm{d}i_{\mathrm{a1}}}{dt} + m_{21} \frac{\mathrm{d}i_{\mathrm{a2}}}{dt} + m_{31} \frac{\mathrm{d}i_{\mathrm{a3}}}{dt} + \frac{\mathrm{d}\psi_{\mathrm{ma}}}{dt} \tag{2.128}$$

$$u_{\mathrm{b1}} = i_{\mathrm{b1}} r_1 + L_1 \frac{\mathrm{d}i_{\mathrm{b1}}}{dt} + m_{21} \frac{\mathrm{d}i_{\mathrm{b2}}}{dt} + m_{31} \frac{\mathrm{d}i_{\mathrm{b3}}}{dt} + \frac{\mathrm{d}\psi_{\mathrm{mb}}}{dt} \tag{2.129}$$

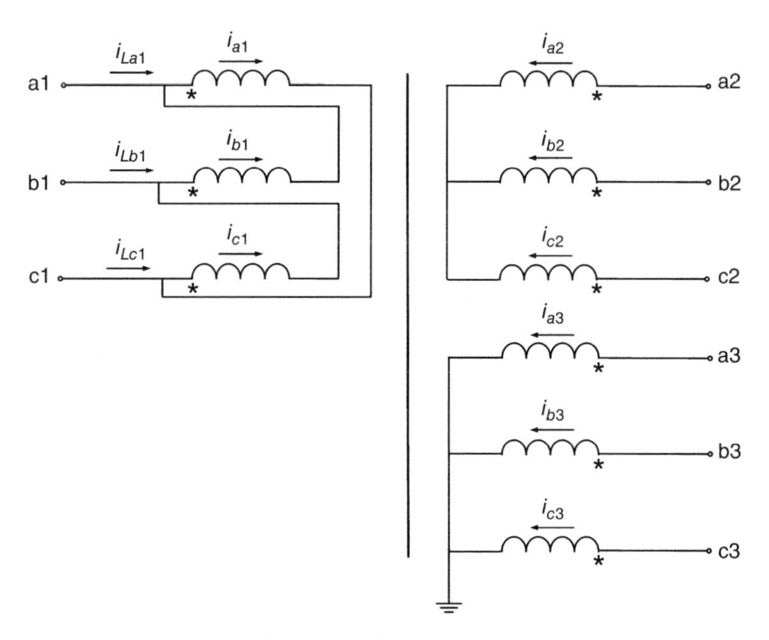

Figure 2.48 A three-winding, three-phase Δ/Y/Y_0 transformer.

$$u_{c1} = i_{c1}r_1 + L_1\frac{di_{c1}}{dt} + m_{21}\frac{di_{c2}}{dt} + m_{31}\frac{di_{c3}}{dt} + \frac{d\psi_{mc}}{dt} \tag{2.130}$$

$$u_{a2} = i_{a2}r_2 + L_2\frac{di_{a2}}{dt} + m_{12}\frac{di_{a1}}{dt} + m_{32}\frac{di_{a3}}{dt} + \frac{d\psi_{ma}}{dt} \tag{2.131}$$

$$u_{b2} = i_{b2}r_2 + L_2\frac{di_{b2}}{dt} + m_{12}\frac{di_{b1}}{dt} + m_{32}\frac{di_{b3}}{dt} + \frac{d\psi_{mb}}{dt} \tag{2.132}$$

$$u_{c2} = i_{c2}r_2 + L_2\frac{di_{c2}}{dt} + m_{12}\frac{di_{c1}}{dt} + m_{32}\frac{di_{c3}}{dt} + \frac{d\psi_{mc}}{dt} \tag{2.133}$$

$$u_{a3} = i_{a3}r_3 + L_3\frac{di_{a3}}{dt} + m_{13}\frac{di_{a1}}{dt} + m_{23}\frac{di_{a2}}{dt} + \frac{d\psi_{ma}}{dt} \tag{2.134}$$

$$u_{b3} = i_{b3}r_3 + L_3\frac{di_{b3}}{dt} + m_{13}\frac{di_{b1}}{dt} + m_{23}\frac{di_{b2}}{dt} + \frac{d\psi_{mb}}{dt} \tag{2.135}$$

$$u_{c3} = i_{c3}r_3 + L_3\frac{di_{c3}}{dt} + m_{13}\frac{di_{c1}}{dt} + m_{23}\frac{di_{c2}}{dt} + \frac{d\psi_{mc}}{dt} \tag{2.136}$$

where the parameters of the secondary and the tertiary sides have been converted to the primary side by the transformer ratio. u_{a1}, u_{b1} and u_{c1} are the voltages of the primary windings. i_{a1}, i_{b1} and i_{c1} are the currents of the primary windings. i_{La1}, i_{Lb1} and i_{Lc1} are the line currents of the primary windings. r_1 is the resistance of the primary windings. L_1 is the self-leakage inductance of the primary windings. u_{a2}, u_{b2} and u_{c2} are the voltages of the secondary windings. i_{a2}, i_{b2} and i_{c2} are the currents of the secondary windings. r_2 is the resistance of the secondary windings. L_2 is the self-leakage inductance of the secondary windings. u_{a3}, u_{b3} and u_{c3} are the voltages of the tertiary windings. i_{a3}, i_{b3} and i_{c3} are the currents of the tertiary windings. r_3 is the resistance of the tertiary windings. L_3 is the self-leakage inductance of the tertiary windings. m_{12} and m_{21} are the mutual leakage inductances between the primary and secondary windings. m_{31} and m_{13} are the mutual leakage inductances between the primary and tertiary windings. m_{32} and m_{23} are the mutual leakage inductances between the secondary and tertiary windings. ψ_{ma}, ψ_{mb} and ψ_{mc} are the mutual flux linkages.

Assume the mutual leakage inductances to be constant and equal during normal operational conditions and inrush currents: $m_{12} = m_{21}$, $m_{13} = m_{31}$, $m_{23} = m_{32}$. A procedure similar to the two-winding transformer provides three groups of ADOAPs:

$$\begin{cases} P_1 = \dfrac{1}{T}\left|\displaystyle\int_0^T \left(u_{ab12}(t)i_{da}(t) - i_{da}^2(t)r_1\right)dt\right| \\ P_2 = \dfrac{1}{T}\left|\displaystyle\int_0^T \left(u_{bc12}(t)i_{db}(t) - i_{db}^2(t)r_1\right)dt\right| \\ P_3 = \dfrac{1}{T}\left|\displaystyle\int_0^T \left(u_{ca12}(t)i_{dc}(t) - i_{dc}^2(t)r_1\right)dt\right| \end{cases} \tag{2.137}$$

$$\begin{cases} P_4 = \dfrac{1}{T}\left|\displaystyle\int_0^T \left(u_{ab23}(t)i_{da}(t) - i_{da}^2(t)r_2\right)dt\right| \\ P_5 = \dfrac{1}{T}\left|\displaystyle\int_0^T \left(u_{bc23}(t)i_{db}(t) - i_{db}^2(t)r_2\right)dt\right| \\ P_6 = \dfrac{1}{T}\left|\displaystyle\int_0^T \left(u_{ca23}(t)i_{dc}(t) - i_{dc}^2(t)r_2\right)dt\right| \end{cases} \tag{2.138}$$

$$\begin{cases} P_7 = \dfrac{1}{T}\left|\displaystyle\int_0^T \left(u_{ab31}(t)i_{da}(t) - i_{da}^2(t)r_3\right)dt\right| \\ P_8 = \dfrac{1}{T}\left|\displaystyle\int_0^T \left(u_{bc31}(t)i_{db}(t) - i_{db}^2(t)r_3\right)dt\right| \\ P_9 = \dfrac{1}{T}\left|\displaystyle\int_0^T \left(u_{ca31}(t)i_{dc}(t) - i_{dc}^2(t)r_3\right)dt\right| \end{cases} \tag{2.139}$$

with

$$\begin{cases} i_{da} = i_{La1} + i_{a2} - i_{b2} + i_{a3} - i_{b3} \\ i_{db} = i_{Lb1} + i_{b2} - i_{c2} + i_{b3} - i_{c3} \\ i_{dc} = i_{Lc1} + i_{c2} - i_{a2} + i_{c3} - i_{a3} \end{cases} \tag{2.140}$$

$$\begin{cases} u_{ab12} = u_{b2} - u_{a2} + u_{a1} - u_{b1} + (i_{a2} - i_{b2})(r_1 + r_2) \\ \quad + \dfrac{x_1 + x_2}{\omega}\dfrac{d(i_{a2} - i_{b2})}{dt} + (i_{a3} - i_{b3})r_1 + \dfrac{x_1}{\omega}\dfrac{d(i_{a3} - i_{b3})}{dt} \\ u_{bc12} = u_{c2} - u_{b2} + u_{b1} - u_{c1} + (i_{b2} - i_{c2})(r_1 + r_2) \\ \quad + \dfrac{x_1 + x_2}{\omega}\dfrac{d(i_{b2} - i_{c2})}{dt} + (i_{b3} - i_{c3})r_1 + \dfrac{x_1}{\omega}\dfrac{d(i_{b3} - i_{c3})}{dt} \\ u_{ca12} = u_{a2} - u_{c2} + u_{c1} - u_{a1} + (i_{c2} - i_{a2})(r_1 + r_2) \\ \quad + \dfrac{x_1 + x_2}{\omega}\dfrac{d(i_{c2} - i_{a2})}{dt} + (i_{c3} - i_{a3})r_1 + \dfrac{x_1}{\omega}\dfrac{d(i_{c3} - i_{a3})}{dt} \end{cases} \tag{2.141}$$

$$\begin{cases} u_{ab23} = u_{b3} - u_{a3} + u_{a2} - u_{b2} + (i_{a3} - i_{b3})(r_2 + r_3) \\ \quad + \dfrac{x_2 + x_3}{\omega}\dfrac{d(i_{a3} - i_{b3})}{dt} + i_{La1}r_2 + \dfrac{x_2}{\omega}\dfrac{di_{La1}}{dt} \\ u_{bc23} = u_{c3} - u_{b3} + u_{b2} - u_{c2} + (i_{b3} - i_{c3})(r_2 + r_3) \\ \quad + \dfrac{x_2 + x_3}{\omega}\dfrac{d(i_{b3} - i_{c3})}{dt} + i_{Lb1}r_2 + \dfrac{x_2}{\omega}\dfrac{di_{Lb1}}{dt} \\ u_{ca23} = u_{a3} - u_{c3} + u_{c2} - u_{a2} + (i_{c3} - i_{a3})(r_2 + r_3) \\ \quad + \dfrac{x_2 + x_3}{\omega}\dfrac{d(i_{c3} - i_{a3})}{dt} + i_{Lc1}r_2 + \dfrac{x_2}{\omega}\dfrac{di_{Lc1}}{dt} \end{cases} \tag{2.142}$$

$$\begin{cases} u_{ab31} = u_{b1} - u_{a1} + u_{a3} - u_{b3} + i_{La1}(r_1 + r_3) \\ \quad + \dfrac{x_1 + x_3}{\omega}\dfrac{di_{La1}}{dt} + (i_{a2} - i_{b2})r_3 + \dfrac{x_3}{\omega}\dfrac{d(i_{a2} - i_{b2})}{dt} \\ u_{bc31} = u_{c1} - u_{b1} + u_{b3} - u_{c3} + i_{Lb1}(r_1 + r_3) \\ \quad + \dfrac{x_1 + x_3}{\omega}\dfrac{di_{Lb1}}{dt} + (i_{b2} - i_{c2})r_3 + \dfrac{x_3}{\omega}\dfrac{d(i_{b2} - i_{c2})}{dt} \\ u_{ca31} = u_{a1} - u_{c1} + u_{c3} - u_{a3} + i_{Lc1}(r_1 + r_3) \\ \quad + \dfrac{x_1 + x_3}{\omega}\dfrac{di_{Lc1}}{dt} + (i_{c2} - i_{a2})r_3 + \dfrac{x_3}{\omega}\dfrac{d(i_{c2} - i_{a2})}{dt} \end{cases} \tag{2.143}$$

where x_1, x_2 and x_3 are the short-circuit reactances of the primary, secondary and tertiary windings, respectively. These values can be obtained from the transformer manufacturer.

$$\begin{cases} x_1 = \omega(L_1 - m_{12} - m_{13} + m_{23}) \\ x_2 = \omega(L_2 - m_{12} - m_{23} + m_{13}) \\ x_3 = \omega(L_3 - m_{13} - m_{23} + m_{12}) \end{cases} \tag{2.144}$$

If the three groups of ADOAPs are all less than a preset threshold, the relay determines that there is an inrush current and rejects the tripping. Otherwise, the relay determines that an internal fault has occurred.

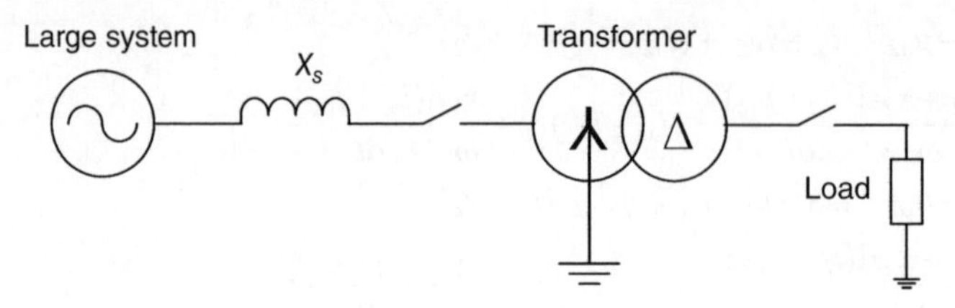

Figure 2.49 Experimental system.

Table 2.10 Parameters of the transformer used in the test.

Rated capacity	30 kVA
Rated voltage ratio	1732.05/380 V
Rated current ratio	10/45.58 A
Rated frequency	50 Hz
No load current	1.45%
No load loss	1%
Short circuit voltage	9.0 ~ 15.0%
Short circuit loss	0.35%
Load	0.9 kW

2.2.6.2 Experimental System

To verify the effectiveness of the proposed method, experimental tests have been carried out at the Electrical Power Dynamic Laboratory (EPDL). The experimental transformer is a three-phase, two-winding transformer bank with Y_0/Δ-11 connection, which is fed by a large power system grid, as shown in Figure 2.49. The parameters of the two-winding transformers are given in Table 2.10. Three identical current transformers (CTs) are connected in Δ on the primary side, and another three identical CTs are connected in Y on the secondary side of the power transformer.

The experiments provide samples of three phase voltages and differential currents when the transformer is energized or when a fault occurs or when both occur simultaneously. In order to test various features of the algorithm, a total of 162 cases have been divided into three main categories: 56 cases for inrush conditions only, 52 cases for simultaneous internal fault and inrush conditions, and 54 cases for faulty conditions only. Different switching and clearing instants for inrush currents, as well as different faults and short circuit turn ratios for internal faults were considered in the tests. The measured data were used as input to the developed algorithm to identify its response.

Figures 2.50–2.54 show some examples of the experimental test results: the differential currents and the resulting analysis. The ADOAPs are calculated just after the relay starts up for one cycle. In addition, the threshold is set to 5 W and fairly good results were obtained in all 162 cases. This is represented by the dashed line.

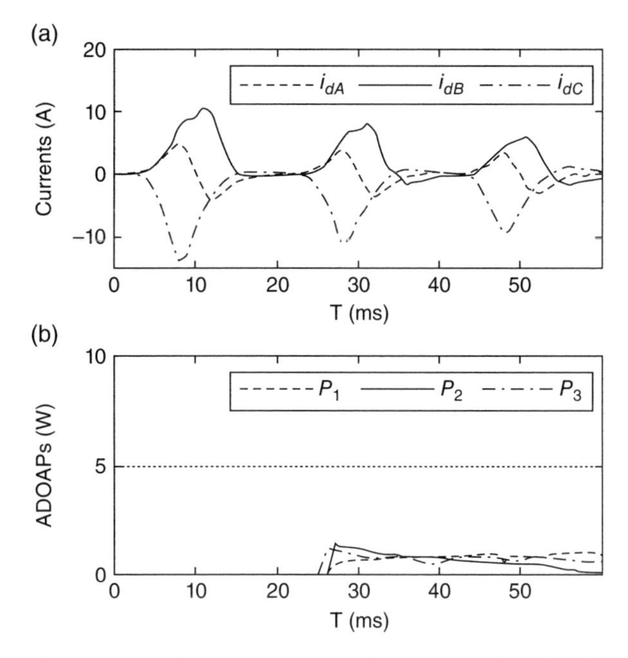

Figure 2.50 Differential currents and experimental results when the transformer is energized. (a) Differential currents. (b) ADOAPs.

2.2.6.3 Testing Results and Analysis

A. Responses to Different Inrush Conditions

A total of 56 test cases were carried out in this situation. The inrush current waveform is a function of the different core residual magnetization and the switching instant, so the waveforms of the inrush current are different from each other. However, the ADOAPs calculated by using Equation (2.127) present identical results due to eliminating flux linkages. An example taken from these cases is given in Figure 2.50, where the differential currents of three phases and their ADOAPs are shown in Figures 2.50(a) and (b), respectively. We can see that the calculated ADOAPs of three phases are negligible and only have a small variation resulting from the measurement and calculation errors, as shown in Figure 2.50(b). According to the criterion of the two-winding, three-phase transformer, the protection will be blocked.

B. Responses to Simultaneous Fault and Inrush Conditions

When the transformer is energized with an internal fault, an inrush current may occur and this would affect the differential current waveforms of the fault phases. This has been verified by a total of 52 cases with simultaneous inrush currents and internal fault currents.

As an example, Figure 2.51(a) shows this condition, which is obtained by switching in the transformer bank with no load and a turn-to-ground fault in phase B. Non-fault phase C presents the magnetizing current and fault phases A and B show little distortion. Using Equation (2.127), we can calculate the

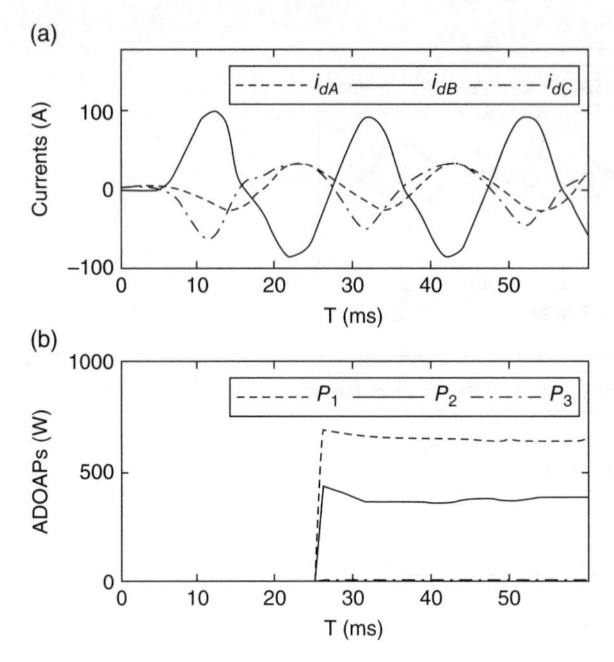

Figure 2.51 Differential currents and experimental results when the transformer is switched with no load and a turn-to-ground fault in phase B. (a) Differential currents. (b) ADOAPs.

ADOAPs of three phases, as shown in Figure 2.51(b). The ADOAP of phase C is close to zero, whereas the ADOAPs of phases A and B are very noticeable.

Compared with Figure 2.51(a), Figure 2.52(a) shows differential currents with more severe distortion, where the transformer bank is energized with no load and a 2.4% turn-to-turn fault (minimum ratio of turns provided by the experimental transformer) occurs in phase A.

After analysis in the frequency domain, we find that the magnitudes of the second harmonic in fault phases A and C are greater than those of some magnetizing inrush currents. Consequently, the commonly employed conventional differential protection technique based on the second harmonic will have difficulty in distinguishing between an internal fault and an inrush current. However, the ADOAPs of fault phases show much higher amplitudes than the threshold in Figure 2.52(b), which makes the relay determine that there is an internal fault and it lets the relay trip.

These results are in accordance with the practical state of the transformer bank. In a total of 52 cases, identical results verify that the proposed technique is sensitive and reliable to discriminate internal faults from inrush currents when simultaneous inrush currents and faults occur in the transformer.

C. Responses to Internal Fault Conditions Only

Data from a total of 54 cases were used to calculate the ADOAPs based on Equation (2.127). In all of the 54 cases, the ADOAPs of faulty phases are noticeable, whereas the ADOAPs of non-faulty phases are negligible. Two examples are shown in Figures 2.53 and 2.54. One is a turn-to-ground fault in phase B (the same

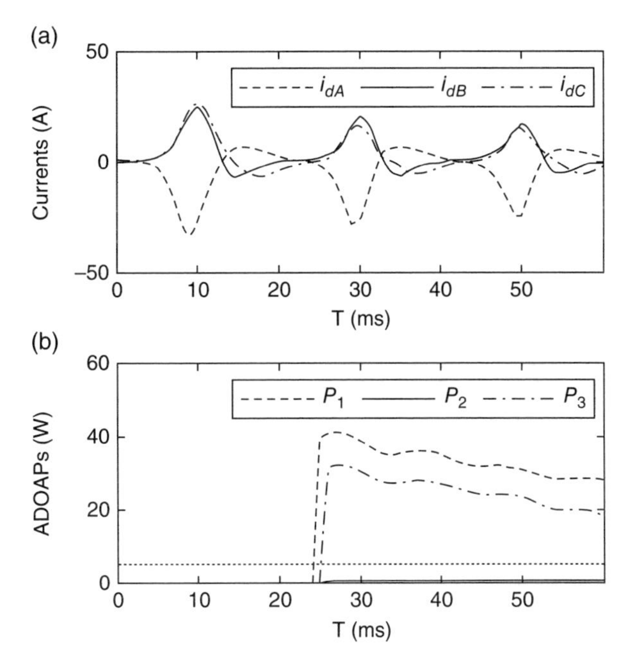

Figure 2.52 Differential currents and experimental results when the transformer is energized with a 2.4% turn-to-turn fault in phase A. (a) Differential currents. (b) ADOAPs.

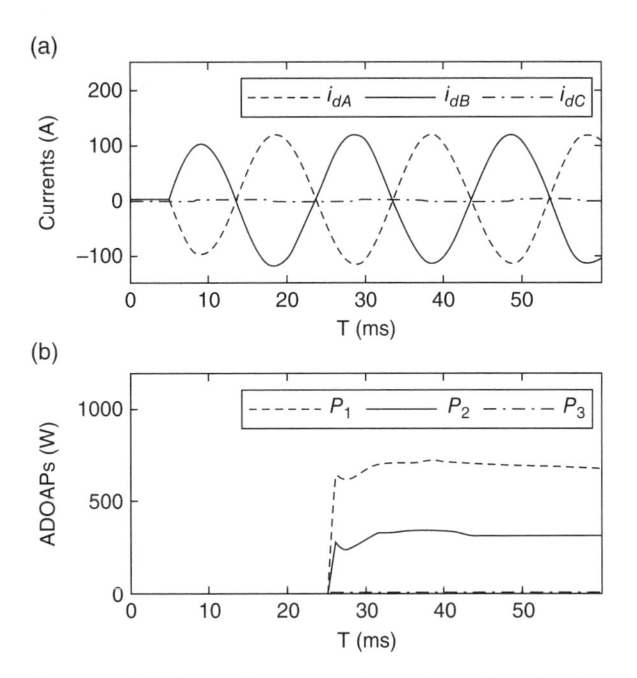

Figure 2.53 Differential currents and experimental results when a turn-to-ground internal fault occurs in phase B. (a) Differential currents. (b) ADOAPs.

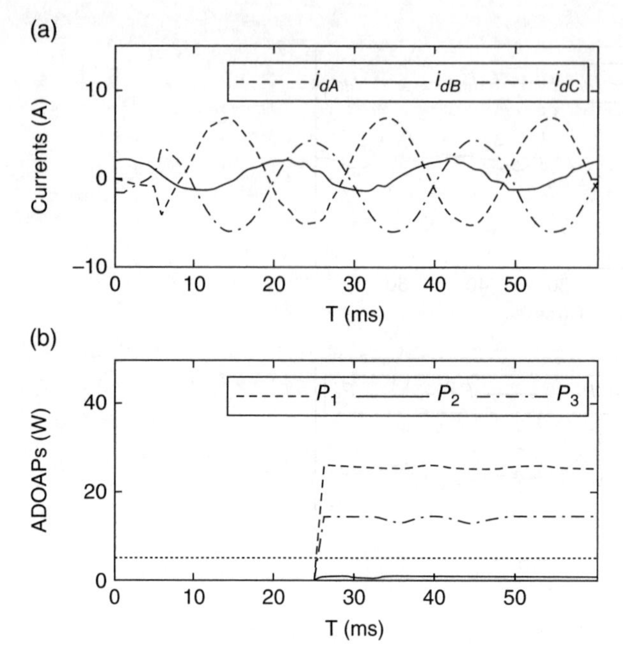

Figure 2.54 Differential currents and experimental results when a 2.4% turn-to-turn fault occurs in phase A. (a) Differential currents. (b) ADOAPs.

fault location as the example shown in Figure 2.51), and the other one is a 2.4% turn-to-turn internal fault in phase A (the same fault location as the example shown in Figure 2.52), where the differential currents of faulty phases become sufficiently small to be comparable with the nominal value.

In Figures 2.53(b) and 2.54(b), we find that the ADOAPs of non-fault phases are close to zero, but the ADOAPs of fault phases are all larger than the threshold. Moreover, the ADOAPs in Figures 2.53(b) and 2.54(b) present similar results to those in Figures 2.51(b) and 2.52(b), respectively. These results prove the accuracy of the calculated ADOAPs and the sensitivity of the method for identifying internal faults.

2.3 Transmission Line Protection

New transmission line protection schemes are introduced in this section. Sections 2.3.1 and 2.3.2 utilize voltage phase comparison and fault supplementary impedance to construct different protection criteria respectively, eliminating the impact of transition resistance [12–13]. Section 2.3.3 introduces a new location method for multiple inter-line and grounded faults of double-circuit transmission lines based on compensation voltage. Section 2.3.4 constructs overload identification criteria and can identify overload and symmetrical fault accurately combined with a conventional distance protection III segment. Section 2.3.5 introduces a novel

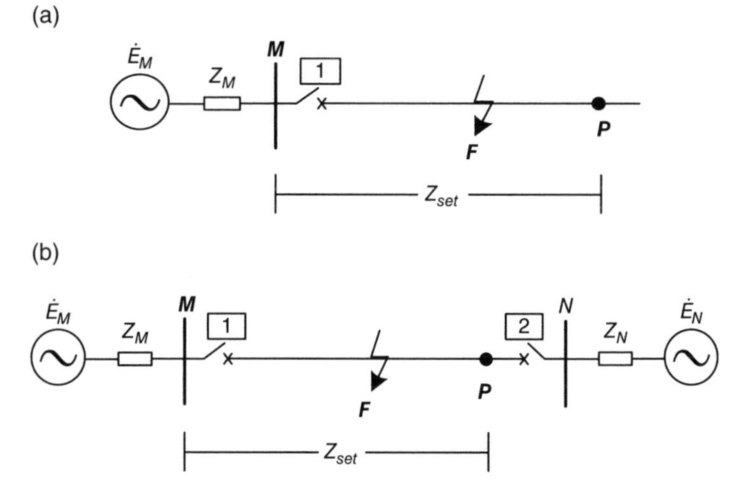

Figure 2.55 Analysis models. (a) Single-source system. (b) Double-source system.

fault-phase selection scheme utilizing fault-phase selection factors and solves the malfunction of distance protection caused by false phase selection [14].

2.3.1 Line Protection Scheme for Single-Phase-to-Ground Faults Based on Voltage Phase Comparison

2.3.1.1 Basic Principles

In the analysis models shown in Figure 2.55, the system voltages of the system on the M side and the system on the N side are $\dot{E}_M$ and $\dot{E}_N$, and the corresponding equivalent impedances are Z_M and Z_N. Take the line protection on the M side, for example. The setting impedance is Z_{set} and the protection range is MP.

A. Analysis on the Voltage Phasor Plane

In the single-source system shown in Figure 2.55(a), suppose that a single-phase-to-ground fault occurs at point F via a fault resistance. The voltage phasor diagram is shown in Figure 2.56, where $\dot{U}_{\text{m}}$ and $\dot{I}_{\text{m}}$ are the measured voltage and current at the relaying point. $\dot{U}_{\text{f}}$ is the voltage at the fault point. When the fault occurs at the same point via different fault resistances, $\dot{U}_{\text{f}}$ will move along the arc with $\dot{E}_M$ as the string. Since the current at the relaying point is equal to the current at the fault point, $\dot{I}_{\text{m}}$ has the same phase angle. as $\dot{U}_{\text{f}}$. φ_{ui} is the angular difference between the measured voltage and measured current at the relaying point. φ_{line} is the line impedance angle. $\dot{U}_p$ is the calculated voltage at the setting point P, which is called the compensatory voltage. With the measured voltage $\dot{U}_{\text{m}}$ as reference, the phase angle of $\dot{U}_{\text{f}}$ is

$$\varphi_{uf} = \arg\left(\frac{\dot{U}_{\text{f}}}{\dot{U}_{\text{m}}}\right) = \arg\left(\frac{\dot{I}_{\text{m}} R_g}{\dot{U}_{\text{m}}}\right) = \varphi_{ui} \tag{2.145}$$

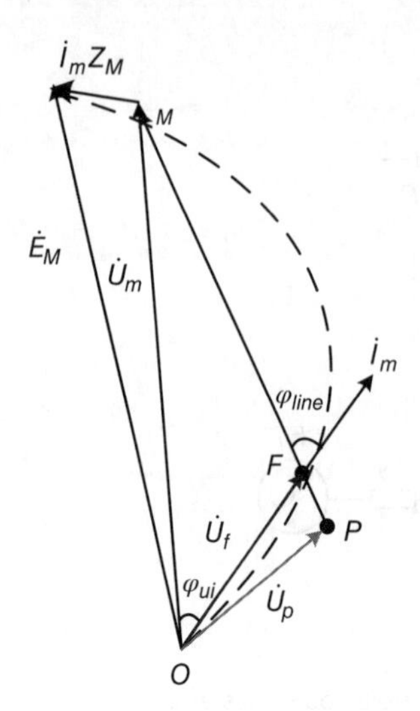

Figure 2.56 Voltage phasor diagram when a single-phase-to-ground fault occurs in the single-source system.

Table 2.11 Measured voltage and current.

Fault type		Measured voltage $\dot{U}_m$	Measured current $\dot{I}_m$
Single-phase-to-ground fault	AG	$\dot{U}_A$	$\dot{I}_A + K \times 3\dot{I}_0$
	BG	$\dot{U}_B$	$\dot{I}_B + K \times 3\dot{I}_0$
	CG	$\dot{U}_C$	$\dot{I}_C + K \times 3\dot{I}_0$

The compensatory voltage is expressed as

$$\dot{U}_p = \dot{U}_m - \dot{I}_m Z_{set} \tag{2.146}$$

In the protection scheme proposed here, Z_{set} applies the setting impedance of distance protection zone I. The measured voltage $\dot{U}_m$ and measured current $\dot{I}_m$ of different single-phase-to-ground faults are shown in Table 2.11, where K is the zero sequence current compensation coefficient.

With the measured voltage $\dot{U}_m$ as reference, the phase angle of $\dot{U}_p$ is

$$\varphi_{up} = \arg\left(\frac{\dot{U}_p}{\dot{U}_m}\right) = \arg\left(\frac{\dot{U}_m - \dot{I}_m Z_{set}}{\dot{U}_m}\right) \tag{2.147}$$

In the double-source system shown in Figure 2.54(b), suppose a single-phase-to-ground fault occurs at point F via the fault resistance. The voltage phasor diagram is shown in Figure 2.56, where $\dot{U}_{M|0|}$, $\dot{U}_{N|0|}$ and $\dot{U}_{f|0|}$ are the voltage at bus M, bus N and the fault point when the system is in normal operation. $\dot{U}_m$ and $\dot{I}_m$ are the measured voltage and current at the relaying point on the M side. $\dot{I}_f$ is the current at the fault point. Due to the contribution current from the opposite side, an angular difference φ_{if} exists between $\dot{I}_m$ and $\dot{I}_f$. $\dot{U}_f$ is the voltage at the fault

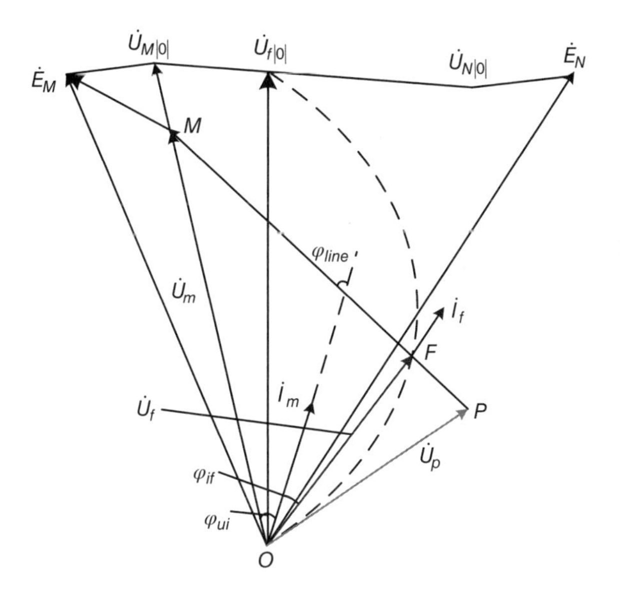

Figure 2.57 Voltage phasor diagram when a single-phase-to-ground fault occurs in the double-source system.

point. When a fault occurs at the same point via different fault resistances, $\dot{U}_\mathrm{f}$ will move along the arc with $\dot{U}_{\mathrm{f}|0|}$ as the string. φ_{ui} is the angular difference between the measured voltage and measured current at the relaying point. φ_{line} is the line impedance angle. $\dot{U}_p$ is the compensatory voltage.

The angular difference between the measured current $\dot{I}_\mathrm{m}$ and the fault point voltage $\dot{U}_\mathrm{f}$ can be expressed as

$$\varphi_{if} = \arg\left(\frac{\dot{I}_\mathrm{m}}{\dot{U}_\mathrm{f}}\right) = \arg\left(\frac{\dot{I}_\mathrm{m}}{\dot{I}_\mathrm{f}R_g}\right) = \arg\left(\frac{\dot{I}_\mathrm{m}}{\dot{I}_\mathrm{f}}\right) \tag{2.148}$$

where R_g is the fault resistance.

With the measured voltage $\dot{U}_\mathrm{m}$ as reference, the phase angle of $\dot{U}_\mathrm{f}$ is

$$\varphi_{uf} = \arg\left(\frac{\dot{U}_\mathrm{m}}{\dot{U}_\mathrm{f}}\right) = \arg\left(\frac{\dot{U}_\mathrm{m}}{\dot{I}_\mathrm{m}}\right) + \arg\left(\frac{\dot{I}_\mathrm{m}}{\dot{I}_\mathrm{f}}\right) = \varphi_{ui} + \varphi_{if} \tag{2.149}$$

It can be seen from Equation (2.149) that, in the double-source system, in order to calculate the phase angle of the fault point voltage, the angular difference between the measured current at the relaying point and the fault point φ_{if} must be calculated first.

B. Calculation of the Angular Difference Between the Measured Current and the Fault Point Voltage

Since the fault point current is immeasurable, it is difficult to calculate the angular difference between the measured current and the fault point current φ_{if} directly. Considering that the phase of sequence current at the relaying point is

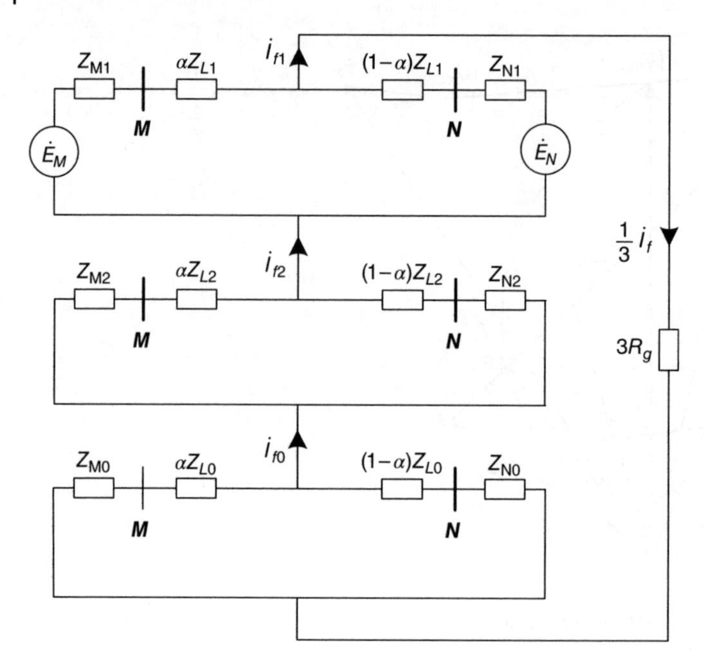

Figure 2.58 Compound sequence network of a single-phase-to-ground fault.

approximately the same as the phase of sequence current at the fault point, φ_{if} can be calculated indirectly as follows.

For the double-source system in Figure 2.55(b), supposing that a single-phase-to-ground fault occurs, the compound sequence network is as shown in Figure 2.58. Z_{M1}, Z_{M2} and Z_{M0} are the equivalent positive, negative and zero sequence impedance of the system on the M side. Z_{N1}, Z_{N2} and Z_{N0} are the equivalent positive, negative and zero sequence impedance of the system on the N side. Z_{L1}, Z_{L2} and Z_{L0} are the positive, negative and zero sequence impedances of line MN. α represents the percentage of distance from the fault point to the relaying point in the whole line length. R_g is the fault resistance. $\dot{I}_{\mathrm{f1}}$, $\dot{I}_{\mathrm{f2}}$ and $\dot{I}_{\mathrm{f0}}$ are the positive, negative and zero sequence current at the fault point.

It can be seen that, when a single-phase-to-ground fault occurs in the system, the fault point current has the same phase as the fault point sequence current. In a high-voltage system, the phase of sequence current at the relaying point is approximately the same as the phase of sequence current at the fault point, i.e. the phase of the fault point current can be equivalent to the phase of sequence current at the relaying point. Compared with zero sequence current, the negative sequence current at the relaying point is closer to the fault point current in phase. Thus, here the phase of the negative sequence current at the relaying point is used to approximate the phase of the negative sequence current at the fault point. Therefore, the angular difference between the measured current and the fault point voltage φ_{if} can be expressed as

$$\varphi_{\mathrm{if}} \approx \arg\left(\frac{\dot{I}_{\mathrm{m}}}{\dot{I}_{\mathrm{m2}}}\right) \tag{2.150}$$

where $\dot{I}_{\mathrm{m2}}$ is the negative sequence current at the relaying point.

C. Protection Scheme and Its Implementation

Whether in a single-source system or a double-source system, when the fault point F is within the protection setting range, the fault point voltage is ahead of the compensatory voltage in phase, i.e. $\varphi_{uf} < \varphi_{up}$. When the fault point F is on the protection setting boundary, the fault point voltage is the same as the compensatory voltage in phase, i.e. $\varphi_{uf} = \varphi_{up}$. When the fault point F is outside the protection setting range, the fault point voltage lags behind the compensatory voltage in phase, i.e. $\varphi_{uf} > \varphi_{up}$. According to the phase relationship between the fault point voltage and the compensatory voltage, the following line protection criterion is formed:

$$\begin{cases} \varphi_{uf} \leq \varphi_{up} & \text{in-zone fault} \\ \varphi_{uf} > \varphi_{up} & \text{out-of-zone fault} \end{cases} \tag{2.151}$$

where φ_{uf} and φ_{up} are the phase angles of the fault point voltage and the compensatory voltage, with the measured voltage as the reference value.

Use the jump-value of current as the startup component. When the maximum value of the phase current mutation variables exceeds the threshold value, single-phase-to-ground line protection starts. When the maximum value of the phase current jump-values exceeds the threshold value, the protection criterion will start up. Half- and quarter-wave Fourier filters are used to calculate the phase angles of the fault point voltage and the compensatory voltage. Thus, the speed of the proposed protection algorithm is effectively improved. If the operation criterion is satisfied, line protection operates; otherwise, line protection does not operate. The flow chart of the line protection scheme for a single-phase-to-ground fault is shown in Figure 2.59, where i_{mA}, i_{mB} and i_{mC} are the current mutation variables of phases A, B and C. I_ε is the rated line current when the line is in normal operation. Take 0.1 I_ε as the threshold value of the current jump-value startup criterion.

2.3.1.2 Simulation Verification

A two-machine system, as shown in Figure 2.55(b), is established in RTDS for simulation tests. The line parameters of line MN are: $R_1 = 0.021\ \Omega/\text{km}$, $X_1 = 0.281\ \Omega/\text{km}$, $C_1 = 500\ \text{M}\Omega{*}\text{m}$; $R_0 = 0.115\ \Omega/\text{km}$, $X_0 = 0.719\ \Omega/\text{km}$, $C_0 = 800\ \text{M}\Omega{*}\text{m}$; the whole line length is 300 km. The impedance of system E_M is: $Z_{\mathrm{S1}} = 4.264 + \mathrm{j}45.15\Omega$, $Z_{\mathrm{S0}} = 0.6 + \mathrm{j}9.09\Omega$, $E_M = 525\angle\delta$ kV. The impedance of system E_N is: $Z_{\mathrm{R1}} = 8.0 + 59.65\Omega$, $Z_{\mathrm{R0}} = 2.0 + 7.47\Omega$, $E_N = 525\angle 0°$ kV. To verify the correctness and feasibility of the proposed scheme, take the protection at bus M for example, the setting range of which is 80% of the whole length of line MN. In the simulation process, suppose that a fault occurs at $t = 0.5$ s.

A. Simulation of Different Fault Resistances

When $\delta = \pm 20°$, supposing that a phase-A-to-ground fault occurs at 180 km from bus M via different fault resistances, simulation results of the proposed scheme are shown in Figures 2.60 and 2.61.

When $\delta = 20°$, bus M is the sending end; when $\delta = -20°$, bus M is the receiving end. It can be seen that, whether bus M is the sending end or the receiving end, the proposed scheme is not affected by the fault resistance. No matter how big the

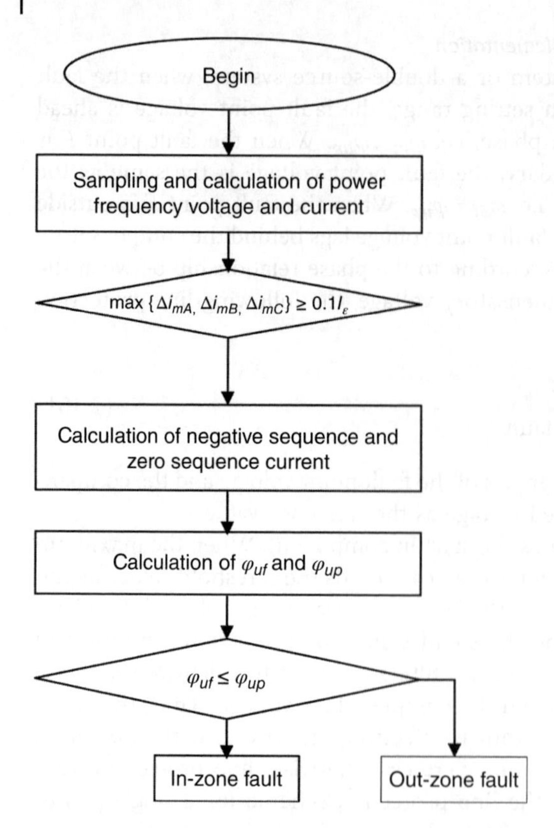

Figure 2.59 Flowchart of the line protection scheme for a single-phase-to-ground fault.

fault resistance is, the in-zone fault can be identified correctly. In addition, the protection operation time is shorter than 14 ms, which satisfies the requirement of line protection on operational time [15].

B. Simulation of Different Fault Locations

When $\delta = \pm 20°$, supposing that a phase-A-to-ground fault occurs via 300 Ω fault resistance at different locations on line MN, simulation results of the proposed scheme are shown in Figure 2.62 and Figure 2.63.

When $\delta = 20°$, bus M is the sending end; when $\delta = -20°$, bus M is the receiving end. It can be seen that, whether bus M is the sending end or the receiving end, when the fault location is within 225 km from bus M (i.e. within 95% of the protection range), the proposed scheme can identify in-zone fault correctly within 14 ms. When the fault location is 255 km from bus M (i.e. outside 105% of the protection range), the proposed scheme can identify out-of-zone fault correctly. Therefore, the proposed scheme satisfies the requirement of line protection on setting value error and operational time.

C. Simulation of a Non-Linear, High-Impedance Fault

In order for the simulation results to better conform to engineering practice, simulation tests concerning non-linear, high-impedance grounding faults are added in the revised paper using the HIF model mentioned in reference [16]. When $\delta = 20°$,

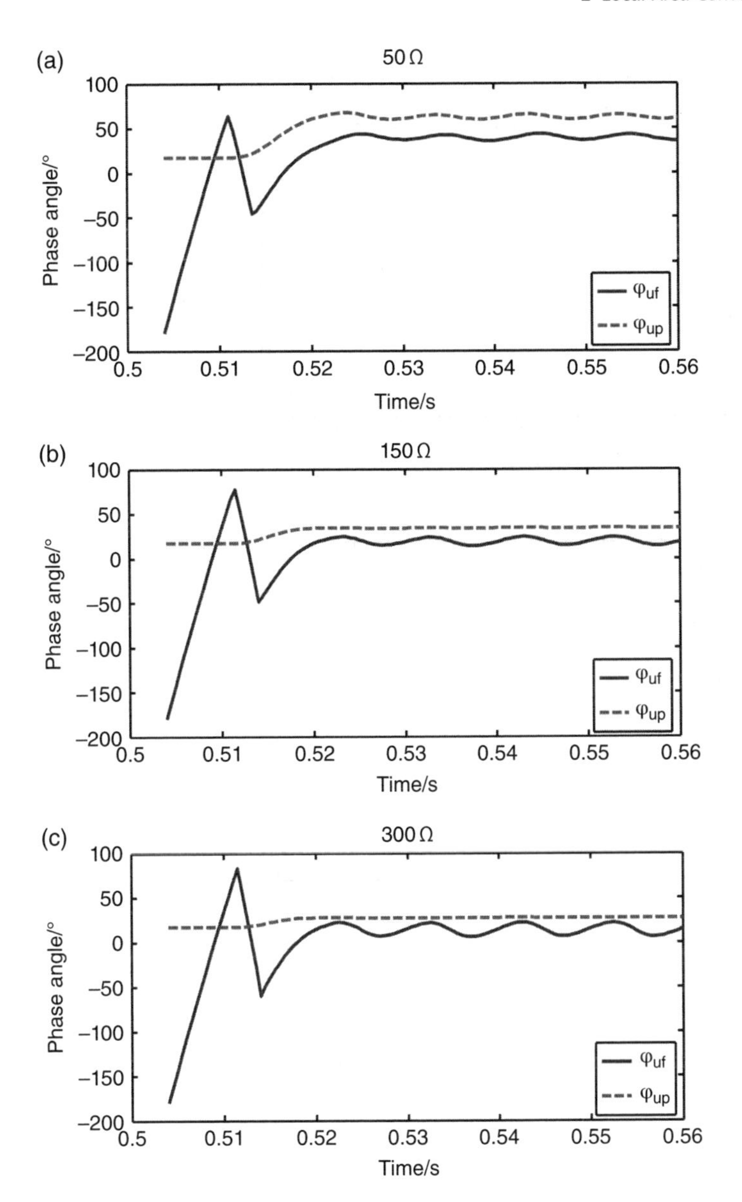

Figure 2.60 Simulation results of a phase-A-to-ground fault occurring at 180 km from bus *M* via different fault resistances ($\delta = 20°$).

supposing a phase-A-to-ground fault occurs at different locations, the simulation results of the *M* side are as shown in Figure 2.64.

It can be seen that, when a fault occurs via non-linear fault resistance, both φ_{uf} and φ_{up} undergo a dynamic variation period. However, the proposed scheme is still able to identify in-zone and out-zone faults correctly. When the fault location is within 95% of the protection range, the proposed scheme can identify an in-zone fault correctly within 13 ms. When the fault location is outside 105% of

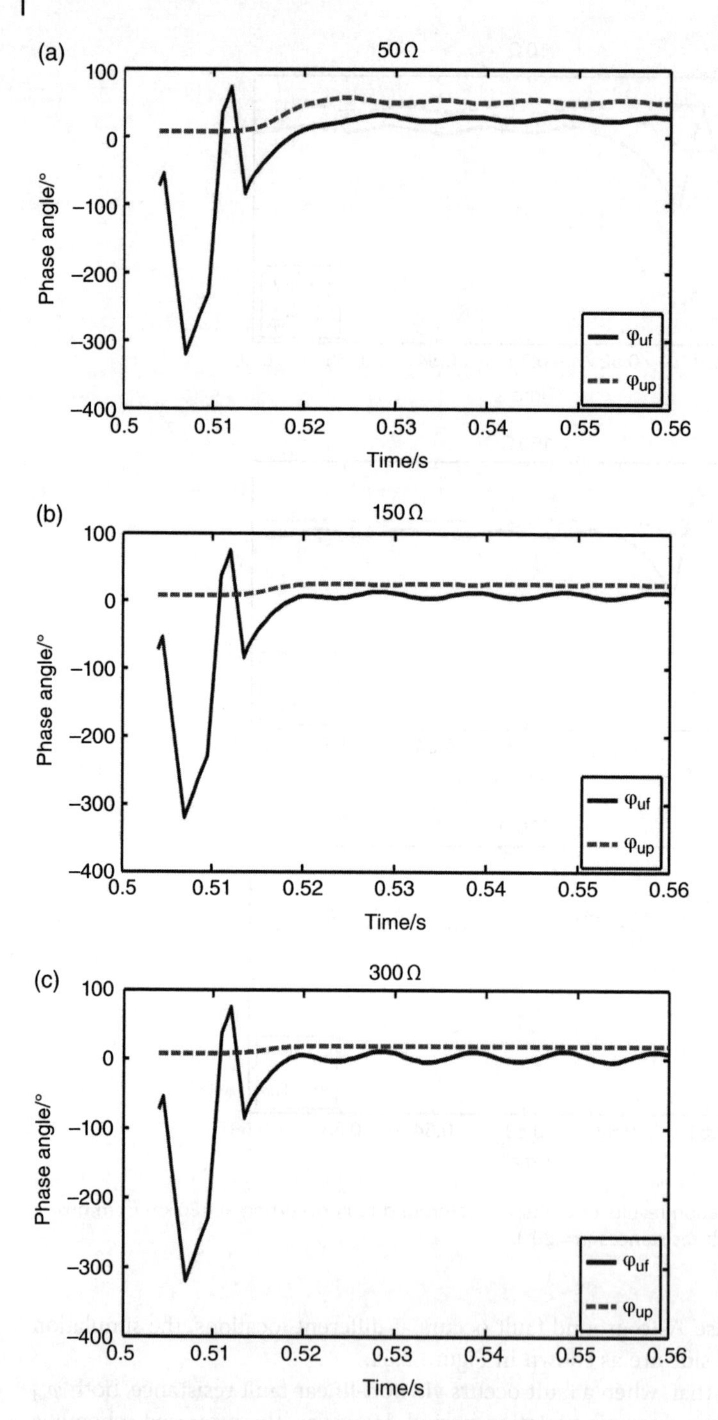

Figure 2.61 Simulation results of a phase-A-to-ground fault occurring at 180 km from bus *M* via different fault resistances ($\delta = -20°$).

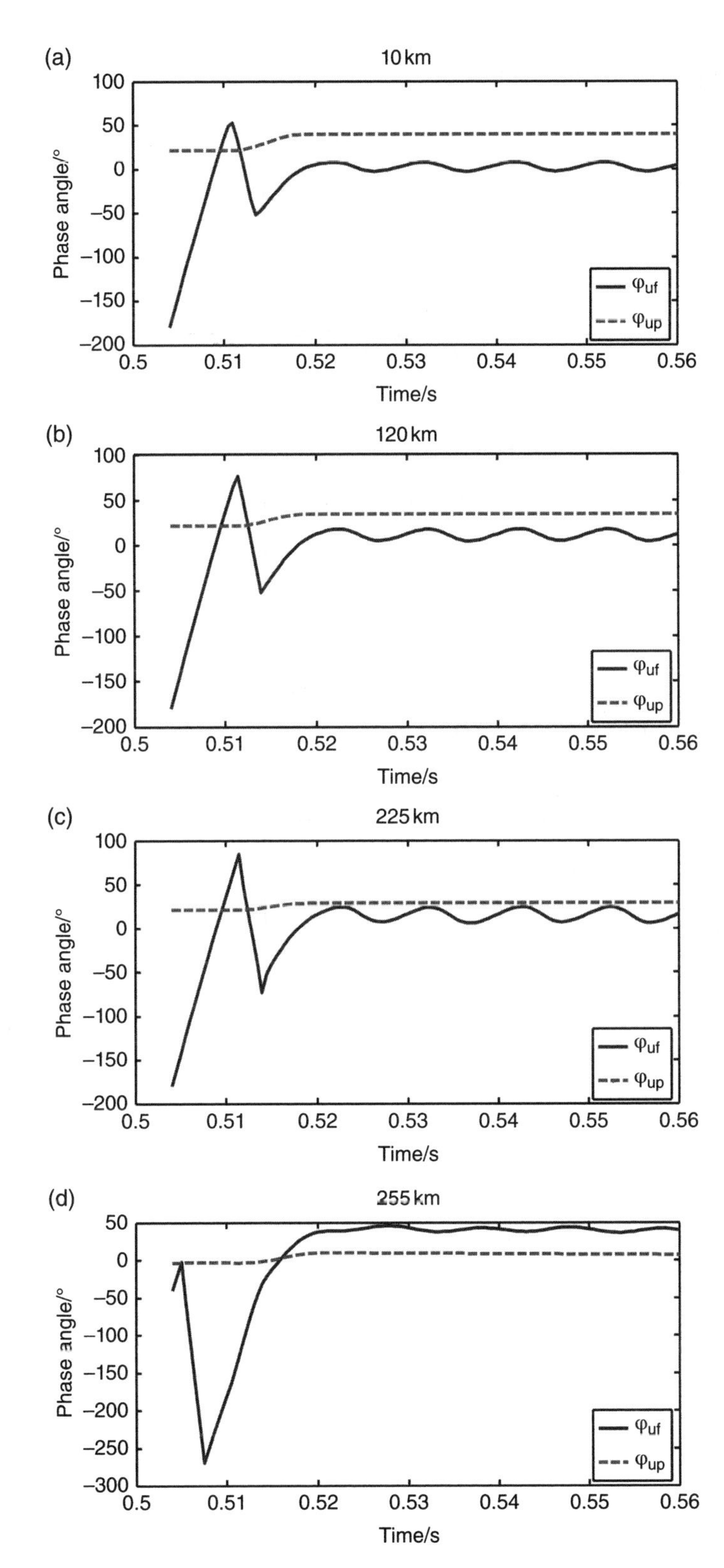

Figure 2.62 Simulation results of a phase-A-to-ground fault occurring via 300 Ω fault resistance at different distances from bus M ($\delta = 20°$).

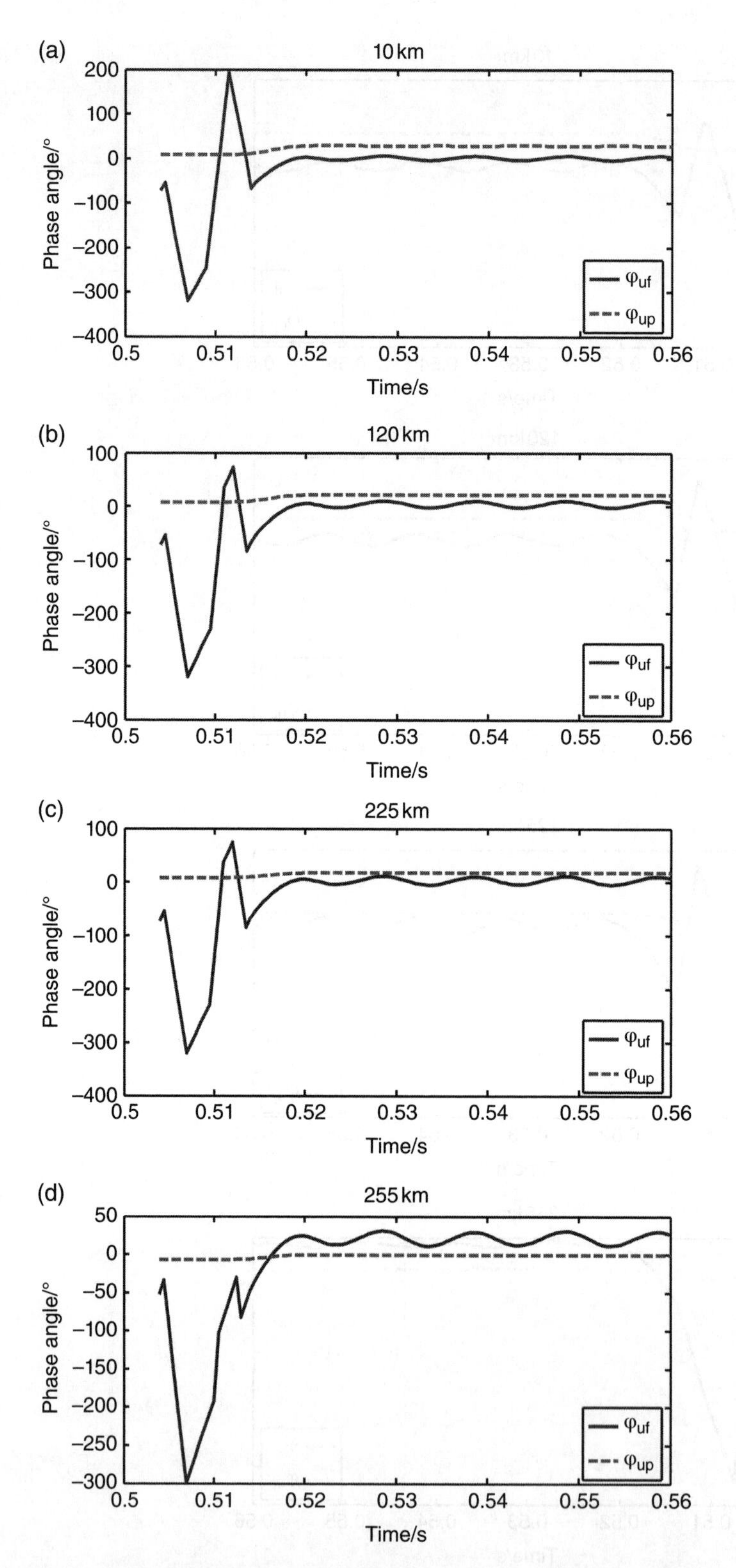

Figure 2.63 Simulation results of a phase-A-to-ground fault occurring via 300 Ω fault resistance at different distances from bus M ($\delta = -20°$).

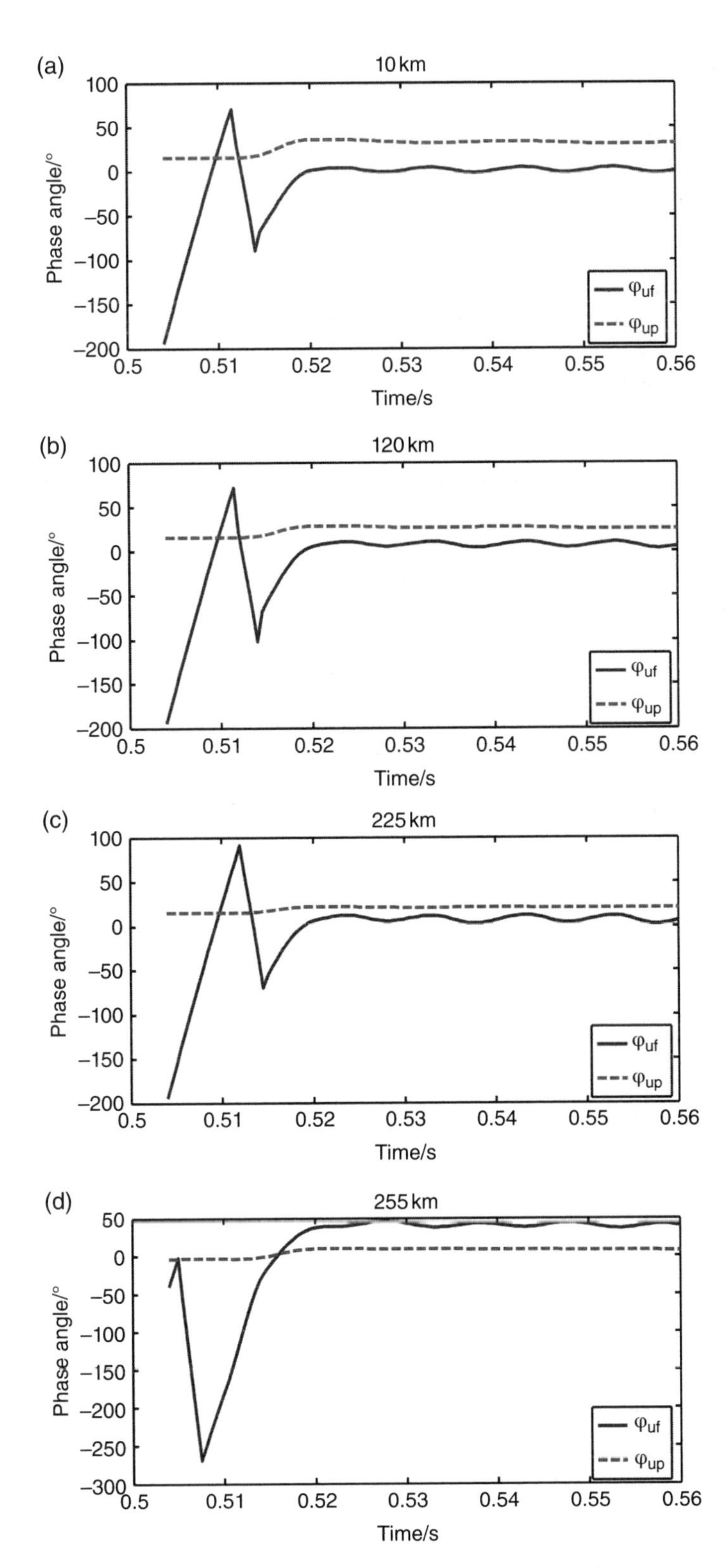

Figure 2.64 Simulation results of a non-linear HIF at different distances from bus M ($\delta = 20°$).

the protection range, the proposed scheme can correctly identify it as an out-zone fault. Thus, the setting error and operational time requirements of grounding fault protection are both satisfied.

D. Simulation of Different System Operating Conditions

The angular difference between the voltages at two line ends reflects the system operating condition. The bigger the angular difference, the more heavily loaded the line. When the system is in different operating conditions, supposing that a phase-A-to-ground fault occurs at different locations via different fault resistances, the simulation results of the proposed scheme are as shown in Table 2.12. In Table 2.12, '√' means that the protection operates, and '×' means

Table 2.12 Simulation results of a phase-A-to-ground fault occurring under different system operating conditions.

		***M*-side system voltage phase angle**			
Fault resistance	**Fault location**	$\delta = 10°$	$\delta = -10°$	$\delta = 30°$	$\delta = -30°$
5/Ω	225/km	√	√	√	√
	*240/km	√	√	√	×
	255/km	×	×	×	×
50/Ω	225/km	√	√	√	√
	*240/km	√	√	√	×
	255/km	×	×	×	×
100/Ω	225/km	√	√	√	√
	*240/km	√	×	√	×
	255/km	×	×	×	×
150/Ω	225/km	√	√	√	√
	*240/km	√	×	√	×
	255/km	×	×	×	×
200/Ω	225/km	√	√	√	√
	*240/km	√	×	×	×
	255/km	×	×	×	×
250/Ω	225/km	√	√	√	√
	*240/km	√	×	×	×
	255/km	×	×	×	×
300/Ω	225/km	√	√	√	√
	*240/km	×	×	×	×
	255/km	×	×	×	×

* *Note:* 240 km is the setting range boundary of line protection.

Table 2.13 Errors of the proposed method for different fault resistance under different conditions.

	Fault resistance	Fault location	*M*-side system voltage phase angle			
			$\delta = 10°$	$\delta = -10°$	$\delta = 30°$	$\delta = -30°$
Errors of proposed method (%)	5/Ω	225/km	2.13	1.96	1.12	2.31
	50/Ω	225/km	1.82	2.02	2.56	2.33
	100/Ω	225/km	2.11	1.88	2.25	2.60
	150/Ω	225/km	1.94	2.07	3.02	1.45
	200/Ω	225/km	2.39	2.08	1.11	2.71
	250/Ω	225/km	1.63	1.99	2.05	2.31
	300/Ω	225/km	2.22	1.86	2.42	1.75

that the protection does not operate. Comparative analysis shows that the simulation results are similar to those of $\delta = \pm 20°$, i.e. the proposed scheme is not affected by the system operating condition. No matter how big the fault resistance, the protection operates when a fault occurs in the zone, and it does not operate when a fault occurs out of the zone. Thus, the correctness and feasibility of the proposed scheme is further proved.

Shown in Table 2.13 are the errors of the proposed scheme when a fault occurs via different fault resistances in different system operating modes. It can be seen that the proposed method is not affected by the fault resistance. Higher fault resistance will not cause bigger error or poorer performance.

To verify the performance of the proposed scheme in the case of weak feed end fault, suppose that a phase-A-to-ground fault occurs at different distances from bus N via different linear fault resistances and non-linear fault resistances under different system operating conditions. The simulation results are shown in Table 2.14. Comparison with the simulation results in Table 2.12 shows that the proposed scheme is not affected by the weak feed end fault. When a fault occurs within the protection range of the weak feed end, the protection operates reliably. When a fault occurs outside the protection range, the protection does not operate.

E. Simulation of Different System Source Impedances

Simulation tests concerning different source impedances are made in this section. When $\delta = 20°$, suppose that a phase-A-to-ground fault occurs at 225 km from bus N with different source impedances; simulation results of the proposed scheme are shown in Figure 2.65.

It can be seen that variation of the source impedance does not affect the operational performance of the proposed scheme. Even when the source impedance increases to 50 times the original value, the proposed scheme can still identify in-zone fault reliably within 14 ms.

Table 2.14 Simulation results of phase-A-to-ground fault occurring under weak feed conditions.

Weak-infeed conditions	Fault location	*N*-side system voltage phase angle			
		$\delta = 10°$	$\delta = -10°$	$\delta = 30°$	$\delta = -30°$
5/Ω	225/km	√	√	√	√
	***240**/km	√	√	√	√
	255/km	×	×	×	×
50/Ω	225/km	√	√	√	√
	***240**/km	√	√	×	√
	255/km	×	×	×	×
100/Ω	225/km	√	√	√	√
	***240**/km	√	×	√	√
	255/km	×	×	×	×
150/Ω	225/km	√	√	√	√
	***240**/km	√	×	√	×
	255/km	×	×	×	×
200/Ω	225/km	√	√	√	√
	***240**/km	√	√	×	×
	255/km	×	×	×	×
250/Ω	225/km	√	√	√	√
	***240**/km	√	×	×	×
	255/km	×	×	×	×
300/Ω	225/km	√	√	√	√
	***240**/km	×	×	×	×
	255/km	×	×	×	×
Non-linear	225/km	√	√	√	√
	***240**/km	√	√	×	√
	255/km	×	×	×	×

* *Note:* 240 km is the setting range boundary of distance protection.

F. Comparative Analysis of the Proposed Scheme and the Other Line Protection Scheme

In order to further verify the performance of the proposed scheme, a comparison is made between the proposed scheme and the distance protection scheme in reference [17]. Suppose that an *N*-side system is the weak feed end with a voltage phase angle $\delta = -20°$, and a phase-A-to-ground fault occurs at different locations

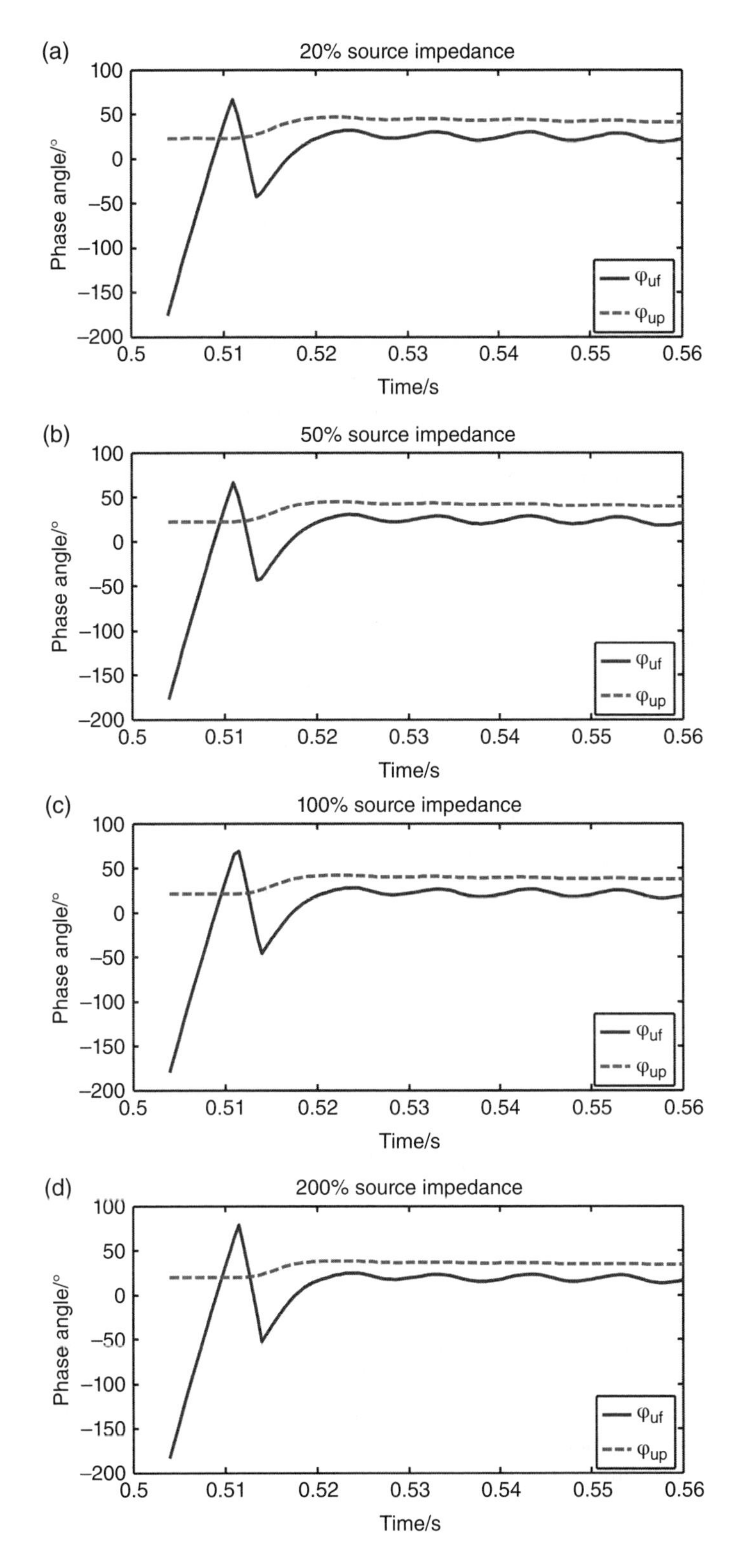

Figure 2.65 Simulation results of non-linear HIF under different source impedances for bus *N* ($\delta = 20°$).

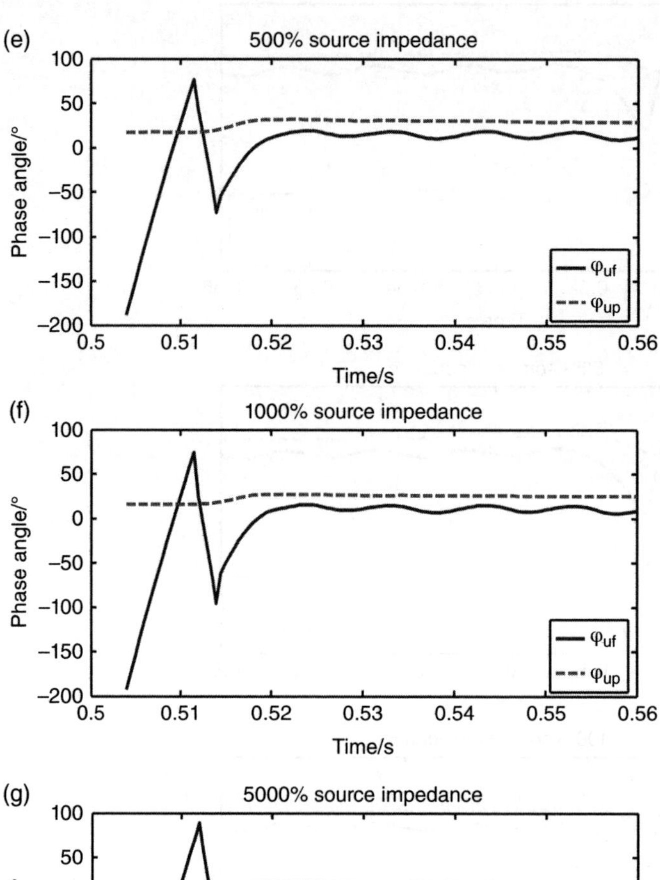

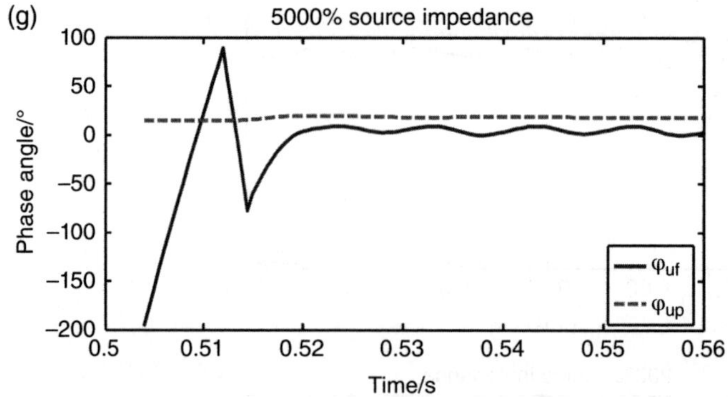

Figure 2.65 (*Continued*)

via non-linear fault resistances. For the proposed scheme, the distance relay applies the polygonal operation characteristics [15]. When the measured impedance falls into the operation zone, distance protection operates; otherwise, distance protection does not operate. When a phase-A-to-ground fault occurs at different distances from bus N, the trajectory of the measured impedance is as shown in Figure 2.66.

It can be seen that, for the distance protection scheme in reference [17], when the fault location is within 225 km of bus N (i.e. the fault occurs within 95% of the protection range), the measured impedance may fall outside the operational zone and thus distance protection refuses to operate. When the fault location is at 255 km from bus N (i.e. the fault occurs outside 105% of the protection range), the measured impedance may fall inside the operational zone and thus distance protection malfunctions.

According to the above simulation analysis, the distance protection scheme in reference [17] is strongly affected by the source impedance. It may refuse to operate when a fault occurs within the protection range of the weak feed end, and may malfunction when a fault occurs outside the protection range. However, the proposed scheme is not affected by the source impedance and could operate correctly and reliably no matter where the fault occurs (on the strong feed end or weak feed end), as can be seen from Figure 2.65 and Table 2.14.

2.3.2 Adaptive Distance Protection Scheme Based on the Voltage Drop Equation

2.3.2.1 The Definition of Compensation Voltage Adaptive Distance Protection Based on Phasor Analysis

In the distance protection analysis model shown in Figure 2.55, the system voltages of side M and side N are $\dot{E}_{\mathrm{M}}$ and $\dot{E}_{\mathrm{N}}$, and the equivalent impedances are Z_{M} and Z_{N} respectively. Taking protection R1 at bus M for example, the protection zone is MP, and the setting value is Z_{set}. F is the fault point, and Z represents the positive sequence impedance of the line length from F to the relaying point R1.

When a fault occurs on the line, the following relationship exists between the measured voltage $\dot{U}_{\mathrm{m}}$ and the measured current $\dot{I}_{\mathrm{m}}$:

$$\dot{U}_{\mathrm{m}} = \dot{I}_{\mathrm{m}} Z + \dot{U}_{\mathrm{f}} \tag{2.152}$$

where $\dot{U}_{\mathrm{f}}$ is the voltage at the fault point.

According to Equation (2.152), the measured impedance at the relaying point is

$$Z_{\mathrm{m}} = Z + \frac{\dot{U}_{\mathrm{f}}}{\dot{I}_{\mathrm{m}}} = Z + \frac{\dot{I}_{\mathrm{f}} R_g}{\dot{I}_{\mathrm{m}}} \tag{2.153}$$

where $\dot{I}_{\mathrm{f}}$ is the current at the fault point. R_g is the transition resistance. It can be seen that the measured impedance Z_{m} contains not only the real fault impedance Z, but also a supplementary impedance $\frac{\dot{I}_{\mathrm{f}} R_g}{\dot{I}_{\mathrm{m}}}$, which might cause distance protection to malfunction.

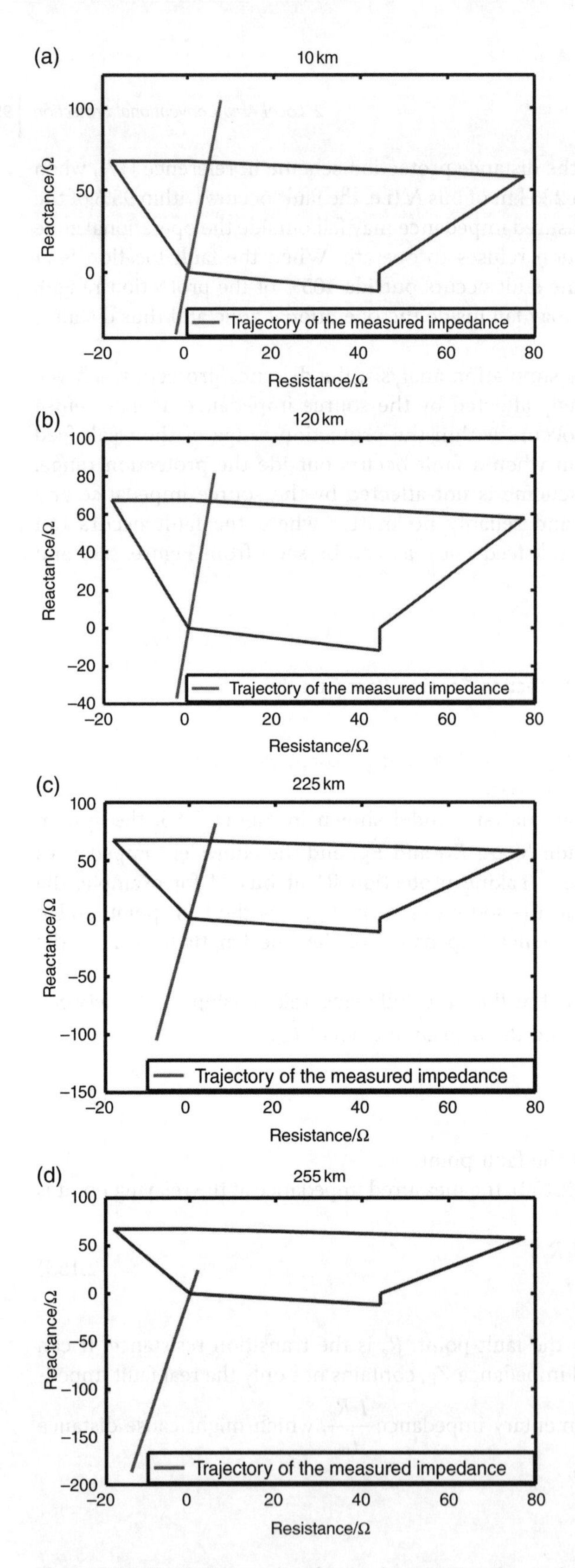

Figure 2.66 Simulation results of a non-linear HIF under weak-infeed conditions at different distances from bus N ($\delta = -20°$).

Table 2.15 Measured voltage and measured current in different fault types.

Fault type	$\dot{U}_m$	$\dot{I}_m$
phase-A-to-ground fault	$\dot{U}_A$	$\dot{I}_A + K \times 3\dot{I}_0$
phase-B-to-ground fault	$\dot{U}_B$	$\dot{I}_B + K \times 3\dot{I}_0$
phase-C-to-ground fault	$\dot{U}_C$	$\dot{I}_C + K \times 3\dot{I}_0$

Note: K represents the zero sequence current compensation coefficient.

In the adaptive distance protection scheme, the choice of measured voltage $\dot{U}_m$ and measured current $\dot{I}_m$ in different fault types is shown in Table 2.15.

A. Single-Source System

In the single-source system shown in Figure 2.55(a), when a fault occurs on the line, the voltage and current phasor distribution is as shown in Figure 2.56. $\overrightarrow{\boldsymbol{OM}}$ is the measured voltage phasor $\dot{U}_m$ at R1. $\overrightarrow{\boldsymbol{OF}}$ is the voltage at the fault point $\dot{U}_f$, and when the fault occurs at the same location via different transition resistances, $\overrightarrow{\boldsymbol{OF}}$ will move along the arc with $\dot{E}_M$ as the chord. $\overrightarrow{\boldsymbol{MF}}$ is the voltage drop from the fault point F to bus M, the modulus of which is $|\overrightarrow{\boldsymbol{MF}}| = |\dot{I}_m \times Z|$. The perpendicular foot of $\boldsymbol{O}$ on phasor $\overrightarrow{\boldsymbol{MF}}$ is $\boldsymbol{A}$. The angular difference between $\overrightarrow{\boldsymbol{OA}}$ and $\overrightarrow{\boldsymbol{OM}}$ is $\varphi = \varphi_{ui} + 90° - \varphi_{line}$.

The following relationships exist in Figure 2.56:

$$\begin{cases} |\overrightarrow{\boldsymbol{OA}}| = |\dot{U}_m \cos\varphi| \\ |\overrightarrow{\boldsymbol{MA}}| = |\dot{U}_m \sin\varphi| \\ |\overrightarrow{\boldsymbol{FA}}| = |\dot{U}_m \cos\varphi \tan(90° - \varphi_{line})| \\ |\overrightarrow{\boldsymbol{MF}}| = |\overrightarrow{\boldsymbol{MA}}| - |\overrightarrow{\boldsymbol{FA}}| = |\dot{U}_m \sin\varphi| - |\dot{U}_m \cos\varphi \tan(90° - \varphi_{line})| \end{cases} \tag{2.154}$$

Applying $|\overrightarrow{\boldsymbol{MF}}| = |\dot{I}_m \times Z|$ to Equation (2.154), the voltage drop equation can be obtained:

$$|\dot{I}_m||Z| = |\dot{U}_m \sin\varphi| - |\dot{U}_m \cos\varphi \tan(90° - \varphi_{line})| \tag{2.155}$$

Multiply both sides of Equation (2.155) by $\left|\dfrac{\cos(90^\circ - \varphi_{line})}{\dot{I}_m}\right|$, so that

$$|Z\cos(90^\circ - \varphi_{line})| = |Z_m(\sin\varphi\cos(90^\circ - \varphi_{line}) - \cos\varphi\sin(90^\circ - \varphi_{line}))| \tag{2.156}$$

According to the trigonometric function, Equation (2.156) can also be expressed as

$$|Z\sin\varphi_{line}| = |Z_m\sin\varphi_{ui}| \tag{2.157}$$

Transform Equation (2.157) to the impedance plane form:

$$|Z| = |Z_m|\left|\frac{\sin\varphi_{ui}}{\sin\varphi_{line}}\right| \tag{2.158}$$

Consider that $|Z| = \dfrac{Z}{\cos\varphi_{line} + j\sin\varphi_{line}}$, $|Z_m| = \dfrac{Z_m}{\cos\varphi_{ui} + j\sin\varphi_{ui}}$, Equation (2.158) can also be expressed as

$$Z = Z_m\left|\frac{\sin\varphi_{ui}}{\sin\varphi_{line}}\right|\frac{\cos\varphi_{line} + j\sin\varphi_{line}}{\cos\varphi_{ui} + j\sin\varphi_{ui}} = (R + jX)\left|\frac{\sin\varphi_{ui}}{\sin\varphi_{line}}\right| \tag{2.159}$$

where Z_m is the measured impedance. φ_{line} is the impedance angle of the protected line. φ_{ui} is the angular difference between the measured voltage and measured current. $R = \text{Real}\left(Z_m\dfrac{\cos\varphi_{line} + j\sin\varphi_{line}}{\cos\varphi_{ui} + j\sin\varphi_{ui}}\right)$, $X = \text{Imag}\left(Z_m\dfrac{\cos\varphi_{line} + j\sin\varphi_{line}}{\cos\varphi_{ui} + j\sin\varphi_{ui}}\right)$.

In view of the wide application of the impedance relay with quadrilateral characteristic in power systems, such as the quadrilateral characteristic shown in Figure 2.67, value setting of the proposed protection scheme is carried out. In Figure 2.67, R_{set} is the resistance setting value, and X_{set} is the reactance setting

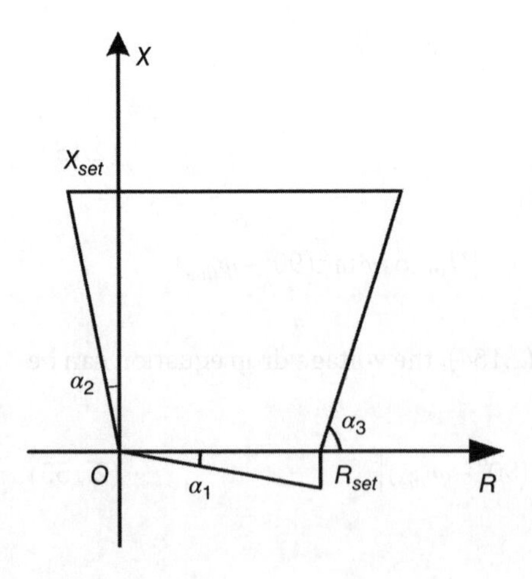

Figure 2.67 Impedance relay with quadrilateral characteristic.

value. When the fault point is inside the protection zone, the impedance Z will fall in the operational zone and the protection will operate. When the fault point is outside the protection zone, the impedance Z will fall outside the operational zone and the protection will not operate.

In the quadrilateral characteristic, the operating equations of quadrant I are

$$\begin{cases} R\left|\dfrac{\sin\varphi_{ui}}{\sin\varphi_{line}}\right| - X\left|\dfrac{\sin\varphi_{ui}}{\sin\varphi_{line}}\right|\cot\alpha_3 \le R_{\text{set}} \\ X\left|\dfrac{\sin\varphi_{ui}}{\sin\varphi_{line}}\right| \le X_{\text{set}} \end{cases} \tag{2.160}$$

The operating equations of quadrant II are

$$\begin{cases} R\left|\dfrac{\sin\varphi_{ui}}{\sin\varphi_{line}}\right| \ge -X\left|\dfrac{\sin\varphi_{ui}}{\sin\varphi_{line}}\right|\tan\alpha_2 \\ X\left|\dfrac{\sin\varphi_{ui}}{\sin\varphi_{line}}\right| \le X_{\text{set}} \end{cases} \tag{2.161}$$

The operating equations of quadrant IV are

$$\begin{cases} R\left|\dfrac{\sin\varphi_{ui}}{\sin\varphi_{line}}\right| \le R_{\text{set}} \\ X\left|\dfrac{\sin\varphi_{ui}}{\sin\varphi_{line}}\right| \ge -R\left|\dfrac{\sin\varphi_{ui}}{\sin\varphi_{line}}\right|\tan\alpha_1 \end{cases} \tag{2.162}$$

If Equations (2.160)–(2.162) are all satisfied at the same time, then it is identified that the fault is inside the protection zone and the protection operates. Otherwise, the protection does not operate.

Define $k_{\text{m}} = \left|\dfrac{\sin\varphi_{line}}{\sin\varphi_{ui}}\right|$ as the adaptive setting coefficient. Thus, in the single-source system, the operating equations of the adaptive distance protection with quadrilateral characteristic can be expressed as

$$\begin{cases} R - X\cot\alpha_3 \le k_{\text{m}}R_{\text{set}} \\ X \le k_{\text{m}}X_{\text{set}} \end{cases} \tag{2.163}$$

$$\begin{cases} R \ge -X\tan\alpha_2 \\ X \le k_{\text{m}}X_{\text{set}} \end{cases} \tag{2.164}$$

$$\begin{cases} R \le k_{\text{m}}R_{\text{set}} \\ X \ge -R\tan\alpha_1 \end{cases} \tag{2.165}$$

where R and X are adaptive protection measured values, and $k_{\text{m}}R_{\text{set}}$ and $k_{\text{m}}X_{\text{set}}$ are adaptive protection setting values.

On the other hand, none of the parameters in the operating equations are related to the source impedance. Therefore, in a single-source system, the adaptive distance protection criterion is not affected by a change in the source impedance.

B. Double-Source System

In the double-source system shown in Figure 2.55(b), when a fault occurs on the line, the voltage and current phasor distribution is as shown in Figure 2.68. $\dot{U}_{M|0|}$, $\dot{U}_{N|0|}$ and $\dot{U}_{f|0|}$ are the voltages at bus M, bus N and the fault point respectively when the system is in a normal operating state. $\dot{I}_m$, $\dot{I}_n$ and $\dot{I}_f$ are the current at bus M, bus N and the fault point respectively, and $\dot{I}_f = \dot{I}_m + \dot{I}_n$. Due to the contribution of the opposite side current $\dot{I}_n$, there is a deviation angle ψ between $\dot{I}_m$ and $\dot{I}_f$. When the fault occurs at the same location via different transition resistances, $\overrightarrow{OF}$ will move along the arc with $\dot{U}_{f|0|}$ as the chord. C is the intersection of the measured current $\dot{I}_m$ and the extended line of $\overrightarrow{MF}$. The angular difference between $\overrightarrow{OA}$ and $\overrightarrow{OM}$ is $\varphi = \varphi_{ui} + 90° - \varphi_{line}$.

According to Figure 2.68, the following relationships exist:

$$\begin{cases} |\overrightarrow{OA}| = |\dot{U}_m \cos\varphi| \\ |\overrightarrow{MA}| = |\dot{U}_m \sin\varphi| \\ |\overrightarrow{FA}| = |\dot{U}_m \cos\varphi \tan(90° - \varphi_{line} - \psi)| \\ |\overrightarrow{MF}| = |\overrightarrow{MA}| - |\overrightarrow{FA}| = |\dot{U}_m \sin\varphi| - |\dot{U}_m \cos\varphi \tan(90° - \varphi_{line} - \psi)| \end{cases} \tag{2.166}$$

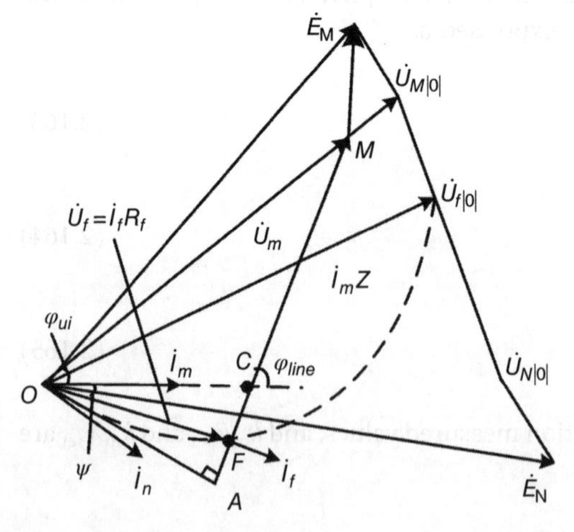

Figure 2.68 Voltage and current vector in the case of a double-source system fault.

Apply $|\overrightarrow{MF}| = |\dot{I}_\mathrm{m} \times Z|$ to Equation (2.166), so that the voltage drop equation can be obtained:

$$\begin{aligned} |\dot{I}_\mathrm{m}||Z| &= |\dot{U}_\mathrm{m}\sin\varphi| - |\dot{U}_\mathrm{m}\cos\varphi\tan(90^\circ - \varphi_{line} - \psi)| \\ &= \left|\dot{U}_\mathrm{m}\frac{\sin(\varphi_{ui} + \psi)}{\cos(90^\circ - \varphi_{line} - \psi)}\right| \end{aligned} \tag{2.167}$$

Divide both sides of Equation (2.167) by the amplitude of the measured current $|\dot{I}_\mathrm{m}|$, so that

$$|Z| = |Z_\mathrm{m}|\left|\frac{\sin(\varphi_{ui} + \psi)}{\sin(\varphi_{line} + \psi)}\right| \tag{2.168}$$

Considering that $|Z| = \dfrac{Z}{\cos\varphi_{line} + \mathrm{j}\sin\varphi_{line}}$, $|Z_\mathrm{m}| = \dfrac{Z_\mathrm{m}}{\cos\varphi_{ui} + \mathrm{j}\sin\varphi_{ui}}$, Equation (2.168) can also be expressed as

$$Z = Z_\mathrm{m}\left|\frac{\sin(\varphi_{ui} + \psi)}{\sin(\varphi_{line} + \psi)}\right|\frac{\cos\varphi_{line} + \mathrm{j}\sin\varphi_{line}}{\cos\varphi_{ui} + \mathrm{j}\sin\varphi_{ui}} = (R + \mathrm{j}X)\left|\frac{\sin(\varphi_{ui} + \psi)}{\sin(\varphi_{line} + \psi)}\right| \tag{2.169}$$

where Z_m is the measured impedance. φ_{line} is the impedance angle of the protected line. φ_{ui} is the angular difference between the measured voltage and measured current. ψ is the angular difference between the measured current and fault current, $R = \mathrm{Real}\left(Z_\mathrm{m}\dfrac{\cos\varphi_{line} + \mathrm{j}\sin\varphi_{line}}{\cos\varphi_{ui} + \mathrm{j}\sin\varphi_{ui}}\right)$, $X = \mathrm{Imag}\left(Z_\mathrm{m}\dfrac{\cos\varphi_{line} + \mathrm{j}\sin\varphi_{line}}{\cos\varphi_{ui} + \mathrm{j}\sin\varphi_{ui}}\right)$.

In order to realize adaptive distance protection in the double-source system, the deviation angle ψ between the measured current $\dot{I}_\mathrm{m}$ and the fault current $\dot{I}_\mathrm{f}$ needs to be calculated first. Since the fault current cannot be measured, the deviation angle ψ is calculated according to the measured value of the sequence current at the relaying point.

When a fault occurs in the double-source system shown in Figure 2.55(b), the negative sequence network is as shown in Figure 2.69. α is the percentage of the distance from F to bus M in the whole line length. Z_{L2} is the line negative sequence impedance. Z_{M2} and Z_{N2} are the equivalent negative sequence impedances of the system on side M and side N respectively.

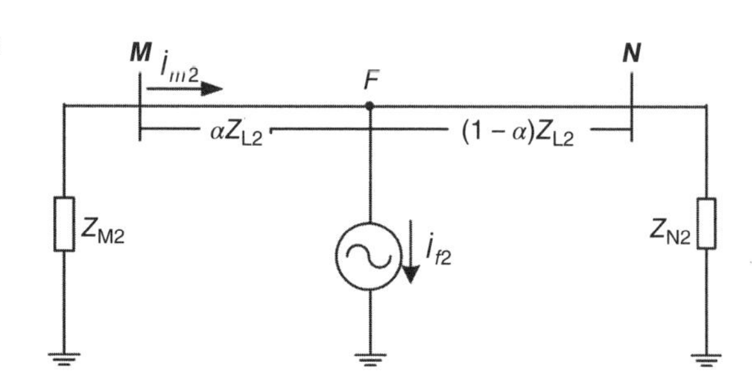

Figure 2.69 Negative sequence network of fault system.

According to Figure 2.69, the following relationship exists between the negative sequence fault current $\dot{I}_{f2}$ and the measured negative sequence current $\dot{I}_{m2}$:

$$\dot{I}_{m2} = C_{m2}\dot{I}_{f2} = \frac{Z_{N2} + (1-\alpha)Z_{L2}}{Z_{M2} + Z_{N2} + Z_{L2}}\dot{I}_{f2} \tag{2.170}$$

where C_{m2} is the negative sequence current distribution coefficient.

Consider that in an EHV/UHV system, the system impedance angle and line impedance angle are approximately the same [18], thus the negative sequence current distribution coefficient C_{m2} can be taken as a real number. Therefore, the angular difference between the negative sequence current at the fault point and the negative sequence current at the relaying point is approximately 0, i.e.

$$\arg\left(\frac{\dot{I}_{m2}}{\dot{I}_{f2}}\right) \approx 0° \tag{2.171}$$

When a single-phase grounding fault occurs,

$$\dot{I}_{f} = 3\dot{I}_{f2} \tag{2.172}$$

Therefore, the deviation angle of the measured current $\dot{I}_m$ from the fault current $\dot{I}_f$ can be expressed as

$$\psi = \arg\left(\frac{\dot{I}_m}{\dot{I}_f}\right) = \arg\left(\frac{\dot{I}_m}{\dot{I}_{f2}}\right) \approx \arg\left(\frac{\dot{I}_m}{\dot{I}_{m2}}\right) \tag{2.173}$$

Similarly to the single-source system, define $k_a = \left|\frac{\sin(\varphi_{line} + \psi)}{\sin(\varphi_{ui} + \psi)}\right|$ as the adaptive setting coefficient. Thus, in the double-source system, the operating equations of the adaptive distance protection with quadrilateral characteristic can be expressed as

$$\begin{cases} R - X\cot\alpha_3 \le k_a R_{set} \\ X \le k_a X_{set} \end{cases} \tag{2.174}$$

$$\begin{cases} R \ge -X\tan\alpha_2 \\ X \le k_a X_{set} \end{cases} \tag{2.175}$$

$$\begin{cases} R \le k_a R_{set} \\ X \ge -R\tan\alpha_1 \end{cases} \tag{2.176}$$

where R and X are adaptive protection measured values, and $k_a R_{set}$ and $k_a X_{set}$ are adaptive protection setting values.

On the other hand, none of the parameters in the operating equations are related to the source impedance. Therefore, a change in the source impedance does not affect the correctness and effectiveness of the proposed adaptive distance protection scheme.

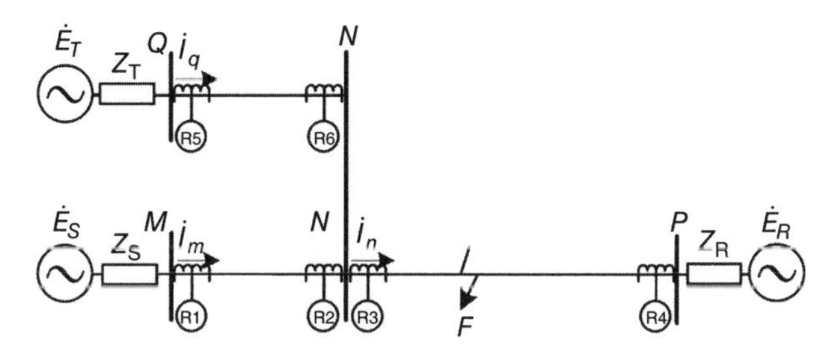

Figure 2.70 Analysis model.

C. *Three-Source System*

In the system shown in Figure 2.70, $\dot{E}_S$, $\dot{E}_R$ and $\dot{E}_T$ are equivalent power sources at buses *M*, *P* and *Q*, respectively, and Z_S, Z_R and Z_T are the corresponding equivalent impedances. The measured currents at relays R1, R3 and R5 are $\dot{I}_m$, $\dot{I}_n$ and $\dot{I}_q$ respectively. Taking R1 for example, it is proved below that when a fault occurs in the downstream (adjacent) line NP, the proposed scheme will not malfunction.

When a fault occurs at point *F* on line NP, the impedance from the fault location *F* to relay R1 is

$$Z = Z_{MN} + \frac{\dot{I}_n}{\dot{I}_m}\alpha Z_{NP} = Z_{MN} + K_b \alpha Z_{NP} \tag{2.177}$$

where Z_{MN} and Z_{NP} are the impedances of line MN and line NP. α is the percentage of the distance between fault point *F* and bus *N* in the whole line length. K_b is the branch coefficient:

$$K_b = \frac{\dot{I}_n}{\dot{I}_m} = 1 + \frac{\dot{I}_q}{\dot{I}_m} = 1 + \frac{Z_S + Z_{MN}}{Z_T + Z_{QN}} \tag{2.178}$$

Consider that in an EHV/UHV system, the system impedance angle and line impedance angle are approximately the same, thus K_b is approximately a real number. Thus, the reactance part of impedance *Z* is

$$\text{Imag}(Z) = X_{MN} + K_b \alpha X_{NP} \geq X_{MN} > X_{set} \tag{2.179}$$

where X_{MN} and X_{NP} are the reactances of line MN and line NP. X_{set} is the setting value of the reactance, $X_{set} = (80 \sim 90\%)\, X_{MN}$.

According to Equation (2.169), Imag(*Z*) can be calculated:

$$\text{Imag}(Z) = X\left|\frac{\sin(\varphi_{ui} + \psi)}{\sin(\varphi_{line} + \psi)}\right| \tag{2.180}$$

Combine Equations (2.179) and (2.180), so that

$$X \geq X_{set}\left|\frac{\sin(\varphi_{line} + \psi)}{\sin(\varphi_{ui} + \psi)}\right| = k_a X_{set} \tag{2.181}$$

where k_a is the adaptive setting coefficient.

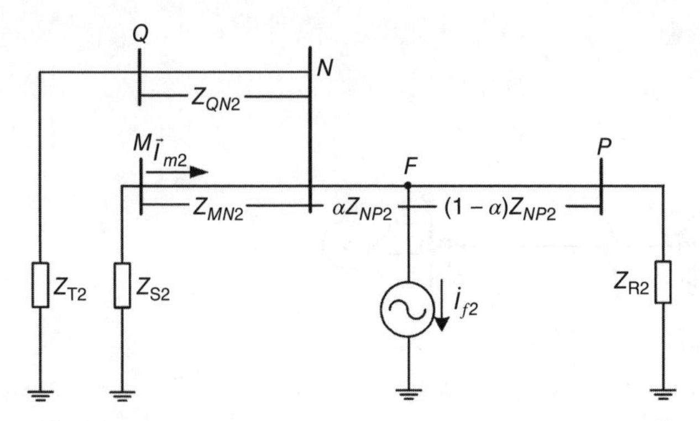

Figure 2.71 Negative sequence network.

It can be seen that Equation (2.181) does not meet the operating equation of the adaptive distance protection (shown in Equation (2.174)), thus the protection will not operate.

In addition, when a fault occurs on the downstream line NP, the formulas to calculate the deviation angle ψ shown in Equations (2.170)–(2.173) are still applicable. This is proved as follows.

In the analysis model shown in Figure 2.70 when a fault occurs on line NP, the negative sequence fault network is shown in Figure 2.71. The relationship between the negative sequence current at the relaying point and the negative sequence current at the fault point in Equation (2.170) becomes

$$\dot{I}_{\mathrm{m2}} = C_{\mathrm{m2}}\dot{I}_{\mathrm{f2}} = \frac{Z_{\mathrm{T2}} + Z_{\mathrm{Q2}}}{\dfrac{(Z_{\mathrm{S2}} + Z_{\mathrm{MN2}})*(Z_{\mathrm{T2}} + Z_{\mathrm{QN2}})}{Z_{\mathrm{S2}} + Z_{\mathrm{QN2}} + Z_{\mathrm{T2}} + Z_{\mathrm{QN2}}}} * \frac{Z_{\mathrm{R2}} + (1-\alpha)Z_{\mathrm{NP2}}}{\dfrac{(Z_{\mathrm{S2}} + Z_{\mathrm{MN2}})*(Z_{\mathrm{T2}} + Z_{\mathrm{QN2}})}{Z_{\mathrm{S2}} + Z_{\mathrm{QN2}} + Z_{\mathrm{T2}} + Z_{\mathrm{QN2}}} + Z_{\mathrm{NP2}} + Z_{\mathrm{R2}}}\dot{I}_{\mathrm{f2}} \tag{2.182}$$

where $\dot{I}_{\mathrm{m2}}$ is the negative sequence current at the relaying point, $\dot{I}_{\mathrm{f2}}$ is the negative sequence current at the fault point. C_{m2} is the negative sequence current distribution coefficient. α is the percentage of the distance between fault point F and bus N in the whole line length. Z_{MN2}, Z_{QN2} and Z_{NP2} are the negative sequence impedances of lines MN, QN and NP respectively. Z_{S2}, Z_{T2} and Z_{R2} are the equivalent negative sequence impedances of the systems at bus M, bus T and bus N.

Consider that in an EHV/UHV system, the system impedance angle and line impedance angle are approximately the same, thus in the case of a downstream line fault, C_{m2} can still be taken as a real number, and the formulas shown in Equations (2.171)–(2.173) are still true.

2.3.2.2 Simulation and Verification

In order to verify and evaluate the adaptive distance protection scheme, a series of faults with different transition resistances are simulated at several locations along the transmission line. The adaptive protection setting values and measured values are calculated according to the full cycle Fourier algorithm with a sampling frequency of 1000 Hz. Taking protection R1 at bus *M* for example, the reliability of the adaptive distance protection is verified. In the simulation process, the protection zone of distance protection is set to cover 80% of the whole line length, and the fault time is $t = 0.5$ s.

A. Simulation of a Single-Source Power System

The single-source power system shown in Figure 2.55(a) is established in RTDS for simulation, where the rated power is 100 MVA, the rated voltage is 220 kV, the fundamental frequency is 50 Hz, and the line length is 150 km. The equivalent system parameters are $Z_{M1} = 4.3578 + j49.8079\ \Omega$, and $Z_{M0} = 1.1 + j16.6\ \Omega$. The parameters of the transmission line are $Z_{L1} = 0.029 + j0.362\ \Omega/\text{km}$, and $Z_{L0} = 0.255 + j0.971\ \Omega$.

When a single-phase grounding fault occurs at 70 km from bus *M* via different transition resistances, simulation results of the adaptive distance protection at one cycle after the fault are as shown in Figure 2.72, where the solid line is the setting boundary of the adaptive impedance relay with quadrilateral characteristic, and the circles represent the adaptive protection measured values. It can be seen that the proposed scheme is immune to the transition resistance. When a fault occurs via different transition resistances, the adaptive protection measured values all fall inside the operational zone, so that the protection is able to operate correctly. The operational time of the adaptive distance protection is about 20 ms, which meets the requirement of distance protection on the operational time.

When a single-phase grounding fault occurs via 100 Ω transition resistances at different locations, simulation results of the adaptive distance protection at one cycle after the fault are as shown in Figure 2.73, where the solid line is the setting boundary of the adaptive impedance relay with quadrilateral characteristic, and the circles represent the adaptive protection measured values. It can be seen that, when the fault location is less than 114 km away from bus *M* (i.e. when the fault location is within 95% of the protection range), the adaptive protection measured values all fall inside the operational zone, so that the adaptive distance protection operates. When the fault location is more than 126 km away from bus *M* (i.e. when the fault location falls 105% outside the protection range), the adaptive protection measured values fall outside the operational zone, so that the adaptive distance protection does not operate. The requirement of distance protection on the setting value error is met.

B. Simulation of a Double-Source Power System

The double-source power system shown in Figure 2.55(b) is established in a real time digital simulator (RTDS) for simulation, where the rated power is 500 MVA, the rated voltage is 500 kV, the fundamental frequency is 50 Hz, and the line length is 400 km. The equivalent system parameters are: $Z_{M1} = 1.05 + j6.0\ \Omega$, $Z_{M0} = 0.06 + j7.8\ \Omega$, $Z_{N1} = 1.28 + j9.0\ \Omega$, $Z_{N0} = 6.4 + j18.2\ \Omega$. The parameters of the transmission line are: $r_1 = 0.01839\ \Omega/\text{km}$, $l_1 = 0.8376\ \text{mH/km}$, $c_1 = 500.1314$ MΩ.m; $r_0 = 0.1417\ \Omega/\text{km}$, $l_0 = 1.91943\ \text{mH/km}$, $c_0 = 800.5217$ MΩ.m.

When a single-phase grounding fault occurs at 200 km from bus *M* via different transition resistances, simulation results of the adaptive distance protection at one cycle after the fault are as shown in Figure 2.74, where the solid line is the setting

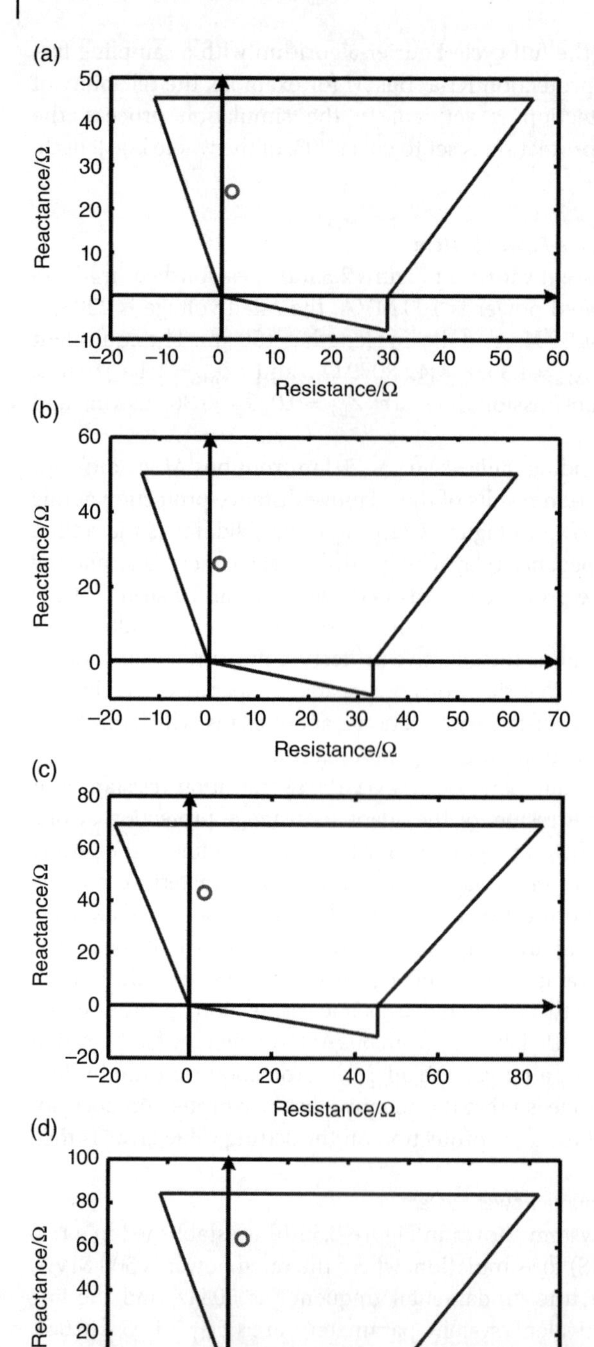

Figure 2.72 Simulation results of the adaptive distance protection for a single-source power system when a phase-A-to-ground fault occurs via different transition resistances at 70 km away from relay R1. (a) 0 Ω. (b) 10 Ω. (c) 50 Ω. (d) 100 Ω.

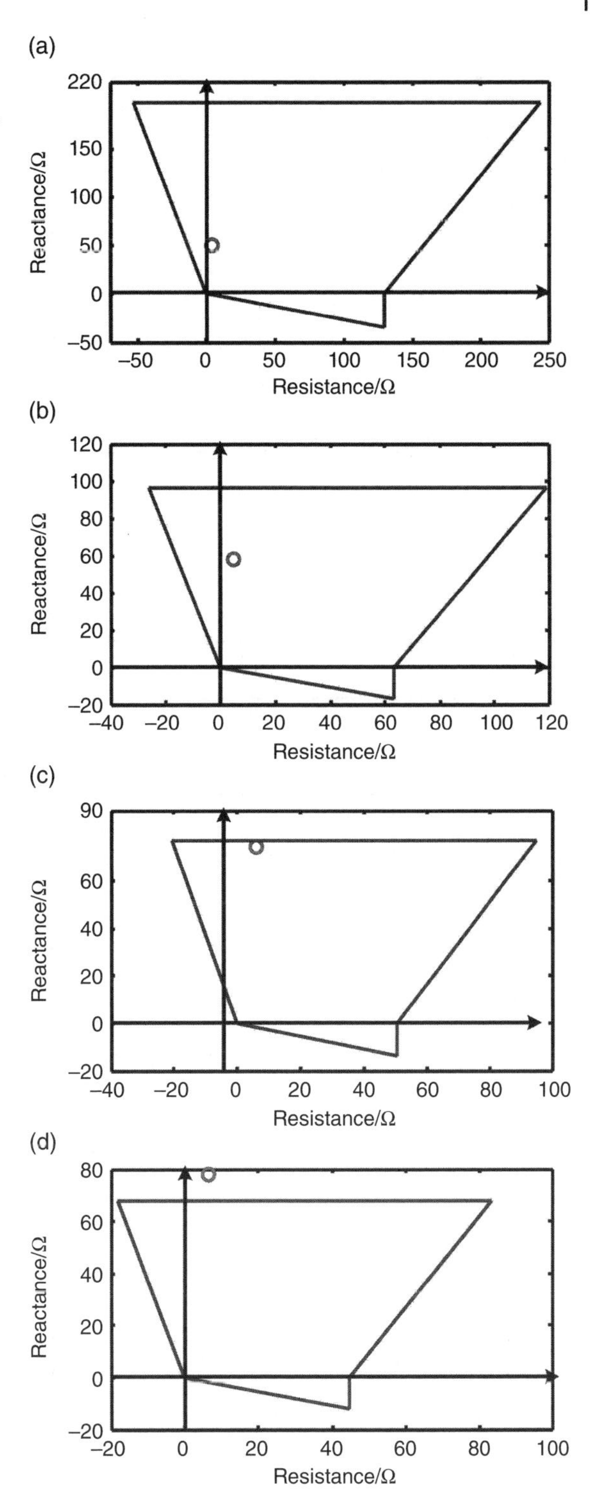

Figure 2.73 Simulation results of the adaptive distance protection for a single-source power system when a phase-A-to-ground fault occurs via 100 Ω transition resistance at different points away from relay R1. (a) 5 km. (b) 50 km. (c) 114 km. (d) 126 km.

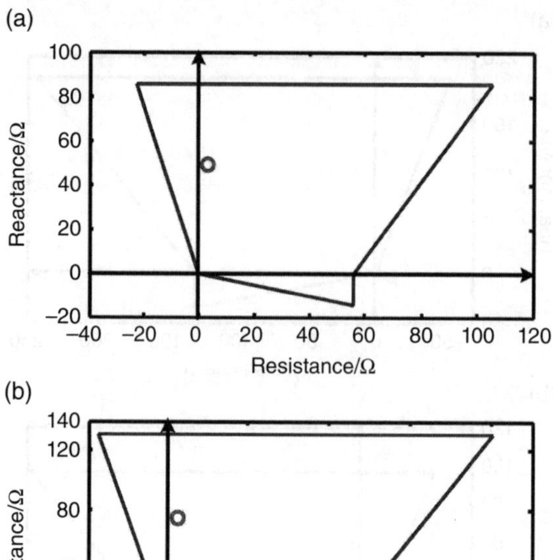

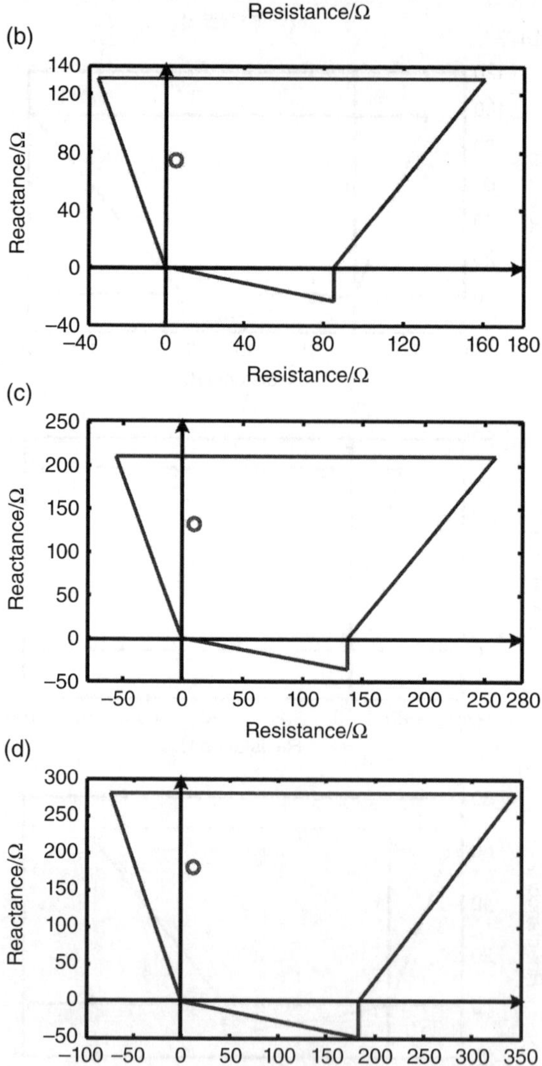

Figure 2.74 Simulation results of the adaptive distance protection for a double-source power system when a phase-A-to-ground fault occurs via different transition resistances at 200 km away from relay R1. (a) 0 Ω. (b) 50 Ω. (c) 150 Ω. (d) 300 Ω.

boundary of the adaptive impedance relay with quadrilateral characteristic, and the circles represent the adaptive protection measured values. It can be seen that, after the fault occurs, the adaptive protection measured values all fall inside the operational zone, so that the protection is able to operate correctly no matter how big the transition resistance is. The operational time of the adaptive distance protection is about 20 ms, which meets the requirement of distance protection on the operational time.

When a single-phase grounding fault occurs via 300 Ω transition resistances at different locations, simulation results of the adaptive distance protection at one cycle after the fault are as shown in Figure 2.75, where the solid line is the setting boundary of the adaptive impedance relay with quadrilateral characteristic, and the circles represent the adaptive protection measured values. It can be seen that, when a fault occurs within 300 km of bus M, the adaptive protection measured values all fall inside the operational zone, so that the adaptive distance protection operates. When a fault occurs at 340 km of bus M, the adaptive protection measured values fall outside the operational zone, so that the adaptive distance protection does not operate. The setting value error is less than 5% around the protection setting.

When the fault occurs on the downstream line, the proposed adaptive distance protection criterion is still valid.

C. *Simulation of a Three-Source Power System*

In the test system shown in Figure 2.70, when a high resistance fault occurs at different locations on line NP, the simulation results of the adaptive distance protection at one cycle after the fault are as shown in Figure 2.76. It can be seen that, in the case of a downstream line fault, the adaptive protection measured value falls outside the operational zone, thus the distance protection will not operate. In the test system, the line length of line NP is 200 km, line parameters are: $R_1 = 0.02083\ \Omega/\text{km}$, $L_1 = 0.8948\ \text{mH/km}$, $C_1 = 0.0129\ \mu\text{F/km}$; $R_0 = 0.1148\ \Omega/\text{km}$, $L_0 = 2.2886\ \text{mH/km}$, $C_0 = 0.00523\ \mu\text{F/km}$. The line length of line QN is 280 km, line parameters are: $R_1 = 0.0133\ \Omega/\text{km}$, $L_1 = 0.8469\ \text{mH/km}$, $C_1 = 0.0139\ \mu\text{F/km}$; $R_0 = 0.3080\ \Omega/\text{km}$, $L_0 = 2.5941\ \text{mH/km}$, $C_0 = 0.0098\ \mu\text{F/km}$. Parameters of the system on the Q side are: $Z_{T1} = 1.05 + j43.14\ \Omega$, $Z_{T0} = 0.6 + j29.7\ \Omega$. The other parameters are the same as those in the double-source simulation system.

D. *Comparison with Conventional Distance Protection*

In order to show the superiority of the proposed scheme, a comparison is made between the proposed scheme and the conventional distance protection scheme with quadrilateral characteristics.

In the single-source simulation system, when a single-phase grounding fault occurs at 70 km from bus M via different transition resistances, the measured impedance trajectory of the conventional distance protection after the fault is shown in Figure 2.77. It can be seen that, when the transition resistance is 10 Ω or 50 Ω, the measured value falls in the operational zone, and the conventional distance protection operates correctly. However, when the transition resistance is 100 Ω, the measured value falls outside the operational zone, and the conventional distance protection fails to operate.

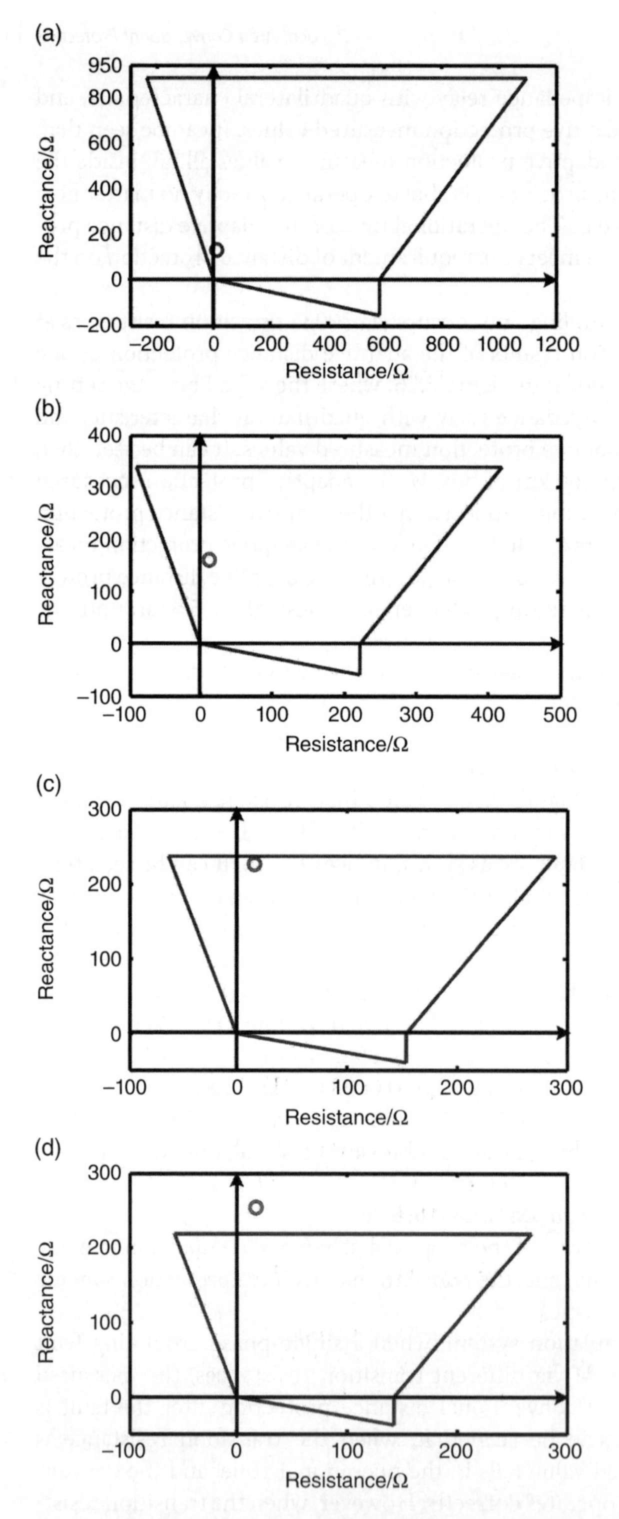

Figure 2.75 Simulation results of the adaptive distance protection for a double-source power system when a phase-A-to-ground fault occurs via 300 Ω transition resistance at different points away from relay R1. (a) 50 km. (b) 150 km. (c) 300 km. (d) 340 km.

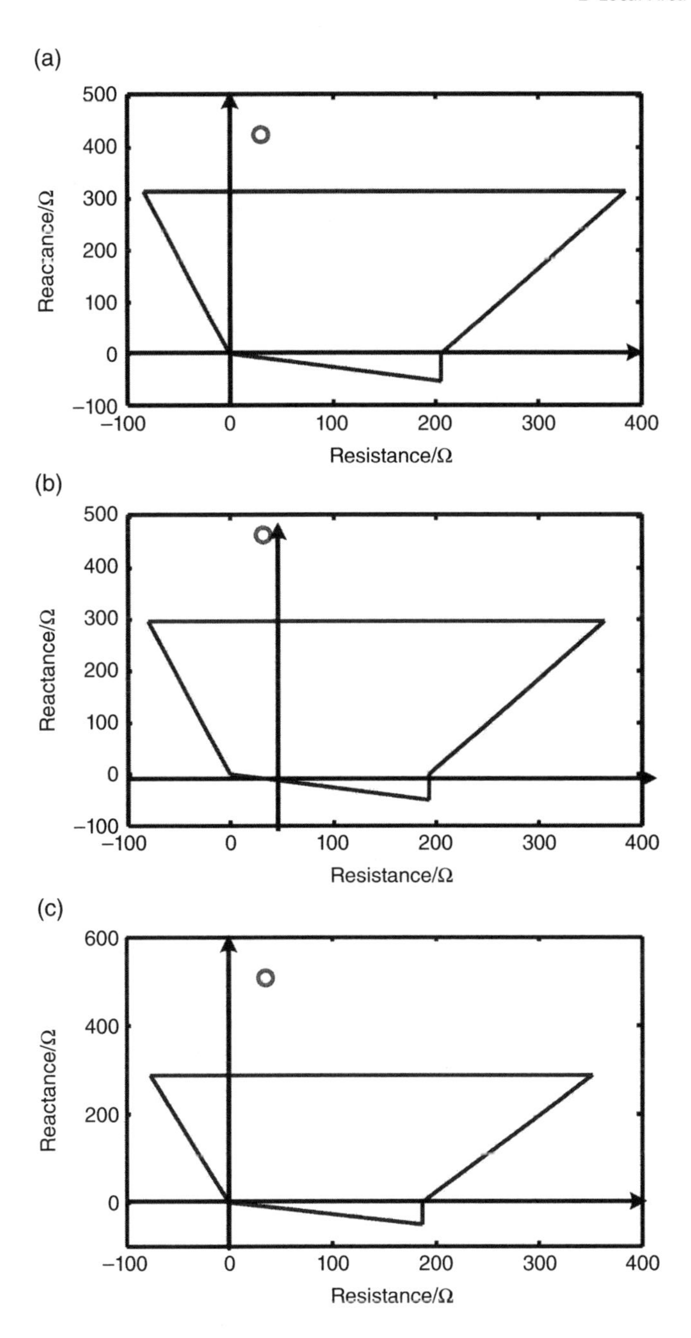

Figure 2.76 Simulation results of the adaptive distance protection when a fault occurs via 300 Ω transition resistance at different distances away from bus *N*. (a) 0 km. (b) 40 km. (c) 80 km.

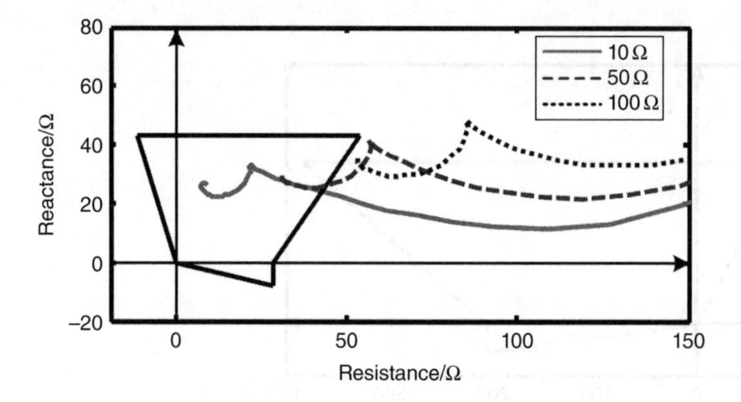

Figure 2.77 Measured impedance trajectory of the conventional distance protection for a single-source power system when a phase-A-to-ground fault occurs at 70 km from bus *M* via different transition resistances.

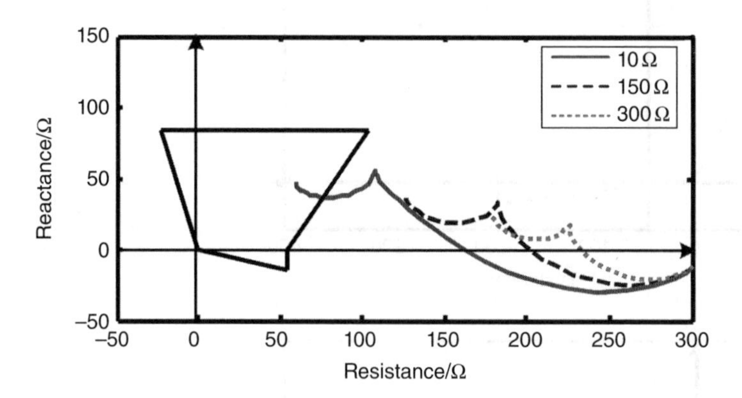

Figure 2.78 Measured impedance trajectory of the conventional distance protection for a double-source power system when a phase-A-to-ground fault occurs at 200 km from bus *M* via different transition resistances.

In the double-source simulation system, when a single-phase grounding fault occurs at 200 km from bus *M* via different transition resistances, the measured impedance trajectory of conventional distance protection after the fault is shown in Figure 2.78. It can be seen that, when the transition resistance is 50 Ω, the measured value falls in the operational zone, and the conventional distance protection operates correctly. However, when the transition resistance is 150 Ω or 300 Ω, the measured value does not fall in the operational zone, and the conventional distance protection will refuse to operate.

However, according to Figures 2.72 and 2.74, the proposed distance protection scheme could operate correctly no matter how big the transition resistance was. Thus, the proposed scheme has better immunity to transition resistance than the conventional scheme.

2.3.3 Location Method for Inter-Line and Grounded Faults of Double-Circuit Transmission Lines Based on Distributed Parameters

2.3.3.1 Definition of Compensation Voltage

The system diagram of a double-circuit line is shown in Figure 2.79. The generators at bus M and N represent equivalent transmission systems, the equivalent impedances being Z_{MS} and Z_{NS}, respectively. CT and PT represent the current transformer and potential transformer, respectively.

Voltage and current in symmetric coupled double-circuit transmission lines can be decoupled through matrix $\boldsymbol{T}$, where matrix $\boldsymbol{T}$ is ($\alpha = e^{j120°}$), resulting in the six-sequence network of the distributed parameter model:

$$\boldsymbol{T} = \frac{1}{6}\begin{bmatrix} 1 & \alpha & \alpha^2 & 1 & \alpha & \alpha^2 \\ 1 & \alpha^2 & \alpha & 1 & \alpha^2 & \alpha \\ 1 & 1 & 1 & 1 & 1 & 1 \\ 1 & \alpha & \alpha^2 & -1 & -\alpha & -\alpha^2 \\ 1 & \alpha^2 & \alpha & -1 & -\alpha^2 & -\alpha \\ 1 & 1 & 1 & -1 & -1 & -1 \end{bmatrix} \tag{2.183}$$

When a fault occurs on the double-circuit transmission line, the electric quantities of each sequence network are as shown in Figure 2.80, where the voltage and current of each sequence network satisfy

$$\begin{bmatrix} \dot{U}_{\mathrm{mi}} \\ \dot{I}_{\mathrm{mi}} \end{bmatrix} = \begin{bmatrix} \cosh(\gamma_i l_{\mathrm{mf}}) & z_{ci}\sinh(\gamma_i l_{\mathrm{mf}}) \\ \sinh(\gamma_i l_{\mathrm{mf}})/z_{ci} & \cosh(\gamma_i l_{\mathrm{mf}}) \end{bmatrix} \begin{bmatrix} \dot{U}_{\mathrm{mf}i} \\ \dot{I}_{\mathrm{mf}i} \end{bmatrix} \tag{2.184}$$

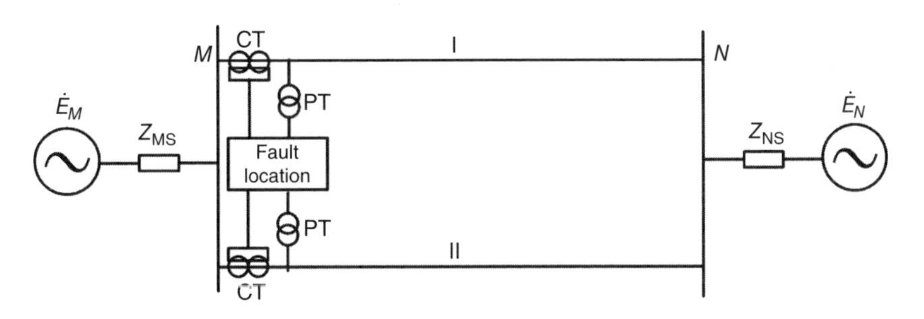

Figure 2.79 System diagram of a double-circuit line.

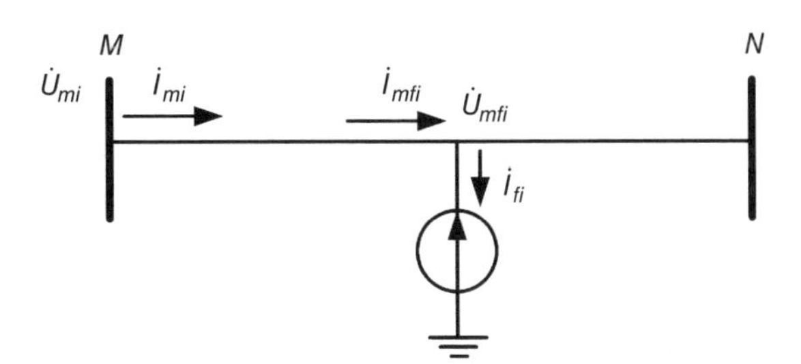

Figure 2.80 The electrical quantity of each sequence network.

where $i = T1, T2, T0, F1, F2, F0$ represents a co-rotating positive sequence network, a co-rotating negative sequence network, a co-rotating zero sequence network, a reverse positive sequence network, a reverse negative sequence network and a reverse zero sequence network, respectively. $\dot{U}_{mi}$ and $\dot{I}_{mi}$ are the sequence voltage and sequence current, measured from terminal M in each sequence network, respectively. $\dot{U}_{mfi}$ is the sequence voltage at the fault in each sequence network. $\dot{I}_{mfi}$ is the sequence short circuit current provided by terminal M at the fault in each sequence network. $\dot{I}_{fi}$ is the sequence short circuit current at the fault in each sequence network. l_{mf} is the distance from M to the fault in each sequence network. z_{ci} and γ_i are the characteristic impedance and propagation coefficients.

Adding up the six-sequence voltages, $\dot{U}_{mi}$ measured from terminal M in each sequence network in Equation (2.184), yields

$$\begin{aligned}\dot{U}_{m} &= \frac{\dot{U}_{mfT1}}{\cosh(\gamma_{T1}l_{mf})} + z_{cT1}\tanh(\gamma_{T1}l_{mf})\dot{I}_{mT1} + \frac{\dot{U}_{mfT2}}{\cosh(\gamma_{T2}l_{mf})} \\ &+ z_{cT2}\tanh(\gamma_{T2}l_{mf})\dot{I}_{mT2} + \frac{\dot{U}_{mfT0}}{\cosh(\gamma_{T0}l_{mf})} + z_{cT0}\tanh(\gamma_{T0}l_{mf})\dot{I}_{mT0} \\ &+ \frac{\dot{U}_{mfF1}}{\cosh(\gamma_{F1}l_{mf})} + z_{cF1}\tanh(\gamma_{F1}l_{mf})\dot{I}_{mF1} + \frac{\dot{U}_{mfF2}}{\cosh(\gamma_{F2}l_{mf})} \\ &+ z_{cF2}\tanh(\gamma_{F2}l_{mf})\dot{I}_{mF2} + \frac{\dot{U}_{mfF0}}{\cosh(\gamma_{F0}l_{mf})} + z_{cF0}\tanh(\gamma_{F0}l_{mf})\dot{I}_{mF0} \\ &= \frac{\dot{U}_{mf}}{\cosh(\gamma_{1}l_{mf})} + z_{c1}\tanh(\gamma_{1}l_{mf})\left[\dot{I}_{m} + k_T(l_{mf})\dot{I}_{mT0} + k_F(l_{mf})\dot{I}_{mF0}\right]\end{aligned} \tag{2.185}$$

where $\dot{U}_m$ and $\dot{I}_m$ are the phase voltage and phase current measured from terminal M. $\dot{U}_{mf}$ is the phase voltage at the fault, z_{c1} and γ_1 are the characteristic impedance and propagation coefficients in the co-rotating positive sequence network. $k_T(l_{mf})$ and $k_F(l_{mf})$ are the co-rotating zero sequence compensation coefficient and reverse zero sequence compensation coefficient, which are written as:

$$\begin{aligned}k_T(l_{mf}) &= \frac{z_{cT0}\sinh(\gamma_{T0}l_{mf}) - z_{c1}\sinh(\gamma_{1}l_{mf})}{z_{c1}\sinh(\gamma_{1}l_{mf})} \\ &+ \frac{Z_{mT0}\cosh(\gamma_{T0}l_{mf}) - Z_{mT0}\cosh(\gamma_{1}l_{mf})}{z_{c1}\sinh(\gamma_{1}l_{mf})}\end{aligned} \tag{2.186}$$

$$k_F(l_{mf}) = \frac{z_{cF0}\sinh(\gamma_{F0}l_{mf}) - z_{c1}\sinh(\gamma_{1}l_{mf})}{z_{c1}\sinh(\gamma_{1}l_{mf})} \tag{2.187}$$

It can be seen from Equation (2.185) that if a subtraction is conducted between two phase voltages on a single circuit of the transmission line, the co-rotating zero sequence current and reverse zero sequence current will cancel each other out. So the relationship between the phase-to-phase voltage on this circuit

$\dot{U}_{\mathrm{m}\varphi\varphi}$, the phase-to-phase voltage at the fault $\dot{U}_{\mathrm{mf}\varphi\varphi}$ and phase-to-phase current $\dot{I}_{\mathrm{m}\varphi\varphi}$ can be written as

$$\dot{U}_{\mathrm{m}\varphi\varphi} = \frac{\dot{U}_{\mathrm{mf}\varphi\varphi}}{\cosh(\gamma_1 l_{\mathrm{mf}})} + z_{c1}\tanh(\gamma_1 l_{\mathrm{mf}})\dot{I}_{\mathrm{m}\varphi\varphi} \tag{2.188}$$

Similarly, if a subtraction is conducted between two inter-line phase voltages, only the co-rotating zero sequence currents will cancel each other out. So the relationship between inter-line phase-to-phase voltage $\dot{U}_{\mathrm{m}\varphi I\varphi II}$, the inter-line phase-to-phase voltage at the fault $\dot{U}_{\mathrm{mf}\varphi I\varphi II}$ and the inter-line phase-to-phase current $\dot{I}_{\mathrm{m}\varphi I\varphi II}$ can be written as

$$\dot{U}_{\mathrm{m}\varphi I\varphi II} = \frac{\dot{U}_{\mathrm{mf}\varphi I\varphi II}}{\cosh(\gamma_1 l_{\mathrm{mf}})} + z_{c1}\tanh(\gamma_1 l_{\mathrm{mf}})\cdot\left[\dot{I}_{\mathrm{m}\varphi I\varphi II} + 2k_F(l_{\mathrm{mx}})\dot{I}_{\mathrm{m}F0}\right] \tag{2.189}$$

The compensation voltage $\dot{U}_{op}(l_{\mathrm{mx}})$ which has a distance l_{mx} from terminal M can be defined as

$$\dot{U}_{op}(l_{\mathrm{mx}}) = \dot{U}\cosh(\gamma_1 l_{\mathrm{mx}}) - z_{c1}\sinh(\gamma_1 l_{\mathrm{mx}})\dot{I}' \tag{2.190}$$

where the value of $\dot{I}'$ is related to the type of the voltage $\dot{U}$. As shown in Equation (2.191), if $\dot{U}$ represents the single-phase voltage $\dot{U}_{\mathrm{m}}$ on a single circuit of the transmission line, then $\dot{I}'$ represents the sum of single-phase current on this circuit $\dot{I}_{\mathrm{m}}$, the co-rotation zero sequence current and reverse zero sequence current. If $\dot{U}$ represents the phase-to-phase voltage on a single circuit $\dot{U}_{\mathrm{m}\varphi\varphi}$, then $\dot{I}'$ represents the phase-to-phase current $\dot{I}_{\mathrm{m}\varphi\varphi}$. If $\dot{U}$ represents inter-line phase-to-phase voltage $\dot{U}_{\mathrm{m}\varphi I\varphi II}$, then $\dot{I}'$ represents the sum of inter-line phase-to-phase current $\dot{I}_{\mathrm{m}\varphi I\varphi II}$ and reverse zero sequence current.

$$\dot{I}' = \begin{cases} \dot{I}_{\mathrm{m}} + k_T(l_{\mathrm{mx}})\dot{I}_{\mathrm{m}T0} + k_F(l_{\mathrm{mx}})\dot{I}_{\mathrm{m}F0} & \dot{U} = \dot{U}_{\mathrm{m}} \\ \dot{I}_{\mathrm{m}\varphi\varphi} & \dot{U} = \dot{U}_{\mathrm{m}\varphi\varphi} \\ \dot{I}_{\mathrm{m}\varphi I\varphi II} + 2k_F(l_{\mathrm{mx}})\dot{I}_{mF0} & \dot{U} = \dot{U}_{\mathrm{m}\varphi I\varphi II} \end{cases} \tag{2.191}$$

2.3.3.2 Voltage Phase Estimation at the Fault Point

The relationship between reverse positive/negative/zero sequence current at terminal M $\dot{I}_{\mathrm{m}F012}$ and the reverse positive/negative/zero sequence current at the fault $\dot{I}_{\mathrm{f}F012}$ can be represented by

$$\frac{\dot{I}_{\mathrm{f}F012}}{\dot{I}_{\mathrm{m}F012}} = \cosh(\gamma_{F012} l_{\mathrm{mf}})\cdot\left\{1 + \frac{\tanh(\gamma_{F012} l_{\mathrm{mf}})}{\tanh[\gamma_{F012}(l_{\mathrm{mn}} - l_{\mathrm{mf}})]}\right\} \tag{2.192}$$

where l_{mn} is the distance between the two terminals of the line.

The curve for the angle of $\dot{I}_{\mathrm{f}F1}/\dot{I}_{\mathrm{m}F1}$ with fault distance increasing is shown in Figure 2.81. Considering that the parameters in the negative sequence network and positive sequence network are the same, the curve for the angle of $\dot{I}_{\mathrm{f}F2}/\dot{I}_{\mathrm{f}F2}$ with fault distance increasing should be the same as the curve for the

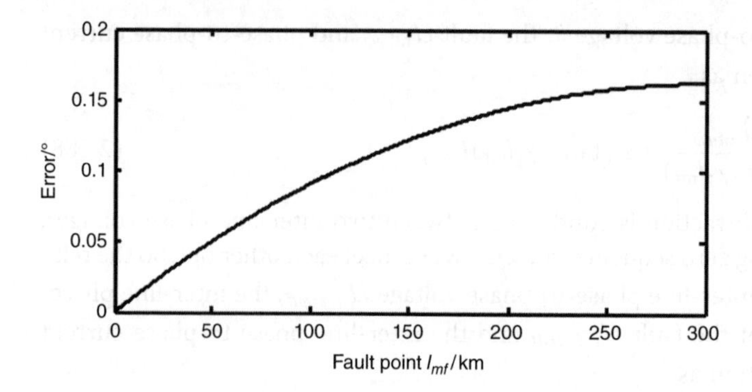

Figure 2.81 Angle of $\dot{I}_{mF1}/\dot{I}_{fF1}$ with fault distance increasing.

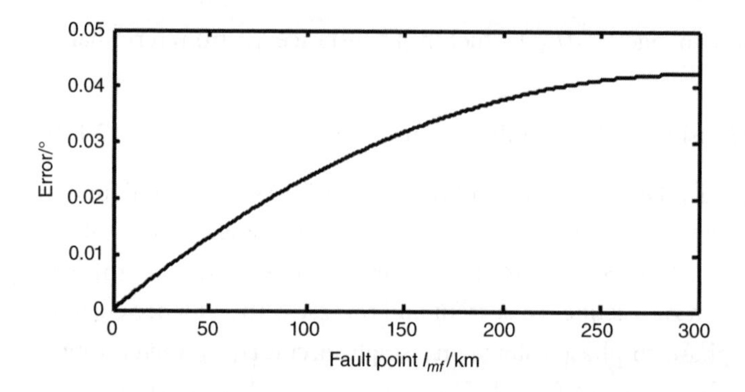

Figure 2.82 Angle of $\dot{I}_{fF0}/\dot{I}_{mF0}$ with fault distance increasing.

angle of $\dot{I}_{mF1}/\dot{I}_{fF1}$. Similarly, the curve for the angle of $\dot{I}_{fF0}/\dot{I}_{mF0}$ with fault distance increasing is as shown in Figure 2.82. It can be seen that the maximum error is less than 0.2°.

The phase of $\dot{I}_{fF1}$ and $\dot{I}_{fF2}$ at the fault is estimated by the phase of $\dot{I}_{mF1}$ and $\dot{I}_{mF2}$ respectively, as follows:

$$\begin{cases} \dot{I}_{fF1} \approx k_1(l_{mf})\dot{I}_{mF1} \\ \dot{I}_{fF2} \approx k_2(l_{mf})\dot{I}_{mF2} \end{cases} \tag{2.193}$$

where $k_1(l_{mf}) = k_2(l_{mf}) = \text{abs}(\dot{I}_{fF1}/\dot{I}_{mF1})$.

It can be seen from Equation (2.193) that the phase of the reverse positive and negative sequence current at the fault can be estimated by the phase of the reverse positive and negative sequence current measured at terminal M:

$$\begin{aligned} \arg(a_1\dot{I}_{fF1} + a_2\dot{I}_{fF1}) &\approx \arg\left[a_1k_1(l_{mf})\dot{I}_{mF1} + a_2k_2(l_{mf})\dot{I}_{mF2}\right] \\ &= \arg\left[k_1(l_{mf})(a_1\dot{I}_{mF1} + a_2\dot{I}_{mF2})\right] \\ &= \arg\left[(a_1\dot{I}_{mF1} + a_2\dot{I}_{mF2})\right] \end{aligned} \tag{2.194}$$

where a_1 and a_2 are arbitrary real numbers.

Selecting IA as the special phase, the six-phase current $I_{\mathrm{fAT},F}$ at the fault boundary is decoupled into six-sequence current $I_{\mathrm{fAI,II}}$ through matrix $\boldsymbol{T}$:

$$I_{\mathrm{fA}T,F} = \boldsymbol{T} I_{\mathrm{fAI,II}} \tag{2.195}$$

where $I_{\mathrm{fA}T,F} = \left[\dot{I}_{\mathrm{fA}T1}\ \ \dot{I}_{\mathrm{fA}T2}\ \ \dot{I}_{\mathrm{fA}T0}\ \ \dot{I}_{\mathrm{fA}F1}\ \ \dot{I}_{\mathrm{fA}F2}\ \ \dot{I}_{\mathrm{fA}F0}\right]^{\mathrm{T}}$,
$I_{\mathrm{fAI,II}} = \left[\dot{I}_{\mathrm{fAI}}\ \ \dot{I}_{\mathrm{fBI}}\ \ \dot{I}_{\mathrm{fCI}}\ \ \dot{I}_{\mathrm{fAII}}\ \ \dot{I}_{\mathrm{fBII}}\ \ \dot{I}_{\mathrm{fCII}}\right]^{\mathrm{T}}$.

According to the boundary conditions of two-line, two-phase fault, three-line, two-phase fault and three-line, three-phase fault and Equation (2.195), the relationship between the phases of voltage and reverse sequence current at the fault is deduced.

A. *Two-Line, Two-Phase Fault*

1) Two-line, two-phase inter-line phase-to-phase fault – When an IBIIC inter-line ungrounded fault occurs, the current condition at the fault boundary is

$$I_{\mathrm{fAI,II}} = \left[0\ \ \dot{I}_{\mathrm{fBI}}\ \ 0\ \ 0\ \ 0\ \ -\dot{I}_{\mathrm{fBI}}\right]^{\mathrm{T}} \tag{2.196}$$

Substitute Equation (2.196) into Equation (2.195) and the relationship between the reverse positive sequence current and phase current at the fault can be written as

$$\dot{I}_{\mathrm{fA}F1} = \frac{1}{6}\left(\alpha\dot{I}_{\mathrm{fBI}} - \alpha^2\dot{I}_{\mathrm{fCII}}\right) = \frac{1}{6}\left(\alpha\dot{I}_{\mathrm{fBI}} + \alpha^2\dot{I}_{\mathrm{fBI}}\right) = -\frac{1}{6}\dot{I}_{\mathrm{fBI}} \tag{2.197}$$

According to the fault type, the relationship between the phase of voltage between two fault phases (circuit I phase B and circuit II phase C) $\dot{U}_{\mathrm{fBICII}}$ and the phase of the reverse positive sequence current $\dot{I}_{\mathrm{fA}F1}$ is

$$\arg\left(\dot{U}_{\mathrm{fBICII}}\right) = \arg\left(\dot{I}_{\mathrm{fBI}}\right) = \arg\left(\dot{I}_{\mathrm{fA}F1}\right) + 180^\circ \tag{2.198}$$

2) Two-line, two-phase inter-line grounded fault – When an IBIIC inter-line grounded fault (IBIIC-G) occurs, the current condition at the fault boundary is

$$I_{\mathrm{fAI,II}} = \left[0\ \ \dot{I}_{\mathrm{fBI}}\ \ 0\ \ 0\ \ 0\ \ \dot{I}_{\mathrm{fCII}}\right]^{\mathrm{T}} \tag{2.199}$$

Substitute Equation (2.199) into Equation (2.195) and the relationship between the reverse zero sequence current and the phase current at the fault can be written as

$$\dot{I}_{\mathrm{fA}F0} = \frac{1}{6}\left(\dot{I}_{\mathrm{fBI}} - \dot{I}_{\mathrm{fCII}}\right) \tag{2.200}$$

According to the fault type, the relationship between the phases of voltage between the two fault phases $\dot{U}_{\mathrm{fBICII}}$ and the reverse zero sequence $\dot{I}_{\mathrm{fA}F0}$ is

$$\arg\left(\dot{U}_{\mathrm{fBICII}}\right) = \arg\left(\dot{I}_{\mathrm{fBI}} - \dot{I}_{\mathrm{fCII}}\right) = \arg\left(\dot{I}_{\mathrm{fA}F0}\right) \tag{2.201}$$

B. Three-Line, Two-Phase Fault

1) Three-line, two-phase inter-line phase-to-phase fault – When an IBCIIC inter-line ungrounded fault occurs, the current conditions at the fault boundary are

$$\begin{cases} \dot{I}_{\text{fAI,II}} = \begin{bmatrix} 0 & \dot{I}_{\text{fBI}} & \dot{I}_{\text{fCI}} & 0 & 0 & \dot{I}_{\text{fCII}} \end{bmatrix}^{\text{T}} \\ \dot{I}_{\text{fBI}} + \dot{I}_{\text{fCI}} + \dot{I}_{\text{fCII}} = 0 \end{cases} \tag{2.202}$$

Substitute Equation (2.202) into Equation (2.195) and the reverse positive sequence current $\dot{I}_{\text{fAF1}}$ at the fault and reverse negative sequence current $\dot{I}_{\text{fAF2}}$ are

$$\begin{cases} \dot{I}_{\text{fAF1}} = \dfrac{1}{6}\left(\alpha \dot{I}_{\text{fBI}} + \alpha^2 \dot{I}_{\text{fCI}} - \alpha^2 \dot{I}_{\text{fCII}}\right) \\ \dot{I}_{\text{fAF2}} = \dfrac{1}{6}\left(\alpha^2 \dot{I}_{\text{fBI}} + \alpha \dot{I}_{\text{fCI}} - \alpha \dot{I}_{\text{fCII}}\right) \end{cases} \tag{2.203}$$

Eliminate $\dot{I}_{\text{fCI}}$ and $\dot{I}_{\text{fCII}}$ in Equation (2.203) and the relationship between the reverse positive current, reverse negative current and phase current can be written as

$$\dot{I}_{\text{fAF1}} - \alpha \dot{I}_{\text{fAF2}} = \frac{1}{6}(\alpha - 1)\dot{I}_{\text{fBI}} \tag{2.204}$$

According to the fault type, the relationship between the phases of the phase-to-phase voltage on circuit I at the fault $\dot{U}_{\text{fBCI}}$, the reverse positive sequence and the reverse negative sequence is

$$\arg\left(\dot{U}_{\text{fBCI}}\right) = \arg\left(\dot{I}_{\text{fBI}}\right) = \arg\left(\frac{\dot{I}_{\text{fAF1}} - \alpha \dot{I}_{\text{fAF2}}}{\alpha - 1}\right) \tag{2.205}$$

2) Three-line, two-phase inter-line grounded fault – When an IBCIIC inter-line grounded fault (IBCIIC-G) occurs, the relationship between the fault current in phase B, the reverse positive sequence current and the reverse negative sequence current is still satisfied. The phase of the voltage in phase B circuit I at the fault $\dot{U}_{\text{fBI}}$ can be represented by the phase of $\dot{I}_{\text{fBI}}$:

$$\arg\left(\dot{U}_{\text{fBI}}\right) = \arg\left(\dot{I}_{\text{fBI}}\right) = \arg\left(\frac{\dot{I}_{\text{fAF1}} - \alpha \dot{I}_{\text{fAF2}}}{\alpha - 1}\right) \tag{2.206}$$

3) Three-line, two-phase non-inter-line single-phase grounded fault – If an ungrounded fault occurs between phases B and C on circuit I simultaneously with a grounded fault on phase C on circuit II (IBC-IICG), the relationship represented by Equation (2.204) is still satisfied. The phase of the phase-to-phase voltage on circuit I at the fault $\dot{U}_{\text{fBCI}}$ can be represented by the phase of $\dot{I}_{\text{fBI}}$:

$$\arg\left(\dot{U}_{\text{fBCI}}\right) = \arg\left(\dot{I}_{\text{fBI}}\right) = \arg\left(\frac{\dot{I}_{\text{fAF1}} - \alpha \dot{I}_{\text{fAF2}}}{\alpha - 1}\right) \tag{2.207}$$

4) Three-line, two-phase inter-line single-phase grounded fault – If an IBIIC inter-line ungrounded fault occurs simultaneously with a grounded fault on phase C on circuit I (IBIIC-ICG), the relationship represented by Equation (2.204) is still satisfied. The phase of the phase-to-phase voltage (phase B on circuit I and phase C on circuit II) at the fault $\dot{U}_{\text{fBICII}}$ is represented by the phase of $\dot{I}_{\text{fBI}}$:

$$\arg\left(\dot{U}_{\text{fBICII}}\right) = \arg\left(\dot{I}_{\text{fBI}}\right) = \arg\left(\frac{\dot{I}_{\text{fAF1}} - \alpha\dot{I}_{\text{fAF2}}}{\alpha - 1}\right) \tag{2.208}$$

C. Three-Line, Three-Phase Fault

1) Three-line, three-phase inter-line phase-to-phase fault – When an IBCIIA inter-line ungrounded fault occurs, the current condition at the fault boundary is

$$I_{\text{fAI,II}} = \begin{bmatrix} 0 & \dot{I}_{\text{fBI}} & \dot{I}_{\text{fCI}} & \dot{I}_{\text{fAII}} & 0 & 0 \end{bmatrix}^{\text{T}} \tag{2.209}$$

Substitute Equation (2.209) into Equation (2.195) and the reverse positive sequence current and reverse negative sequence current can be written as

$$\begin{cases} \dot{I}_{\text{fAF1}} = \dfrac{1}{6}\left(\alpha\dot{I}_{\text{fBI}} + \alpha^2\dot{I}_{\text{fCI}} - \dot{I}_{\text{fAII}}\right) \\ \dot{I}_{\text{fAF2}} = \dfrac{1}{6}\left(\alpha^2\dot{I}_{\text{fBI}} + \alpha\dot{I}_{\text{fCI}} - \dot{I}_{\text{fAII}}\right) \end{cases} \tag{2.210}$$

Eliminate $\dot{I}_{\text{fAII}}$ in Equation (2.210) and the relationship between reverse positive sequence current, reverse negative sequence current and phase current is

$$\dot{I}_{f\text{AF1}} - \dot{I}_{\text{fAF2}} = \frac{1}{6}\left(\alpha - \alpha^2\right)\left(\dot{I}_{\text{fBI}} - \dot{I}_{\text{fCI}}\right) = \frac{\text{j}\sqrt{3}}{6}\left(\dot{I}_{\text{fBI}} - \dot{I}_{\text{fCI}}\right) \tag{2.211}$$

According to the fault type, the relationship between the phases of the phase-to-phase voltage on circuit I at the fault $\dot{U}_{\text{fBCI}}$, the reverse positive sequence current and the reverse negative sequence current can be written as

$$\arg\left(\dot{U}_{\text{fBCI}}\right) = \arg\left(\dot{I}_{\text{fBI}} - \dot{I}_{\text{fCI}}\right) = \arg\left(\dot{I}_{\text{fAF1}} - \dot{I}_{\text{fAF2}}\right) - 90^\circ \tag{2.212}$$

2) Three-line, three-phase inter-line grounded fault – When an IBCIIA inter-line grounded fault (IBCIIA-G) occurs, the phase relationship deduced from the IBCIIA three-line, three-phase inter-line phase-to-phase fault represented by Equation (2.212) is still satisfied.
3) Three-line, three-phase non-inter-line single phase grounded fault – If an ungrounded fault occurs between phases B and C on circuit I simultaneously with a grounded fault on phase A on circuit II (IBC-IIAG), Equation (2.209) is satisfied together with

$$\dot{I}_{\text{fBI}} + \dot{I}_{\text{fCI}} = 0 \tag{2.213}$$

Substitute Equation (2.213) into Equation (2.195) and the relationship between the reverse positive sequence current, reverse negative sequence current and phase current can be written as

$$\dot{I}_{\text{fAF1}} - \dot{I}_{\text{fAF2}} = \frac{1}{3}(\alpha - \alpha^2)\dot{I}_{\text{fBI}} = \frac{\text{j}\sqrt{3}}{3}\dot{I}_{\text{fBI}} \tag{2.214}$$

According to the fault type, the relationship between the phases of phase-to-phase voltage on circuit I at the fault $\dot{U}_{\text{fBCI}}$, the reverse positive sequence current and the reverse negative sequence current is

$$\arg(\dot{U}_{\text{fBCI}}) = \arg(\dot{I}_{\text{fBI}}) = \arg(\dot{I}_{\text{fAF1}} - \dot{I}_{\text{fAF2}}) - 90° \tag{2.215}$$

The relationship between the phase of the voltage at the fault and the phase of its corresponding reverse sequence current is deduced under different fault types. With the help of the phase relationship between measured reverse sequence current and reverse sequence current at the fault, the phase of the voltage at the fault can be calculated.

2.3.3.3 Estimation of the Voltage Amplitude at the Fault Point

After the phase of the voltage at the fault is obtained, the amplitude of the voltage at the fault can be calculated, together with the voltage and current at terminal M.

According to the relationship shown in Figure 2.83, the amplitude of the voltage at the fault can be written as

$$U_{\text{f}} = \left|\frac{\dot{U}\sin(\varphi_L - \theta)}{\sin(\pi - \varphi_L - \gamma)}\right| \tag{2.216}$$

where U_{f} is the single phase voltage at the fault $\dot{U}_{\text{mf}}$, or the phase-to-phase voltage $\dot{U}_{\text{mf}\varphi\varphi}$ or the inter-line phase-to-phase voltage $\dot{U}_{\text{mf}\varphi I\varphi II}$. φ_L is the impedance angle of the line. θ is the angle between $\dot{U}$ and $\dot{I}'$, and γ is the angle between $\dot{I}'$ and $\dot{U}_{\text{f}}$.

2.3.3.4 Fault Location Method

According to the compensation voltage $\dot{U}_{op}(l_{\text{m}x})$ and the estimated voltage at the fault $\dot{U}_{\text{f}}$, a relationship equation can be deduced from Equations (2.185) and (2.193):

$$\frac{\dot{U}_{op}(l_{\text{m}x}) - \dot{U}_{\text{f}}\cosh(\gamma_1 l_{\text{m}x})}{\dot{I}'} = z_{c1}\sinh[\gamma_1(l_{\text{mf}} - l_{\text{m}x})] \tag{2.217}$$

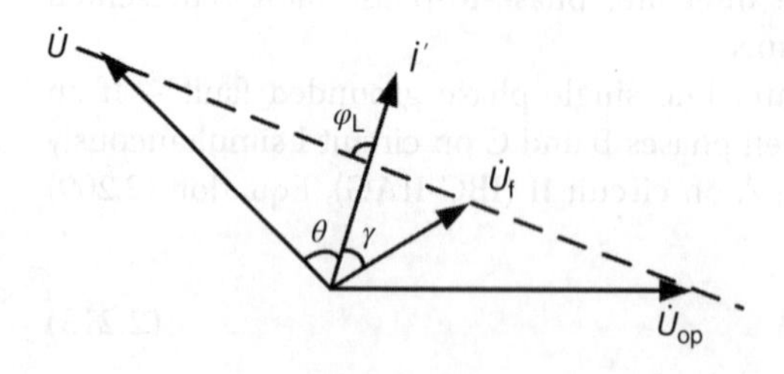

Figure 2.83 The estimated voltage amplitude of the fault point.

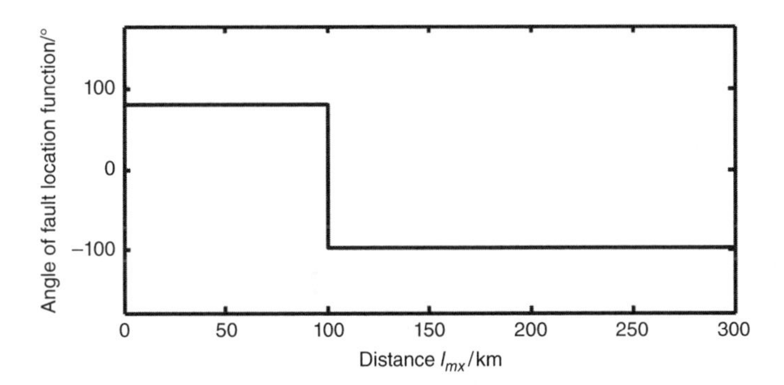

Figure 2.84 Phase characteristics analysis of the fault location function when the fault happened at 100 km.

The fault location function can be constructed as

$$f(l_{\mathrm{mx}}) = \frac{\dot{U}_{op}(l_{\mathrm{mx}}) - \dot{U}_{\mathrm{f}} \cosh(\gamma_1 l_{\mathrm{mx}})}{\dot{I}'} \tag{2.218}$$

Supposing the fault distance l_{mf} is 100 km, the phase characteristics analysis of the fault location function is as shown in Figure 2.84.

It can be seen from Figure 2.84 that when $l_{\mathrm{m}x} < l_{\mathrm{mf}}$, the phase of the fault location function is nearly 90°. When $l_{\mathrm{m}x} > l_{\mathrm{mf}}$, the phase of the fault location function is nearly –90°. When $l_{\mathrm{m}x} = l_{\mathrm{mf}}$, the phase of the fault location function is 0°. Thus, the mutation of the phase of the fault location function occurs at fault distance l_{mf}, and the phase characteristic curve has only one mutation point. So the location of the fault can be identified by solving the phase mutation point of the fault location function as follows:

1) Divide the line into n equal segments and calculate the phase of the fault location function at each equal diversion point.
2) Identify the segment which l_{mf} belongs to by the fact that the phase characteristic of the fault location function at one of the two equal diversion points at each end of this segment that contains the fault is larger than 0, and the phase characteristic of the fault location function is less than 0 at the other point.
3) Calculate the phase of the fault location function point by point with a step length of Δl within the segment to which l_{mf} belongs, starting from the smaller boundary point. Use $l_{\mathrm{f}} = l_{\mathrm{f+}} - 0.5\Delta l$ as the distance between terminal M and the fault, where $l_{\mathrm{f+}}$ is the first reference point that corresponds to a phase less than 0.

2.3.3.5 Simulations and Analysis

A 330 kV, 300 km two-terminal system is built here using PSCAD/EMTDC for simulation analysis, shown in Figure 2.85. δ represents the phase angle difference between the two equivalent power sources at terminals M and N on this line. The amplitudes of these two sources are 1.05 p.u. and 1 p.u., respectively. The full-

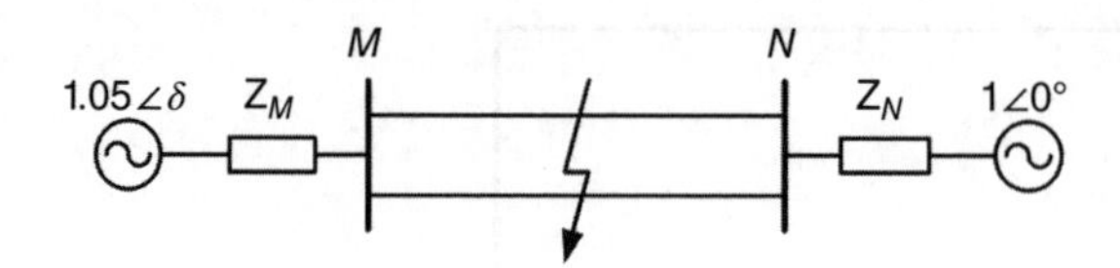

Figure 2.85 Simulation model of the double-circuit lines.

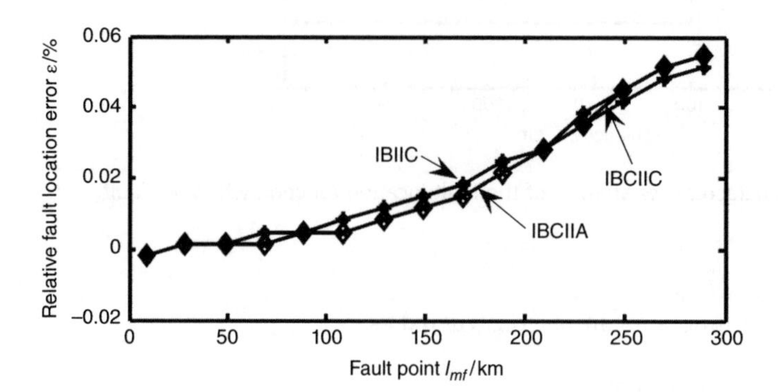

Figure 2.86 Influence of fault position and fault type on fault location for the ungrounded inter-line fault.

cycle Fourier correlation is used as the filter algorithm with a sampling frequency 2000 Hz.

The parameters for the two systems on each side of the line are: $Z_{M1} = 1.0515 + j43.1749\ \Omega$, $Z_{M0} = 0.6 + j29.0911\ \Omega$, $Z_{N1} = 26 + j44.9185\ \Omega$, $Z_{N0} = 20 + j37.4697\ \Omega$.

Single-circuit transmission line positive (negative) sequence parameters: $R_1 = 0.05468\ \Omega/\text{km}$, $L_1 = 1.0264\ \text{mH/km}$, $C_1 = 0.01095\ \mu\text{F/km}$.

Single-circuit transmission line zero sequence parameters: $R_0 = 0.2931\ \Omega/\text{km}$, $L_0 = 3.9398\ \text{mH/km}$, $C_0 = 0.05473\ \mu\text{F/km}$.

Double-circuit transmission line zero sequence mutual impedance parameters: $R_{m0} = 0.2385\ \Omega/\text{km}$, $L_{m0} = 2.6274\ \text{mH/km}$, $C_{m0} = 0.00026\ \mu\text{F/km}$.

The relative fault location error is defined as:

$$\varepsilon = \frac{\text{calculated fault distance} - \text{actual fault distance}}{\text{faulted section length}} \times 100\% \tag{2.219}$$

A. Influence of the Fault Position

The influence of fault location difference on fault location accuracy under two-line, two-phase inter-line ungrounded fault, three-line, two-phase inter-line ungrounded fault, and three-line, three-phase inter-line ungrounded fault is shown in Figure 2.86. It can be seen that the relative fault location error reaches its maximum at 290 km, and the maximum relative errors are 0.055%, 0.0517% and 0.055%, respectively, which are all less than 0.1%.

Figure 2.87 shows the influence of fault location difference on fault location accuracy under two-line, two-phase inter-line grounded fault. Figure 2.88 shows the influence of fault location difference on fault location accuracy under

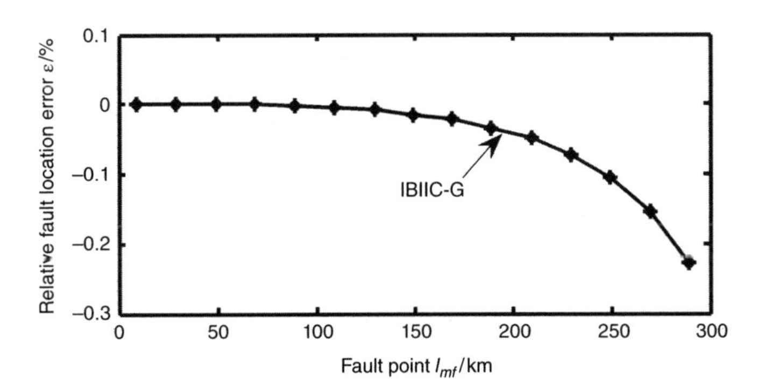

Figure 2.87 Influence of transition position on fault location for the grounded inter-two-line fault.

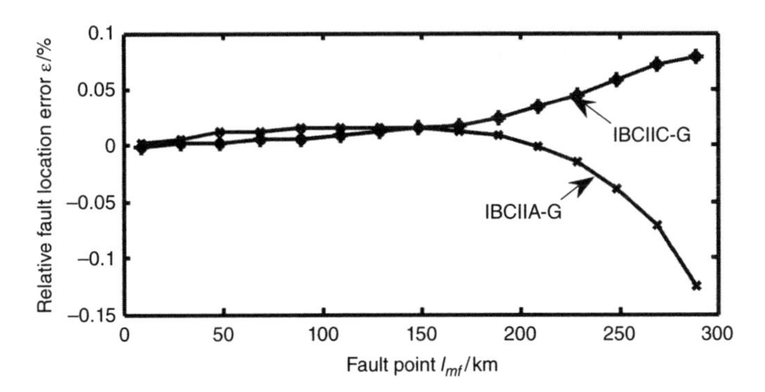

Figure 2.88 Influence of fault position and fault type on fault location for the grounded inter-three-line fault.

three-line, two-phase inter-line grounded fault and three-line, three-phase inter-line grounded fault. From Figures 2.87 and 2.88 it can be seen that the relative fault location error increases with the fault distance under all fault types. The maximum absolute values of relative fault location error are 0.2283%, 0.0783% and 0.125%, respectively, all less than 0.3%.

Figure 2.89 shows the influence of fault location difference on fault location accuracy under three-line, two-phase non-inter-line single-phase grounded fault, three-line, two-phase inter-line single-phase grounded fault and three-line, three-phase non-inter-line single-phase grounded fault. It can be seen that the maximum relative fault location errors are 0.0783%, 0.0783% and 0.055%, respectively, all less than 0.1%.

B. Influence of the Transition Resistance

When a two-line, two-phase inter-line grounded fault occurs at 290 km from the measuring point, the influence of transition resistance on the fault location accuracy is as shown in Figure 2.90. It can be seen that the relative fault location error

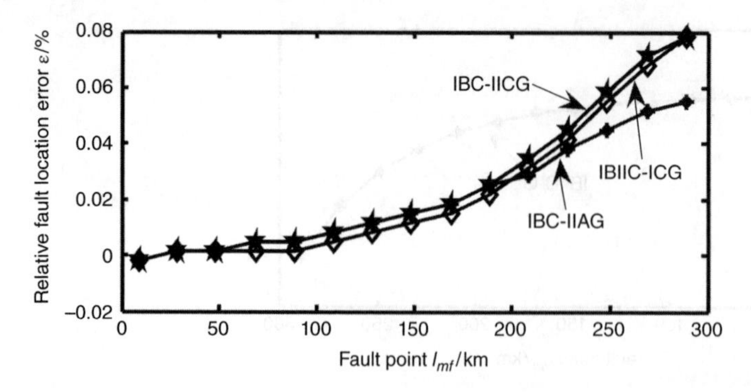

Figure 2.89 Influence of fault position and fault type on fault location for the three-line fault with single-phase grounded.

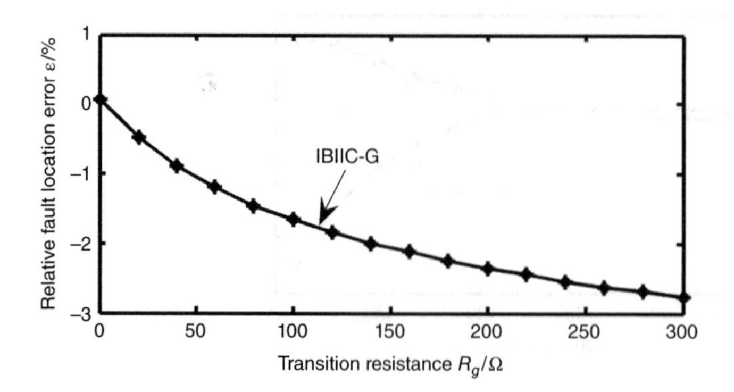

Figure 2.90 Influence of transition resistance on fault location for the grounded inter-two-line fault.

reaches its maximum when the transition resistance is 300 Ω, and its maximum absolute value is 2.7717%.

Figure 2.91 shows the influence of transition resistance on fault location accuracy under three-line, two-phase inter-line grounded fault and three-line, three-phase inter-line grounded fault respectively when the fault is 290 km from the measuring point. Similarly, the absolute value of the relative fault location error reaches its maximum of 1.2683% and 1.6183%, respectively, when transition resistance is 300 Ω.

Figure 2.92 shows the influence of transition resistance on fault location accuracy under three-line, two-phase non-inter-line single-phase grounded fault, three-line, two-phase inter-line single-phase grounded fault and three-line, three-phase non-inter-line single-phase grounded faults, respectively, when the fault is 290 km from the measuring point. It can be seen that the maximum relative fault location errors are 0.095%, 0.215% and 0.055%, respectively, and the maximum absolute value of relative error is only 2.7717%, which is less than 3%.

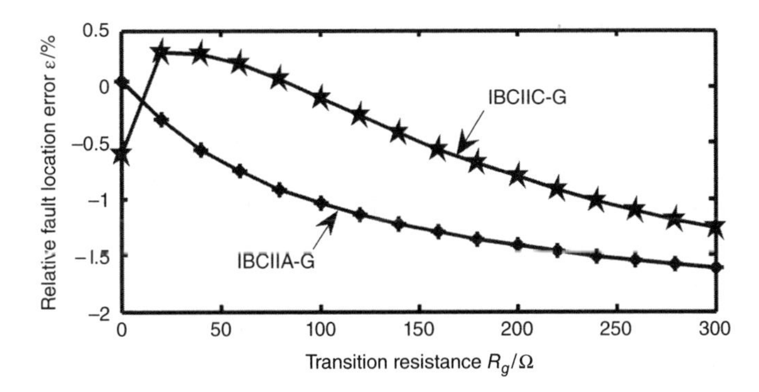

Figure 2.91 Influence of transition resistance and fault type on fault location for the grounded inter-three-line fault.

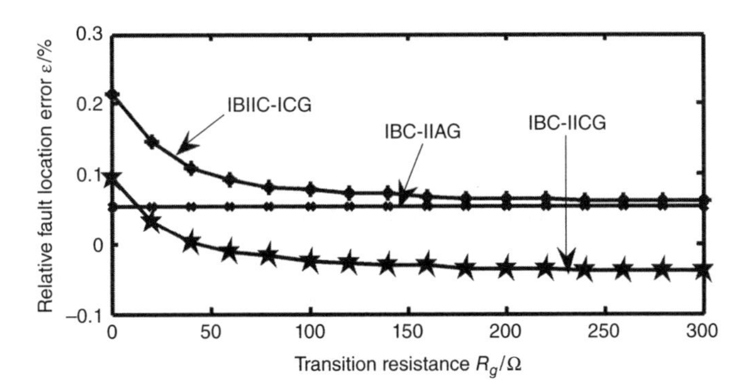

Figure 2.92 Influence of transition resistance and fault type on fault location for the three-line fault with single-phase grounded.

C. *Influence of the Source Impedance*

When a fault occurs at 230 km from the bus via 10 Ω transition resistance, the relative fault location errors obtained by varying the local, remote, and local and remote source impedances, respectively are shown in Tables 2.16–2.18. The values of the impedances in local and remote line terminals have been set equal to 20%, 50%, 100%, 200% and 500% of the actual system values. It can be seen that source impedance variations have an insignificant influence on the accuracy of the proposed method. This is in full accordance with the fault location function of the proposed method, which does not include the source impedances.

D. *Influence of Measurement Errors*

Current phasors and voltage phasors used as input data in the algorithm are obtained through measuring and filtering processes. This makes it necessary to analyse the effects of measurement errors on the accuracy of the proposed method. To test the impact of instrument transformers on the fault location accuracy, suppose that a two-line, two-phase inter-line grounding fault (IBIIC-G), a

Table 2.16 Influence of source impedance in terminal *M* on the accuracy of the method.

	Source impedance variation at *M*				
Fault type	**20%**	**50%**	**100%**	**200%**	**500%**
IBIIC	0.54	0.52	0.54	0.53	0.54
IBCIIC	0.21	0.20	0.21	0.21	0.21
IBCIIA	0.18	0.18	0.18	0.18	0.18
IBIC-G	−0.09	−0.10	−0.11	−0.11	−0.13
IBCIIC-G	0.03	0.04	0.05	0.06	0.06
IBCIIA-G	−0.04	−0.04	−0.04	−0.04	−0.05
IBC-IICG	0.05	0.05	0.06	0.07	0.07
IBC-IIAG	0.04	0.04	0.05	0.05	0.06
IBIIC-ICG	0.02	0.03	0.03	0.03	0.04

Table 2.17 Influence of source impedance in terminal *N* on the accuracy of the method.

	Source impedance variation at *N*				
Fault type	**20%**	**50%**	**100%**	**200%**	**500%**
IBIIC	0.53	0.53	0.54	0.53	0.53
IBCIIC	0.20	0.20	0.20	0.21	0.21
IBCIIA	0.18	0.19	0.18	0.18	0.18
IBIC-G	−0.10	−0.11	−0.12	−0.13	−0.13
IBCIIC-G	0.04	0.05	0.05	0.06	0.06
IBCIIA-G	−0.03	−0.04	−0.04	−0.04	−0.05
IBC-IICG	0.05	0.06	0.06	0.07	0.07
IBC-IIAG	0.04	0.04	0.05	0.05	0.05
IBIIC-ICG	0.02	0.03	0.03	0.04	0.04

three-line, two-phase inter-line grounding fault (IBCIIC-G) and a three-line, three-phase non-inter-line single-phase grounding fault (IBC-IIAG) occur separately, and the influence of the fault position and transition resistance on the fault location accuracy is shown in Figure 2.93. It can be seen that, when an IBIIC-G fault occurs at different locations on the line via different transition resistances, the absolute value of the maximum relative ranging error is 2.9167%. When an IBCIIC-G fault occurs at different locations on the line via different transition resistances, the absolute value of the maximum relative ranging error of the proposed method is 2.1132%. When an IBC-IIAG fault occurs at different locations

Table 2.18 Influence of source impedance in terminal M and N on the accuracy of the method.

	Source impedance variation at M and N				
Fault type	**20%**	**50%**	**100%**	**200%**	**500%**
IBIIC	0.54	0.53	0.54	0.53	0.54
IBCIIC	0.21	0.21	0.21	0.21	0.21
IBCIIA	0.18	0.18	0.18	0.19	0.19
IBIC-G	−0.09	−0.12	−0.11	−0.11	−0.12
IBCIIC-G	0.04	0.04	0.05	0.06	0.06
IBCIIA-G	−0.04	−0.04	−0.04	−0.05	−0.05
IBC-IICG	0.05	0.05	0.06	0.06	0.07
IBC-IIAG	0.04	0.04	0.05	0.05	0.05
IBIIC-ICG	0.02	0.03	0.03	0.03	0.04

on the line via different transition resistances, the absolute value of the maximum relative ranging error of the proposed method is 2.4891%.

E. Comparison with Another Method

In order to evaluate the proposed method, a comparison is made between the proposed method and the method that utilizes two-end unsynchronized current

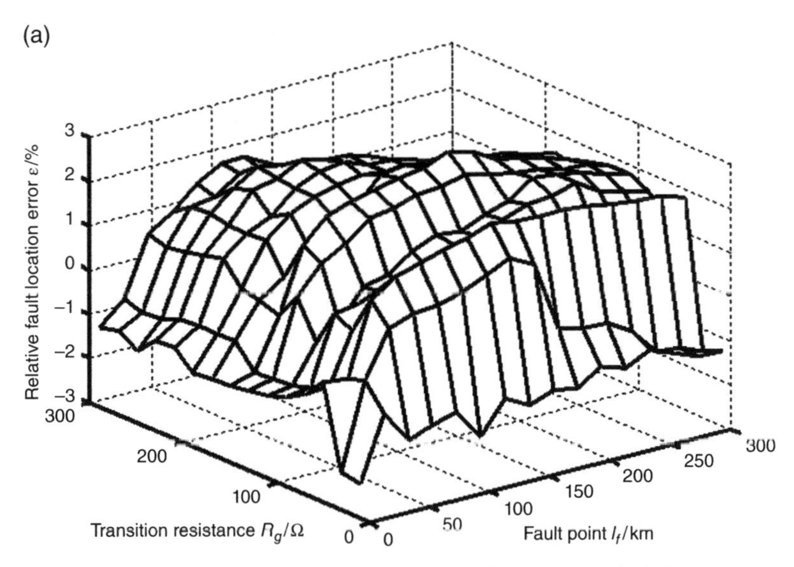

Figure 2.93 Influence of transition resistance and fault point on fault location considering the measuring transformer models. (a) Two-line, two-phase inter-line grounded fault (IBICG). (b) Three-line, two-phase inter-line grounded fault (IBCIIC-G). (c) Three-line, three-phase non-inter-line single-phase grounded fault (IBC-IIAG).

(b)

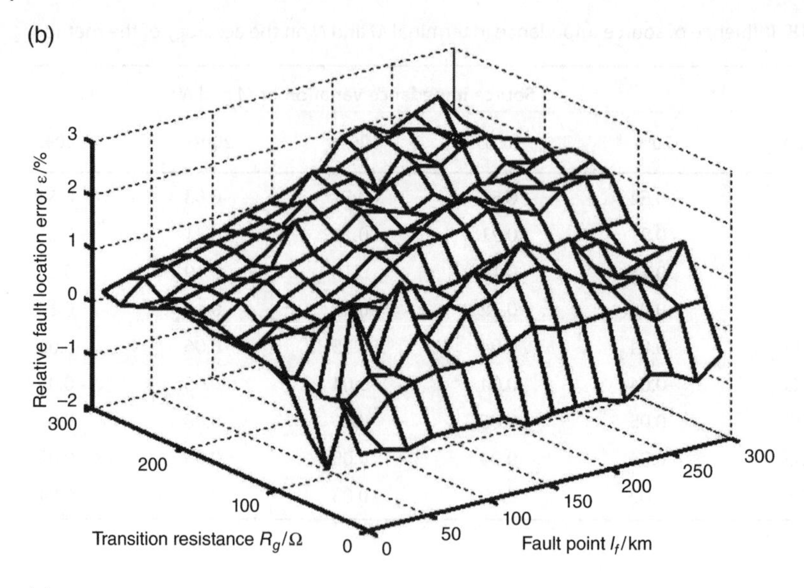

(c)

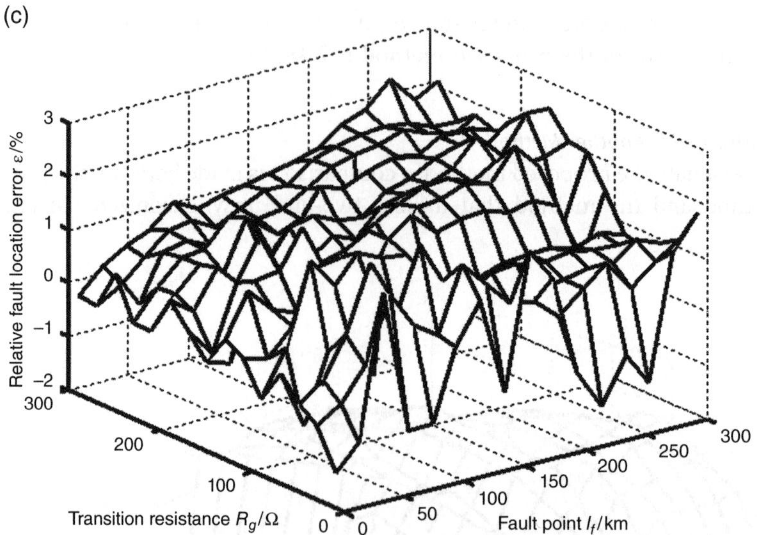

Figure 2.93 (*Continued*)

measurements in reference [19]. If a two-line, two-phase inter-line grounding fault (IBIIC-G), a three-line, two-phase inter-line grounding fault (IBCIIC-G) and a three-line, three-phase non-inter-line single-phase grounding fault (IBC-IIAG) occur separately, the influence of the fault position and transition resistance on fault location accuracy of the existing method is as shown in Figure 2.94.

It can be seen from Figure 2.94 that when an IBIIC-G fault occurs at different locations on the line via different transition resistances, the maximum relative error of the existing method is 2.6630%. When an IBCIIC-G fault occurs at

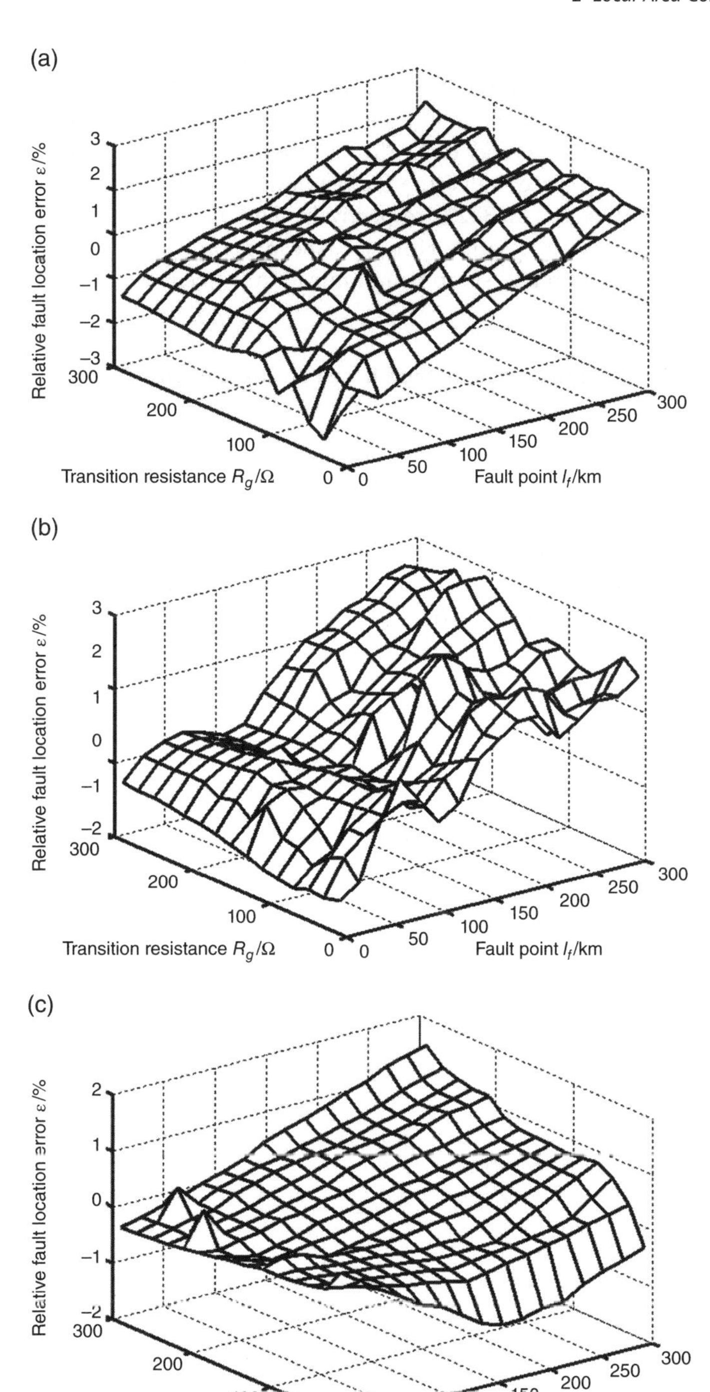

Figure 2.94 Influence of transition resistance and fault point on fault location of the existing method. (a) Two-line, two-phase inter-line grounded fault (IBICG). (b) Three-line, two-phase inter-line grounded fault (IBCIIC-G). (c) Three-line, three-phase non-inter-line single-phase grounded fault (IBC-IIAG).

different locations on the line via different transition resistances, the maximum relative error of the existing method is 0.9012%. When an IBC-IIAG fault occurs at different locations on the line via different transition resistances, the maximum relative error of the existing method is 1.4851%. When using the proposed method to locate the above three faults, the corresponding maximum relative errors are 2.7717%, 0.8343% and 1.2342%, respectively. Therefore, the accuracy of the two methods is nearly the same. However, the proposed method only requires single-end information and so no communication is needed.

2.3.4 Adaptive Overload Identification Method Based on the Complex Phasor Plane

2.3.4.1 Adaptive Overload Identification Criterion Based on the Complex Phasor Plane

Figure 2.95 shows a two-source power system. M is the sending end and N is the receiving end. The protection relays are installed at both ends of line MN. The line impedance is Z_L and the system impedances at the two ends are Z_S and Z_R. When there is no fault, the voltage and current phasor diagram is as shown in Figure 2.96.

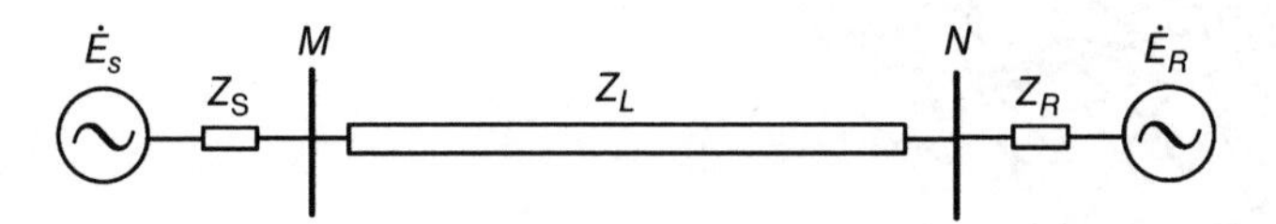

Figure 2.95 A two-source power system.

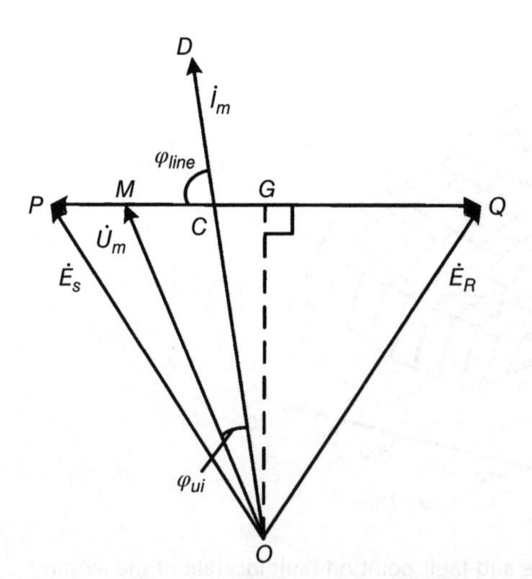

Figure 2.96 Voltage and current phasor diagram without fault.

Looking at the protection at bus M for example, $\overrightarrow{OM}$ represents the measured voltage phasor at bus M; $\dot{U}_{\mathrm{m}}\cdot\overrightarrow{OD}$ represents the measured current phasor at bus M; $\dot{I}_{\mathrm{m}}\cdot\angle$MOC is the angular difference between the measured voltage phasor and the measured current phasor; $\varphi_{ui}\cdot\overrightarrow{OP}$ and $\overrightarrow{OQ}$ represent the emf phasors of the two sources, $\dot{E}_S$ and $\dot{E}_R$. Thus, $\overrightarrow{PQ}$ is the voltage drop phasor of the two-source system. G is the lowest voltage point in the system. The angular difference between the measured current phasor and the voltage drop phasor $\overrightarrow{PQ}$ is the line impedance angle, φ_{line}. The angular difference between $\overrightarrow{OG}$ and $\overrightarrow{OM}$ is $\varphi = \varphi_{ui} + 90^\circ - \varphi_{line}$. In Figure 2.96, there are the following relationships:

$$\begin{cases} |\overrightarrow{OG}| = |\dot{U}_{\mathrm{m}}| \cos\varphi \\ |\overrightarrow{MG}| = |\dot{U}_{\mathrm{m}}|\sin(\varphi_{ui} + 90^\circ - \varphi_{line}) \\ |\overrightarrow{CG}| = |\dot{U}_{\mathrm{m}}|\cos(\varphi_{ui} + 90^\circ - \varphi_{line})\tan(90^\circ - \varphi_{line}) \\ |\overrightarrow{CM}| = |\overrightarrow{MG}| - |\overrightarrow{CG}| = |\dot{U}_{\mathrm{m}}|\sin(\varphi_{ui} + 90^\circ - \varphi_{line}) \\ \qquad - |\dot{U}_{\mathrm{m}}|\cos(\varphi_{ui} + 90^\circ - \varphi_{line})\tan(90^\circ - \varphi_{line}) \end{cases} \tag{2.220}$$

In the case of a three-phase fault, the fault point is grounded via the arc resistance (this is because most three-phase faults are caused by lightning, and an electric arc will form at the fault point). As shown in Figure 2.97, a three-phase fault occurs at F on line MN. The voltage equation of the fault loop is

$$\dot{U}_{\mathrm{m}} = \dot{U}_{\mathrm{arc}} + \dot{I}_{\mathrm{m}}Z_d \tag{2.221}$$

where $\dot{U}_{\mathrm{arc}}$ is the voltage drop on the arc resistance. $\dot{U}_{\mathrm{m}}$ and $\dot{I}_{\mathrm{m}}$ are the measured voltage and current at the relaying point. Z_d is the positive sequence impedance of the line length from the fault point to the relaying point.

The voltage and current phasor diagram in the case of a three-phase fault is shown in Figure 2.98. $\overrightarrow{OC}$ represents the arc voltage phasor $\dot{U}_{\mathrm{arc}}$, which is in the same direction as the measured current phasor $\dot{I}_{\mathrm{m}}$. $\overrightarrow{OB}$ is the measured voltage phasor at M, $\dot{U}_{\mathrm{m}}$. $\angle BOC$ is the angular difference between the measured voltage phasor and the measured current phasor, φ_{ui}. $\overrightarrow{CB}$ is the voltage drop phasor from

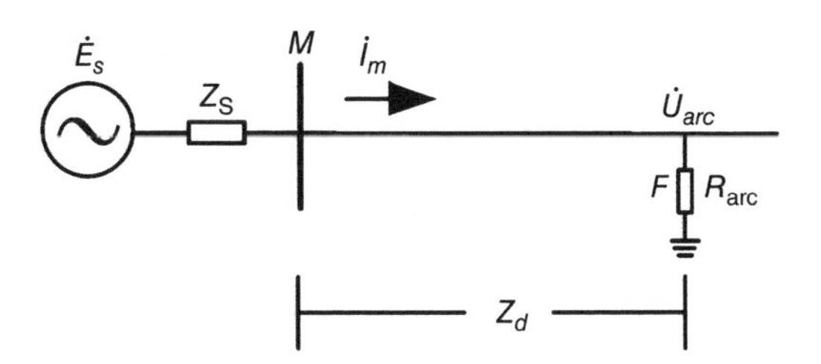

Figure 2.97 Three-phase fault at F.

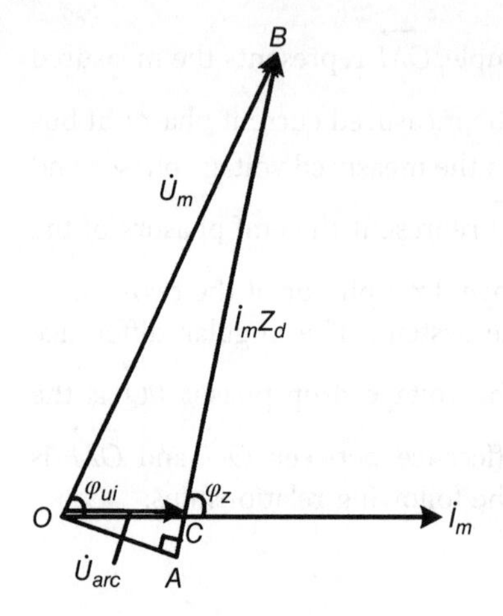

Figure 2.98 Voltage and current phasor diagram in the case of a three-phase fault.

the fault point to bus M, and the angle between $\overrightarrow{CB}$ and the measured current phasor is φ_z which is equal to φ_{line}. Similarly, A is the perpendicular foot of the origin O on $\overrightarrow{CB}$, and the angular difference between $\overrightarrow{OA}$ and $\overrightarrow{OB}$ is $\varphi = \varphi_{ui} + 90° - \varphi_{line}$. In Figure 2.98, there are the following relationships:

$$\begin{cases} |\overrightarrow{OA}| = |\dot{U}_{\mathrm{m}}| \ \cos\varphi \\ |\overrightarrow{BA}| = |\dot{U}_{\mathrm{m}}|\sin(\varphi_{ui} + 90° - \varphi_{line}) \\ |\overrightarrow{CA}| = |\dot{U}_{\mathrm{m}}|\cos(\varphi_{ui} + 90° - \varphi_{line})\tan(90° - \varphi_{line}) \\ |\overrightarrow{CB}| = |\overrightarrow{BA}| - |\overrightarrow{CA}| = |\dot{U}_{\mathrm{m}}|\sin(\varphi_{ui} + 90° - \varphi_{line}) \\ \qquad - |\dot{U}_{\mathrm{m}}|\cos(\varphi_{ui} + 90° - \varphi_{line})\tan(90° - \varphi_{line}) \end{cases} \tag{2.222}$$

When the fault occurs, we have

$$|\overrightarrow{CB}| = |\dot{I}_{\mathrm{m}} \times Z_d| \tag{2.223}$$

When there is no fault, the amplitude of the voltage drop between the virtual point (a certain point in the network) and the measuring point on the M side $|\overrightarrow{CM}|$ is:

$$|\overrightarrow{CM}| = |\dot{I}_{\mathrm{m}} \times Z_x| \tag{2.224}$$

where Z_x is the impedance of the line length between the virtual point and the measuring point on the M side. When there is no fault, the currents on the two ends of the line have the same direction. Therefore, in Figure 2.96, the angular difference between $\overrightarrow{OC}$ and the voltage phasor on the opposite line end is positive. The virtual point corresponding to $\overrightarrow{OC}$ is outside line MN. Thus, Z_x is a virtual impedance with bigger amplitude than Z_{MN}.

Note that $|\overrightarrow{CB}|$ in Equation (2.222) and $|\overrightarrow{CM}|$ in Equation (2.220) have the same expression. Applying Equation (2.223) to Equation (2.222) and Equation (2.224) to Equation (2.220), and replacing Z_d in Equation (2.223) and Z_x in Equation (2.224) with a unified Z, the virtual voltage drop phasor equation can be obtained:

$$|\dot{I}_{\mathrm{m}} \times Z| = |\dot{U}_{\mathrm{m}}|\sin(\varphi_{ui} + 90^\circ - \varphi_{line}) - |\dot{U}_{\mathrm{m}}|\cos(\varphi_{ui} + 90^\circ - \varphi_{line})\tan(90^\circ - \varphi_{line}) \tag{2.225}$$

Change the phasor relationship in Equation (2.225) into an amplitude relationship:

$$|\dot{I}_{\mathrm{m}}||Z| = |\dot{U}_{\mathrm{m}}|\sin(\varphi_{ui} + 90^\circ - \varphi_{line}) - |\dot{U}_{\mathrm{m}}|\cos(\varphi_{ui} + 90^\circ - \varphi_{line})\tan(90^\circ - \varphi_{line}) \tag{2.226}$$

Divide both sides of Equation (2.226) by the amplitude of the measured current phasor $|\dot{I}_{\mathrm{m}}|$:

$$|Z| = |Z_{\mathrm{m}}|(\sin(\varphi_{ui} + 90^\circ - \varphi_{line}) - \cos(\varphi_{ui} + 90^\circ - \varphi_{line})\tan(90^\circ - \varphi_{line})) \tag{2.227}$$

And then multiply both sides of Equation (2.227) by $\cos(90^\circ - \varphi_{line})$:

$$\begin{aligned}|Z|\cos(90^\circ - \varphi_{line}) = |Z_{\mathrm{m}}| \times (&\sin(\varphi_{ui} + 90^\circ - \varphi_{line})\cos(90^\circ - \varphi_{line}) \\ &- \cos(\varphi_{ui} + 90^\circ - \varphi_{line})\sin(90^\circ - \varphi_{line}))\end{aligned} \tag{2.228}$$

According to the characteristic that the sine function is an odd function and the cosine function is an even function, the $\sin(90^\circ - \varphi_{line})$ and $\cos(90^\circ - \varphi_{line})$ in Equation (2.228) can be replaced by $-\sin(\varphi_{line} - 90^\circ)$ and $\cos(\varphi_{line} - 90^\circ)$, respectively:

$$\begin{aligned}|Z|\cos(90^\circ - \varphi_{line}) = |Z_{\mathrm{m}}| \times (&\sin(\varphi_{ui} + 90^\circ - \varphi_{line})\cos(\varphi_{line} - 90^\circ) \\ &+ \cos(\varphi_{ui} + 90^\circ - \varphi_{line})\sin(\varphi_{line} - 90^\circ))\end{aligned} \tag{2.229}$$

The items on the right side of Equation (2.229) can be merged according to trigonometric formulas:

$$|Z|\cos(90^\circ - \varphi_{line}) = |Z_{\mathrm{m}}|\sin\varphi_{ui} \tag{2.230}$$

The impedance plane form of Equation (2.230) is:

$$|Z| = \frac{\sin\varphi_{ui}}{\cos(90^\circ - \varphi_{line})}|Z_{\mathrm{m}}| \tag{2.231}$$

Based on the above analysis, when there is a symmetrical fault within the protection range, then $Z = Z_d$ and $Z < Z_{MN}$, while in the case of overload or normal operation, $Z = Z_x$ and $Z > Z_{MN}$. This difference of Z in the cases of overload and three-phase fault forms the basis for the adaptive overload identification criterion, i.e.

$$\begin{cases} |Z| < |Z_{MN}| & \text{fault} \\ |Z| > |Z_{MN}| & \text{overload} \end{cases} \tag{2.232}$$

Applying Equation (2.231) to Equation (2.232):

$$\begin{cases} \dfrac{\sin\varphi_{ui}}{\cos(90^\circ-\varphi_{line})}|Z_\mathrm{m}|<|Z_{MN}| \text{ fault} \\ \dfrac{\sin\varphi_{ui}}{\cos(90^\circ-\varphi_{line})}|Z_\mathrm{m}|>|Z_{MN}| \text{ overload} \end{cases} \tag{2.233}$$

Inequality (2.233) can be transformed to

$$\begin{cases} |Z_\mathrm{m}|<\dfrac{\cos(90^\circ-\varphi_{line})}{\sin\varphi_{ui}}|Z_{MN}| \text{ fault} \\ |Z_\mathrm{m}|>\dfrac{\cos(90^\circ-\varphi_{line})}{\sin\varphi_{ui}}|Z_{MN}| \text{ overload} \end{cases} \tag{2.234}$$

Replace the relevant expressions in Inequality (2.234) with coefficients, so that the adaptive overload identification criterion can be obtained:

$$\begin{cases} |Z_\mathrm{m}|<k_l k_\mathrm{m}|Z_{MN}| \text{ fault} \\ |Z_\mathrm{m}|>k_l k_\mathrm{m}|Z_{MN}| \text{ overload} \end{cases} \tag{2.235}$$

where $k_\mathrm{m}=\dfrac{\cos(90^\circ-\varphi_{line})}{\sin\varphi_{ui}}$ is the adaptive setting coefficient and k_m is able to adjust according to real-time voltage and current phasor distribution characteristics. φ_{line} is the impedance angle of the protected line. φ_{ui} is the angular difference between the measured voltage and current. Z_m is the measured impedance. Z_{MN} is the impedance boundary value of the set line length. k_l is the reliability coefficient, which can extend the protection range considerably and improve the transition resistance tolerance ability; usually $k_l = 1.3$. The adaptive overload identification criterion in Inequality (2.235) displayed in the impedance plane is a full impedance circle with the origin as the centre, and the setting value $Z_\mathrm{set} = k_l k_\mathrm{m} Z_{MN}$ as the radius, shown in Figure 2.99 (the small circle). When there is a forward fault, the protection is switched on. When there is a reverse fault, the adaptive setting coefficient k_m is set to zero, so that the adaptive overload identification criterion is switched off. The time processing of the algorithm is 0.685 ms, so it can meet the requirements of protection systems. The CPU used is the Intel core i7 processor Q720 (1.60 GHz).

The operational characteristic of traditional distance protection zone-III is a directional circle set according to the minimum load impedance in the normal operating state, as shown in Figure 2.99 (the big circle) [20]. Applying the adaptive overload identification criterion to traditional distance protection zone-III, supposing that the system is symmetrical, when the measured impedance falls in the operational circle of traditional distance protection zone-III, the adaptive overload identification criterion is switched on.

1) If the measured impedance falls in the full impedance circle representing the overload identification criterion, then it is identified as a three-phase fault within the protection range, and the protection will operate reliably.
2) If the measured impedance falls outside the full impedance circle representing the overload identification criterion, then it is identified as an overload (no fault), and the protection will be blocked.

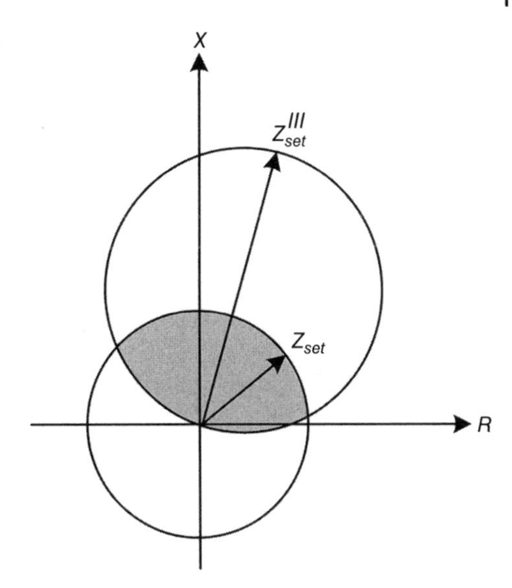

Figure 2.99 Protection operational zone.

It can be seen that, as an effective supplement to the traditional distance protection zone-III, the overload identification criterion is able to improve its overload identification ability, thus protection malfunction caused by accidental overloading can be avoided, as well as blackouts. The operational characteristic of distance protection zone-III supplemented by the adaptive overload identification criterion is shown in Figure 2.99, where Z_{set}^{III} and Z_{set} are the setting values of traditional distance protection zone-III and the adaptive overload identification criterion, respectively. The shaded intersection of the two impedance circles is the protection operational zone.

2.3.4.2 Simulation Verification

The IEEE 10-machine, 39-bus system is used here for simulation tests, as shown in Figure 2.100. In the normal operating state, the current on line 13 is relatively small (1.86 + j0.02 p.u.), and the current on its neighbouring lines is relatively big (e.g. the current on line 11 is 4.21 + j0.92 p.u.). Therefore, line 13 is vulnerable to accidental overloading caused by the cutting of neighbouring lines. Relay R8 on line 13 close to bus 8 is chosen here for an anti-overload capability test. The protection range of the chosen protection extends to the whole line length of the downstream line, i.e. the maximum line impedance of line 11. Considering the branch coefficient, the amplitude of the maximum line impedance is 0.0138 p.u. ($|Z_{MN}|$), and the line impedance angle φ_{line} is 85.86°. The setting value of the adaptive overload identification criterion Z_{set} could be calculated according to Inequality (2.235). The traditional distance protection zone-III applies the directional circle characteristic, where the setting impedance $Z_{\text{set}}^{III} = 0.1390 + j0.5188$ p.u. The operating equation is

$$\left|Z_{\text{m}} - \frac{1}{2}Z_{\text{set}}^{III}\right| \leq \left|\frac{1}{2}Z_{\text{set}}^{III}\right| \tag{2.236}$$

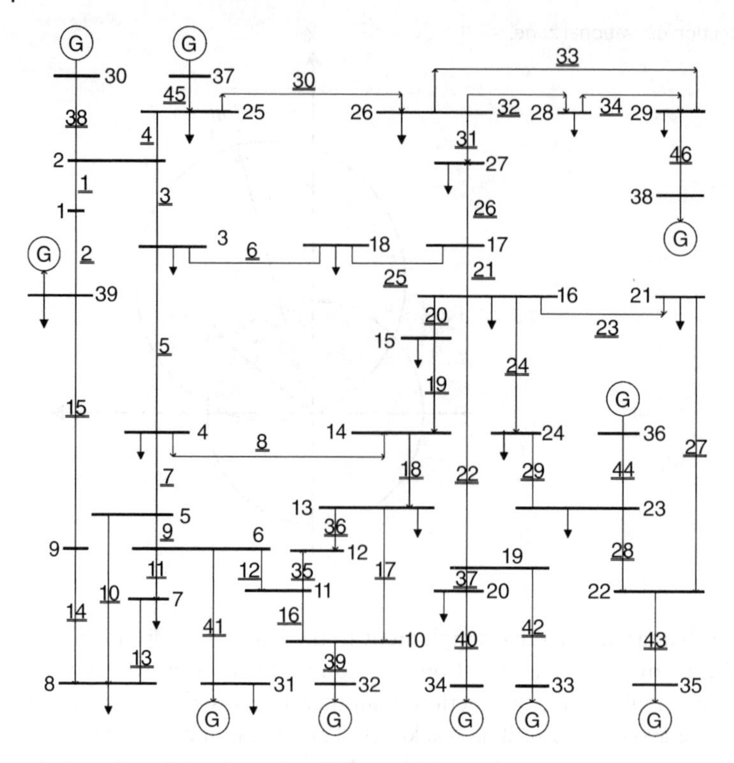

Figure 2.100 IEEE 10-machine, 39-bus system.

The following are some requirements for overload identification:

1) When a fault occurs in the protection range, the protection should operate reliably.
2) When an accidental overload occurs on the line, the protection should be able to identify the overload and not operate.
3) When a fault occurs on an overloaded line, the protection should be able to identify the fault within the protection range and operate reliably.

Based on the above requirements, the protection anti-overload capability test includes the following four stages:

Stage 1: Before $t = 0.3$ s, when the system is in the normal operating state, the protection zone-III should not operate.

Stage 2: When a three-phase fault occurs on the neighbouring line at $t = 0.3$ s and is cut by the primary protection at $t = 0.36$ s, the protection zone-III should start up but not operate reliably as the backup protection in this process. Thus, requirement (1) may be met.

Stage 3: After the fault on the neighbouring line is cut at $t = 0.36$ s and before a three-phase fault occurs on the tested line at $t = 0.5$ s, the protection zone-III should be able to identify the overload on the tested line and not malfunction. Thus, requirement (2) may be met.

Table 2.19 Operation status of relay R8 when line 11 is in fault.

Protection type	Normal operation (before $t = 0.3$ s)	Line 11 fault ($t = 0.3 - 0.36$ s)	Line 13 overload ($t = 0.36 - 0.5$ s)	Line 13 fault (after $t = 0.5$ s)
Traditional distance protection zone- III	No operation	Operation	Operation	Operation
Adaptive overload identification criterion	No operation	Operation	No operation	Operation
Integrated	No operation	Operation	Overload	Operation

Stage 4: After a fault occurs on the tested line at $t = 0.5$ s while it is overloaded, the protection zone-III should be able to identify the fault within the protection range and operate reliably. Thus, requirement (3) may be met.

The neighbouring lines of line 13 are line 11, line 10 and line 14. In the following, simulation tests are carried out on each line according to the above four stages.

A. *Simulation of Line 11*

When line 11 is in fault, the operational status of the relay 8 is shown in Table 2.19. In Stage 1, when the system is in the normal operating state, the measured impedance of relay R8 is $Z_{m1} = -0.3769 + j0.3952$ p.u., which is outside the protection range of traditional distance protection zone-III. In Stage 2, a three-phase fault occurs on line 11 at 50% of the line length at $t = 0.3$ s. The measured impedance changes to $Z_{m2} = 0.0017 + j0.0091$ p.u., which falls in the protection range of traditional distance protection zone-III. Because the system is symmetrical, the adaptive overload identification criterion will start, with $|Z_{set2}| = 0.0211$ p.u. Since $|Z_{m2}| < |Z_{set2}|$, the measured impedance also falls in the full-impedance circle of the overload identification criterion. Therefore, it is identified as a three-phase fault within the protection range and relay R8 will operate. In Stage 3, the primary protection on line 11 operates to cut the fault at $t = 0.36$ s. Then, power flow will be transferred, causing the current on line 13 to increase, and thus overload will result. The measured impedance now changes to $Z_{m2} = 0.1829 + j0.1565$ p.u., which falls in the protection range of traditional distance protection zone-III. Similarly, because the system is symmetrical, the adaptive overload identification criterion will start, with $|Z_{set2}| = 0.0319$ p.u. However, since $|Z_{m3}| > |Z_{set3}|$, the measured impedance falls outside the full-impedance circle of the overload identification criterion. Therefore, it is identified as overload and relay R8 will not operate. In Stage 4, a three-phase fault occurs on line 13 at $t = 0.5$ s. The measured impedance now changes to $Z_{m4} = 0.0002 + j0.0023$ p.u., which is still in the protection range of traditional distance protection zone-III. Because the system is still symmetrical, the adaptive overload identification criterion will start, with

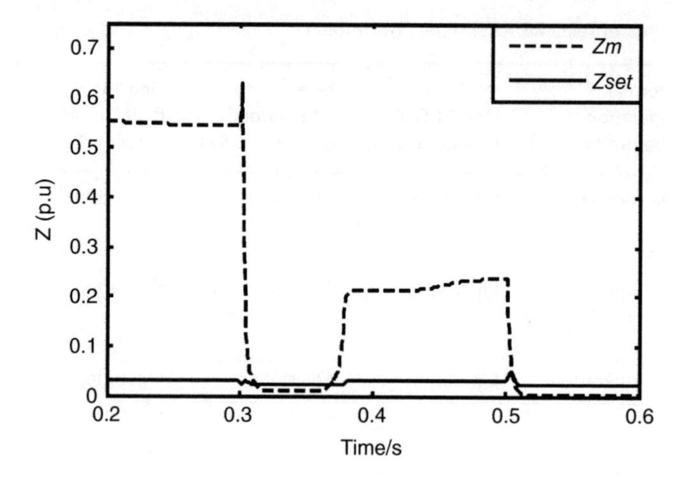

Figure 2.101 Z_m and Z_{set} variation diagram of relay R8 when line 11 is in fault.

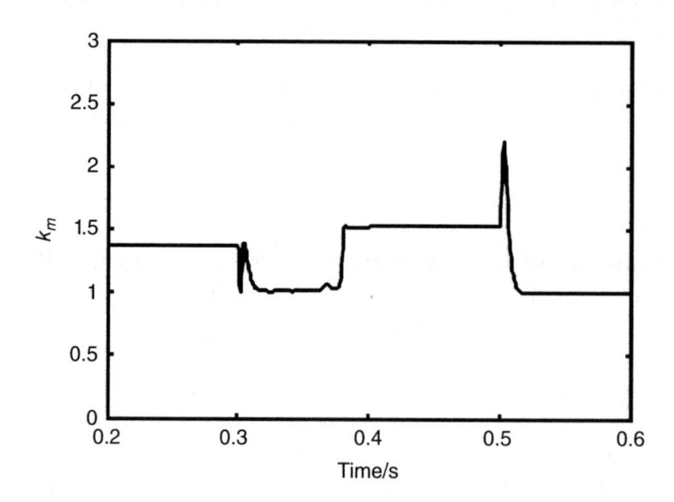

Figure 2.102 k_m variation diagram of relay R8 when line 11 is in fault.

$|Z_{set4}| = 0.0208$ p.u. Since $|Z_{m4}| < |Z_{set4}|$, the measured impedance falls in the full-impedance circle of the overload identification criterion. Therefore, it is identified as a three-phase fault within the protection range and relay R8 will operate.

In the above simulation test, variation of the measured impedance Z_m and setting impedance Z_{set} of the adaptive overload identification criterion is shown in Figure 2.101. It can be seen that the adaptive overload identification criterion is able to identify the in-zone faults in Stage 2 and Stage 4, and the overload in Stage 3 reliably.

According to the setting formula of the adaptive overload identification criterion, $Z_{set} = k_l k_m Z_{MN}$, the variation of the setting value Z_{set} depends on the adaptive coefficient k_m. Therefore, the variation trend of the setting value Z_{set} is the same as that of the adaptive coefficient k_m, which is shown in Figure 2.102. It can be seen

Table 2.20 Operational status of relay R8 when line 10 is in fault.

Protection type	Normal operation (before $t = 0.3$ s)	Line 10 fault ($t = 0.3 - 0.36$ s)	Line 13 overload ($t = 0.36 - 0.5$ s)	Line 13 fault (after $t = 0.5$ s)
Traditional distance protection zone- III	No operation	No operation	No operation	Operation
Adaptive overload identification criterion	No operation	No operation	No operation	Operation
Integrated	No operation	No operation	No operation	Operation

that, when line 13 is overloaded in Stage 3, k_m increases automatically to adjust the setting value of the overload criterion, so that protection malfunction can be prevented.

B. Simulation of Line 10

When line 10 is in fault, the operational status of relay R8 is shown in Table 2.20. In Stage 1, when the system is in the normal operating state, the measured impedance of relay R8 is $Z_{m1} = -0.5745 + j0.3801$ p.u., which is outside the protection range of traditional distance protection zone-III. In Stage 2, a three-phase fault occurs on line 10 at $t = 0.3$ s and the measured impedance changes to $Z_{m2} = -0.0033 + j0.0115$ p.u. Since this is a backward line fault, the measured impedance does not fall in the protection range of traditional distance protection zone-III. Therefore, relay R8 will not operate. In Stage 3, the primary protection on line 10 operates to cut the fault at $t = 0.36$ s. Then, power flow will be transferred, causing the current on line 13 to increase, and thus overload will result. The measured impedance now changes to $Z_{m3} = -0.2775 + j0.0505$ p.u. Since the impedance angle is too large, the measured impedance does not fall in the protection range of traditional distance protection zone-III. Therefore, relay R8 will not operate. In Stage 4, a three-phase fault occurs on line 13 at $t = 0.5$ s. The measured impedance now changes to $Z_{m4} = 0.0002 + j0.0023$ p.u., which falls in the protection range of traditional distance protection zone-III. Because the system is symmetrical, the adaptive overload identification criterion will start, with $|Z_{set4}| = 0.0208$ p.u. Since $|Z_{m4}| < |Z_{set4}|$, the measured impedance falls in the full-impedance circle of the overload identification criterion. Therefore, it is identified as a three-phase fault within the protection range and relay R8 will operate reliably.

In the above simulation test, variation of the measured impedance Z_m and setting impedance Z_{set} of the adaptive overload identification criterion is shown in Figure 2.103. It can be seen that the adaptive overload identification criterion is able to identify the in-zone fault in Stage 4 reliably, and block the protection in the non-fault cases of Stage 1 and Stage 3, and in the case of out-of-zone fault in Stage 2.

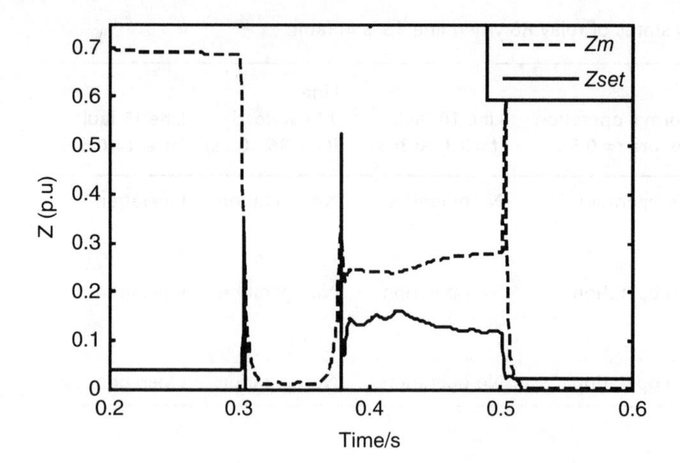

Figure 2.103 Z_m and Z_{set} variation diagram of relay 8 when line 10 is in fault.

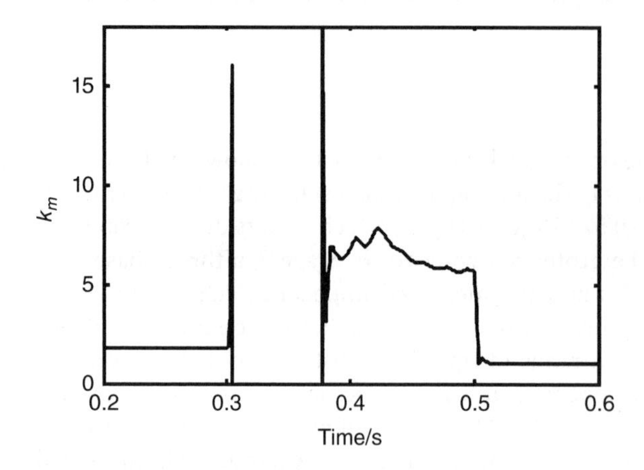

Figure 2.104 k_m variation diagram of relay 8 when line 10 is in fault.

The variation of the adaptive coefficient k_m is shown in Figure 2.104. It can be seen that, in Stage 2 when a backward line fault occurs, k_m is set to zero so that the overload criterion is blocked. In Stage 3 when the tested line is overloaded, k_m increases automatically to adjust the setting value of the criterion, so that protection malfunction can be prevented.

C. *Simulation of Line 14*

When line 14 is in fault, the operational status of relay R8 is shown in Table 2.21. In Stage 1, when the system is in the normal operating state, the measured impedance of relay R8 is $Z_{m1} = -0.5173 + j0.4164$ p.u., which is outside the protection range of traditional distance protection zone-III. In Stage 2, a three-phase fault occurs on line 14 at $t = 0.3$ s and the measured impedance changes to $Z_{m1} = -0.0089 - j0.0361$ p.u. Since this is a backward line fault, the measured impedance does not fall in the protection range of traditional distance protection zone-III.

Table 2.21 Operational status of relay R8 when line 14 is in fault.

Protection type	Normal operation (before $t = 0.3$ s)	Line 14 fault ($t = 0.3 - 0.36$ s)	Line 13 overload ($t = 0.36 - 0.5$ s)	Line 13 fault (after $t = 0.5$ s)
Traditional distance protection zone-III	No operation	No operation	No operation	Operation
Adaptive overload identification criterion	No operation	No operation	No operation	Operation
Integrated	No operation	No operation	No operation	Operation

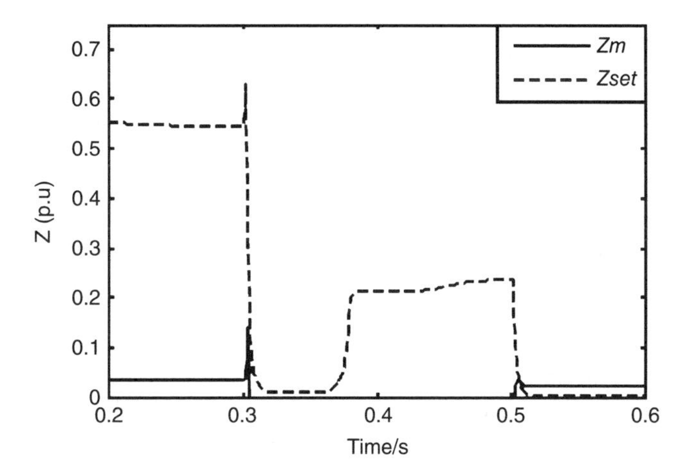

Figure 2.105 Z_m and Z_{set} variation diagram of relay R8 when line 14 is in fault.

Therefore, relay R8 will not operate. In Stage 3, the primary protection on line 10 operates to cut the fault at $t = 0.36$ s. Then, the power flow will be transferred, causing the current on line 13 to increase, and thus overload will result. The measured impedance now changes to $Z_{m3} = -0.3422 + j0.0465$ p.u. Since the overloaded power flow is opposite to the forward direction of relay R8, the measured impedance does not fall in the protection range of traditional distance protection zone-III. Therefore, relay R8 will not operate. In Stage 4, a three-phase fault occurs on line 13 at $t = 0.5$ s. The measured impedance now changes to $Z_{m4} = 0.0002 + j0.0023$ p.u., which falls in the protection range of traditional distance protection zone-III. Because the system is symmetrical, the adaptive overload identification criterion will start, with $|Z_{set4}| = 0.0208$ p.u. Since $|Z_{m4}| < |Z_{set4}|$, the measured impedance falls in the full-impedance circle of the overload identification criterion. Therefore, it is identified as a three-phase fault within the protection range and relay R8 will operate reliably.

In the above simulation test, variation of the measured impedance Z_m and setting impedance Z_{set} of the adaptive overload identification criterion is shown in Figure 2.105. It can be seen that the adaptive overload identification criterion is able to identify the in-zone fault in Stage 4 reliably, and block the protection in the

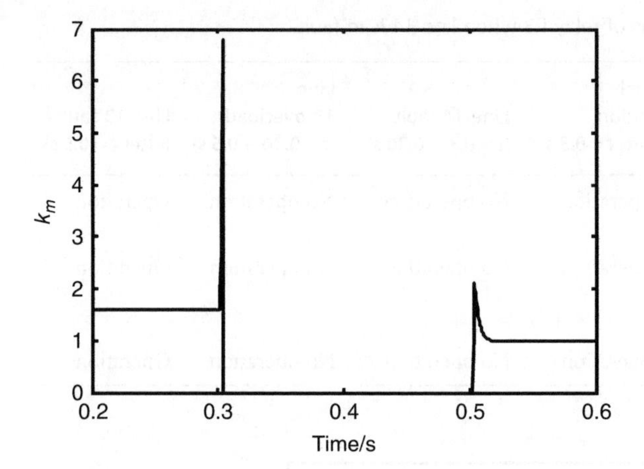

Figure 2.106 k_{m} variation diagram of relay R8 when line 14 is in fault.

non-fault cases of Stage 1 and Stage 3 and in the case of backward out-of-zone fault in Stage 2.

The variation of the adaptive coefficient k_{m} is shown in Figure 2.106. It can be seen that, in Stage 2 when a backward fault occurs and in Stage 3 when the tested line is backward overloaded, k_{m} is set to be zero so that the overload criterion is blocked. In Stage 4, when an in-zone fault occurs on the tested line, k_{m} increases so that the criterion starts and identifies the fault.

It can be seen from the above simulation results that the adaptive overload identification criterion is effective in distinguishing three-phase fault from overload. When a three-phase, in-zone fault occurs, the protection operates reliably, and in the cases of overload, normal operation and out-of-zone fault, the protection does not operate.

D. Comparison Between the Proposed Method and the Existing Method

In order to verify the efficiency of the proposed method, a comparison is made between the proposed method and the method mentioned in [21], where the load limiting characteristics (shown in Figure 2.107) are used to deal with the malfunction of distance protection caused by power flow transference. When the measured impedance of the relay falls in the load impedance region (the shaded part), the distance protection zone-III will be blocked. On the basis of a report from the North American Electric Reliability Corporation, the impedance angle of load blocking is set to 30°.

Relay R6 on line 9 close to bus 6 is taken to verify the efficiency of the method mentioned in [21]. A three-phase fault occurs on line 11 at $t = 0.3$ s and is cut at $t = 0.36$ s in Figure 2.100. The current of line 9 increases due to the effect of power flow transference. The measured impedance Z_{m1} of relay R6 falls in the protection range of distance protection zone-III, where Z_{m1} is 0.1004 + j0.0514 p.u. and Z_{set}^{III} is 0.0600 + j0.2239 p.u., respectively. However, Z_{m1} is in the load impedance region and relay R6 will be blocked, which is able to avoid the unreasonable removal of line 9 caused by power flow transference. When a three-phase fault occurs via 50 Ω transition resistance in the middle of line 9 at $t = 0.5$ s, relay R6 refuses to

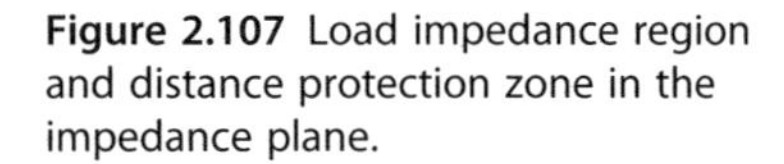
Figure 2.107 Load impedance region and distance protection zone in the impedance plane.

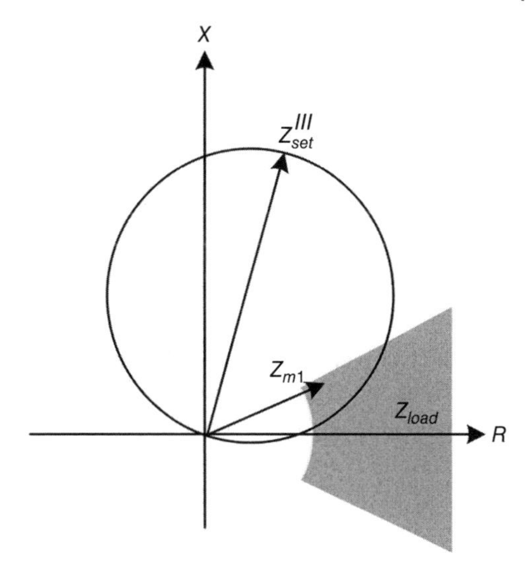

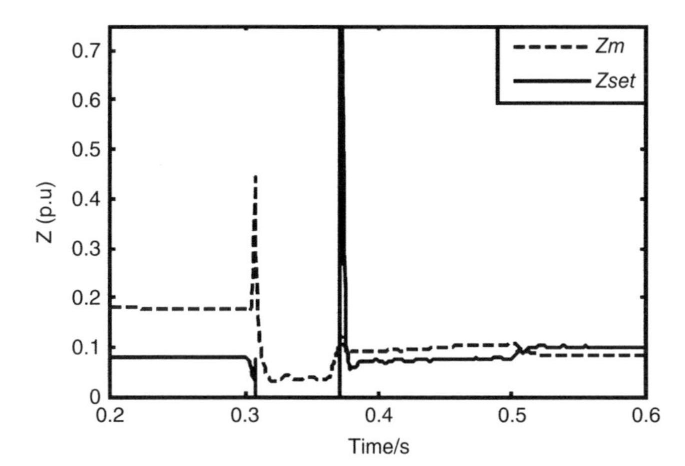

Figure 2.108 Z_m and Z_{set} variation diagram of relay R6 when line 9 is in fault.

operate since the measured impedance Z_{m2} is outside the protection range of distance protection zone-III, where Z_{m2} is 0.0805 + j0.0255 p.u. On the other hand, relay R6 is able to operate accurately by using the proposed scheme. The variation of the measured impedance Z_m and setting impedance Z_{set} of the adaptive overload identification criterion is shown in Figure 2.108.

On the basis of the above analysis, the method in [21] is able to identify overload correctly, but may lose the function of backup protection in the case of overload and symmetrical fault coexisting. However, the proposed method could be applied to traditional distance protection zone-III as an effective supplement to improve its backup protection performance.

In view of asymmetrical flow transference caused by the open-phase status, which may lead to the unreasonable removal of a non-faulty line, there is a need to study the overload identification scheme for asymmetrical flow transference.

2.3.5 Novel Fault Phase Selection Scheme Utilizing Fault Phase Selection Factors

2.3.5.1 Fault Phase Selection Based on FPSFs

To simplify the analysis, a dual-source power system shown in Figure 2.109 is used as the model here. The system voltages of side M and side N are $\dot{E}_M$ and $\dot{E}_N$, and the equivalent impedances are Z_{MS} and Z_{NS} respectively.

Suppose a fault occurs at point F on line MN. The sequence network is shown in Figure 2.110, where Z_{MSi} and Z_{NSi} represent the equivalent sequence impedances of the system on side M and side N, respectively. $\dot{I}_{Mi}$ and $\dot{I}_{Ni}$ represent the sequence currents on side M and side N, respectively. $\dot{I}_{Fi}$ represents the sequence current at the fault location F. R_g is the transition resistance. $i = 1, 2, 0$.

The sequence current distribution coefficient on side M is defined as:

$$C_i = \frac{\dot{I}_{Mi}}{\dot{I}_{Fi}} = \frac{Z_{NSi} + (1-p)Z_{MNi}}{Z_{MSi} + Z_{MNi} + Z_{NSi}} \tag{2.237}$$

where Z_{MNi} is the sequence impedance of line MN and p represents the percentage of the distance from fault location F to bus M in the line length. For a high-voltage system, it can be approximately viewed that C_1, C_2 and C_0 are all real values. Meanwhile, the positive sequence and negative sequence impedances are approximately the same, thus $C_1 = C_2$.

The fault phase currents at the relaying point of bus M are

$$\begin{cases} \dot{I}_{Ma} = C_1\dot{I}_{Fa1} + C_2\dot{I}_{Fa2} + C_0\dot{I}_{Fa0} \\ \dot{I}_{Mb} = \alpha^2 C_1\dot{I}_{Fa1} + \alpha C_2\dot{I}_{Fa2} + C_0\dot{I}_{Fa0} \\ \dot{I}_{Mc} = \alpha C_1\dot{I}_{Fa1} + \alpha^2 C_2\dot{I}_{Fa2} + C_0\dot{I}_{Fa0} \end{cases} \tag{2.238}$$

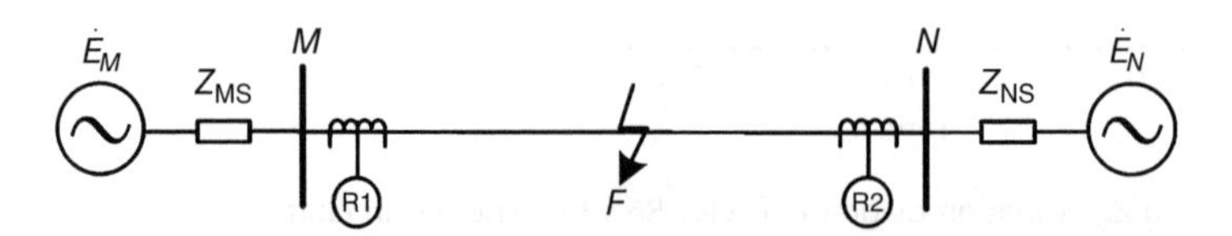

Figure 2.109 Model of a dual-source power system.

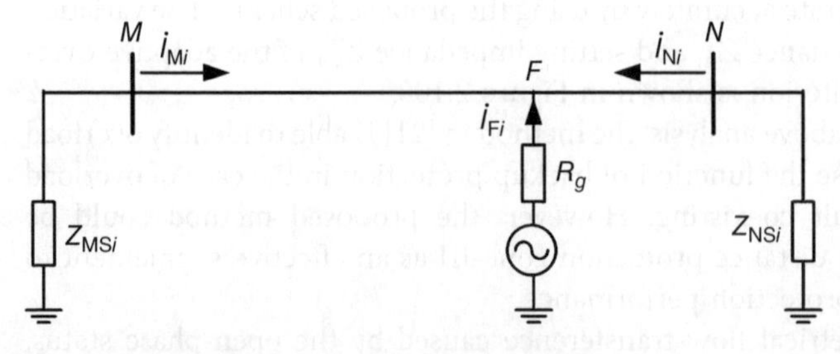

Figure 2.110 Sequence network of internal fault.

where $\alpha = e^{j120^\circ}$. $\dot{I}_{Ma}$, $\dot{I}_{Mb}$ and $\dot{I}_{Mc}$ are the fault currents of phases A, B and C measured at bus M. $\dot{I}_{Fa1}$, $\dot{I}_{Fa2}$ and $\dot{I}_{Fa0}$ are the sequence components of the fault current of phase A at the fault location F.

According to the proportion relation between the fault current of each phase and the fault current difference of the other two phases at the relaying point, the FPSFs are defined as

$$S_1 = \frac{|\dot{I}_{Ma}|}{|\dot{I}_{Mb} - \dot{I}_{Mc}|}, S_2 = \frac{|\dot{I}_{Mb}|}{|\dot{I}_{Mc} - \dot{I}_{Ma}|}, S_3 = \frac{|\dot{I}_{Mc}|}{|\dot{I}_{Ma} - \dot{I}_{Mb}|} \tag{2.239}$$

where (S_1, S_2, S_3) is a group of FPSFs.

A. Characteristics of FPSFs for Different Fault Types

1) Single-phase grounding fault – When a phase-A-to-ground fault occurs, the fault currents of phase B and C at the fault point F are both zero, i.e. $\dot{I}_{Fb} = \dot{I}_{Fc} = 0$. While it can be gleaned from fault analysis that the sequence components of the fault current of phase A meet the following relationship:

$$\dot{I}_{Fa1} = \dot{I}_{Fa2} = \dot{I}_{Fa0} \tag{2.240}$$

Applying Equation (2.240) to Equation (2.238) results in

$$\begin{cases} \dot{I}_{Ma} = (C_1 + C_2 + C_0)\dot{I}_{Fa1} \\ \dot{I}_{Mb} = (\alpha^2 C_1 + \alpha C_2 + C_0)\dot{I}_{Fa1} \\ \dot{I}_{Mc} = (\alpha C_1 + \alpha^2 C_2 + C_0)\dot{I}_{Fa1} \end{cases} \tag{2.241}$$

Thus, the FPSFs can be gained as follows:

$$S_1 = \frac{|\dot{I}_{Ma}|}{|\dot{I}_{Mb} - \dot{I}_{Mc}|} = \frac{\sqrt{3}}{3} \frac{|C_1 + C_2 + C_0|}{|C_1 - C_2|} \tag{2.242}$$

$$S_2 = \frac{|\dot{I}_{Mb}|}{|\dot{I}_{Mc} - \dot{I}_{Ma}|} = \frac{1}{3} \frac{|C_1^2 + C_2^2 + C_0^2 - C_1C_2 - C_1C_0 - C_2C_0|}{|C_1^2 + C_2^2 + C_1C_2|} \tag{2.243}$$

$$S_3 = \frac{|\dot{I}_{Mc}|}{|\dot{I}_{Ma} - \dot{I}_{Mb}|} = \frac{1}{3} \frac{|C_1^2 + C_2^2 + C_0^2 - C_1C_2 - C_1C_0 - C_2C_0|}{|C_1^2 + C_2^2 + C_1C_2|} \tag{2.244}$$

Consider that C_1 is approximately equal to C_2, thus the denominator of the expression on the right side of Equation (2.242) is nearly zero. Therefore, S_1 is very big. Regardless of the values that C_1, C_2 and C_0 may take, the relationship between S_2 and S_3 is that $S_2 = S_3$.

2) Phase-to-phase fault – When a phase-B-to-C fault occurs, the fault currents of phases B and C at the fault point F are equal in amplitude and opposite in direction, and the fault current of phase A is zero, i.e. $\dot{I}_{Fb} = -\dot{I}_{Fc}$, $\dot{I}_{Fa} = 0$. The sequence components of the fault current of phase A are

$$\begin{cases} \dot{I}_{Fa1} = -\dot{I}_{Fa2} \\ \dot{I}_{Fa0} = 0 \end{cases} \tag{2.245}$$

Applying Equations (2.245) to Equations (2.238) results in

$$\begin{cases} \dot{I}_{\text{Ma}} = C_1\dot{I}_{\text{Fa1}} + C_2\dot{I}_{\text{Fa2}} + C_0\dot{I}_{\text{Fa0}} = (C_1 - C_2)\dot{I}_{\text{Fa1}} \\ \dot{I}_{\text{Mb}} = \alpha^2 C_1\dot{I}_{\text{Fa1}} + \alpha C_2\dot{I}_{\text{Fa2}} + C_0\dot{I}_{\text{Fa0}} = (\alpha^2 C_1 - \alpha C_2)\dot{I}_{\text{Fa1}} \\ \dot{I}_{\text{Mc}} = \alpha C_1\dot{I}_{\text{Fa1}} + \alpha^2 C_2\dot{I}_{\text{Fa2}} + C_0\dot{I}_{\text{Fa0}} = (\alpha C_1 - \alpha^2 C_2)\dot{I}_{\text{Fa1}} \end{cases} \tag{2.246}$$

Thus, the FPSFs can be gained as follows:

$$S_1 = \frac{|\dot{I}_{\text{Ma}}|}{|\dot{I}_{\text{Mb}} - \dot{I}_{\text{Mc}}|} = \frac{\sqrt{3}}{3}\frac{|C_1 - C_2|}{|C_1 + C_2|} \tag{2.247}$$

$$S_2 = \frac{|\dot{I}_{\text{Mb}}|}{|\dot{I}_{\text{Mc}} - \dot{I}_{\text{Ma}}|} = \frac{|-(C_1 - C_2) - j\sqrt{3}(C_1 + C_2)|}{|-(C_1 - C_2) + j\sqrt{3}(C_1 + C_2)|} = 1 \tag{2.248}$$

$$S_3 = \frac{|\dot{I}_{\text{Mc}}|}{|\dot{I}_{\text{Ma}} - \dot{I}_{\text{Mb}}|} = \frac{|-(C_1 - C_2) + j\sqrt{3}(C_1 + C_2)|}{|(C_1 - C_2) + j\sqrt{3}(C_1 + C_2)|} = 1 \tag{2.249}$$

Consider that C_1 is approximately equal to C_2, so the numerator of the expression on the right side of Equation (2.242) is nearly zero. Therefore, S_1 is nearly zero. In addition, whatever values C_1 and C_2 may take, the relationship between S_2 and S_3 is that $S_2 = S_3 = 1$.

3) Phase-to-phase grounding fault – When a phase-BC-to-ground fault occurs, the fault current of phase A at the fault point F is zero. The sequence components of the fault current of phase A are

$$\begin{cases} \dot{I}_{\text{Fa2}} = -\dfrac{Z_{0\Sigma}}{Z_{2\Sigma} + Z_{0\Sigma}}\dot{I}_{\text{Fa1}} \\ \dot{I}_{\text{Fa0}} = -\dfrac{Z_{2\Sigma}}{Z_{2\Sigma} + Z_{0\Sigma}}\dot{I}_{\text{Fa1}} \end{cases} \tag{2.250}$$

where $Z_{2\Sigma}$ and $Z_{0\Sigma}$ are the negative and zero sequence integrated impedance, respectively.

Applying Equations (2.250) to Equations (2.238) results in

$$\begin{cases} \dot{I}_{\text{Ma}} = [(Z_{2\Sigma} + Z_{0\Sigma})C_1 - Z_{0\Sigma}C_2 - Z_{2\Sigma}C_0]\dfrac{\dot{I}_{\text{Fa1}}}{Z_{2\Sigma} + Z_{0\Sigma}} \\ \dot{I}_{\text{Mb}} = [\alpha^2(Z_{2\Sigma} + Z_{0\Sigma})C_1 - \alpha Z_{0\Sigma}C_2 - Z_{2\Sigma}C_0]\dfrac{\dot{I}_{\text{Fa1}}}{Z_{2\Sigma} + Z_{0\Sigma}} \\ \dot{I}_{\text{Mc}} = [\alpha(Z_{2\Sigma} + Z_{0\Sigma})C_1 - \alpha^2 Z_{0\Sigma}C_2 - Z_{2\Sigma}C_0]\dfrac{\dot{I}_{\text{Fa1}}}{Z_{2\Sigma} + Z_{0\Sigma}} \end{cases} \tag{2.251}$$

Thus, the FPSFs can be gained as follows:

$$S_1 = \frac{|\dot{I}_{\text{Ma}}|}{|\dot{I}_{\text{Mb}} - \dot{I}_{\text{Mc}}|} = \frac{\sqrt{3}}{3}\frac{|Z_{2\Sigma}(C_1 - C_0) + Z_{0\Sigma}(C_1 - C_2)|}{|Z_{2\Sigma}C_1 + Z_{0\Sigma}(C_1 + C_2)|} \tag{2.252}$$

$$S_2 = \frac{|\dot{I}_{\text{Mb}}|}{|\dot{I}_{\text{Mc}} - \dot{I}_{\text{Ma}}|} = \frac{\left|\left(\frac{1}{2}\text{A} - \frac{\sqrt{3}}{2}\text{B}\right) + j\left(\frac{1}{2}\text{C} + \frac{\sqrt{3}}{2}\text{D}\right)\right|}{\left|\left(\frac{3}{2}\text{A} + \frac{\sqrt{3}}{2}\text{B}\right) + j\left(\frac{3}{2}\text{C} - \frac{\sqrt{3}}{2}\text{D}\right)\right|} \tag{2.253}$$

$$S_3 = \frac{|\dot{I}_{\text{Mc}}|}{|\dot{I}_{\text{Ma}} - \dot{I}_{\text{Mb}}|} = \frac{\left|\left(\frac{1}{2}\text{A} + \frac{\sqrt{3}}{2}\text{B}\right) + j\left(\frac{1}{2}\text{C} - \frac{\sqrt{3}}{2}\text{D}\right)\right|}{\left|\left(\frac{3}{2}\text{A} - \frac{\sqrt{3}}{2}\text{B}\right) + j\left(\frac{3}{2}\text{C} + \frac{\sqrt{3}}{2}\text{D}\right)\right|} \tag{2.254}$$

where $\text{A} = R_{2\Sigma}(\text{C}_1 + 2\text{C}_0) + R_{0\Sigma}(\text{C}_1 - \text{C}_2)$, $\text{B} = X_{2\Sigma}\text{C}_1 + X_{0\Sigma}(\text{C}_1 + \text{C}_2)$, $\text{C} = X_{2\Sigma}(\text{C}_1 + 2\text{C}_0) + X_{0\Sigma}(\text{C}_1 - \text{C}_2)$, $\text{D} = R_{2\Sigma}\text{C}_1 + R_{0\Sigma}(\text{C}_1 + \text{C}_2)$, and A, B, C and D are all real values.

Whatever values C_1, C_2 and C_0 may take, it can be seen from Equations (2.252)–(2.254) that S_1, S_2 and S_3 differ from each other. Because of this, the amplitudes of the fault currents can be used to identify phase-to-phase-to-ground faults.

According to Equations (2.251),

$$\frac{|\dot{I}_{\text{Mb}}|}{|\dot{I}_{\text{Mc}}|} = \frac{\left|\text{C}_1\left(-\alpha^2 + j\sqrt{3}\frac{Z_{0\Sigma}}{Z_{2\Sigma}}\right) + \text{C}_0\right|}{\left|\text{C}_1\left(-\alpha - j\sqrt{3}\frac{Z_{0\Sigma}}{Z_{2\Sigma}}\right) + \text{C}_0\right|} = 1 \tag{2.255}$$

$$\frac{|\dot{I}_{\text{Ma}}|}{|\dot{I}_{\text{Mb}}|} = \frac{\left|(\text{C}_1 - \text{C}_0) + \frac{Z_{0\Sigma}}{Z_{2\Sigma}}(\text{C}_1 - \text{C}_2)\right|}{\left|\text{C}_1\left(-\alpha^2 + j\sqrt{3}\frac{Z_{0\Sigma}}{Z_{2\Sigma}}\right) + \text{C}_0\right|} < \frac{\left|\frac{\text{C}_0}{\text{C}_1} - 1\right|}{\left|\frac{\text{C}_0}{\text{C}_1} - \alpha^2\right|} \tag{2.256}$$

Let $Y = \frac{\left|\frac{\text{C}_0}{\text{C}_1} - 1\right|}{\left|\frac{\text{C}_0}{\text{C}_1} - \alpha^2\right|}$, then Y can be written as a function of x:

$$Y = \begin{cases} \dfrac{1 - x}{\sqrt{x^2 + x + 1}} & 0 \le x \le 1 \\ \dfrac{x - 1}{\sqrt{x^2 + x + 1}} & x > 1 \end{cases} \tag{2.257}$$

where $x = \frac{\text{C}_0}{\text{C}_1}$.

Consider that C_1, C_2 and C_0 can all be approximately viewed as positive real values. Therefore, x can also be seen as a positive real value. According to Equation (2.257), when $x = 0$ or $x \to +\infty$, Y has the maximum value 1; when $0 < x < +\infty$, Y is always below 1. Thus, for phase-BC-to-ground faults, it is always true that

$$|\dot{I}_{\text{Ma}}| < |\dot{I}_{\text{Mb}}| = |\dot{I}_{\text{Mc}}| \tag{2.258}$$

4) Three-phase fault – When a three-phase fault occurs, the fault currents of phase A, B and C at the fault point F are all the same as the positive sequence fault current of phase A, i.e. $\dot{I}_{\mathrm{Fa}} = \dot{I}_{\mathrm{Fb}} = \dot{I}_{\mathrm{Fc}} = \dot{I}_{\mathrm{Fa1}}$. Applying this expression to Equations (2.238) results in

$$\begin{cases} \dot{I}_{\mathrm{Ma}} = \mathrm{C}_1 \dot{I}_{\mathrm{Fa1}} \\ \dot{I}_{\mathrm{Mb}} = \alpha^2 \mathrm{C}_1 \dot{I}_{\mathrm{Fa1}} \\ \dot{I}_{\mathrm{Mc}} = \alpha \mathrm{C}_1 \dot{I}_{\mathrm{Fa1}} \end{cases} \tag{2.259}$$

Thus, the FPSFs can be gained as follows:

$$S_1 = S_2 = S_3 = \frac{\sqrt{3}}{3} \tag{2.260}$$

B. Fault Phase Selection Criterion

Concerning the different characteristics of FPSFs in different fault types, the fault phase selection criterion is established.

Criterion A: As shown in Inequalities (2.261), when one factor in the group of FPSFs (S_1,S_2,S_3) exceeds the threshold ε_1, and the absolute value of the difference between the other two factors in the group divided by the lesser of the two remains below the threshold ε_2, a single-phase grounding fault can be determined. The fault phase is that corresponding to the maximum factor in the group of FPSFs.

$$\begin{cases} \max(S_1, S_2, S_3) \geq \varepsilon_1 \\ \min\left(\left| \dfrac{S_1 - S_2}{\min(S_1, S_2)} \right|, \left| \dfrac{S_2 - S_3}{\min(S_2, S_3)} \right|, \left| \dfrac{S_3 - S_1}{\min(S_3, S_1)} \right| \right) \leq \varepsilon_2 \end{cases} \tag{2.261}$$

where $\varepsilon_1 = 50$, $\varepsilon_2 = 0.1$.

Criterion B: As shown in Inequalities (2.262), when the minimum factor in the group of FPSFs (S_1,S_2,S_3) does not surpass the threshold ε_3, and the differences between the other two factors and one are both smaller than the threshold ε_4, a phase-to-phase fault can be determined, and the selected phase is the one corresponding to the minimum factor in the group.

$$\begin{cases} \min(S_1, S_2, S_3) \leq \varepsilon_3 \\ |S_{\mathrm{m1}} - 1| \leq \varepsilon_4 \\ |S_{\mathrm{m2}} - 1| \leq \varepsilon_4 \end{cases} \tag{2.262}$$

where $\varepsilon_3 = 0.1$, $\varepsilon_4 = 0.2$. S_{m1} and S_{m2} are the other two factors in the FPSFs group which have higher values.

Criterion C: As shown in Inequality (2.263), when the zero sequence current exceeds the threshold ε_0, and the absolute value of the difference between any two factors in the group divided by the lesser of the two surpasses the threshold ε_5, a phase-to-phase-to-ground fault can be determined. Meanwhile, according to Equation (2.258), the fault phases are the two with higher fault currents.

$$\min\left(\left|\frac{S_1-S_2}{\min(S_1,S_2)}\right|,\left|\frac{S_2-S_3}{\min(S_2,S_3)}\right|,\left|\frac{S_3-S_1}{\min(S_3,S_1)}\right|\right)>\varepsilon_5 \tag{2.263}$$

where $\varepsilon_0 = 30$ A, $\varepsilon_5 = 0.2$.

Criterion D: As shown in Inequality (2.264), when the absolute value of the difference between any two factors in the group divided by the lesser of the two does not surpass the threshold ε_6, a three-phase fault can be determined.

$$\max\left(\left|\frac{S_1-S_2}{\min(S_1,S_2)}\right|,\left|\frac{S_2-S_3}{\min(S_2,S_3)}\right|,\left|\frac{S_3-S_1}{\min(S_3,S_1)}\right|\right)<\varepsilon_6 \tag{2.264}$$

where $\varepsilon_6 = 0.1$.

2.3.5.2 Performance Analysis

In the algorithm flow shown in Figure 2.111, phasor measurements of the current mutation variable are used to select the fault phases. When the system operates in the stable and steady state, the current mutation at the relaying point is nearly zero. Once a fault occurs, the current mutation value increases significantly. The starting threshold value is set to $i_{start} = 0.13\,I_e$ (the operating value of the mutation phase selector is usually set to $i_{start} = 0.1 \sim 0.15\,I_e$ in real relay protection devices), where I_e represents the rated current. When the maximum current mutation value of different phases exceeds i_{start}, the fault phase selection scheme embedded in digital protective relaying (DPR) starts.

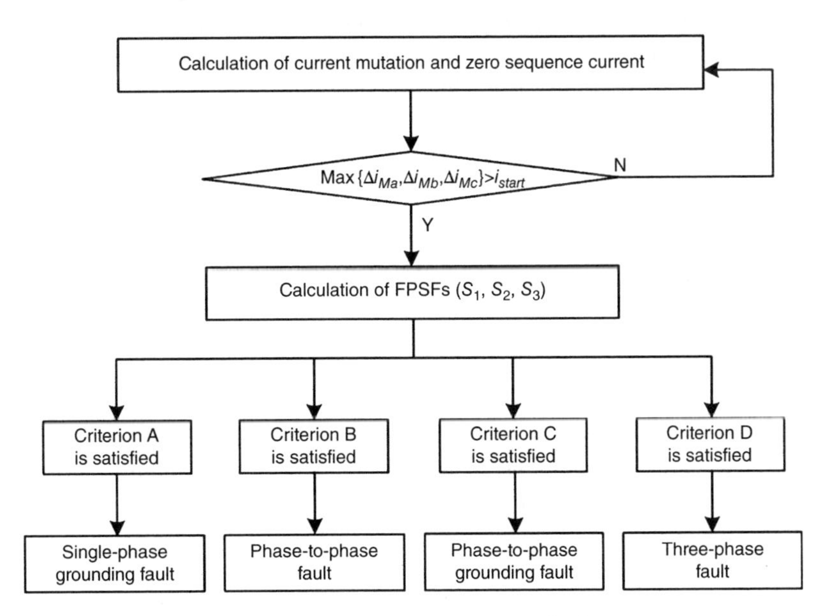

Figure 2.111 Flowchart of the fault phase selection method.

The proposed method has the following characteristics:

1) The proposed method is unaffected by the transition resistance and fault location, and is applicable to fault phase selection on the weak-infeed side. According to Equation (2.237), the current distribution coefficients are not related to the transition resistance. Thus, however big the transition resistance is, the proposed method is able to select the fault phase correctly. Also, although the current distribution coefficients are affected by the system impedance, line impedance and fault location, no matter what values C_1, C_2 and C_0 take, the characteristics that FPSFs demonstrate in different fault types always remain the same. Therefore, the proposed method is not affected by the fault location, and is applicable to fault phase selection on the weak-infeed side (the weak-infeed side means that the system impedance is large).
2) The proposed method is unaffected by the fault initial phase angle. According to analysis of different fault types, for single-phase-to-ground faults, phase-to-phase faults and three-phase faults, the FPSFs are only related to the current distribution coefficients; for a two-phase-to-ground fault, the FPSFs have to do with the current distribution coefficients and the negative and zero sequence integrated impedances. Since the current distribution coefficients and the negative and zero sequence integrated impedances are not related to the fault instant, the proposed method is not affected by the fault initial phase angle.
3) The proposed method is applicable to fault identification of a developing fault. A developing fault refers to a single-phase-to-ground fault developing into a two-phase-to-ground fault (including the former single phase) at the same fault location. Comparative analysis shows that the phase selection criterion of a single-phase-to-ground fault (Criterion A) is obviously different from the phase selection criterion of a two-phase-to-ground fault (Criterion C). Thus, in the case of a developing fault, the proposed method has good selectivity.
4) The proposed method is applicable to fault phase selection of double circuit lines. For double circuit transmission lines on the same tower, if any type of fault occurs on one line, the phase electrical variables on two lines at the fault location are independent of each other, thus the relationship between sequence currents is the same as in a single circuit transmission line. Therefore, the proposed method is applicable to fault phase selection of double circuit lines.
5) The proposed method is not affected by TCSC and UPFC, and is applicable to the rectifier side in an AC/DC hybrid system. When TCSC or UPFC is installed on the transmission line, the positive sequence and negative sequence impedances are still the same, thus $C_1 \approx C_2$. In addition, the FPSFs are not related to the values of the current distribution coefficients. Therefore, the proposed method is not affected by TCSC and UPFC. For an AC/DC hybrid system, when a short circuit fault occurs on the rectifier-side AC line, usually there is no commutation failure, thus the phase/sequence components 'felt' at two ends of the fault line are the same as those in an ordinary AC line. In this case, false identification will not happen. On the other hand, when a short circuit fault occurs on the inverter-side AC line, due to the influence of commutation failure, the positive and negative sequence impedances of the backside system 'felt' by the inverter-side relay protection during the fault transient can be very different. In this case, false identification may result.

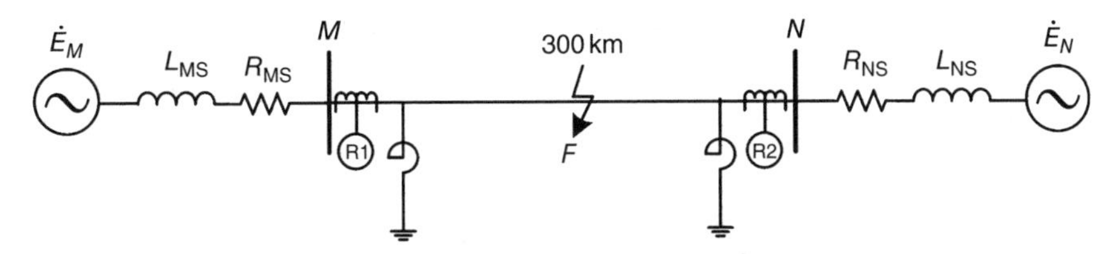

Figure 2.112 Simulation model of a high-voltage transmission line.

2.3.5.3 Simulation Verification and Analysis

With reference to the Beijing–Tianjin–Tangshan 500 kV UHV transmission line parameters, a 300 km transmission line model with distributed parameters is constructed here using PSCAD/EMTDC, shown in Figure 2.112. The system rated power $S_N = 100$ MVA, rated voltage $U_N = 500$ kV, fundamental frequency $f = 50$ Hz. System parameters are: $L_{M0} = 0.0926$ H, $R_{M0} = 0.6\,\Omega$, $L_{M1} = 0.13743$ H, $R_{M1} = 1.0515\,\Omega$, $L_{N0} = 0.11927$ H, $R_{N0} = 20\,\Omega$, $L_{N1} = 0.14298$ H, $R_{N1} = 26\,\Omega$. Parameters for the transmission line are: $r_1 = 0.02083\,\Omega$/km; $l_1 = 0.8948$ mH/km; $C_1 = 0.0129\,\mu$F/km; $r_0 = 0.1148\,\Omega$/km; $l_0 = 2.2886$ mH/km; $C_0 = 0.00523\,\mu$F/km. The compensation degree of the shunt reactor is set to 70% with $L_L = 7.4803$ H, $L_N = 2.4934$ H. The full-wave Fourier algorithm is used to extract the phasor with a sampling frequency 2000 Hz.

The fault phase selector is installed at bus M. Five locations for fault simulation are chosen, i.e. d_1, d_2, d_3, d_4 and d_5, their distances from bus M being 0 km, 75 km, 150 km, 225 km and 300 km, respectively. Since the statistical maximum grounding resistance for 500 kV lines is 300 Ω, the fault resistance for high-resistance faults is set to be 300 Ω here.

A. *Simulation of Fault Resistance*

In real system operation, a fault usually occurs via a particular transition resistance, which will weaken the fault characteristics of the power grid and exert an unfavourable influence on the fault phase selection. The different types of faults via fault resistance, including single-phase grounding fault, phase-to-phase grounding fault, phase-to-phase fault, and three-phase fault, are simulated to estimate the influence of the fault resistance on the performance of the proposed method.

When $t = 0.40$ s, a phase-A-to-ground fault occurs at the middle point of the line via a 300 Ω fault resistance. After the fault occurs, the fault phase selector immediately starts to calculate the FPSFs. As shown in Figure 2.113, S_1 increases rapidly, while S_2 and S_3 remain approximately the same. According to Criterion A, a phase-A-to-ground fault can be determined. The time taken to make such a judgement is 4 ms.

When $t = 0.40$ s, a phase-B-to-C fault occurs at the middle point of the line via a 300 Ω fault resistance. The variation curves of the FPSFs are as shown in Figure 2.114. It can be seen that S_1 is nearly zero, while S_2 and S_3 are close to 1. Therefore, according to Criterion B, a phase-B-to-C fault can be determined. The time used for the selection process is 2.5 ms.

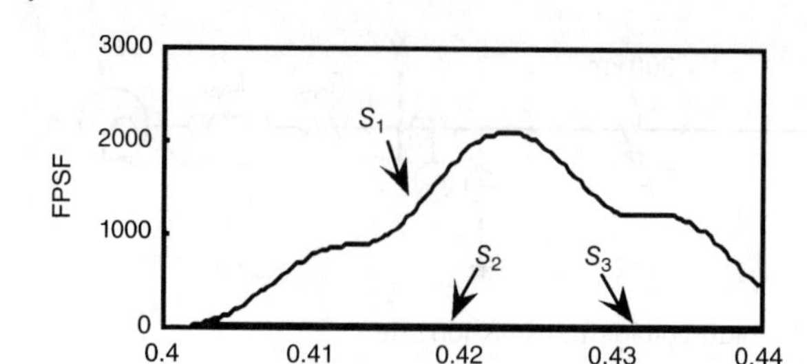

Figure 2.113 FPSFs when a phase-A-to-ground fault occurs via a 300 Ω fault resistance.

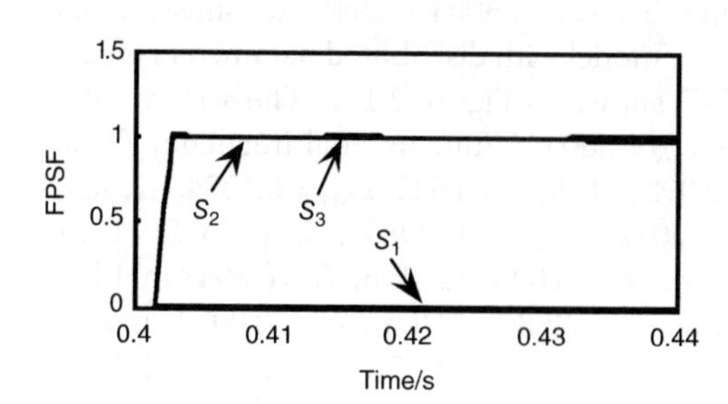

Figure 2.114 FPSFs when a phase-B-to-C fault occurs via a 300 Ω fault resistance.

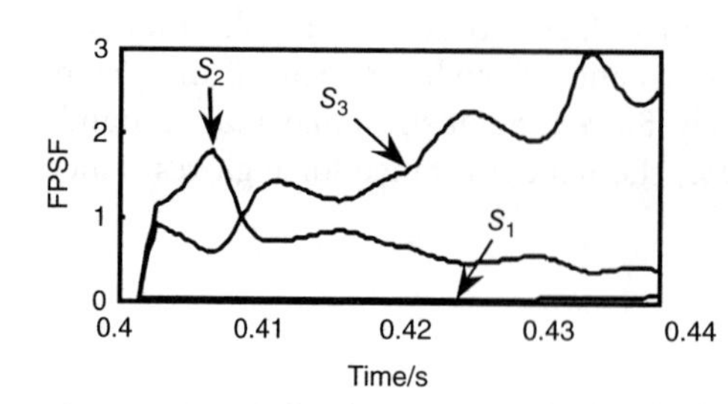

Figure 2.115 FPSFs when a phase-BC-to-ground fault occurs via a 300 Ω fault resistance.

When $t = 0.40$ s, a phase-BC-to-ground fault occurs at the middle point of the line via a 300 Ω fault resistance. The FPSF curves are as shown in Figure 2.115. It can been seen that S_1, S_2 and S_3 are different from each other, and the zero sequence current is relatively large. However, it can be seen that the fault currents of phases B and C are higher than the fault current of phase A. Thus, according to Criterion C, a phase-BC-to-ground fault can be determined. The time for identification is 2.5 ms.

When $t = 0.40$ s, a three-phase symmetrical fault occurs at the middle point of the line. The FPSF curves are as shown in Figure 2.116. It can be seen that S_1, S_2 and S_3 are very close to one another. In accordance with Criterion D, a three-phase fault can be determined.

For different types of faults, the values of the FPSFs at a cycle after the fault phase selector starts and the selection results are shown in Table 2.22, where AG, BC, BCG and ABC represent phase-A-to-ground, phase-B-to-C, phase-BC-to-ground, and three-phase fault, respectively. It can be seen that for either metallic or high-resistance faults, accurate identification of the fault type and

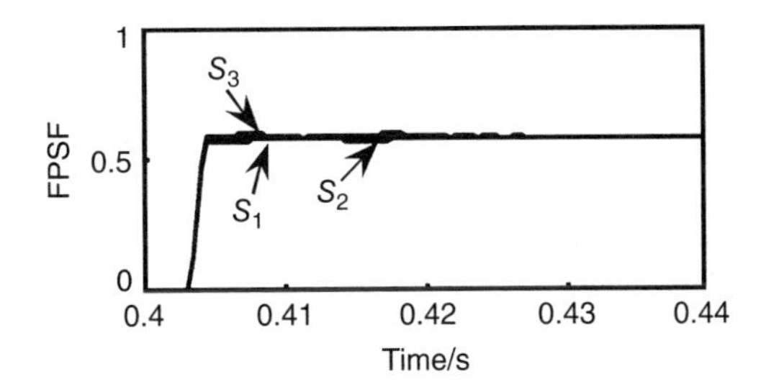

Figure 2.116 FPSFs when a three-phase fault occurs via a 300 Ω fault resistance.

Table 2.22 Results of fault phase identification at the middle point of the line via different fault resistances.

Fault type	Fault resistance(Ω)	Fault phase selection factors (FPSFs)			Selection result
		S_1	S_2	S_3	
AG	0	3292.3	0.0348	0.0349	AG
	300	1820.6	0.0145	0.0151	AG
BC	0	0.0000	1.0000	1.0000	BC
	300	0.0001	1.0002	0.9998	BC
BCG	0	0.0013	0.8906	1.1095	BCG
	300	0.0059	0.6932	1.4436	BCG
ABC	0	0.5724	0.5751	0.5774	ABC
	300	0.5740	0.5773	0.5723	ABC

selection of the fault phases is possible with the proposed method. This method is highly sensitive to all types of fault, especially for phase-to-phase grounding faults, when the fault resistance changes from 0 Ω to 300 Ω. In this case, the FPSFs do not vary much; rather it remains significantly distinct from the cases of the other fault types. Thus, Criterion C will never lose its applicability in determining a phase-to-phase grounding fault.

B. Simulation of Fault Inception Angle

The fault inception angle, especially for the voltage zero-crossing situation, is one of the most frequent problems found in fault phase selection schemes whose performance is based on travelling waves or on superimposed quantities due to the small travelling waves generated by the fault. When faults occur at the middle point of the line at a wide range of fault inception angles varying between 0° and 180°, the effect of the fault inception angle on the proposed method is simulated. Table 2.23 shows the values of the FPSFs at a cycle after the fault phase selector starts and the selection results. It can be seen that the proposed method has satisfactory performance for identifying the fault type and phase occurring at different fault inception angles.

Table 2.23 The influence of fault inception angle on the fault phase identification results.

		Fault phase selection factors (FPSFs)			
Fault inception angle	**Fault type**	S_1	S_2	S_3	**Selection result**
0°	AG	1956.5	0.0185	0.0177	AG
	BC	0.0002	0.9994	1.0006	BC
	BCG	0.0118	0.8291	1.1826	BCG
	ABC	0.5719	0.5708	0.5723	ABC
45°	AG	1873.6	0.0192	0.0184	AG
	BC	0.0002	0.9992	1.0008	BC
	BCG	0.0134	0.7523	1.2995	BCG
	ABC	0.5745	0.5763	0.5756	ABC
90°	AG	1905.2	0.0197	0.0188	AG
	BC	0.0002	0.9992	1.0008	BC
	BCG	0.0114	0.6510	1.5142	BCG
	ABC	0.5747	0.5752	0.5763	ABC
135°	AG	1808.2	0.0223	0.0208	AG
	BC	0.0002	0.9992	1.0008	BC
	BCG	0.0104	0.8167	1.2071	BCG
	ABC	0.5751	0.5751	0.5785	ABC
180°	AG	1892.6	0.0234	0.0219	AG
	BC	0.0003	0.9987	1.0013	BC
	BCG	0.0155	0.9438	1.0299	BCG
	ABC	0.5790	0.5781	0.5741	ABC

C. *Simulation of Different Fault Locations*

The fault phase selector should be able to accurately distinguish the fault phase when a fault occurs at any location in the protection area. In order to evaluate the proposed method, a series of faults of different types with 300 Ω fault resistance were simulated at several locations along the transmission line. Table 2.24 shows the values of the FPSFs at a cycle after the fault phase selector starts and the selection results. It can be seen that, for faults at different locations along the line, the fault type and phase can be identified reliably and with a relatively high sensitivity using the proposed selection method.

D. *Simulation of a Developing Fault*

The post-fault power system is under abnormal operating conditions, where it is easy for a developing fault to occur. If the phase selector cannot identify the fault phases quickly and accurately, malfunction of the protection may result.

Table 2.24 The influence of fault positions on the fault phase identification results.

Fault location	Fault type	Fault phase selection factors (FPSFs)			Selection result
		S_1	S_2	S_3	
d1	AG	3834.9	0.0040	0.0037	AG
	BC	0.0000	1.0000	1.0000	BC
	BCG	0.0025	0.6616	1.0416	BCG
	ABC	0.5799	0.5657	0.5635	ABC
d2	AG	2725.3	0.0068	0.0065	AG
	BC	0.0000	1.0000	1.0000	BC
	BCG	0.0032	0.6583	1.5219	BCG
	ABC	0.5780	0.5676	0.5685	ABC
d4	AG	1217.0	0.0861	0.0868	AG
	BC	0.0002	1.0000	1.0000	BC
	BCG	0.0396	0.7701	1.2555	BCG
	ABC	0.5753	0.5773	0.5794	ABC
d5	AG	586.10	0.3193	0.3206	AG
	BC	0.0003	1.0000	1.0000	BC
	BCG	0.2264	1.0626	0.6907	BCG
	ABC	0.5774	0.5782	0.5750	ABC

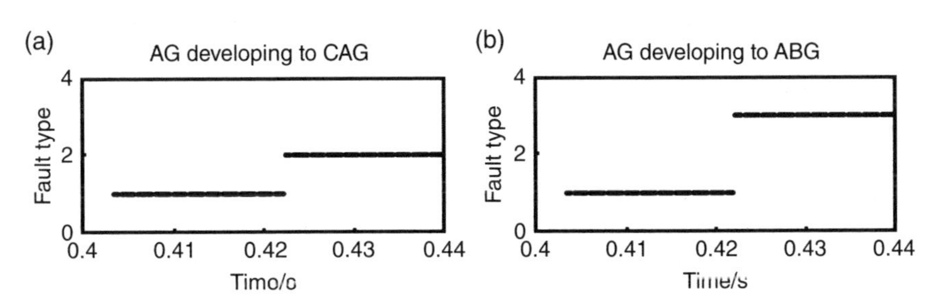

Figure 2.117 Results of fault phase identification with a developing fault.

Therefore, research on a phase selector applicable to developing faults is a key issue for high-voltage line protection. Take phase-A-to-ground fault, for example; the applicability of the proposed method to developing faults is verified.

When $t = 0.40$ s, a phase-A-to-ground fault occurs at the middle point of the line. After a cycle, the fault develops into a phase-CA-to-ground fault and a phase-AB-to-ground fault. Selection results of these developing faults using the proposed method are as shown in Figure 2.117. The fault type codes are shown in Table 2.25.

Table 2.25 Fault type codes.

Numbering	Selection result
0	The proposed method does not start
1	A-to-ground fault (AG)
2	CA-to-ground fault (CAG)
3	AB-to-ground fault (ABG)

As can be seen from Figure 2.117, 4 ms after the fault occurs, a phase-A-to-ground fault can be accurately determined according to Criterion A. After a cycle, the single-phase fault develops into a phase-CA-to-ground fault and a phase-AB-to-ground fault, and this developing fault is correctly identified using Criterion C within 5 ms.

E. Simulation of the Weak-Infeed Side

The weak-infeed system often has relatively low values of fault current availability. Under the weak-infeed condition, the fault phase selector may not have enough sensitivity for fault detection. By reducing the capacity of the generator on side M by 100 times, the generator on side N can be equivalent to an infinite bus while the generator on side M becomes the weak-infeed system, and the performance of the proposed method is studied. When a fault occurs at different locations along the line via different fault resistances, the values of the FPSFs at a cycle after the fault, the zero sequence current and the phase selection results are shown in Table 2.26.

It can be seen from Table 2.26 that it is convenient for the proposed method to identify different kinds of faults sensitively on the weak-infeed side within one cycle of time. Meanwhile, it is unaffected by the fault location and fault resistance. Therefore, the adaptability of the proposed method in phase selection on the weak-infeed system is proved.

F. Simulation of the Effects of TCSC and UPFC

TCSC and UPFC make the control of large-scale interconnected power grids faster, more frequent, continuous and flexible, thus improving the power grids' stability level. However, the application of these techniques also affects the relay protection. For example, the distance protection may refuse to operate in the case of an in-zone fault and malfunction in the case of a backward fault. To examine the influence of TCSC and UPFC on the proposed method, simulation tests were carried out based on the system model shown in Figure 2.112.

Suppose that TCSC is installed at bus N and it compensates according to 30% of the line reactance. When a fault occurs at the midpoint of line MN, the phase selection results are as shown in Figure 2.118. When UPFC is installed on line MN, if there is a fault at the midpoint of the line, the phase selection results are as shown in Figure 2.119. It can be seen from Figures 2.118 and 2.119 that the proposed method is not affected by the installation of TCSC and UPFC. It is still able to identify the fault type and fault phase quickly and accurately, with high sensitivity.

Table 2.26 Results of fault phase identification at the weak-infeed side.

Fault location	Fault type	Fault resistance (Ω)	Fault phase selection factors (FPSFs)			Zero sequence current i_0/A	Selection result
			S_1	S_2	S_3		
d1	AG	0	25110	0.3358	0.3358	—	AG
		300	3307.1	0.3979	0.3976	—	AG
	BC	0	0.0000	1.0001	0.9999	—	BC
		300	0.0000	1.0002	0.9999	—	BC
	BCG	0	0.2406	0.9002	1.6515	832.74	BCG
		300	0.2378	0.9389	1.3584	814.84	BCG
	ABC	0	0.5662	0.5721	0.5718	—	ABC
		300	0.5589	0.5694	0.5758	—	ABC
d3	AG	0	3242.1	0.3499	0.3499	—	AG
		300	1810.4	0.3110	0.3106	—	AG
	BC	0	0.0000	1.0001	0.9999	—	BC
		300	0.0001	1.0003	0.9997	—	BC
	BCG	0	0.0267	0.9058	1.0975	395.23	BCG
		300	0.1088	0.5206	1.7229	299.02	BCG
	ABC	0	0.5781	0.5758	0.5698	—	ABC
		300	0.5764	0.5782	0.5771	—	ABC
d5	AG	0	8046.8	0.3025	0.3026	—	AG
		300	834.89	0.2941	0.2951	—	AG
	BC	0	0.0000	1.0001	0.9999	—	BC
		300	0.0003	1.0011	0.9989	—	BC
	BCG	0	0.1571	0.8901	0.9224	186.58	BCG
		300	0.2040	0.8698	0.8415	105.91	BCG
	ABC	0	0.5681	0.5717	0.5698	—	ABC
		300	0.5701	0.5728	0.5766	—	ABC

G. Simulation of Double-Circuit Lines

A high-voltage parallel transmission line has the advantages of high transmission capacity as well as great economic benefits. It is able to improve the security and stability of the power system and has been widely applied in recent years. In order to minimize the power outage of double-circuit lines on the same tower when a fault occurs, it is required that the fault phase selector identify the fault phases correctly. To examine the suitability of the proposed method for parallel lines, an extra transmission line is added between bus M and bus N on the basis of Figure 2.112, so that a parallel line results (see Figure 2.120).

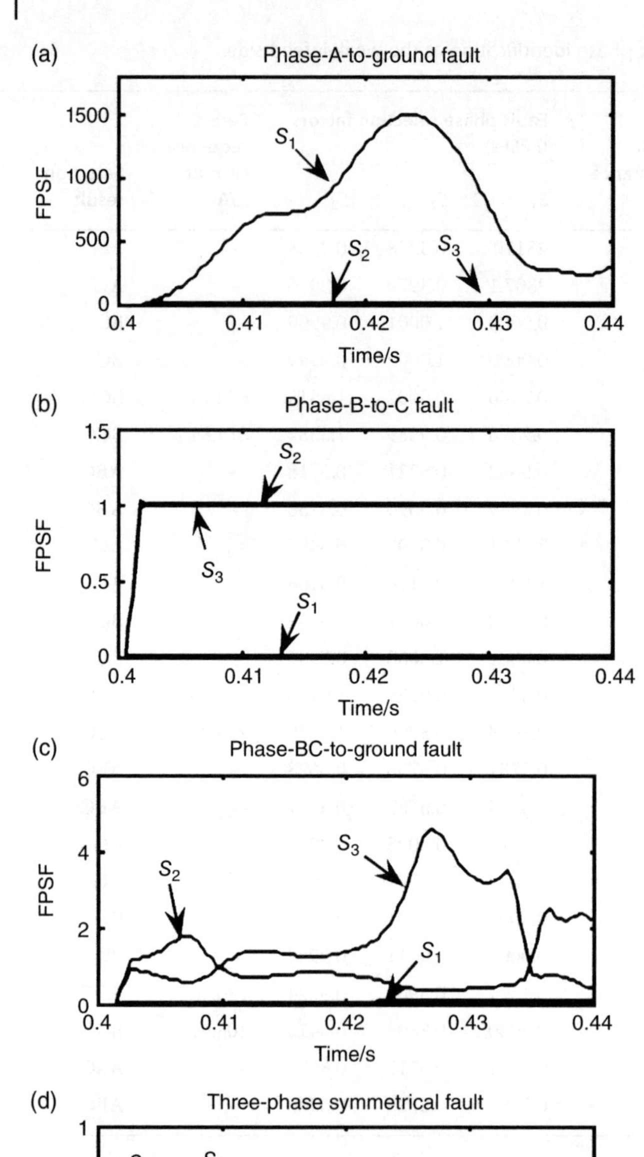

Figure 2.118 FPSFs when TCSC is installed at bus *N*.

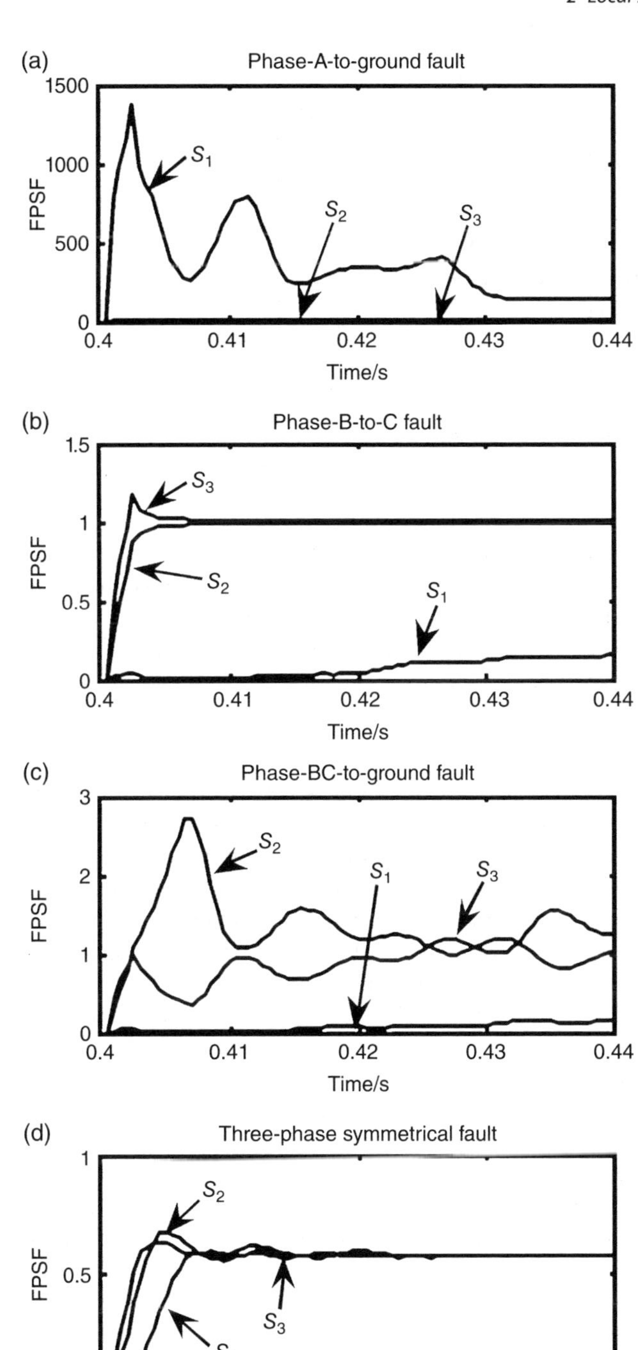

Figure 2.119 FPSFs when UPFC is installed on line *MN*.

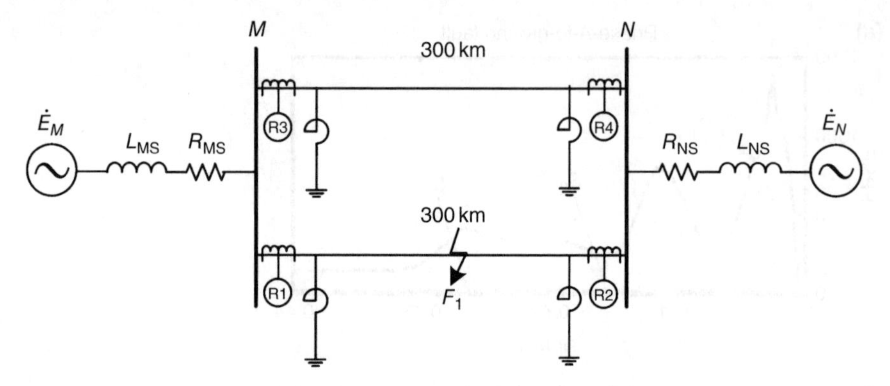

Figure 2.120 Double high-voltage transmission line.

When different types of faults occur at F_1 (which is 150 km from bus M) via a 300 Ω fault resistance, the phase selection results at relay R1 are as shown in Figure 2.121. It can be seen that, in the case of a double-circuit line, the proposed method is highly reliable in identifying different types of faults and the fault phases, unaffected by the fault resistance.

H. Simulation of an AC–DC Hybrid System

High-voltage, direct current (HVDC) transmission is widely applied due to its obvious advantage in long-distance, large-capacity power transmission and large power grid interconnection. Compared with a traditional AC power system, when a fault occurs, the AC–DC hybrid system will exhibit many new electrical characteristics, which will have an unfavourable influence on the operating behaviour of AC system relay protection. Simulation experiments have been carried out in the laboratory based on the CIGRÉ benchmark HVDC system (see Figure 2.122) established in the real time digital simulator (RTDS). The system parameters can be found in [14].

When a metallic phase-A-to-ground fault and a phase-A-to-ground fault via fault resistance (300 Ω) occur at the midpoint of line PQ, the instantaneous phase currents at the relaying point of bus P are as shown in Figure 2.123, and the corresponding phase selection results are as shown in Figure 2.124. The instantaneous phase currents at the relaying point of bus Q are as shown in Figure 2.125, and the corresponding phase selection results are as shown in Figure 2.126. It can be seen from Figures 2.123–2.126 that when the fault occurs on the AC line at the rectifier side, the proposed method is highly reliable in identifying the fault type and fault phase, unaffected by the fault resistance.

In the case of an inverter-side AC line fault, suppose that a metallic phase-A-to-ground fault and a phase-A-to-ground fault via fault resistance (300 Ω) occur at the midpoint of line MN; the instantaneous phase currents at the relaying point of bus M are as shown in Figure 2.127, and the corresponding phase selection results are as shown in Figure 2.128. The instantaneous phase currents at the relaying point of bus N are as shown in Figure 2.129, and the corresponding phase

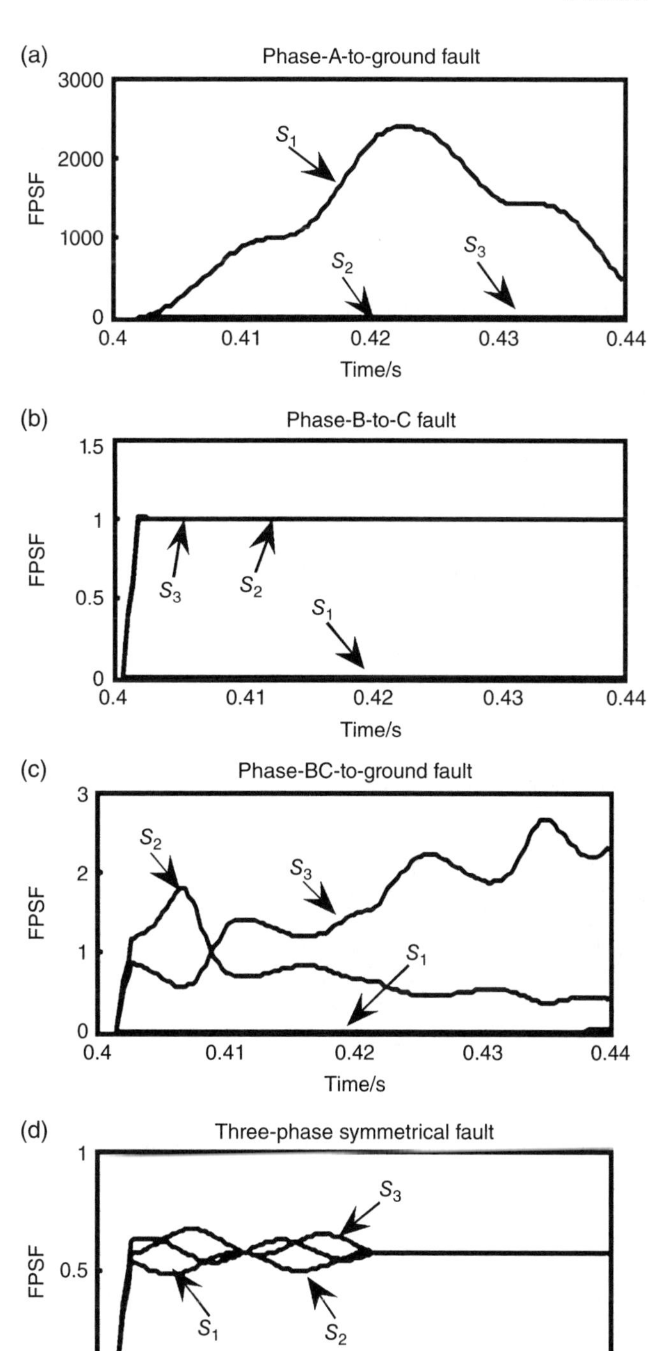

Figure 2.121 FPSFs when different types of faults occur at F_1.

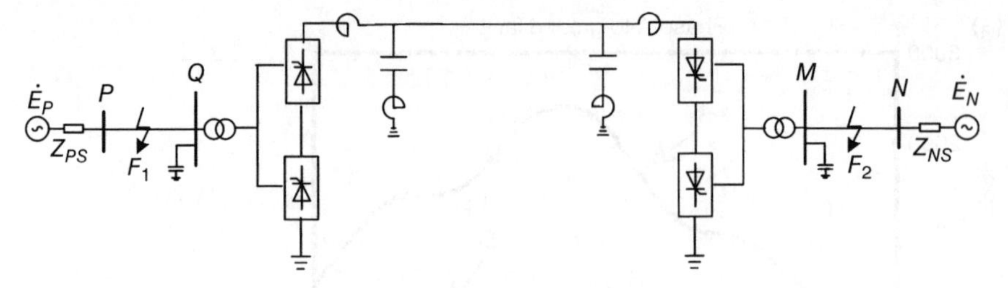

Figure 2.122 Single-line diagram of the CIGRÉ benchmark HVDC system.

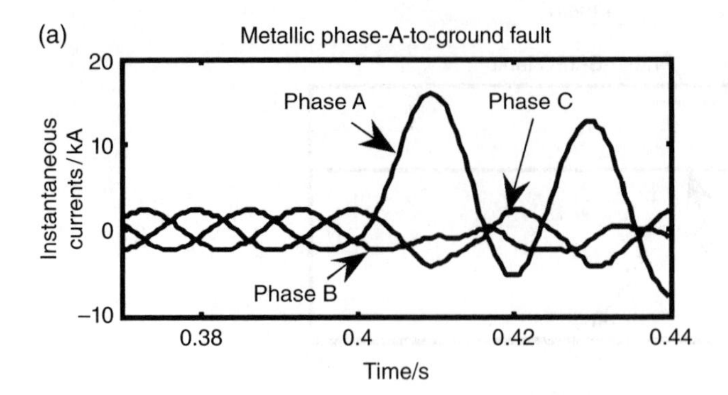

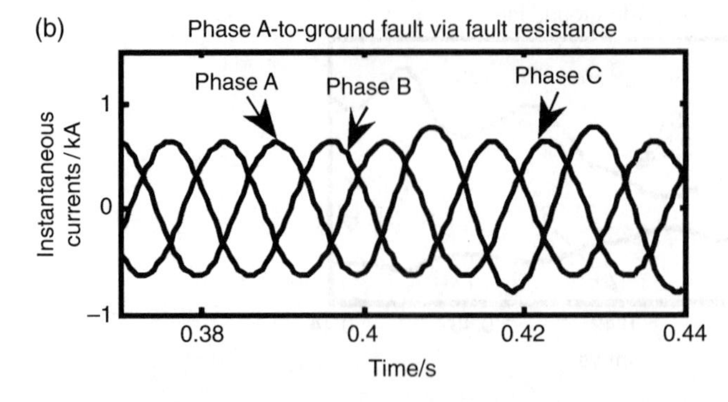

Figure 2.123 Phase currents at the relaying point of bus *P* when line *PQ* is in fault.

selection results are as shown in Figure 2.130. It can be seen from Figures 2.127–2.130 that, when a metallic fault occurs on the inverter-side AC line, both line ends are able to identify the fault type and fault phase accurately. However, when a fault via fault resistance occurs, due to the influence of the DC system, the fault type is mistakenly identified as a three-phase fault at bus *M*, while bus *N* fails to identify the fault phase.

According to the above simulation analysis, in the AC–DC hybrid system, the proposed method is able to identify faults on the rectifier-side AC line accurately. However, false identification may happen with the proposed method for faults on the inverter-side AC line.

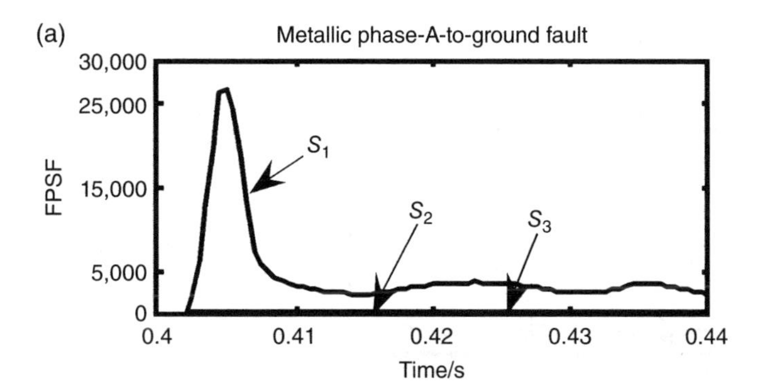

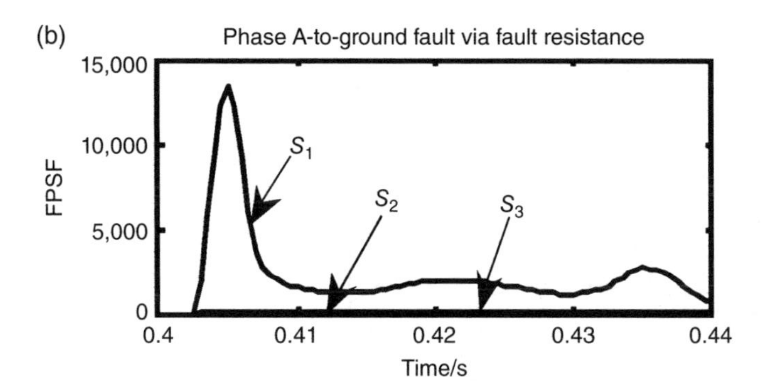

Figure 2.124 FPSFs at the relaying point of bus *P* when line *PQ* is in fault.

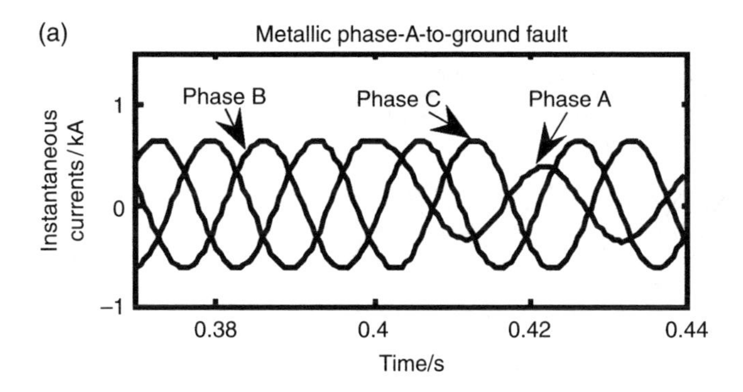

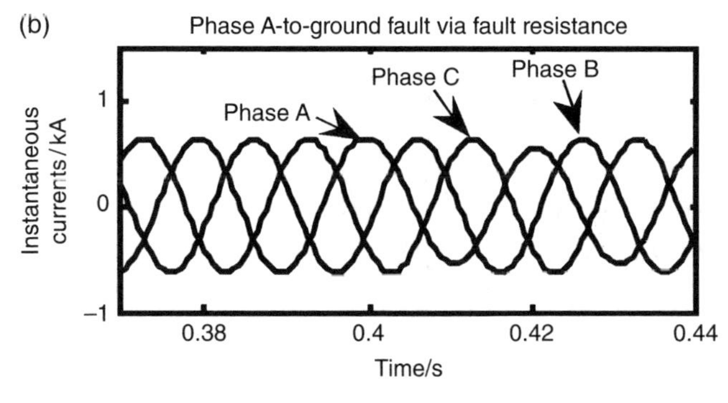

Figure 2.125 Phase currents at the relaying point of bus *Q* when line *PQ* is in fault.

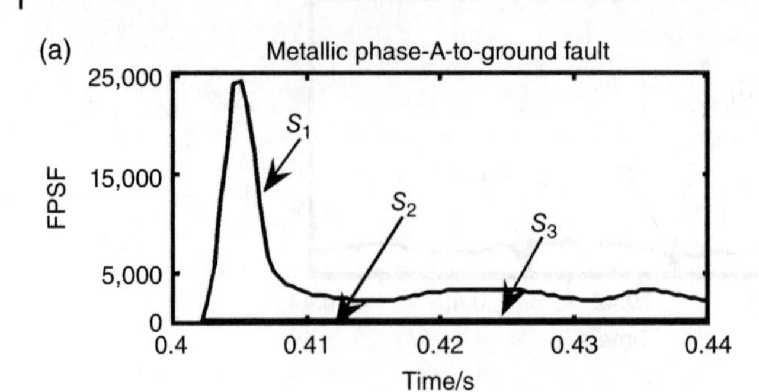

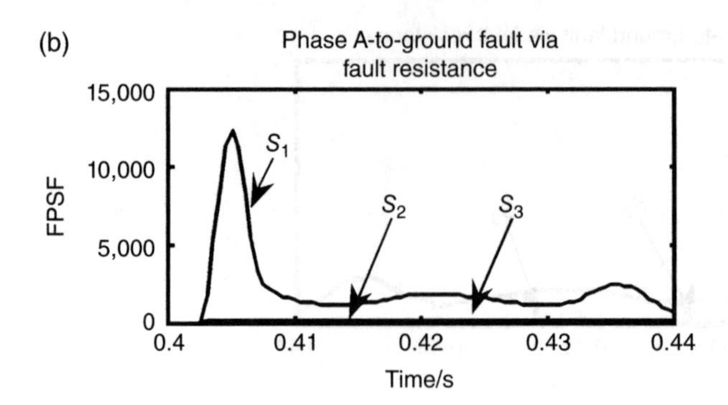

Figure 2.126 FPSFs at the relaying point of bus Q when line PQ is in fault.

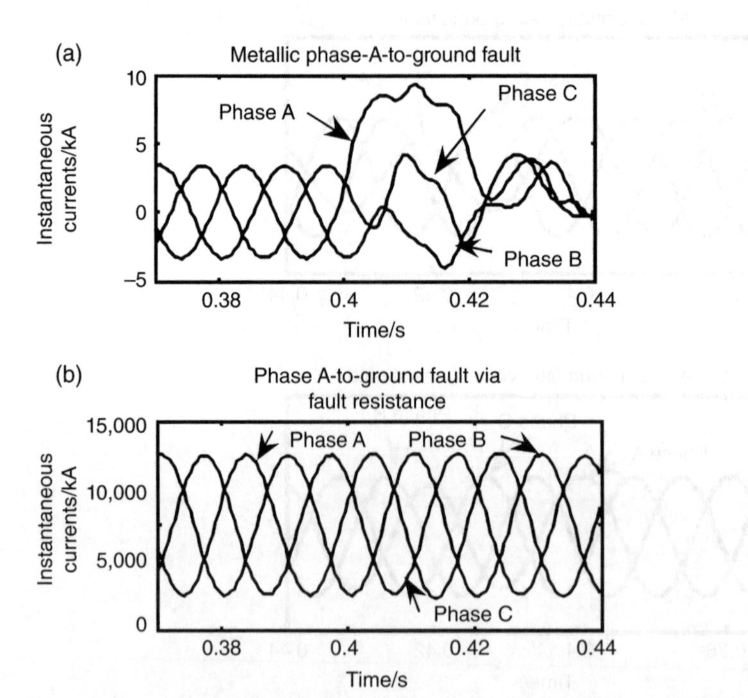

Figure 2.127 Phase currents at the relaying point of bus M when line MN is in fault.

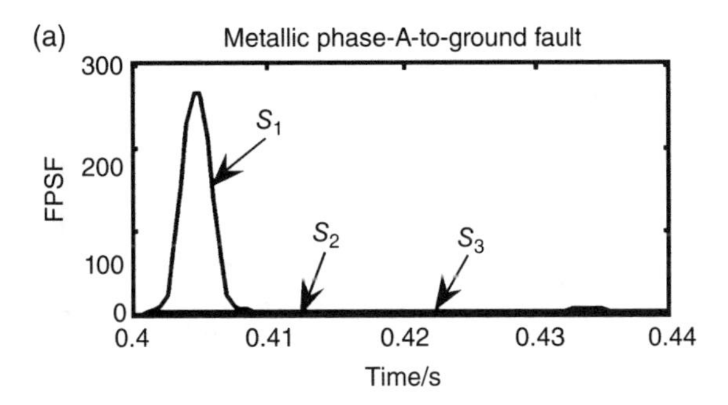

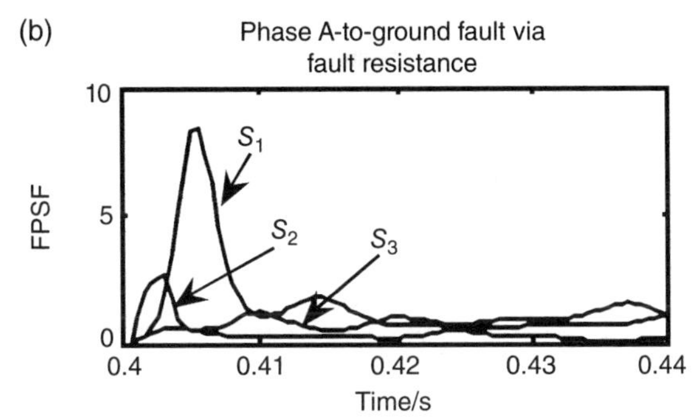

Figure 2.128 FPSFs at the relaying point of bus M when line MN is in fault.

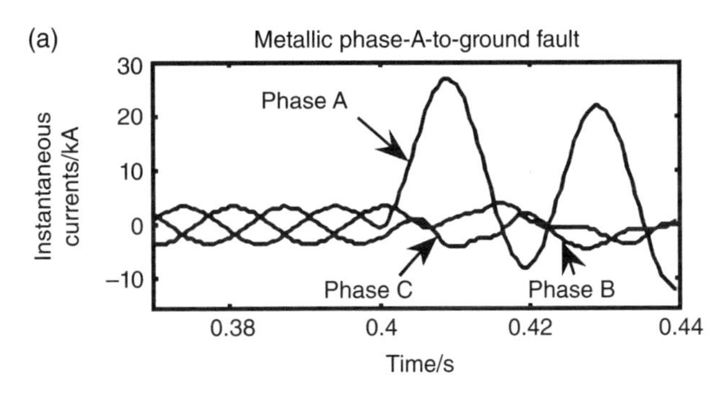

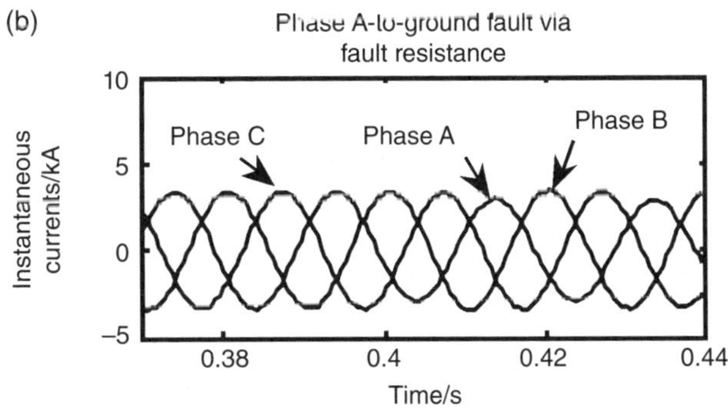

Figure 2.129 Phase currents at the relaying point of bus N when line MN is in fault.

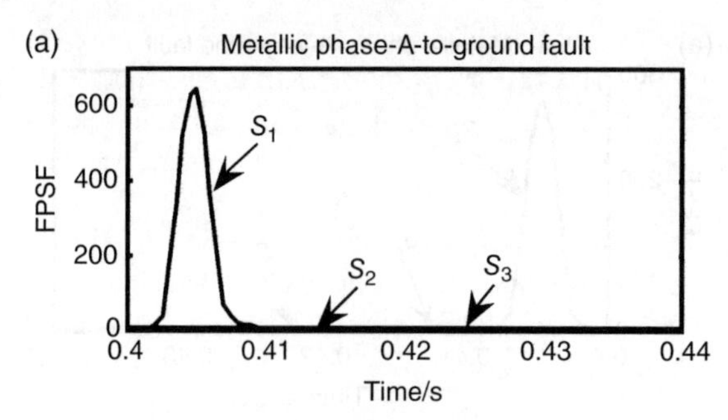

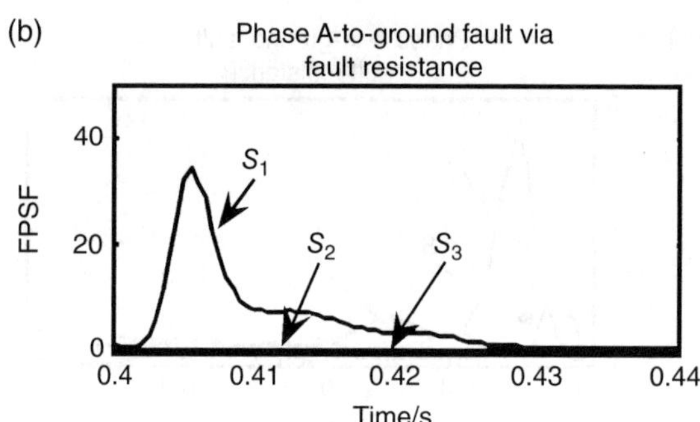

Figure 2.130 FPSFs at the relaying point of bus *N* when line *MN* is in fault.

I. Comparisons with Other Fault Phase Selection Methods

Digital high-voltage line protection in China mainly applies the phase current difference mutation phase selector after the protection starts. When a fault occurs in the system, the amplitude characteristic of the phase current difference variation is used to select the fault phases.

In the simulation model shown in Figure 2.112, suppose a phase-A-to-ground fault (AG) occurs at the midpoint of line MN at $t = 0.40$ s, and develops into a phase CA-to-ground fault (CAG) and a phase AB-to-ground fault (ABG) separately after one cycle. The variation curves of phase current difference mutations $|\dot{I}_{\mathrm{Mab}}|$, $|\dot{I}_{\mathrm{Mbc}}|$ and $|\dot{I}_{\mathrm{Mca}}|$ are as shown in Figure 2.131. It can be seen that it takes 15 ms for the traditional phase current difference mutation phase selection scheme to identify the developing fault. However, according to Figure 2.117, it takes only 5 ms for the proposed phase selection scheme to identify the fault, with high sensitivity and accuracy.

Suppose that the generator on the *M* side is the weak-infeed side of the system. When a three-phase fault occurs at different locations on line MN, the phase selection results of the traditional phase current difference mutation method are shown in Table 2.27. It can be seen that false selection may occur with the traditional phase current difference mutation method. For example, when a fault occurs at d3, it is falsely identified as a phase-A-to-B fault, and when a fault occurs at d5, it is falsely identified as a phase-B-to-C fault. On the other hand, according to Table 2.26, in phase selection on the weak-infeed side, the proposed method is able to identify different types of line faults correctly, unaffected by the fault location and the fault resistance.

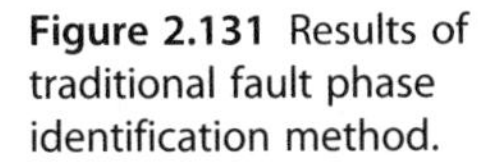
Figure 2.131 Results of traditional fault phase identification method.

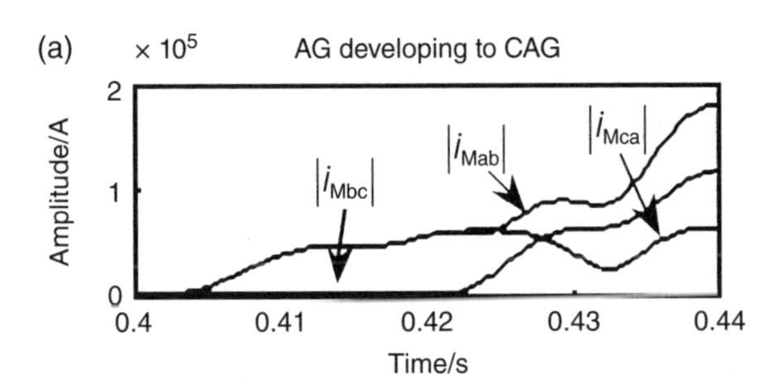

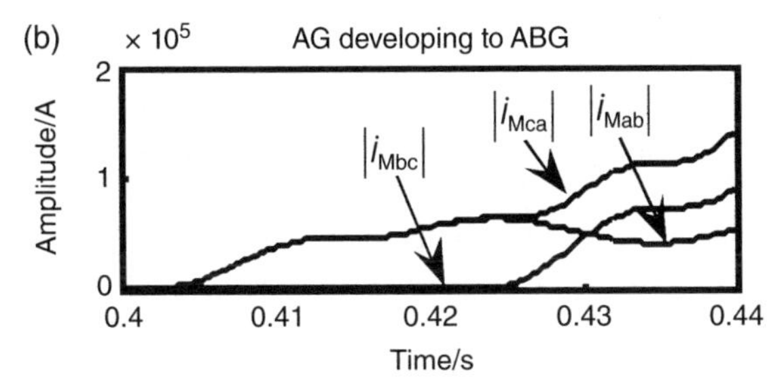

Table 2.27 Simulation results with different fault locations when a three-phase fault occurs at line MN.

Fault location	$\lvert i_{Mab} \rvert$/A	$\lvert i_{Mbc} \rvert$/A	$\lvert i_{Mca} \rvert$/A
d1	4202.9	3771.1	4146.0
d3	4129.5	3353.7	3728.1
d5	2956.7	3740.0	3083.4

Table 2.28 Comparison with other fault phase selection methods.

Conditions	Number	Methods	Failed fault phase selection	Accuracy
$t_A = 0^0, d = 100$ km, R_g is from 10 Ω to 300 Ω (step by 10 Ω)	30	WE	26	13%
		WMM	12	60%
		FPSFs	0	100%

t_A = Fault inception angle of phase A; d = Distance from fault point to bus M; R_g = fault resistance.

Further tests have been conducted to perform comparisons between FPSFs and other fault phase selection methods, such as the wavelet energy (WE) and wavelet modulus maxima (WMM), which are both based on wavelet transforms (WT). The tests under different conditions show that the worst case for fault phase selection is the low-transient condition. Some of the worst-case classification results are shown in Table 2.28. It can be seen that, due to the very low transient magnitude, the accuracies of the fault phase selections based on WE and WMM are 13% and 60%, respectively. However, the proposed method is unaffected by fault inception time and fault resistance, and the accuracy is 100%.

2.4 Summary

The problems of conventional transformer protection and transmission line protection are analysed in this chapter. On this basis, new schemes of transformer and line protection widely different from traditional ones are introduced, making up for the disadvantages of conventional protection and laying a solid foundation for hierarchical protection. With regard to transformer protection, new transformer main protection schemes are constructed with the help of voltage and current information, significantly enhancing the protection sensitivity efficiently when no-load switching at faults and avoiding malfunction in the case of TA saturation due to external faults. As for transmission line protection, new distance protection schemes are constructed, avoiding the impact of transition resistance, line overload and false phase selection and realizing one-terminal fault location double-circuit transmission faults.

References

1 Phadke, A. G. and Thorp, J. S. (1983) A new computer-based flux-restrained current-differential relay for power transformer protection, *IEEE Trans. Power Appl. Syst.*, **102** (11), 3624–3629.
2 Rahman, M. A. and Jeyasurya, B. (1988) A state-of-the-art review of transformer protection algorithms, *IEEE Trans. Power Del.*, **3** (2), 534–544.
3 Ma, J., Wang, Z. P., Xu, Y. and Wu, J. (2006) A novel adaptive algorithm to identify inrush using mathematical morphology, *Proc. IEEE 2006 Power System Conference and Exposition*, 1020–1028.
4 Ma, J., Wang, Z., Zheng, S. *et al.* (2013) A new algorithm to discriminate internal fault current and inrush current utilizing feature of fundamental current, *Canadian Journal of Electrical & Computer Engineering*, **36** (36), 26–31.
5 Ma, J., Wang, Z., Yang, Q. *et al.* (2011) Identifying transformer inrush current based on normalized grille curve, *IEEE Transactions on Power Delivery*, **26** (2), 588–595.
6 Ma, J. and Wang, Z. (2007) A novel algorithm for discrimination between inrush currents and internal faults based on equivalent instantaneous leakage inductance. *Power Engineering Society General Meeting, IEEE*, 1–8.
7 Ma, J., Wang, Z., Yang, Q. *et al.* (2010) A two terminal network-based method for discrimination between internal faults and inrush currents, *Power Delivery IEEE Transactions*, **25** (3), 1599–1605.
8 Matheron, G. (1974) *Random sets and integral geometry*, John Wiley & Sons, Inc., New York.
9 Serra, J. (1982) *Image Analysis and Mathematical Morphology*, Academic Press, New York.
10 Zhang, J. (2004) *Principles of Protective Relaying Based on Microcomputers*, Water Conservancy and Hydropower Publishing House, Beijing, China.

11 Ye, D. (1994) *Electro-mechanics,*Tianjin Science and Technology Publishing House, Tianjin, China.
12 Ma, J., Ma, W., Qiu, Y. *et al.* (2015) An adaptive distance protection scheme based on the voltage drop equation. *IEEE Transactions on Power Delivery,* **30** (4), 1–1.
13 Ma, J., Shi, Y. X., Ma, W. and Wang, Z. P. (2015) Location method for interline and grounded faults of double-circuit transmission lines based on distributed parameters, *IEEE Transactions on Power Delivery,* **30** (3), 1307–1316.
14 Ma, J., Ma, W., Zhu, X. S. *et al.* (2015) A novel fault phase selection scheme utilizing fault phase selection factors, *Electric Power Components and Systems,* **43** (5), 491–507.
15 Stanley, H. H. and Arun, G. (2008) *Power system relaying,* Third edition. John Wiley & Sons Inc., New York.
16 Mahari, A. and Seyedi, H. (2015) High impedance fault protection in transmission lines using a WPT-based algorithm, *International Journal of Electrical Power and Energy Systems,* **67**.
17 Xu, Z. Y., Xu, G., Ran, L. *et al.* (2010) A new fault-impedance algorithm for distance relaying on a transmission line, *IEEE Transactions on Power Delivery,* **25** (3), 1384–1392.
18 State Grid Electric Power Dispatching and Communication Center (2012) *Relay protection training materials of State Grid Corporation of China,* China Electric Power Press, Beijing.
19 Apostolopoulos, C. A. and Korres, G. N. (2011) A novel fault-location algorithm for double-circuit transmission lines without utilizing line parameters, *IEEE Trans. Power Del.,* **26** (3), 1467–1478.
20 Phadke, A. G. and Thorp, J. S. (2009) *Computer Relaying for Power Systems,* Second edition., John Wiley & Sons, Inc., New York.
21 Novosel, D., Bartok, G., Henneberg, G. *et al.* (2010) IEEE PSRC report on performance of relaying during wide-area stressed conditions, *IEEE Trans Power Del.,* **25** (1), 3–16.

3

Local Area Protection for Renewable Energy

3.1 Introduction

In recent years, intermittent renewable energy sources – wind power being the most obvious – have been developing rapidly, the proportion of which in the power grid is getting bigger and bigger. Thus, traditional fault analysis and calculation models and relay protection technology based on synchronous generators is facing several new challenges, one of which is that the fault characteristics of renewable energy sources are affected by the coupling effect of electromagnetic transient characteristics and the power electronic converter control strategy. Currently, knowledge of the fault characteristics of renewable energy sources is not comprehensive and thorough, and cannot be applied to power grid relay protection. Another challenge is that the short circuit calculation model of renewable energy sources is not clear, thus the adaptability analysis of existing protection in the power grid it is connected to, the protection configuration and setting calculation, and research on new protection principles all lack theoretical support.

In relevant research in China and elsewhere, usually simulation methods are used to reveal the fault characteristics of renewable energy sources and to analyse the adaptability of power grid relay protection. However, the essential causes of the variation in fault characteristics of renewable energy sources are not known, and there is, as yet, no short circuit calculation model of renewable energy established from the theoretical perspective, neither is there any systematic and quantitative evaluation of the existing power grid protection configuration principle, setting calculation method and protection principle and algorithm. Therefore, in a real power grid, the impact of renewable energy is usually neglected, i.e. renewable energy is treated as a load, which will bring hidden risk to the existing power grid protection system. Take the renewable energy plant with centralized integration to the power grid as an example. The setting calculation of protection does not take into account the transient regulation and weak feed characteristics of a renewable energy converter, which will result in problems such as collector line protection malfunction and low sensitivity of outgoing transmission line protection. In addition, for a distributed power integration system which is an effective complement to a centralized grid connection, the protection configuration principle is too simple, and the impact of renewable energy fault current on the setting calculation and equipment safety is not considered.

Hierarchical Protection for Smart Grids, First Edition. Jing Ma and Zengping Wang.

To deal with these problems, this chapter analyses the electromagnetic transient process and converter transient control process of renewable energy when crowbar protection is put into operation and when it is not, reveals the current variation characteristics in the whole process of renewable energy fault, and theoretically derives an accurate calculation formula for renewable energy short circuit current under different types of fault. On this basis, protection schemes for the collector line and outgoing transmission line of centralized renewable energy plant are introduced, which could improve the selectivity and sensitivity of protection for the centralized integration of renewable energy into the power grid. Also, the effect that the integration of distributed renewable energy has on distribution network protection is introduced, and a protection scheme for a distribution network with distributed power sources is proposed, which ensures that protection may operate reliably when large amounts of distributed renewable sources are integrated. In order to further improve the adaptability of protection schemes to the dynamic adjustment of network structure, three islanding detection methods are put forward which could quickly identify the grid connection status of distributed renewable sources. The local renewable energy protection studied in this chapter is of great theoretical value and practical significance in ensuring the safe and reliable operation of power grids with large-scale renewable sources.

3.2 Fault Transient Characteristics of Renewable Energy Sources

In order to study the protection of renewable energy sources integrated into a power grid, the fault transient characteristics of renewable energy sources need to be analysed in depth. According to the relative power of the converter used, renewable energy sources can be divided into the partial power conversion type (e.g. doubly fed induction generator) and the full power conversion type (e.g. permanent magnet direct drive wind turbine and photovoltaic cell components). When a fault occurs, the doubly fed induction generator (DFIG) is not only affected by the coupling effects of stator/rotor electromagnetic transient characteristics and the inverter control strategy, but is also affected by the low voltage ride through (LVRT) characteristics, and thus the fault transient characteristics of the DFIG are especially complicated. Therefore, Section 3.2 mainly analyses the fault characteristics of the DFIG considering the LVRT characteristics when different types of fault occur in the grid, laying the foundation for the study of the protection of renewable energy integrated into the grid [1]. In Section 3.2.1, the basic mathematical model of the DFIG and the LVRT characteristics are introduced. From there, Section 3.2.2 studies the fault characteristics of the DFIG without considering crowbar protection and derives the accurate expressions of the DFIG short circuit currents considering only grid-side converter control and considering both grid-side and rotor-side converter control. In Section 3.2.3, the fault characteristics of the DFIG considering crowbar protection are introduced. By analysing the effects of crowbar resistance and rotor speed on the transient flux, the instantaneous expressions of the DFIG short circuit currents are derived.

3.2.1 Mathematical Model and LVRT Characteristics of the DFIG

3.2.1.1 Mathematical Model of the DFIG

The equivalent circuit of the DFIG system is shown in Figure 3.1, where RSC and GSC represent the rotor-side converter and grid-side converter, respectively. Currently, the AC–DC–AC converter usually uses the pulse width modulation (PWM) control mode to provide the excitation source [2].

The mathematical model of the DFIG under the synchronous rotating coordinate system can be expressed as:

$$\boldsymbol{u}_{\mathbf{s}} = R_{\mathrm{s}}\boldsymbol{i}_{\mathbf{s}} + p\boldsymbol{\psi}_{\mathbf{s}} + \mathrm{j}\omega_1\boldsymbol{\psi}_{\mathbf{s}} \tag{3.1}$$

$$\boldsymbol{u}_{\mathbf{r}} = R_{\mathrm{r}}\boldsymbol{i}_{\mathbf{r}} + p\boldsymbol{\psi}_{\mathbf{r}} + \mathrm{j}\omega\boldsymbol{\psi}_{\mathbf{r}} \tag{3.2}$$

$$\boldsymbol{\psi}_{\mathbf{s}} = L_{\mathrm{s}}\boldsymbol{i}_{\mathbf{s}} + L_{\mathrm{m}}\boldsymbol{i}_{\mathbf{r}} \tag{3.3}$$

$$\boldsymbol{\psi}_{\mathbf{r}} = L_{\mathrm{m}}\boldsymbol{i}_{\mathbf{s}} + L_{\mathrm{r}}\boldsymbol{i}_{\mathbf{r}} \tag{3.4}$$

where $\boldsymbol{u}_{\mathbf{s}}$, $\boldsymbol{u}_{\mathbf{r}}$, $\boldsymbol{i}_{\mathbf{s}}$, $\boldsymbol{i}_{\mathbf{r}}$, $\boldsymbol{\psi}_{\mathbf{s}}$, $\boldsymbol{\psi}_{\mathbf{r}}$ are the voltage, current and flux vectors of the stator and rotor, respectively. ω_1 is the synchronous angular speed. $\omega = \omega_1 - \omega_{\mathrm{r}}$ is the slip angular speed, i.e. the difference between the synchronous angular speed and the rotor angular speed. R_{s} and R_{r} are the stator and rotor resistances of the DFIG, respectively. L_{m}, L_{s} and L_{r} are respectively the equivalent excitation inductance, the stator inductance and the rotor inductance, and $L_{\mathrm{s}} = L_{\mathrm{m}} + L_{\mathrm{ls}}$, $L_{\mathrm{r}} = L_{\mathrm{m}} + L_{\mathrm{lr}}$, where L_{ls} and L_{lr} are the DFIG stator and rotor leakage reactances.

3.2.1.2 LVRT Characteristics of the DFIG

As the capacity of wind power generation steadily increases, for a conventional wind power system, when the grid voltage drops to a certain value, splitting of generators is required. However, generator splitting has a big impact on the power grid, which could endanger the safe and reliable operation of the power grid.

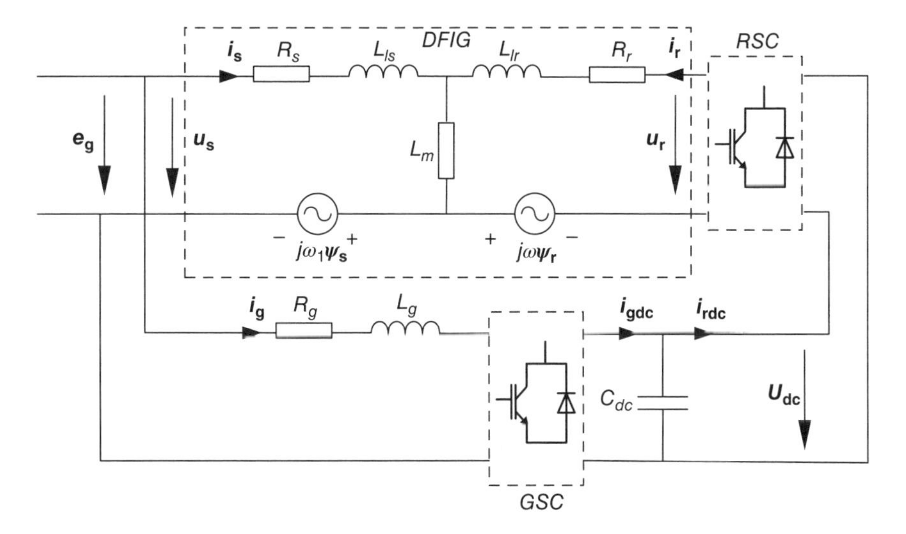

Figure 3.1 Equivalent circuit of the DFIG system.

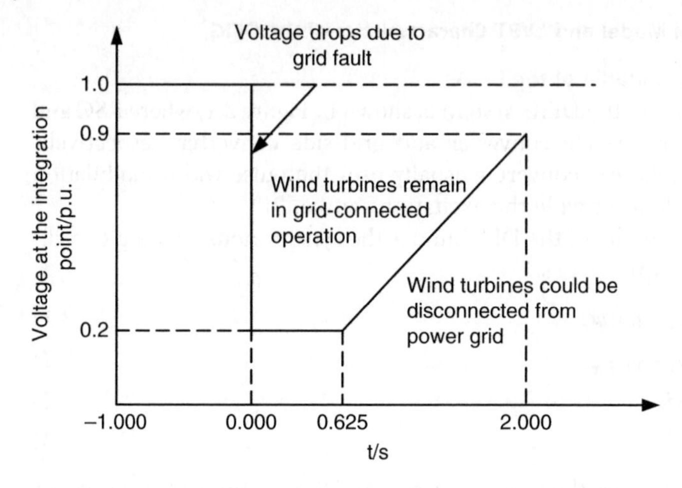

Figure 3.2 Low-voltage operation capability curve of wind turbines.

Therefore, power companies and power grid operators have put forward guidelines for the grid connection of wind turbines, explicitly requiring that wind turbines have the capability of LVRT, i.e. when grid voltage drops to a certain value, wind turbines can remain in grid-connected operation. The 'Technical requirements on the integration of wind farms to the power system' issued in China includes detailed requirements on the LVRT capability of wind farms in China, as shown in Figure 3.2.

Currently, LVRT operation of a DFIG can be realized by adding a hardware circuit. The crowbar circuit is a simple and effective method to realize LVRT operation. The main idea of a crowbar circuit is that, when a DFIG rotor short circuit current or DC bus voltage increases to a preset threshold value, the signal to turn off all power electronic switching devices in the rotor-side converter is issued, so that the DFIG rotor over current and stator current during a grid fault can be suppressed, and the DFIG can remain in grid-connected operation. The crowbar circuit has a big influence on the output characteristics of a DFIG short circuit current, thus the fault transient characteristics of a DFIG when a crowbar is put into operation and when it is not need to be analysed in detail.

3.2.2 DFIG Fault Transient Characteristics When Crowbar Protection Is Not Put into Operation

3.2.2.1 DFIG Fault Transient Characteristics Considering Grid-Side Converter Control

A. Symmetrical Short Circuit Fault

Suppose that a three-phase symmetrical short circuit fault occurs in the grid at $t = t_1$; considering the amplitude drop and phase jump of the DFIG voltage, the DFIG terminal voltage under the synchronous rotating coordinate system can be expressed as:

$$\begin{cases} u_{s1} = U_{sm} e^{j\varphi_1}, t < t_1 \\ u_{s2} = kU_{sm} e^{j\varphi_2}, t \geq t_1 \end{cases} \quad (3.5)$$

where u_{s1} and u_{s2} are the pre- and post-fault DFIG terminal voltages. U_{sm} and φ_1 are the amplitude and initial phase angle of the DFIG terminal voltage when the grid is in steady-state operation. k and φ_2 are the amplitude drop rate and the phase angle of the DFIG terminal voltage after the grid fault. t_1 is the time when the fault occurs.

At the instant of the grid three-phase fault, the stator flux does not vary instantaneously. Thus, according to Equations (3.1) and (3.3), the post-fault stator flux can be shown to be:

$$\psi_s = \psi_{s0} + \psi_{sf} = \frac{(u_{s1} - u_{s2})}{\frac{R_s}{L_s} + j\omega_1} e^{\tau_1 t_1} e^{-\tau_1 t} + \frac{u_{s2} + \frac{R_s}{L_s} L_m i_r}{\frac{R_s}{L_s} + j\omega_1} \quad (3.6)$$

where $\tau_1 = \frac{R_s}{L_s} + j\omega_1$ is the stator transient flux attenuation time constant. The initial value of the transient DC flux ψ_{s0} induced in the stator winding has to do with the amplitude drop rate and phase jump value of the DFIG terminal voltage, the generator parameters and the time of fault occurrence. The steady-state component of the stator flux ψ_{sf} is affected by the rotor current i_r, the stator resistance R_s and the post-fault terminal voltage u_{s2}.

1. DFIG Terminal Voltage Phase Jump Model in the Case of Grid Three-Phase Fault When a three-phase fault occurs in the grid, the phase jump value of the DFIG terminal voltage u_{s2} directly affects the calculation of the stator flux and the short circuit current. Therefore, it is necessary to establish a DFIG transient model considering the terminal voltage phase jump. Suppose the three-phase fault occurs in the grid at $t = t_1$; the fault analysis model is shown in Figure 3.3, where U is the grid voltage and U_s is the DFIG terminal voltage. $L' = \omega_s (L_s - L_m^2 / L_r)$ is the equivalent impedance of the DFIG. Z_l, Z_t and Z_f are respectively the grid transmission line impedance, the transformer transmission impedance and the transition impedance in grounding faults.

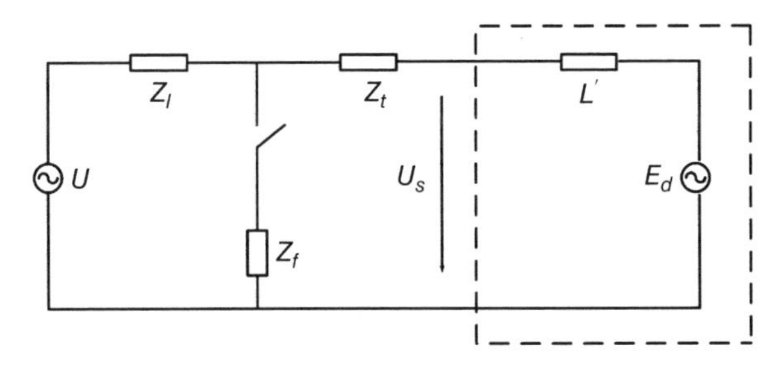

Figure 3.3 Grid fault analysis model.

The 'voltage phase jump' refers to the difference between the phase angle of the DFIG terminal voltage and that of the reference voltage. Before the fault, the grid voltage frequency remains the same. The reference voltage is the pre-fault DFIG terminal voltage. When the fault occurs, the DFIG terminal voltage will deviate from the reference voltage; the phase angle deviation value is called the 'voltage phase jump'. Since the grid-side and rotor-side converters apply the classical vector control strategy, when the voltage phase jump occurs, the directional control of the converters will affect the post-jump phase variation, causing relevant transient processes in the DFIG stator/rotor short circuit currents.

It can be seen from Figure 3.3 that when the grid is in steady-state operation, the DFIG terminal voltage U_{s1} is

$$U_{s1} = \frac{L'}{Z_l + Z_t + L'}U + \frac{Z_l + Z_t}{Z_l + Z_t + L'}E_d \tag{3.7}$$

At the grid fault instant, the DFIG terminal voltage U_{s2} is

$$U_{s2} = \frac{Z_f//(Z_t + L')}{Z_l + Z_f//(Z_t + L')} \cdot \frac{L'}{Z_t + L'}U + \frac{Z_l//Z_f + Z_t}{Z_l//Z_f + Z_t + L'}E_d \tag{3.8}$$

Since the grounding impedance Z_f is relatively small, the stator flux and the induced electromotive force (emf) E_d remain the same at the fault instant; according to Equations (3.7) and (3.8), the variation of the terminal voltage can be put as:

$$\begin{aligned}\Delta U_s = U_{s1} - U_{s2} \approx &\left(\frac{L'}{Z_l + Z_t + L'} - \frac{Z_f}{Z_l + Z_f} \cdot \frac{L'}{Z_t + L'}\right)U \\ &+ \left(\frac{Z_l + Z_t}{Z_l + Z_t + L'} - \frac{Z_f + Z_t}{Z_f + Z_t + L'}\right)E_d\end{aligned} \tag{3.9}$$

The initial value of the DFIG terminal voltage phase jump angle φ_0 can be gained from Equation (3.9). Also, it can be seen from Equation (3.9) that, at the grid fault instant, the DFIG terminal voltage phase jump angle has nothing to do with the time of fault occurrence, but has to do with the system topology, the system operating state, the generator operating parameters and the fault location. In the fault duration, the DFIG terminal voltage phase jump angle is affected by the DFIG rotor-side converter control. The stator flux directional control of the DFIG rotor-side converter is able to realize decoupling control of the DFIG power output by controlling the dq components of the rotor current. The DFIG stator active power and reactive power under the stator flux directional control are

$$\begin{cases} P_s = -\mathrm{Re}\left[\frac{3}{2}U_s I_s\right] \\ Q_s = -\mathrm{Im}\left[\frac{3}{2}U_s I_s\right] = -\frac{3}{2}\mathrm{Re}[\omega_1 \psi_s I_s] \end{cases} \tag{3.10}$$

The relationship between the DFIG stator active and reactive power output and the dq components of the rotor current can be derived from Equations (3.10):

$$\begin{cases} P_{\mathrm{t}} = \dfrac{3L_{\mathrm{m}}}{2L_{\mathrm{S}}}\omega_1\psi_{\mathrm{S}}i_{\mathrm{rq}} \\ Q_{\mathrm{t}} = \dfrac{3L_{\mathrm{m}}^2\omega_1\psi_{\mathrm{S}}}{2L_{\mathrm{S}}}\left(i_{\mathrm{rd}} - \dfrac{\psi_{\mathrm{S}}}{L_{\mathrm{m}}}\right) \\ \varphi_{\mathrm{t}} = \arctan\dfrac{P_{\mathrm{t}}}{Q_{\mathrm{t}}} \\ \varphi_1 = \varphi_0 + \varphi_{\mathrm{t}} \end{cases} \tag{3.11}$$

In the grid fault duration, ψ_{S}, i_{rd} and i_{rq} will vary due to the DFIG rotor-side converter control. By applying the phase jump initial value in Equation (3.9) and the post-fault stator flux and the rotor current in Equations (3.6)–(3.11), the phase jump value in the fault duration φ_1 can be obtained.

2. Control Model of DFIG Grid-Side Converter When a three-phase fault occurs in the grid, the DFIG DC bus voltage U_{dc} will fluctuate. Thus, the grid-side and rotor-side of the DFIG will exchange power. The control of the DFIG grid-side converter on the AC side power will affect the calculation of the stator and rotor short circuit currents.

The DFIG grid-side converter applies double closed-loop control. The voltage outer loop aims to keep the DC bus voltage of the three-phase PWM converter constant. The inner current loop aims to keep the power factor at 1.

Suppose $\boldsymbol{U}_{\mathrm{g}} = u_{\mathrm{gd}} + \mathrm{j}u_{\mathrm{gq}}$ is the grid voltage vector. When the d-axis of the coordinate system is fixed in the direction of the grid voltage vector, $u_{\mathrm{gd}} = |\boldsymbol{U}_{\mathrm{g}}| = U_{\mathrm{g}}$ and $u_{\mathrm{gq}} = 0$. As shown in Figure 3.3, the mathematical model of the DFIG grid-side converter in the synchronous rotating dq coordinate system is:

$$\begin{cases} U_{\mathrm{g}} = R_{\mathrm{g}}I_{\mathrm{g}} + L_{\mathrm{g}}\dfrac{\mathrm{d}I_{\mathrm{g}}}{\mathrm{d}t} + \mathrm{j}\omega_1 L_{\mathrm{g}}I_{\mathrm{g}} + V_{\mathrm{g}} \\ C\dfrac{\mathrm{d}U_{\mathrm{dc}}}{\mathrm{d}t} = \dfrac{P_{\mathrm{r}}}{U_{\mathrm{dc}}} - \dfrac{P_{\mathrm{g}}}{U_{\mathrm{dc}}} \\ U_{\mathrm{g}} = R_{\mathrm{g}}i_{\mathrm{gd}} + L_{\mathrm{g}}\dfrac{\mathrm{d}i_{\mathrm{gd}}}{\mathrm{d}t} - \omega_1 L_{\mathrm{g}}i_{\mathrm{gq}} + v_{\mathrm{gd}} \\ 0 = R_{\mathrm{g}}i_{\mathrm{gq}} + L_{\mathrm{g}}\dfrac{\mathrm{d}i_{\mathrm{gq}}}{\mathrm{d}t} + \omega_1 L_{\mathrm{g}}i_{\mathrm{gd}} + v_{\mathrm{gq}} \end{cases} \tag{3.12}$$

In order to illustrate the control effect of grid-side converter closed-loop control equations on the DC bus voltage and power difference, the comparative curves of DC bus voltage deviation and the power difference between the two sides of the DC capacitance are presented in Figure 3.4 [3], where R_{g} and L_{g} are the inlet line resistance and inductance of the grid-side converter. i_{gd} and i_{gq} are the d-axis and q-axis component of the grid-side converter input current. v_{gd} and v_{gq} are the d-axis and q-axis component of the grid-side converter AC-side voltage. ω_1 is the synchronous rotational speed. U_{dc} is the converter DC-side voltage. C is the DC bus capacitance. P_{g} and P_{r} are the grid-side converter output power and rotor-side converter input power.

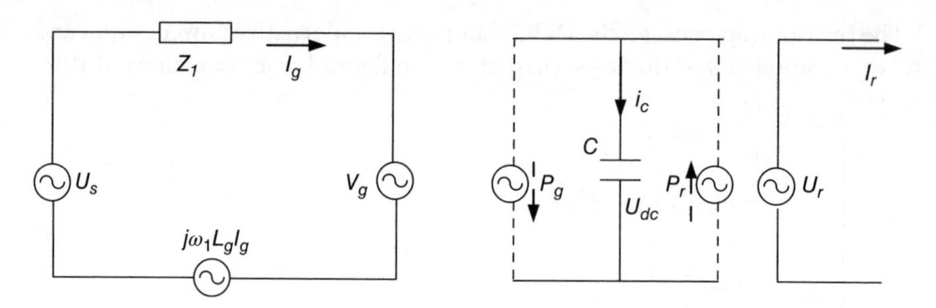

Figure 3.4 Equivalent circuit of the line-side and the rotor-side converter under dq coordinates.

In the DFIG grid-side converter, for the voltage outer loop PI control, the proportion parameter is k_{vp} and the integration parameter is k_{vi}. As for the current inner loop PI control, the proportion parameter is k_{igp} and the integration parameter is k_{igi}. Thus:

$$\begin{cases} C\dfrac{dU_{dc}}{dt} = C\dfrac{dU_{dc}^*}{dt} + k_{vp}\left(U_{dc}^* - U_{dc}\right) + k_{vi}\displaystyle\int\left(U_{dc}^* - U_{dc}\right)dt \\ v'_{gd} = L_g\dfrac{di_{gd}}{dt} = L_g\dfrac{di_{gd}^*}{dt} + k_{igp}\left(i_{gd}^* - i_{gd}\right) + k_{igi}\displaystyle\int\left(i_{gd}^* - i_{gd}\right)dt \\ v'_{gq} = L_g\dfrac{di_{gq}}{dt} = L_g\dfrac{di_{gq}^*}{dt} + k_{igp}\left(i_{gq}^* - i_{gq}\right) + k_{igi}\displaystyle\int\left(i_{gq}^* - i_{gq}\right)dt \end{cases} \quad (3.13)$$

When a three-phase fault occurs in the grid, the DFIG DC bus voltage U_{dc} will not be constant, but it will be affected by the DFIG grid-side converter control parameters. According to Equations (3.12) and (3.13),

$$\begin{cases} C\dfrac{dU_{dc}^*}{dt} + k_{vp}\left(U_{dc}^* - U_{dc}\right) + k_{vi}\displaystyle\int\left(U_{dc}^* - U_{dc}\right)dt = \\ \dfrac{P_r}{U_{dc}^*} - \dfrac{P_g}{U_{dc}^*} \\ P_g = -\dfrac{3u_{gd}v_{gq}}{2\omega_1 L_g} \\ u_r = u_{r1} + u_{rDC} = \dfrac{L_m}{L_s}\left[ku_{s2}s - (1-s)(1-k)u_m e^{j\varphi_1} e^{-\frac{R_S}{L_s}t}\right] \end{cases} \quad (3.14)$$

where U_{dc}^* is the reference value of the DC bus voltage.

In Equations (3.14) the rotor voltage u_r is composed of two parts: u_{r1} which is proportional to the slip s and the post-fault terminal voltage u_{s2}; u_{rDC} which is proportional to $(1 - s)$ and the grid fault degree $(1-k)$. Suppose

$$f(U_{dc}^*) = C\frac{dU_{dc}^*}{dt} + k_{pv}\left(U_{dc}^* - U_{dc}\right) + k_{iv}\int\left(U_{dc}^* - U_{dc}\right)dt \quad (3.15)$$

then the effect of the DFIG grid-side converter control and the DC bus voltage fluctuation ΔU_{dc} on the rotor instantaneous current i_r is

$$i_{\mathrm{rdc}} = \left(U_{\mathrm{dc}}^{*} f\left(U_{\mathrm{dc}}^{*}\right) - \Delta P_{\mathrm{g}}\right) \Big/ u_{\mathrm{r}} \cdot \sin\varphi \tag{3.16}$$

It can be seen from Equation (3.16) that the influence of the DFIG grid-side converter control on the rotor current i_{rdc} has to do with the DC bus voltage fluctuation ΔU_{dc}, the PI control parameters of the DFIG grid-side converter, the fault initial phase angle φ and the voltage drop degree k.

3. DFIG Stator and Rotor Current Analysis in the Case of Grid Three-Phase Fault The DFIG applies classical vector control, with grid voltage directional control for the grid-side converter, and stator flux directional control for the rotor-side converter.

According to Equations (3.2)–(3.4), the 1st-order dynamic expression of the rotor current is

$$\left(\sigma L_{\mathrm{r}} + \frac{\frac{R_{\mathrm{s}} L_{\mathrm{m}}^{2}}{L_{\mathrm{s}}^{2}}}{\tau_1}\right)\frac{\mathrm{d}i_{\mathrm{r}}}{\mathrm{d}t} + \left(R_{\mathrm{r}} + \mathrm{j}\omega\sigma L_{\mathrm{r}} + \frac{\mathrm{j}\omega\frac{R_{\mathrm{s}} L_{\mathrm{m}}^{2}}{L_{\mathrm{s}}^{2}}}{\tau_1}\right) i_{\mathrm{r}} = u_{\mathrm{r}} + e \tag{3.17}$$

where $\sigma = 1 - L_{\mathrm{m}}^{2}/(L_{\mathrm{s}}L_{\mathrm{r}})$ is the magnetic flux leakage parameter of the generator, e is the counter-emf and u_{r} is the AC voltage output of the rotor-side converter.

$$\begin{aligned} e &= e_{\mathrm{p}} + e_{\mathrm{d}} \mathrm{e}^{\tau_1 t_1} \mathrm{e}^{-\tau_1 t} \\ e_{\mathrm{p}} &= -\frac{\mathrm{j}\omega \frac{L_{\mathrm{m}}}{L_{\mathrm{s}}} u_{\mathrm{s}2}}{\tau_1} \\ e_{\mathrm{d}} &= \frac{L_{\mathrm{m}}}{L_{\mathrm{s}}} u_{\mathrm{s}1} - \frac{L_{\mathrm{m}}}{L_{\mathrm{s}}} u_{\mathrm{s}2} + \frac{\mathrm{j}\omega \frac{L_{\mathrm{m}}}{L_{\mathrm{s}}} u_{\mathrm{s}2}}{\tau_1} - \frac{\mathrm{j}\omega \frac{L_{\mathrm{m}}}{L_{\mathrm{s}}} u_{\mathrm{s}1}}{\tau_1} \end{aligned} \tag{3.18}$$

When the grid is in short circuit fault, two processes are involved. The first is the transient process of the generator, i.e. the variation of the air-gap flux and the stator/rotor voltage and current. The second process is the coordination between the grid-side and rotor-side of the DFIG. In Equations (3.18), the counter-emf voltage e characterizes the influence of the instantaneous drop in terminal voltage (i.e. the flux variation) on the rotor current; while u_{r} characterizes the influence of the grid-side and rotor-side control performance on the rotor current.

The rotor voltage vector under the synchronous rotating coordinate system in the case of grid three-phase fault is

$$u_{\mathrm{r}} = k_{\mathrm{p}i}\left(i_{\mathrm{r.ref}} - i_{\mathrm{r}}\right) + k_{\mathrm{i}i}\int\left(i_{\mathrm{r.ref}} - i_{\mathrm{r}}\right)\mathrm{d}t + \mathrm{j}\omega\sigma L_{\mathrm{r}} i_{\mathrm{r}} \tag{3.19}$$

where $k_{\mathrm{p}i}$ and $k_{\mathrm{i}i}$ are the proportion parameter and integration parameter of the current inner loop PI controller. $i_{\mathrm{r.\,ref}}$ is the reference value of the steady-state rotor current.

Considering the influence of the stator flux and rotor-side converter control on the rotor current, derivation of Equation (3.17) will produce the second-order dynamic equation of the rotor current in the case of grid three-phase fault:

$$\frac{\mathrm{d}^2 i_{\mathrm{r}}}{\mathrm{d}t^2} + \lambda_1 \frac{\mathrm{d}i_{\mathrm{r}}}{\mathrm{d}t} + \lambda_2 i_{\mathrm{r}} = \lambda_2 i_{\mathrm{r.ref}} + \lambda_3 \mathrm{e}^{\tau_1 t_1} \mathrm{e}^{-\tau_1 t_1} \tag{3.20}$$

where $\lambda_1 = \left(R_r + k_p + \frac{j\omega \frac{R_s L_m^2}{L_s^2}}{\tau_1} \right) / \left(\sigma L_r + \frac{\frac{R_s L_m^2}{L_s^2}}{\tau_1} \right)$, $\lambda_2 = k_i / \left(\sigma L_r + \frac{\frac{R_s L_m^2}{L_s^2}}{\tau_1} \right)$, $\lambda_3 = (-\tau_1 e_d) / \left(\sigma L_r + \frac{\frac{R_s L_m^2}{L_s^2}}{\tau_1} \right)$.

By solving this second-order differential equation, the post-fault instantaneous expression of the rotor current can be obtained:

$$i_r = i_{r1} + i_{r2} + i_{r3} + i_{rdc} \tag{3.21}$$

$$i_{r1} = \frac{\beta_1 i_{r0} e^{\beta_2 t} - \beta_2 i_{r0} e^{\beta_1 t}}{\beta_1 - \beta_2} \tag{3.22}$$

$$i_{r2} = i_{r \cdot ref} \tag{3.23}$$

$$i_{r3} = \frac{\lambda_3}{(\tau_1^2 - \lambda_1 \tau_1 + \lambda_2)} e^{\tau_1 t_1} e^{-\tau_1 t} \tag{3.24}$$

where $\beta_1 = \frac{-\lambda_1 + \sqrt{\lambda_1^2 - 4\lambda_2}}{2}$, $\beta_2 = \frac{-\lambda_1 - \sqrt{\lambda_1^2 - 4\lambda_2}}{2}$ and i_{rdc} is the influence of the DFIG grid-side converter on the rotor current.

It can be seen from Equation (3.21) that the rotor current contains a steady-state component and two transient components. Transient component i_{r1} is the natural component of the rotor current, which is related to the generator parameters, the rotor-side converter parameters and the steady-state rotor current. The transient component i_{r3} results from the counter-emf, and is related to the terminal voltage drop rate, the rotational speed and the rotor-side control parameters.

According to Equations (3.6) and (3.3), the DFIG stator short circuit current can be obtained:

$$\begin{aligned} i_s &= \frac{1}{L_r}\psi_s - \frac{L_m}{L_r} i_r \\ &= \frac{u_{s1} - u_{s2}}{L_s \tau_1} e^{\tau_1 t_1} e^{-\tau_1 t_1} + \frac{u_{s2}}{L_s \tau_1} + \left(\frac{R_s L_m}{L_s^2 \tau_1} - \frac{L_m}{L_s} \right) i_r \end{aligned} \tag{3.25}$$

When the grid is in three-phase fault, the DFIG stator short circuit current is mainly decided by the stator flux ψ_s and the rotor current i_r. Apply the rotor current to Equation (3.25), so that the expression of the stator short circuit current can be obtained:

$$i_s = i_{s1} + i_{s2} + i_{s3} + i_{s4} + i_{s5} + i_{s6} \tag{3.26}$$

$$i_{s1} = \frac{u_{s2}}{L_s \tau_1} \tag{3.27}$$

$$i_{s2} = \frac{u_{s1} - u_{s2}}{L_s \tau_1} e^{\tau_1 t_1} e^{-\tau_1 t_1} \tag{3.28}$$

where i_{s1} is the periodic component of the stator short circuit current, which is decided by the post-fault terminal voltage. i_{s2} is the transient DC component of the stator short circuit current, which results from the transient DC flux and is related to the terminal voltage drop rate and the time of fault occurrence.

$$i_{s3} = \left(\frac{R_s L_m}{L_s^2 \tau_1} - \frac{L_m}{L_s}\right) i_{r.ref} \tag{3.29}$$

$$i_{s4} = \left(\frac{R_s L_m}{L_s^2 \tau_1} - \frac{L_m}{L_s}\right) \frac{\lambda_3}{(\tau_1^2 - \lambda_1 \tau_1 + \lambda_2)} e^{\tau_1 t_1} e^{-\tau_1 t_1} \tag{3.30}$$

$$i_{s5} = \left(\frac{R_s L_m}{L_s^2 \tau_1} - \frac{L_m}{L_s}\right) \frac{\beta_1 i_{r0} e^{\beta_2 t} - \beta_2 i_{r0} e^{\beta_1 t}}{\beta_1 - \beta_2} \tag{3.31}$$

$$i_{s6} = \left(\frac{R_s L_m}{L_s^2 \tau_1} - \frac{L_m}{L_s}\right) i_{rdc} \tag{3.32}$$

i_{s3} is the periodic component of the stator short circuit current resulting from rotor excitation, which has to do with the steady-state rotor current. i_{s4} is the transient DC component of the stator short circuit current resulting from rotor excitation, which has to do with the terminal voltage drop rate, the rotational speed and the rotor-side control parameters. i_{s5} is the transient component of the stator short circuit current resulting from rotor flux and it has to do with the generator parameters, the rotor-side converter parameters and the steady-state rotor current. i_{s6} characterizes the influence of the grid-side converter control and the DC voltage variation on the rotor current.

The short circuit currents transformed to the three-phase stationary coordinate system become

$$\begin{bmatrix} i_a(t) \\ i_b(t) \\ i_c(t) \end{bmatrix} = \begin{bmatrix} \cos(\omega t + \varphi) & -\sin(\omega t + \varphi) & 1 \\ \cos(\omega t + \varphi + 120^\circ) & -\sin(\omega t + \varphi + 120^\circ) & 1 \\ \cos(\omega t + \varphi - 120^\circ) & \sin(\omega t + \varphi - 120^\circ) & 1 \end{bmatrix} \begin{bmatrix} i_d(t) \\ i_q(t) \\ i_0(t) \end{bmatrix} \tag{3.33}$$

where $i_d(t)$, $i_q(t)$, $i_0(t)$ are respectively the d, q, 0 components of the short circuit current. φ is the voltage phase angle at the fault instant.

B. Asymmetrical Short Circuit Fault

1. Negative Sequence Model of DFIG Under Asymmetrical Grid Fault In the case of an asymmetrical fault, there is a negative sequence voltage. The complex vector expression of the DFIG terminal voltage is

$$\vec{U}_s = \vec{U}_{s+} e^{j\omega_1 t} + \vec{U}_{s-} e^{-j\omega_1 t} \tag{3.34}$$

For the three-phase symmetrical components in the power grid and the DFIG, the sequence components are independent of each other. Thus, the negative sequence vector model of the DFIG under the synchronous rotating coordinate system can be expressed as

$$\vec{U}_{s-} = R_s \vec{I}_{s-} - j\omega_1 \vec{\psi}_{s-} + d\vec{\psi}_{s-}/dt \tag{3.35}$$

$$\vec{U}_{r-} = R_r \vec{I}_{r-} - j(2-s)\omega_1 \vec{\psi}_{r-} + d\vec{\psi}_{r-}/dt \tag{3.36}$$

$$\vec{\psi}_{s-} = L_s \vec{I}_{s-} + L_m \vec{I}_{r-} \tag{3.37}$$

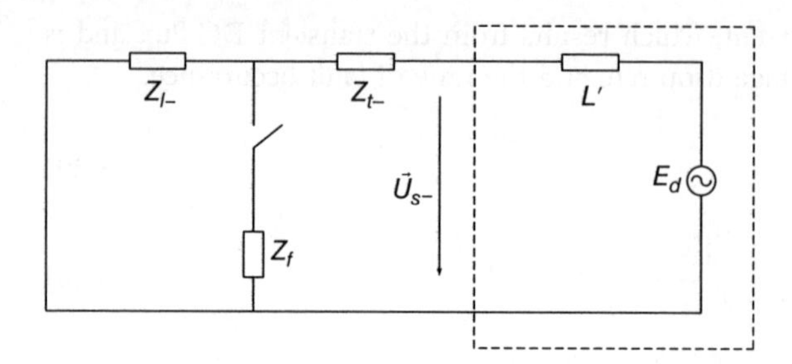

Figure 3.5 Negative sequence analysis model of a DFIG under asymmetrical grid fault.

$$\vec{\psi}_{r-} = L_r\vec{I}_{r-} + L_m\vec{I}_{s-} \tag{3.38}$$

Suppose that the DFIG still applies the control manner in the equilibrium conditions, and the structures of the grid-side and rotor-side converters, as well as the synchronous detection signals remain symmetrical. If the harmonics in rotor voltage caused by the asymmetrical fault are neglected, the negative sequence component of the DFIG terminal voltage is 0 in the normal operating state, and appears at the fault instant. Thus, the DFIG terminal negative sequence voltage can be expressed with a piecewise function as

$$\vec{U}_{s-} = \begin{cases} 0 & t < t_2 \\ \vec{U}_{s-} & t \geq t_2 \end{cases} \tag{3.39}$$

Before the fault occurs, there is no negative sequence component in the DFIG terminal voltage. Therefore, the negative sequence phase angle of the terminal voltage before the fault is 0. After the fault occurs, the stator voltage has an initial phase angle, and the negative sequence circuit is the same as the positive sequence circuit. When an asymmetrical fault occurs in the grid, the phase jump value of the DFIG terminal voltage $\vec{U}_{s-}$ directly affects the calculation of the stator flux and the short circuit current. Therefore, it is necessary to establish a DFIG transient model considering the terminal voltage phase jump. If the asymmetrical fault occurs in the grid at $t = t_1$, the negative sequence fault analysis model is as shown in Figure 3.5, where $\vec{U}_{s-}$ is the DFIG terminal voltage. $L' = \omega_s\left(L_s - L_m^2/L_r\right)$ is the equivalent impedance of the DFIG. Z_{l-}, Z_{t-} and Z_{f-} are respectively the grid transmission line negative sequence impedance, the transformer transmission impedance and the transition impedance in grounding faults.

The initial value of the DFIG terminal voltage phase jump angle φ_{0-} can be obtained. In addition, it can be seen that, at the grid fault instant, the DFIG terminal voltage phase jump angle has nothing to do with the time of the fault occurrence, but has to do with the system topology, the pre-fault system operating state, the generator operating parameters and the fault location.

2. DFIG Stator and Rotor Negative Sequence Currents Calculation The stator negative sequence current can be obtained from Equation (3.37): $\vec{I}_{s-} = \left(\vec{\psi}_{s-} - L_m\vec{I}_{r-}\right)/L_s$.

Apply $\vec{I}_{s-}$ to Equation (3.35), so that the 1st-order differential equation of the negative sequence stator flux is

$$\frac{d\vec{\psi}_{s-}}{dt} + \left(\frac{R_s}{L_s} - j\omega_1\right)\vec{\psi}_s = \vec{U}_{s-} + \frac{R_s L_m}{L_s}\vec{I}_{r-} \tag{3.40}$$

Apply the rotor negative sequence voltage to Equation (3.36), so that the negative sequence stator flux can be calculated:

$$\vec{\psi}_{s-} = \frac{\vec{U}_{s-}}{\tau_2} - \frac{\vec{U}_{s-}}{\tau_2} e^{\tau_2 t_2} e^{-\tau_2 t} \tag{3.41}$$

where $\tau_2 = R_s/L_s - j\omega_1$ is the transient component attenuation time constant of the negative sequence stator flux.

When an asymmetrical fault occurs in the grid, there is a negative sequence component in the terminal voltage. The negative sequence rotor flux can be gained from Equations (3.37) and (3.38):

$$\vec{\psi}_{r-} = \frac{L_m}{L_s}\vec{\psi}_{s-} + \sigma L_r \vec{I}_{r-} \tag{3.42}$$

where σ is the generator leakage inductance coefficient and σL_r is the rotor transient inductance. Apply the negative sequence rotor flux to Equation (3.36), so that the 1st-order differential equation of the rotor negative sequence current can be gained:

$$\begin{aligned} &\sigma L_r \frac{d\vec{I}_{r-}}{dt} + [R_r - j(2-s)\omega_1 \sigma L_r]\vec{I}_{r-} \\ &= \frac{j(2-s)\omega_1 \frac{L_m}{L_s}\vec{U}_{s-}}{\tau_2} - \left[\frac{L_m}{L_s}\vec{U}_{s-} + \frac{j(2-s)\omega_1 \frac{L_m}{L_s}\vec{U}_{s-}}{\tau_2}\right] e^{\tau_2 t_2} e^{-\tau_2 t} \\ &= \vec{U}_p + \vec{U}_d e^{\tau_2 t_2} e^{-\tau_2 t} \end{aligned} \tag{3.43}$$

where $\vec{U}_p$ is the periodic component of the negative sequence counter-emf under the reverse synchronous rotating coordinate system, and $\vec{U}_d$ is the initial value of the transient DC component. The negative sequence counter-emf results from the DFIG terminal negative sequence voltage.

Simplify Equation (3.43) so that the 1st-order differential equation of the rotor negative sequence current becomes

$$\frac{d\vec{I}_{r-}}{dt} + \frac{[R_r - j(2-s)\omega_1 \sigma L_r]}{\sigma L_r}\vec{I}_{r-} = \frac{1}{\sigma L_r}\left(\vec{U}_p + \vec{U}_d e^{\tau_2 t_2} e^{-\tau_2 t}\right) \tag{3.44}$$

Solve the differential equation so that the rotor negative sequence current can be obtained:

$$\begin{aligned} \vec{I}_{r-} &= \left(\frac{\vec{U}_p}{\mu \sigma L_r}\right) + \left(\frac{\vec{U}_d}{(\mu - \tau_2)\sigma L_r} e^{\tau_2 t_2} e^{-\tau_2 t}\right) - \left(\frac{\vec{U}_p}{\mu \sigma L_r} + \frac{\vec{U}_d}{(\mu - \tau_2)\sigma L_r} e^{\tau_2 t_2}\right) e^{-\mu t} \\ &= \vec{I}_{r1-} + \vec{I}_{r2-} + \vec{I}_{r3-} \end{aligned} \tag{3.45}$$

where $\mu = [R_r - j(2-s)\omega_1\sigma L_r]/\sigma L_r$ is the transient component time constant of the rotor negative sequence current.

Transforming the rotor current to the rotor side, it can be seen that the rotor negative sequence current is composed of three parts: the harmonic component $\vec{I}_{r1-}$, the transient periodic component $\vec{I}_{r2-}$ and the transient DC component $\vec{I}_{r3-}$.

Apply $\vec{I}_{r-}$ to Equation (3.44), so that the stator negative sequence short circuit current can be obtained:

$$\begin{aligned}
\vec{I}_{s-} &= \frac{\vec{\psi}_{s-} - L_m\vec{I}_{r-}}{L_s} \\
&= \left(\frac{\vec{U}_{s-}}{L_s\tau_2} - \frac{L_m}{L_s}\frac{\vec{U}_p}{\mu\sigma L_r}\right) + \left(\frac{L_m}{L_s}\frac{\vec{U}_d}{(\mu-\tau_2)} - \frac{\vec{U}_{s-}}{L_s\tau_2}\right)e^{\tau_2 t_2}e^{-\tau_2 t} \\
&\quad + \frac{L_m}{L_s}\left(\frac{\vec{U}_p}{\mu\sigma L_r} + \frac{\vec{U}_d}{(\mu-\tau_2)\sigma L_r}e^{\tau_2 t_2}\right)e^{-\mu t} \\
&= \vec{I}_{s1} + \vec{I}_{s2} + \vec{I}_{s3}
\end{aligned} \tag{3.46}$$

It can be seen from Equation (3.46) that the stator negative sequence current is composed of three parts: (1) the periodic component $\vec{I}_{s1-}$, which is decided by the negative sequence voltage; (2) the transient DC component $\vec{I}_{s2-}$, which results from the negative sequence stator flux and attenuates in time constant τ_2; the initial value of $\vec{I}_{s2-}$ has to do with the post-fault negative sequence voltage; (3) the transient DC component $\vec{I}_{s3-}$, which results from the negative sequence rotor flux and attenuates in time constant μ.

The stator short circuit current can be obtained by adding the positive and negative sequence components, and through Park transformation, the short circuit currents can be transformed to the three-phase stationary coordinate system:

$$\begin{bmatrix} i_a(t) \\ i_b(t) \\ i_c(t) \end{bmatrix} = \begin{bmatrix} \cos(\omega t + \varphi) & -\sin(\omega t + \varphi) & 1 \\ \cos(\omega t + \varphi + 120°) & -\sin(\omega t + \varphi + 120°) & 1 \\ \cos(\omega t + \varphi - 120°) & \sin(\omega t + \varphi - 120°) & 1 \end{bmatrix} \begin{bmatrix} i_d(t) \\ i_q(t) \\ i_0(t) \end{bmatrix} \tag{3.47}$$

where $i_d(t)$, $i_q(t)$, $i_0(t)$ are respectively the d, q, 0 components of the short circuit current. φ is the voltage phase angle at the fault instant.

C. Simulation Verification

1. Simulation System In order to verify that the instantaneous expression of the DFIG short circuit current in the case of a grid three-phase fault derived above is correct, a DFIG transient model was established in PSCAD/EMTDC and dynamic simulation experiments were carried out in the state key laboratory. The line protection and rotor protection are neglected so that the correctness of the current expression and the fault characteristics of the short circuit current can be fully demonstrated.

The simulation model is shown in Figure 3.6. The dynamic simulation experiment parameters are as follows. DFIG: rated power 2 MW, stator rated voltage 690 V, stator resistance 0.0054 p.u., rotor resistance 0.00607 p.u., stator leakage inductance 0.102 p.u., rotor leakage inductance 0.11 p.u., mutual inductance 4.362 p.u. DFIG rotor-side converter current inner loop PI controller: proportion parameter 1.8, integration parameter 0.02. DFIG rotor-side converter voltage outer loop PI controller: proportion parameter 1.4, integration parameter 0.013. DFIG grid-side converter current inner loop PI controller: proportion parameter 1.3, integration parameter 0.1. DFIG grid-side converter voltage outer loop PI controller: proportion parameter 100, integration parameter 0.002.

2. Short Circuit Current Analysis in the Case of Grid Three-Phase Fault Suppose that at $t = 2$s, a three-phase permanent fault occurs at the high-voltage side of the grid transformer, and the DFIG terminal voltage drops to 60% of the rated value. The fault characteristics of the short circuit current and the relevant parameters are analysed as follows.

When the phase jump is not considered, the calculated rotor current is as shown in Figure 3.7(a), which can be seen to be deviating from the simulation results. This is because the phase jump affects the stator flux ψ_s and the counter-emf e, thus neglecting the phase jump will cause the calculated transient component i_{r3} to be smaller than the simulation results. On the other hand, when the phase jump is considered, the rotor current is calculated according to Equation (3.21) and shown in Figure 3.7(b), which can be seen to be consistent with the simulation results in amplitude and the maximum value. Therefore, the proposed calculation method, considering the influence of the DFIG terminal voltage phase jump, is able to describe the characteristics of the rotor short circuit current more accurately.

When the phase jump is not considered, the calculated stator current is as shown in Figure 3.8(a), which can be seen to deviate from the simulation results in both phase and amplitude. This is because the DFIG stator current is jointly decided by the rotor current and the stator flux. In the case of grid fault, the DFIG terminal voltage phase jump affects the calculation of both the rotor current and the stator flux. Meanwhile, it should be noted that due to the DC voltage

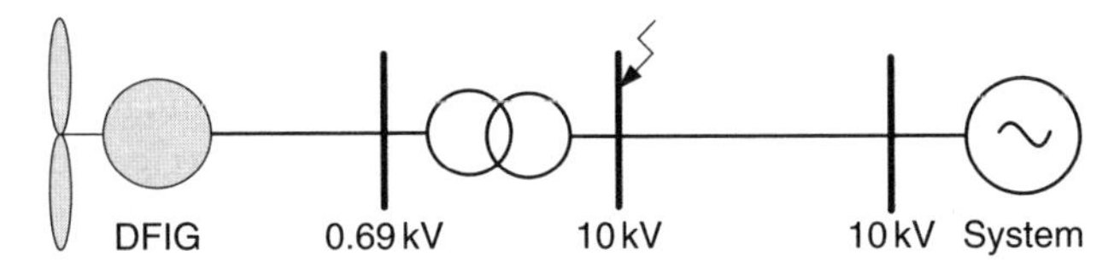

Figure 3.6 Power grid fault.

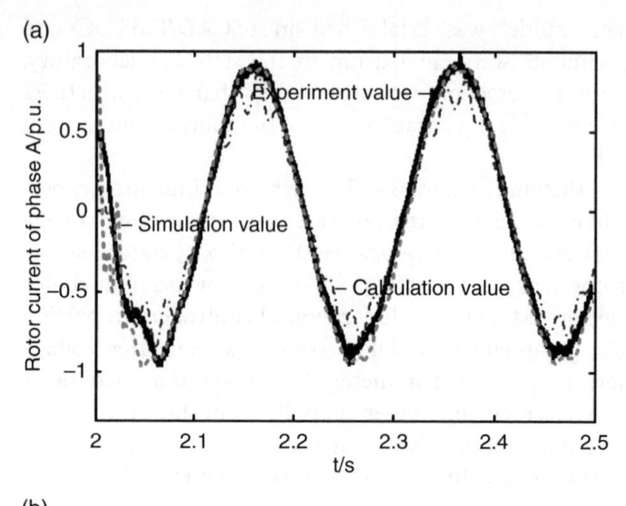

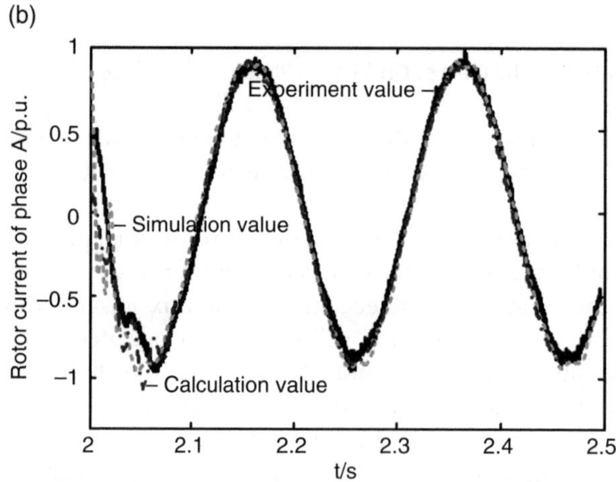

Figure 3.7 Influence of the DFIG terminal voltage phase jump on the rotor current. (a) Rotor current without considering the phase jump. (b) Rotor current considering the phase jump.

fluctuation and the asymmetry in the three-phase parameters, the dynamic simulation result is offset in the first two cycles after the fault and only restores to stability in the third cycle. On the other hand, when the phase jump is considered, the calculated stator current is as shown in Figure 3.8(b), which can be seen to be consistent with the simulation results in phase and amplitude. Therefore, calculation of the stator short circuit current considering a DFIG terminal voltage phase jump is more accurate.

When the DFIG grid-side converter control is not considered, the calculated rotor current is as shown in Figure 3.9(a), which is consistent with the simulation result after the first cycle. However, at the fault instant and in the first cycle, the transient characteristic of the rotor current is not demonstrated. On the other hand, when the DFIG grid-side converter control is considered, the calculated

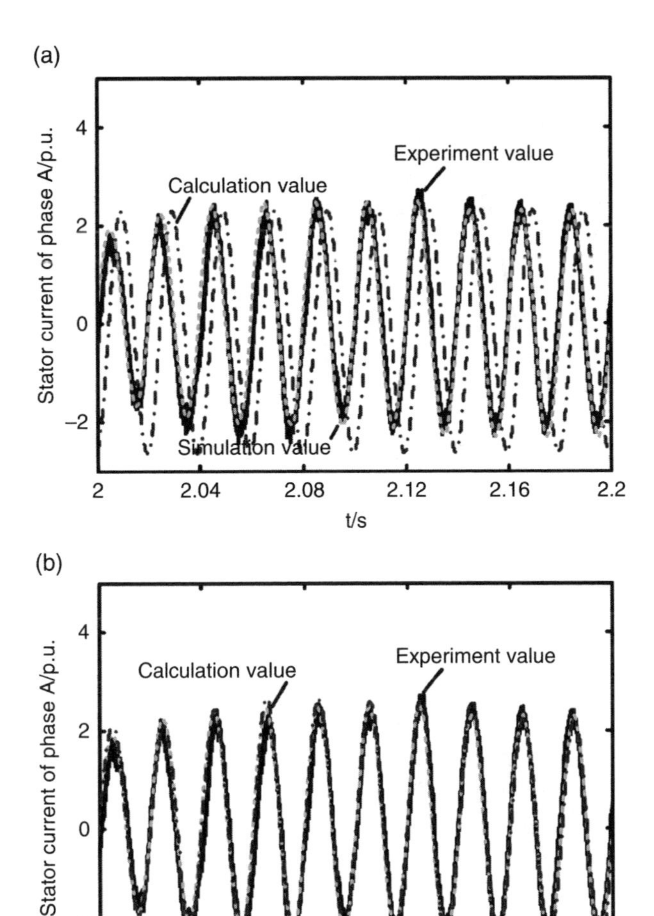

Figure 3.8 Influence of DFIG terminal voltage phase jump on the stator current. (a) Stator current without considering the phase jump. (b) Stator current considering the phase jump.

rotor current is as shown in Figure 3.9(b), which is consistent with the simulation result and the dynamic experiment result. Also, it is able to accurately reflect the transient characteristic of the rotor current at the fault instant and in the first cycle after the fault.

When the DFIG grid-side converter control is not considered, the calculated stator current is as shown in Figure 3.10(a), which has no obvious fluctuation in the fault duration and cannot reflect the transient characteristic of the stator current at the fault instant, but only reflects the post-fault steady-state condition. On the other hand, when the DFIG grid-side converter control is considered, the calculated stator current is as shown in Figure 3.10(b), which is consistent with

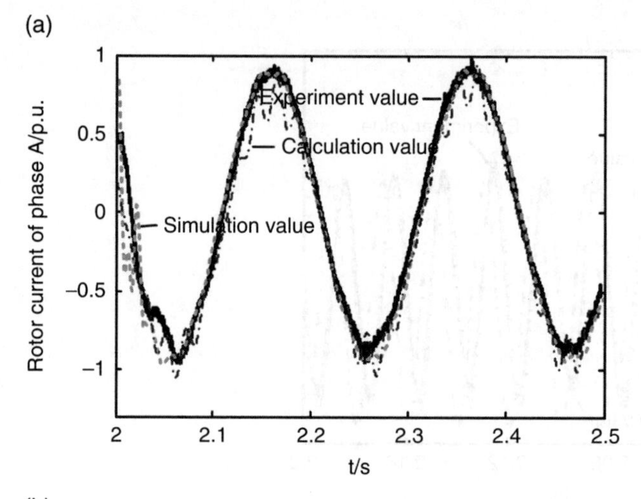

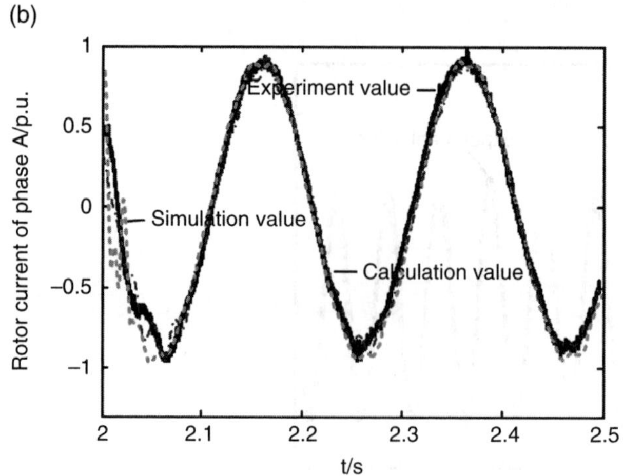

Figure 3.9 Influence of DFIG grid-side converter control on the rotor current. (a) Rotor current without considering the grid-side converter control. (b) Rotor current considering the grid-side converter control.

the simulation result and the dynamic experiment result. Also, it demonstrates obvious attenuation in the transient process, thus it is in line with the characteristic of the stator short circuit current at the fault instant and in the fault duration.

3. DFIG Short Circuit Current Analysis in the Case of Grid Single-Phase Fault Suppose that at $t = 2$ s, a phase-C permanent fault occurs at the high-voltage side of the grid transformer, and the DFIG terminal voltage drops to 70% of the rated value. The fault characteristics of the short circuit currents and relevant parameters are analysed as follows.

When a single-phase fault occurs in the grid, due to the electromagnetic transient characteristic of the DFIG and the influence of DFIG grid-side and rotor-side

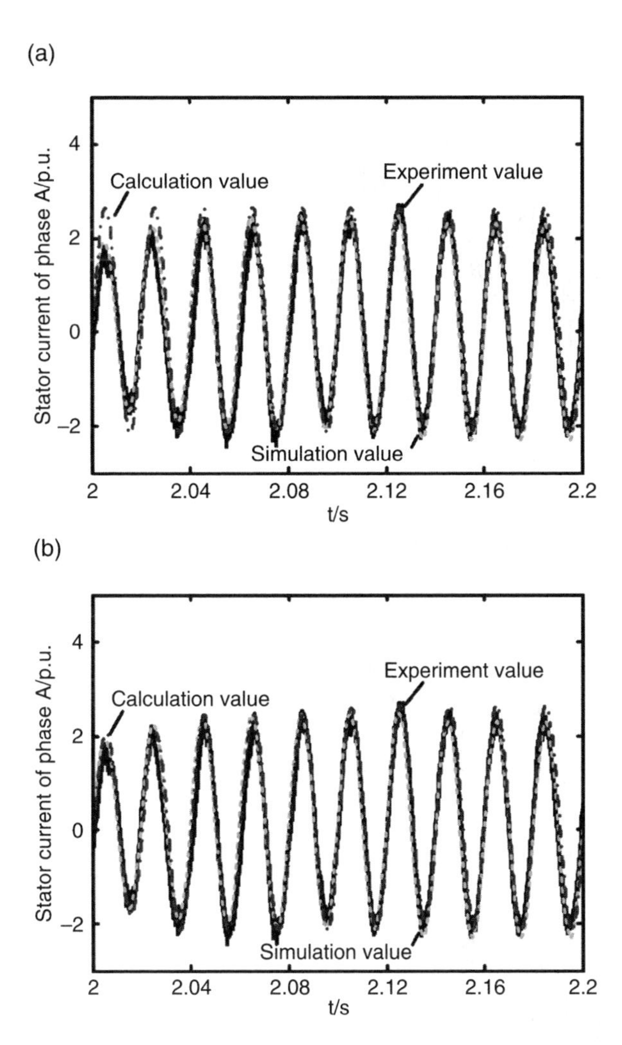

Figure 3.10 Influence of DFIG grid-side converter control on the stator current. (a) Stator current without considering the grid-side converter control. (b) Stator current considering the grid-side converter control.

converter control, the rotor current is no longer sinusoidal. The rotor fault current contains the positive sequence steady-state component, the positive sequence transient component, the negative sequence harmonic component, the negative sequence transient periodic component and the negative sequence transient DC component. The simulation and calculated values of the rotor currents are as shown in Figure 3.11. It can be seen that the calculated values are basically consistent with the PSCAD simulation values in the variation trend of the amplitude and phase, and in the attenuation degree of the transient DC component. However, in the early post-fault stage, the calculated rotor current is not completely consistent with the simulation result in the phase angle. This is

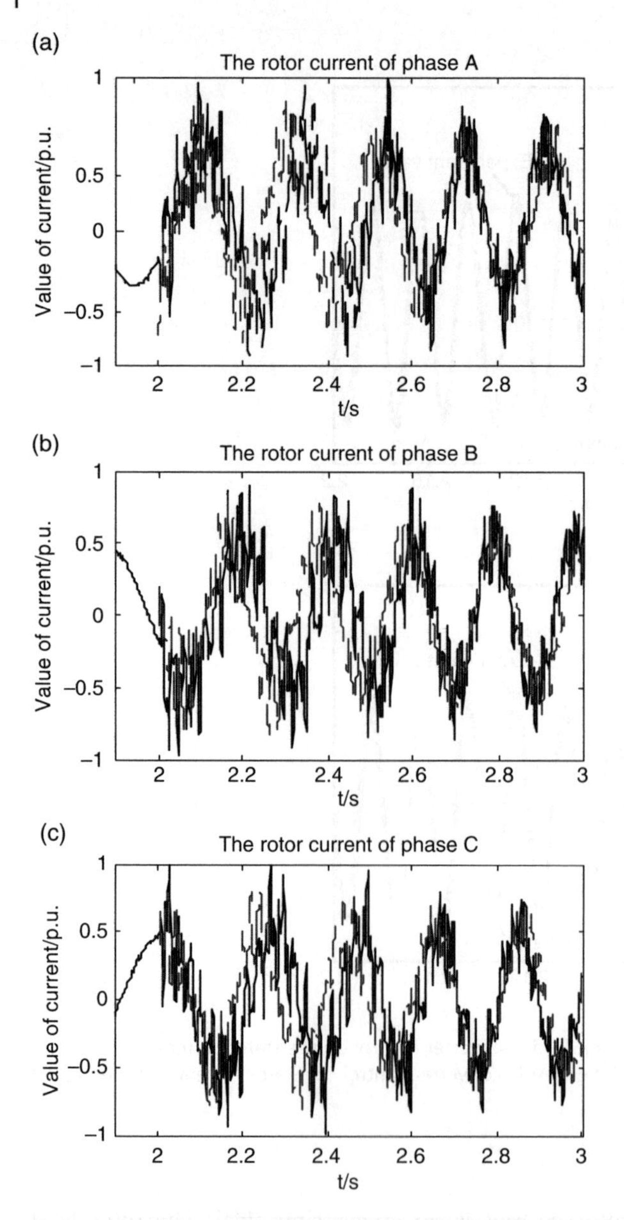

Figure 3.11 DFIG rotor current under grid phase-C short circuit fault. (a) Simulation and calculated values of phase-A rotor current. (b) Simulation and calculated values of phase-B rotor current. (c) Simulation and calculated values of phase-C rotor current.

because the generator electromagnetic torque decreases when the fault occurs, and the rotational speed of the DFIG is unstable at the early post-fault stage. However, after the first four cycles, the phase deviation disappears. Therefore, the rotor current calculation formula derived here is proved correct.

When a single-phase fault occurs in the grid, the variation of DFIG stator current is as shown in Figure 3.12. It can be seen that the fault transient process is very complicated, with transient components such as the attenuating DC component and the negative sequence periodic component. The DFIG grid-side converter control is considered in the calculation, which effectively improves the calculation accuracy. Meanwhile, the terminal voltage phase jump is taken into account, so that the initial phase angle of the stator current at the fault instant can be guaranteed correct. It can be seen from Figure 3.12 that the calculated values of DFIG stator current are basically consistent with the PSCAD simulation values. Therefore, the calculated stator current is able to reflect the transient characteristics and variation trend of a DFIG under grid single-phase fault.

4. DFIG Short Circuit Current Analysis in the Case of Grid Phase-to-Phase Fault Suppose that at $t = 2$ s, a phase-AC permanent fault occurs at the high-voltage side of the grid transformer, and the DFIG terminal voltage drops to 80% of the rated value. The fault characteristic and dynamic process of the DFIG short circuit currents are analysed as follows.

Similar to the case of the single-phase fault, when a phase-to-phase fault occurs in the grid, there are positive and negative sequence transient DC attenuating components in the DFIG rotor current, shown in Figure 3.13. But compared with the single-phase fault, the negative sequence voltage under the phase-to-phase fault is relatively small, thus the negative sequence periodic component and transient DC component of the DFIG rotor current are relatively small.

It can be seen from Figure 3.13 that the calculated values of DFIG rotor current are basically consistent with the PSCAD simulation values in amplitude and phase. The attenuating DC component and periodic component in the calculated rotor current can effectively reflect the actual situation. On the other hand, the phase angle of the calculated current deviates from the simulation result in the first two cycles after the fault. This is because the rotational speed of the DFIG is unstable in the early post-fault stage. However, after the first two cycles, the phase deviation disappears. Therefore, the rotor current calculation formula derived here is able to reflect the transient characteristics and variation trend of the rotor current under grid phase-to-phase fault.

It can be seen from Equation (3.46) that the DFIG stator short circuit current is mainly decided by the stator flux and the rotor fault current. When a phase-AC short circuit fault occurs in the grid, compared with a single-phase fault, the negative sequence periodic component and transient DC component of the DFIG rotor current are smaller, as well as the negative sequence flux, due to the decrease in the negative sequence voltage. Therefore, there are attenuating DC components in the DFIG stator fault current, but the waveform is relatively smooth – see Figure 3.14. It can be seen that the calculated values of the DFIG stator current are basically consistent with the PSCAD simulation values in amplitude and phase. The transient DC components attenuate markedly in the first four cycles after the fault. Thus, the effectiveness of the DFIG stator current calculation formula derived here is verified.

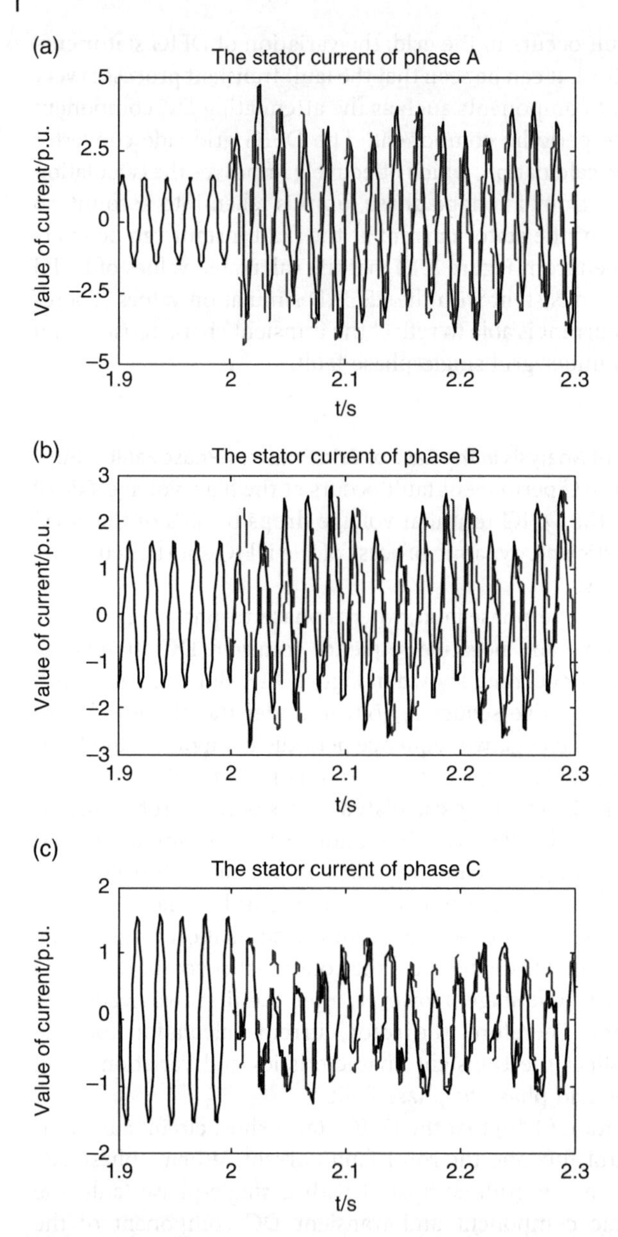

Figure 3.12 DFIG stator current under grid phase-C short circuit fault. (a) Simulation and calculated values of phase-A stator current. (b) Simulation and calculated values of phase-B stator current. (c) Simulation and calculated values of phase-C stator current.

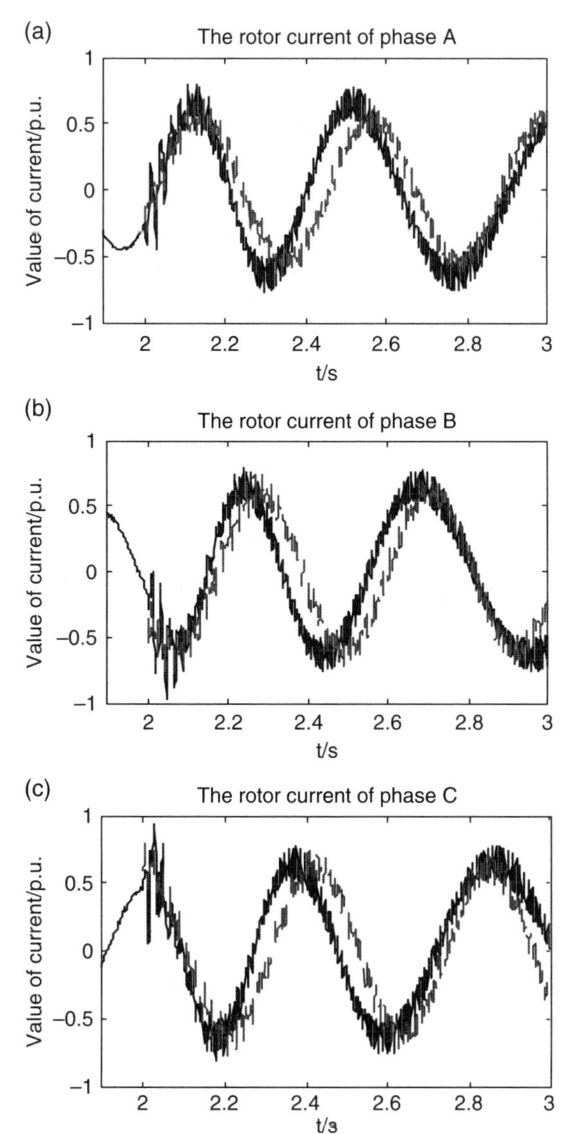

Figure 3.13 DFIG rotor current under grid phase-AC short circuit fault. (a) Simulation and calculated values of phase-A rotor current. (b) Simulation and calculated values of phase-B rotor current. (c) Simulation and calculated values of phase-C rotor current.

3.2.2.2 DFIG Fault Transient Characteristics Considering Both Rotor-Side Control and Grid-Side Converter Control

A. DFIG Short Circuit Characteristics Considering the Converter Transient Regulation

When a fault occurs in the grid, a DFIG fault equivalent network considering the rotor-side converter transient regulation is established, and is decomposed into a normal-operation network and a superimposed network, as shown in Figure 3.15. The fault superimposed network is not only affected by the stator-side superimposed source $\boldsymbol{\Delta u}_{\mathbf{s}} = -A\boldsymbol{u}_{\mathbf{s0}}$ (where A is the terminal voltage drop rate, and $\boldsymbol{u}_{\mathbf{s0}}$ is the steady-state stator voltage vector), but is also affected by the rotor-side fault

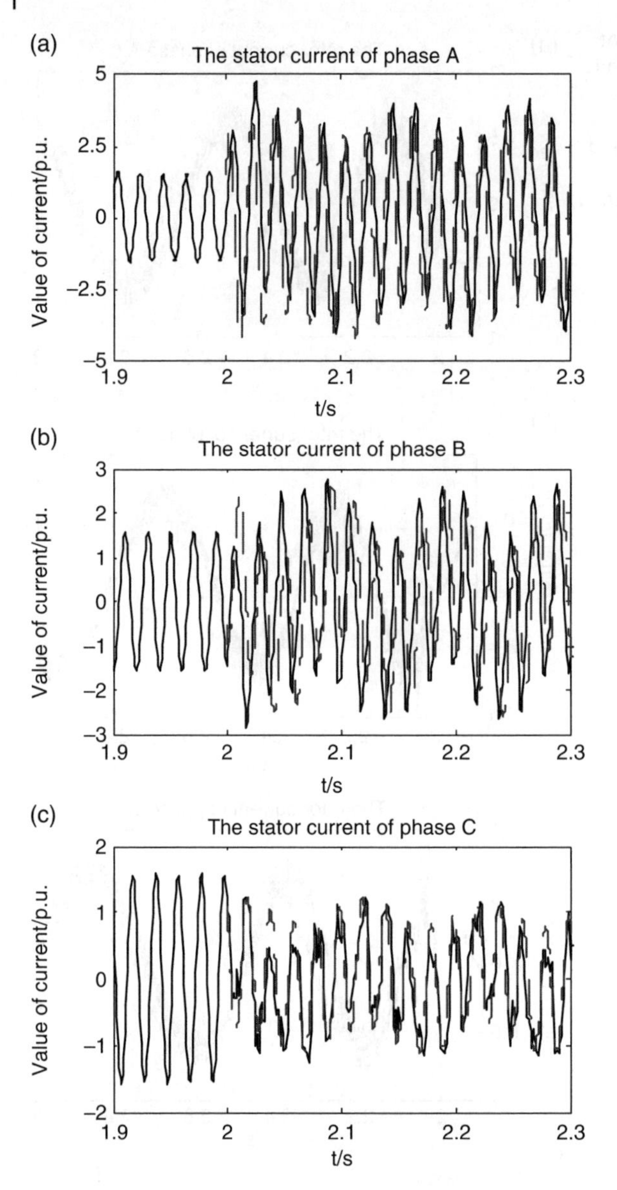

Figure 3.14 DFIG stator current under grid phase-AC short circuit fault (a) Simulation and calculated values of phase-A stator current. (b) Simulation and calculated values of phase-B stator current. (c) Simulation and calculated values of phase-C stator current.

superimposed source resulting from the converter regulation. Neglecting the switch transient, suppose the closed-loop bandwidth of the rotor-side converter current loop is big enough [4], and the converter AC-side voltage can track the reference value faithfully, then the rotor-side fault superimposed source can be expressed as

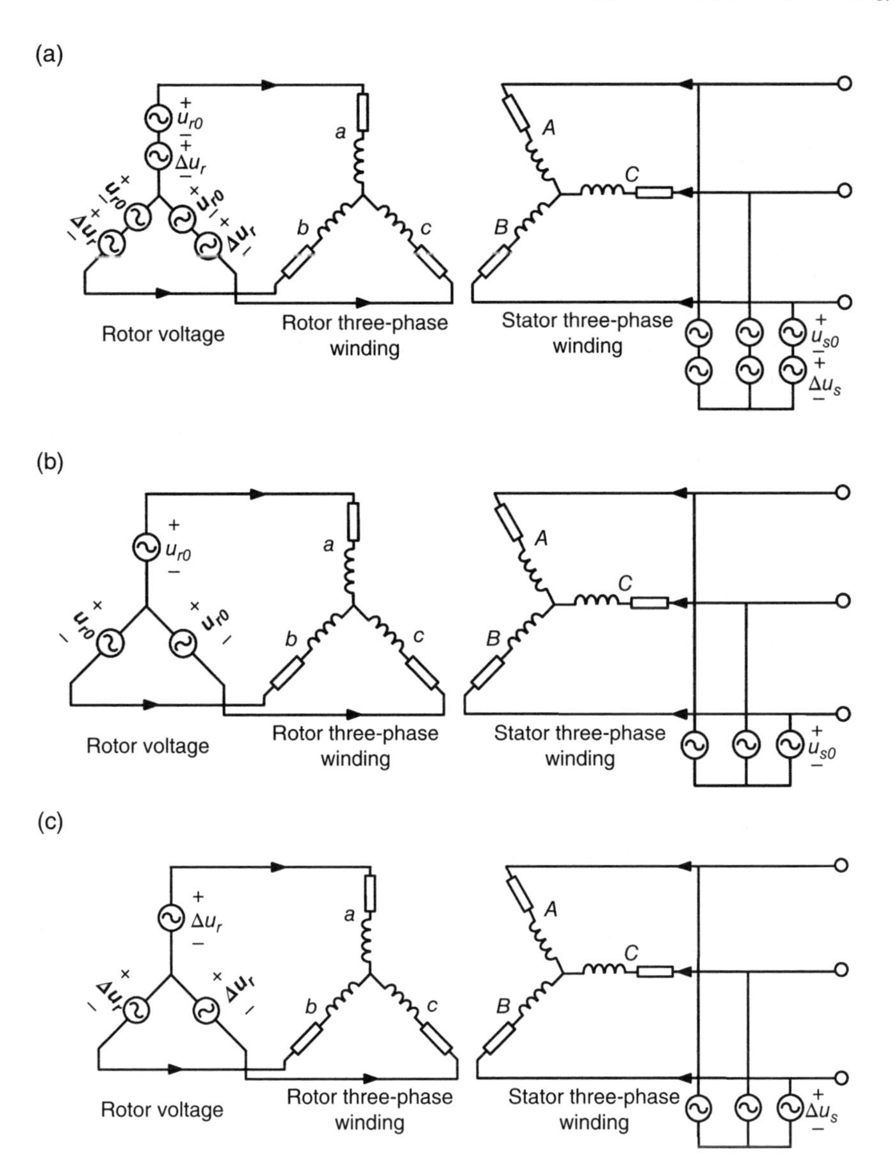

Figure 3.15 DFIG fault equivalent network considering the rotor-side converter control in the case of a three-phase fault. (a) Fault network. (b) Normal-operation network. (c) Superimposed network.

$$\boldsymbol{\Delta u}_\mathbf{r} = k_\mathrm{p}\left(\boldsymbol{\Delta i}_\mathbf{r}^* - \boldsymbol{\Delta i}_\mathbf{r}\right) + k_\mathrm{i}\int\left(\boldsymbol{\Delta i}_\mathbf{r}^* - \boldsymbol{\Delta i}_\mathbf{r}\right)\mathrm{d}t + \mathrm{j}\omega\sigma L_\mathrm{r}\boldsymbol{\Delta i}_\mathbf{r} \tag{3.48}$$

where $\sigma = 1 - L_\mathrm{m}^2/L_\mathrm{s}L_\mathrm{r}$, $\boldsymbol{\Delta i}_\mathbf{r}^*$ is the rotor current reference vector, and $\boldsymbol{\Delta i}_\mathbf{r}$ is the rotor current increment. k_p and k_i are the proportion parameter and integration parameter of the rotor-side converter current inner loop PI controller.

Suppose a symmetrical short circuit fault occurs in the grid at $t = t_1$, the normal-operation network is as shown in Figure 3.15(b), where the stator and rotor currents are

$$\begin{cases} \boldsymbol{i}_{s0} = \left(\dfrac{P_{s.ref} + jQ_{s.ref}}{3\boldsymbol{u}_{s0}/2}\right)^* \\ \boldsymbol{i}_{r0} = \left(\dfrac{\psi_{sm}}{L_m} - \dfrac{2L_s Q_{s.ref}}{3L_m^2 \boldsymbol{u}_{s0}}\right) + j\left(-\dfrac{2L_s P_{s.ref}}{3L_m \boldsymbol{u}_{s0}}\right) \end{cases} \tag{3.49}$$

where ψ_{sm} is the amplitude of the stator flux. $P_{s.\,ref}$ and $Q_{s.\,ref}$ are the reference values of the stator-side active power and reactive power, which can be preset.

The superimposed network is shown in Figure 3.15(c), the mathematical model of which is

$$\begin{cases} \Delta\boldsymbol{u}_s = R_s\Delta\boldsymbol{i}_s + d\Delta\boldsymbol{\psi}_s/dt + j\omega_1\Delta\boldsymbol{\psi}_s \\ \Delta\boldsymbol{u}_r = R_r\Delta\boldsymbol{i}_r + d\Delta\boldsymbol{\psi}_r/dt + js\omega_1\Delta\boldsymbol{\psi}_r \\ \Delta\boldsymbol{\psi}_s = L_s\Delta\boldsymbol{i}_s + L_m\Delta\boldsymbol{i}_r \\ \Delta\boldsymbol{\psi}_r = L_m\Delta\boldsymbol{i}_s + L_r\Delta\boldsymbol{i}_r \end{cases} \tag{3.50}$$

According to Equations (3.50) and the flux conservation law, with the initial value of the stator flux being zero, the stator flux in the superimposed network is

$$\Delta\boldsymbol{\psi}_s = \frac{\Delta\boldsymbol{u}_s - \Delta\boldsymbol{u}_s e^{\tau_1 t_1} e^{-\tau_1 t}}{R_s/L_s + j\omega_1} \tag{3.51}$$

where $\tau_1 = R_s/L_s + j\omega_1$ is the stator transient flux attenuation constant.

Apply Equations (3.48) and (3.51) to the rotor voltage equation in (3.50), so that the rotor current second-order differential equation in the superimposed network is

$$\frac{d^2\Delta\boldsymbol{i}_r}{dt^2} + \beta_1\frac{d\Delta\boldsymbol{i}_r}{dt} + \beta_2\Delta\boldsymbol{i}_r = \beta_2\Delta\boldsymbol{i}_r^* + m_r e^{\tau_1 t_1} e^{-\tau_1 t} \tag{3.52}$$

where $\beta_1 = (R_r + k_p)/(\sigma L_r)$, $\beta_2 = k_i/(\sigma L_r)$, $m_r = L_m\tau_1(1-s)\Delta\boldsymbol{u}_{s0}/(\sigma L_r L_s)$.

Apply the stator flux equation in Equations (3.50)–(3.52), so that the stator current second-order differential equation in the superimposed network is

$$\frac{d^2\Delta\boldsymbol{i}_s}{dt^2} + \beta_1\frac{d\Delta\boldsymbol{i}_s}{dt} + \beta_2\Delta\boldsymbol{i}_s = m_{s1} + m_{s2}e^{-\tau_1 t} \tag{3.53}$$

where $m_{s1} = \frac{k_i\Delta\boldsymbol{u}_{s0} - k_i\tau_1 L_m\Delta\boldsymbol{i}_r^*}{\tau_1\sigma L_r L_s}$, $m_{s2} = \frac{m\Delta\boldsymbol{u}_{s0}}{\tau_1\sigma L_r L_s^2}e^{\tau_1 t_1}$, $m = \tau_1 L_s(R_r + k_p) + (s-1)L_m^2\tau_1^2 - \sigma L_r L_s - k_i L_s$.

Combining Equations (3.52) and (3.53), the stator and rotor current increments in the superimposed network can be calculated:

$$\Delta\boldsymbol{i}_r = \Delta\boldsymbol{i}_{r0} + \Delta\boldsymbol{i}_{rm} + \Delta\boldsymbol{i}_{r1} \tag{3.54}$$

$$\Delta\boldsymbol{i}_s = \Delta\boldsymbol{i}_{s0} + \Delta\boldsymbol{i}_{sm} + \Delta\boldsymbol{i}_{s1} \tag{3.55}$$

In Equations (3.54) and (3.55), $\Delta\boldsymbol{i}_{r0}$ and $\Delta\boldsymbol{i}_{s0}$ are the forced components of the rotor and stator currents, which are only related to the generator parameters, the terminal voltage drop rate and the fault instant.

$$\begin{cases} \boldsymbol{\Delta i}_{\mathrm{r0}} = \boldsymbol{\Delta i}_{\mathrm{r}}^{*} \\ \boldsymbol{\Delta i}_{\mathrm{s0}} = \dfrac{\boldsymbol{\Delta u}_{\mathrm{s0}} - \tau_1 L_{\mathrm{m}} \boldsymbol{\Delta i}_{\mathrm{r}}^{*}}{\tau_1 L_{\mathrm{s}}} \end{cases} \tag{3.56}$$

$\boldsymbol{\Delta i}_{\mathrm{rm}}$ and $\boldsymbol{\Delta i}_{\mathrm{sm}}$ are the rotor and stator transient DC components in the superimposed network, which result from the principle that rotor and stator flux cannot mutate at the fault instant, and have to do with the control parameters of the rotor-side converter.

$$\begin{cases} \boldsymbol{\Delta i}_{\mathrm{rm}} = \dfrac{L_{\mathrm{m}} \tau_1 (1-s) \boldsymbol{\Delta u}_{\mathrm{s0}}}{\sigma L_{\mathrm{r}} L_{\mathrm{s}} (\tau_1^2 - \tau_1 \beta_1 + \beta_2)} \mathrm{e}^{\tau_1 t_1} \mathrm{e}^{-\tau_1 t} \\ \boldsymbol{\Delta i}_{\mathrm{sm}} = \dfrac{m \boldsymbol{\Delta u}_{\mathrm{s0}}}{\tau_1 \sigma L_{\mathrm{r}} L_{\mathrm{s}}^2 (\tau_1^2 - \tau_1 \beta_1 + \beta_2)} \mathrm{e}^{\tau_1 t_1} \mathrm{e}^{-\tau_1 t} \end{cases} \tag{3.57}$$

In Equations (3.54) and (3.55), the transient current natural components $\boldsymbol{\Delta i}_{\mathrm{r1}}$ and $\boldsymbol{\Delta i}_{\mathrm{s1}}$ are current increments in the superimposed network caused by the coupling control of the stator flux and the rotor-side converter. The attenuation time constants α_1 and α_2 are affected by the PI parameters, the terminal voltage drop rate and the control parameters of the rotor-side converter.

$$\begin{cases} \boldsymbol{\Delta i}_{\mathrm{r1}} = \dfrac{\alpha_1 \mathrm{e}^{\alpha_2 t} - \alpha_2 \mathrm{e}^{\alpha_1 t}}{\alpha_1 - \alpha_2} \boldsymbol{i}_{\mathrm{r0}} \\ \boldsymbol{\Delta i}_{\mathrm{s1}} = \dfrac{L_{\mathrm{m}} \boldsymbol{i}_{\mathrm{r0}} (\alpha_2 \mathrm{e}^{\alpha_1 t} - \alpha_1 \mathrm{e}^{\alpha_2 t})}{L_{\mathrm{s}} (\alpha_1 - \alpha_2)} \end{cases} \tag{3.58}$$

$$\begin{cases} \alpha_1 = \dfrac{-(R_{\mathrm{r}} + k_{\mathrm{p}}) + \sqrt{(R_{\mathrm{r}} + k_{\mathrm{p}})^2 - 4k_{\mathrm{i}} \sigma L_{\mathrm{r}}}}{2\sigma L_{\mathrm{r}}} \\ \alpha_1 = \dfrac{-(R_{\mathrm{r}} + k_{\mathrm{p}}) - \sqrt{(R_{\mathrm{r}} + k_{\mathrm{p}})^2 - 4k_{\mathrm{i}} \sigma L_{\mathrm{r}}}}{2\sigma L_{\mathrm{r}}} \end{cases} \tag{3.59}$$

According to Equations (3.49) and (3.54)–(3.59), the rotor and stator short circuit currents in the fault equivalent network considering the rotor-side converter transient regulation can be expressed as

$$\boldsymbol{i}_{\mathrm{r}} = \boldsymbol{i}_{\mathrm{r0}} + \boldsymbol{\Delta i}_{\mathrm{r0}} + \boldsymbol{\Delta i}_{\mathrm{rm}} + \boldsymbol{\Delta i}_{\mathrm{r1}} \tag{3.60}$$

$$\boldsymbol{i}_{\mathrm{s}} = \boldsymbol{i}_{\mathrm{s0}} + \boldsymbol{\Delta i}_{\mathrm{s0}} + \boldsymbol{\Delta i}_{\mathrm{sm}} + \boldsymbol{\Delta i}_{\mathrm{s1}} \tag{3.61}$$

According to Equations (3.60) and (3.61), in the fault equivalent network, the stator and rotor short circuit currents both contain a forced component, a transient DC component and a transient natural component, and have to do with the generator parameters and PI controller parameters.

Based on the above analysis, when a fault occurs in the grid, the DFIG terminal voltage drops instantaneously, resulting in the coupling control of DFIG electromagnetic control and rotor-side converter control. The stator and rotor currents increase instantaneously, causing imbalance in the power exchange on two sides of the DC bus, and fluctuation in the DC bus voltage. The grid-side converter will regulate to suppress the voltage fluctuation, which will further affect the transient characteristics of the stator and rotor currents. Therefore, the study of the DC bus voltage fluctuation characteristic is the foundation of the analysis of grid-side converter transient characteristics.

B. Analysis of DC Bus Voltage Fluctuation Characteristics

Consider that the DFIG rotor-side converter applies a stator flux oriented vector control strategy, i.e. $u_{ds} = 0, u_{qs} = U_s$; according to Equations (3.1) and (3.4), the steady-state stator and rotor fluxes are

$$\begin{cases} \boldsymbol{\psi}_{\mathbf{s}} = \psi_{ds} + j\psi_{qs} = jU_s/\omega_1 \\ \boldsymbol{\psi}_{\mathbf{r}} = \left(L_s L_r - L_m^2\right)\boldsymbol{i}_{\mathbf{r}}/L_s + L_m \boldsymbol{\psi}_{\mathbf{s}}/L_s \end{cases} \tag{3.62}$$

where U_s is the grid voltage amplitude.

In Equations (3.1)–(3.4), considering that $L_{ls} \ll L_m$ and $L_{lr} \ll L_m$, the rotor voltage and grid-side current can be expressed as

$$\boldsymbol{u}_{\mathbf{r}} \approx \boldsymbol{u}_{\mathbf{s}} - j\omega_r \boldsymbol{\psi}_{\mathbf{s}} = jU_s - j\omega_r \boldsymbol{\psi}_{\mathbf{s}} \tag{3.63}$$

$$i_{gq} = \left(1 - \omega_r/\omega_1\right) i_{rq} = s i_{rq} \tag{3.64}$$

In Figure 3.15, neglecting the power loss of reactors and the switching power devices, the grid-side converter input power P_g can be expressed as

$$P_g = \mathrm{Re}\left[3\boldsymbol{u}_{\mathbf{g}}\boldsymbol{i}_{\mathbf{g}}^*\right] \approx 3u_{gq}i_{gq} \tag{3.65}$$

The rotor-side converter input power P_r is

$$P_r = \mathrm{Re}\left[3\boldsymbol{u}_{\mathbf{r}}\boldsymbol{i}_{\mathbf{r}}^*\right] \approx 3\left(u_{sq}i_{rq} - \psi_s i_{rq}\omega_r\right) \tag{3.66}$$

In the grid-side converter, the power storage of the DC bus capacitance is $W = C_{dc}U_{dc}^2/2$, thus the DC bus voltage balance formula is

$$\frac{dW}{dt} = \frac{1}{2}C_{dc}\frac{dU_{dc}^2}{dt} = P_g - P_r = 0 \tag{3.67}$$

According to Equation (3.67), when the DFIG is in steady-state operation, due to the regulatory control of the grid-side converter, the power exchange between the two converters is 0, and the DC bus voltage is constant.

When a short circuit fault occurs in the grid, to facilitate the analysis of the DC bus voltage fluctuation characteristic, the terminal voltage phase jump is neglected. Taking the super-synchronization condition, for example, the post-fault terminal voltage is

$$\boldsymbol{u}_{\mathbf{s}}' = k\boldsymbol{u}_{\mathbf{s}} = jkU_{s0} \tag{3.68}$$

where k is the terminal voltage amplitude drop rate, $k = 1 - A$. U_{s0} is the steady-state terminal voltage amplitude.

According to the flux conservation law and Equations (3.1) and (3.68), the post-fault stator flux can be obtained:

$$\boldsymbol{\psi}_{\mathbf{s}}' = \frac{k\boldsymbol{u}_{\mathbf{s}} + (1-k)\boldsymbol{u}_{\mathbf{s}} e^{\tau_1 t_1} e^{-\tau_1 t}}{j\omega_1} \tag{3.69}$$

The post-fault rotor voltage can be obtained according to Equations (3.2) and (3.69):

$$\boldsymbol{u}_{\mathbf{r}} = jkU_{s0}\left(1 - \frac{\omega_r}{\omega_1}\right) - j(1-k)U_{s0}\frac{\omega_r}{\omega_1} e^{\tau_1 t_1} e^{-\tau_1 t} \tag{3.70}$$

And then, according to Equations (3.67) and (3.69), the post-fault grid-side converter power P_g and the rotor-side converter power P_r are

$$P_g \approx 3kU_{s0} s i_{rq} \tag{3.71}$$

$$P_r = 3kU_{s0} s i_{rq} - 3(1-s)(1-k)U_{s0} i_{rq} e^{\tau_1 t_1} e^{-\tau_1 t} \tag{3.72}$$

According to Equation (3.67), the DC-side power disturbance is

$$\frac{1}{2}C_{dc}\frac{dU_{dc}^2}{dt} = \Delta P_c = 3(1-s)(1-k)U_{s0} i_{rq} e^{\tau_1 t_1} e^{-\tau_1 t} \tag{3.73}$$

Consider that the grid-side converter applies a grid voltage oriented vector control strategy, if the d-axis of the synchronous rotating dq coordinate system is fixed in the direction of the grid voltage vector $\boldsymbol{u}_{\mathbf{s}}$, then the static gain of the DC bus voltage is approximately

$$\left|G_{pedc}(j\omega_1)\right| = \frac{2}{C_{dc}\omega_1} \tag{3.74}$$

According to Equations (3.73) and (3.74), the DC bus voltage disturbance is

$$\Delta U_{dc} = \frac{6(1-k)(1-s)U_{s0}}{C_{dc}\omega_1} i_{rq} e^{\tau_1 t_1} e^{-\tau_1 t} \tag{3.75}$$

It can be seen from Equations (3.73) and (3.75) that the DC-side disturbances are not only related to the terminal voltage drop rate and the generator rotational speed, but are also closely related to the rotor short circuit current. The further the terminal voltage drops, the greater the DC bus voltage fluctuation. Therefore, to suppress the fluctuation of the DC bus voltage, the grid-side converter will regulate to reduce the DC bus voltage difference gradually, until it remains in the vicinity of the reference value.

C. *Analysis of the Grid-Side Converter Control*

Shown in Figure 3.16 is the control model of the grid-side converter. By combining the grid voltage feed-forward control and feedback control, independent control of grid d-axis current and q-axis current can be achieved, as well as the AC-side unit power factor control and DC bus voltage stability control.

According to Equation (3.60), the rotor current q-axis component can be expressed as

$$i_{rq} = I_{rq0} + I_{rqm} e^{-\tau_1 t} + I_{rq1} e^{\alpha_1 t} + I_{rq2} e^{\alpha_2 t} \tag{3.76}$$

And according to Equations (3.75) and (3.76), the post-fault DC bus voltage is

$$U_{dc}(t) = U_{dc.ref} + U_{dc1} e^{-\tau_1 t} + U_{dc2} e^{-2\tau_1 t} + U_{dc3} e^{-\eta_1 t} + U_{dc4} e^{-\eta_2 t} \tag{3.77}$$

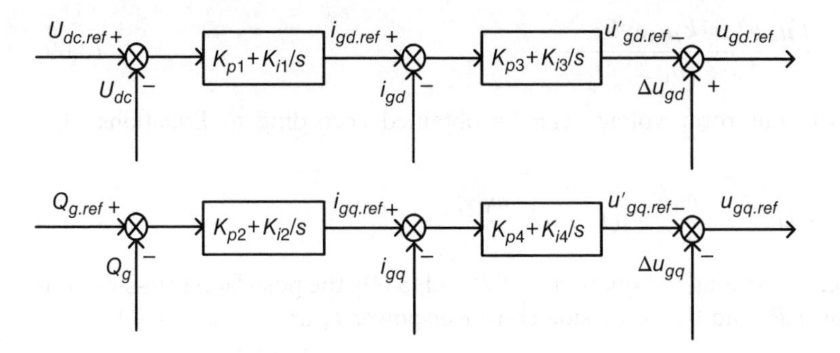

Figure 3.16 Control model of the grid-side converter.

where $m_{\text{dc}} = 6(1-k)(1-s)U_{\text{s0}}\text{e}^{\tau_1 t_1}/(C_{\text{dc}}\omega_1)$, $U_{\text{dc1}} = m_{\text{dc}}I_{\text{rq0}}$, $U_{\text{dc2}} = m_{\text{dc}}I_{\text{rqm}}$, $U_{\text{dc3}} = m_{\text{dc}}I_{\text{rq1}}$, $U_{\text{dc4}} = m_{\text{dc}}I_{\text{rq2}}$, $\eta_1 = \tau_1 - \alpha_1$, $\eta_2 = \tau_1 - \alpha_2$.

After the fault occurs, the grid-side converter input reactive power is

$$\begin{aligned} Q_{\text{g}}(t) &= \text{Im}\left(3\boldsymbol{u}_{\mathbf{s}}\boldsymbol{i}_{\mathbf{g}}^{*}\right) = 3u_{\text{sd}}i_{\text{gq}} = 3kU_{\text{s0}}s\left(I_{\text{rq0}} + I_{\text{rqm}}\text{e}^{-\tau_1 t} + I_{\text{rq1}}\text{e}^{\alpha_1 t} + I_{\text{rq2}}\text{e}^{\alpha_2 t}\right) \\ &= Q_{\text{g0}} + Q_{\text{gm}}\text{e}^{-\tau_1 t} + Q_{\text{g1}}\text{e}^{\alpha_1 t} + Q_{\text{g2}}\text{e}^{\alpha_2 t} \end{aligned} \tag{3.78}$$

where $Q_{\text{g0}} = 3kU_{\text{s0}}sI_{rq0}$, $Q_{\text{gm}} = 3kU_{\text{s0}}sI_{\text{rqm}}$, $Q_{\text{g1}} = 3kU_{\text{s0}}sI_{\text{rq1}}$, $Q_{\text{g2}} = 3kU_{\text{s0}}sI_{\text{rq2}}$.

Applying Laplace transformation to Equations (3.77) and (3.78), the corresponding Laplace equations become

$$\begin{cases} U_{\text{dc}} - U_{\text{dc.ref}} = \left(\dfrac{U_{\text{dc1}}}{s+\tau_1} + \dfrac{U_{\text{dc2}}}{s+2\tau_1} + \dfrac{U_{\text{dc3}}}{s+\eta_1} + \dfrac{U_{\text{dc4}}}{s+\eta_2}\right) \\ Q_{\text{g}} - Q_{\text{g.ref}} = \left(\dfrac{Q_{\text{gm}}}{s+\tau_1} + \dfrac{Q_{\text{g1}}}{s-\alpha_1} + \dfrac{Q_{\text{g2}}}{s-\alpha_2}\right) \end{cases} \tag{3.79}$$

According to Equations (3.79) and the transfer function of the outer loop PI controller in Figure 3.16, the Laplace expressions of the dq components of the grid-side current reference value are

$$\begin{cases} I_{\text{gd.ref}}(s) = \left(K_{\text{p1}} + \dfrac{K_{\text{i1}}}{s}\right)\left(\dfrac{U_{\text{dc1}}}{s+\tau_1} + \dfrac{U_{\text{dc2}}}{s+2\tau_1} + \dfrac{U_{\text{dc3}}}{s+\eta_1} + \dfrac{U_{\text{dc4}}}{s+\eta_2}\right) \\ I_{\text{gq.ref}}(s) = \left(K_{\text{p2}} + \dfrac{K_{\text{i2}}}{s}\right)\left(\dfrac{Q_{\text{gm}}}{s+\tau_1} + \dfrac{Q_{\text{g1}}}{s-\alpha_1} + \dfrac{Q_{\text{g2}}}{s-\alpha_2}\right) \end{cases} \tag{3.80}$$

where K_{p1}, K_{i1} and K_{p2}, K_{i2} are respectively the PI parameters of the grid-side converter voltage outer loop and power outer loop controllers.

Applying Laplace inverse transformation to Equations (3.80), the time-domain expressions of the dq components of the grid-side current reference value are

$$\begin{cases} i_{\text{gd.ref}}(t) = D_0 + D_1\text{e}^{-\tau_1 t} + D_2\text{e}^{-2\tau_1 t} + D_3\text{e}^{-\eta_1 t} + D_4\text{e}^{-\eta_2 t} \\ i_{\text{gq.ref}}(t) = C_{\text{g0}} + C_{\text{gm}}\text{e}^{-\tau_1 t} + C_{\text{g1}}\text{e}^{\alpha_1 t} + C_{\text{g2}}\text{e}^{\alpha_2 t} \end{cases} \tag{3.81}$$

where the calculation formulas for coefficients D_0, D_1, D_2, D_3, D_4 and C_{g0}, C_{gm}, C_{g1}, C_{g2} are shown in Appendix A.

Applying Laplace transformation to Equation (3.81) and combining the current loop in Figure 3.16, the analytical expressions of the dq components of the grid-side voltage reference value can be calculated via the PI controller:

$$\begin{cases} U_{\mathrm{gd.ref}}(s) - \Delta U_{\mathrm{gd}}(s) = -\left(K_{\mathrm{p3}} + \dfrac{K_{\mathrm{i3}}}{s}\right)\left(\dfrac{D_0}{s+\tau_1} + \dfrac{D_2}{s+2\tau_1} + \dfrac{D_3}{s+\eta_1} + \dfrac{D_4}{s+\eta_2}\right) \\ U_{\mathrm{gq.ref}}(s) - \Delta U_{\mathrm{gq}}(s) = -\left(K_{\mathrm{p4}} + \dfrac{K_{\mathrm{i4}}}{s}\right)\left(\dfrac{C_{\mathrm{gm}}}{s+\tau_1} + \dfrac{C_{\mathrm{g1}}}{s-\alpha_1} + \dfrac{C_{\mathrm{g2}}}{s-\alpha_2}\right) \end{cases} \tag{3.82}$$

where K_{p3}, K_{i3} and K_{p4}, K_{i4} are the PI parameters of the grid-side converter d-axis and q-axis current loop controllers.

Applying Laplace inverse transformation to Equations (3.82), the time-domain expressions of the grid-side converter voltage reference values are

$$\begin{cases} u_{\mathrm{gd.ref}}(t) = A_0 + A_1 \mathrm{e}^{-\tau_1 t} + A_2 \mathrm{e}^{-2\tau_1 t} + A_3 \mathrm{e}^{-\eta_1 t} + A_4 \mathrm{e}^{-\eta_2 t} \\ u_{\mathrm{gq.ref}}(t) = B_0 + B_{\mathrm{m}} \mathrm{e}^{-\tau_1 t} + B_1 \mathrm{e}^{\alpha_1 t} + B_2 \mathrm{e}^{\alpha_2 t} \end{cases} \tag{3.83}$$

where the calculation formulas for coefficients A_0, A_1, A_2, A_3, A_4 and B_0, B_m, B_1, B_2 are shown in Appendix B.

It can be seen from Equations (3.83) that when a symmetrical short circuit fault occurs in the grid, the d-axis and q-axis components of the reference voltage output of the grid-side converter controller link contain the following components: DC component which is only affected by the grid-side converter and which attenuates in the time constant τ_1, transient DC component and transient natural component which are jointly affected by the grid-side converter and rotor-side converter. Also, the amplitude of each component has to do with the converter PI parameters, the fault instant, the attenuation time constant, the DC bus voltage oscillation amplitude and the grid-side converter power oscillation amplitude.

D. Analytical Expressions of the Stator and Rotor Short Circuit Currents

Equations (3.81) and (3.83) are both obtained under the synchronous rotating coordinate system. Converting the two equations to the three-phase stationary coordinate system, the grid-side converter voltage and current reference values mainly contain the fundamental frequency component and the second harmonic component.

According to the PWM modulation principle [5], via double PWM modulation, the grid-side converter $2f_1$ current component will produce a corresponding $2f_1$ harmonic current on the rotor side, and the relative angular speed of the rotating magnetic field caused by this harmonic current is $2\omega_1 - \omega_r = (1+s)\omega_1$. This current can be expressed as

$$\boldsymbol{i}_{\mathbf{r}(1+\mathbf{s})} = \frac{U_{\mathrm{dc.ref}}}{2U_{\mathrm{tri}}} f\left(\boldsymbol{U}_{\mathbf{ge}(2\mathbf{f}_1)}\right) \tag{3.84}$$

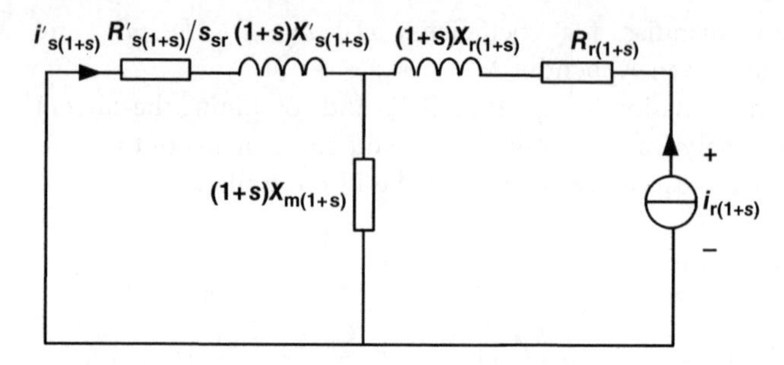

Figure 3.17 Equivalent circuit of the rotor $(1+s)f_1$ current component.

where $\boldsymbol{U}_{\mathbf{ge}(2\mathbf{f}_1)}$ is the voltage component in the grid-side converter AC-side modulation signal whose frequency is $2f_1$. U_{tri} is the carrier signal amplitude.

For the $(1+s)f_1$ current component, due to the coupling between the generator internal windings, a $2\omega_1$ short circuit current will be induced in the stator windings. Suppose that the grid voltage only contains the fundamental frequency component, then the grid side is approximately a short circuit. Converting the stator winding parameters to the rotor side, the equivalent circuit of the rotor $(1+s)f_1$ current component can be obtained, as shown in Figure 3.17.

It can be seen from Figure 3.17 that the second harmonic component of the stator short circuit current is

$$\boldsymbol{i}'_{\mathbf{s}(1+\mathbf{s})} = -\frac{\mathrm{j}(1+s)X_{\mathrm{m}(1+\mathrm{s})}\boldsymbol{i}_{\mathbf{r}(1+\mathbf{s})}}{Z'_{\mathrm{s}(1+\mathrm{S})} + \mathrm{j}(1+s)X_{\mathrm{m}(1+\mathrm{s})}} \tag{3.85}$$

$$\begin{cases} Z'_{\mathrm{s}(1+\mathrm{S})} = R'_{\mathrm{s}(1+\mathrm{s})}/s_{\mathrm{sr}} + \mathrm{j}(1+s)X'_{\mathrm{s}(1+\mathrm{s})} \\ s_{\mathrm{sr}} = 2\omega_1/((1+s)\omega_1) = 2/(1+s) \\ X'_{\mathrm{s}(1+\mathrm{s})} = \omega_1 l'_{\mathrm{ls}(1+\mathrm{s})} \\ X_{\mathrm{m}(1+\mathrm{s})} = \omega_1 l_{\mathrm{m}(1+\mathrm{s})} \end{cases} \tag{3.86}$$

where s_{sr} is the equivalent slip in the equivalent circuit of the rotor $(1+s)f_1$ current component. $X'_{\mathrm{s}(1+\mathrm{s})}$ is the equivalent stator leakage inductance converted to the rotor side. $X_{\mathrm{m}(1+\mathrm{s})}$ is the equivalent mutual inductance.

The generation mechanism of the second harmonic component in the stator short circuit current is as follows. When a fault occurs in the grid, due to the control effect of the stator flux and rotor-side converter, currents containing a forced component, a transient DC component and a transient natural component will be induced in the stator and rotor windings. This will cause the DC bus voltage and the grid-side reactive power to oscillate, thus the grid-side converter will regulate, which will generate second harmonic components in the grid-side converter voltage and current. Then, via double PWM modulation, a $(1+s)f_1$ harmonic current component will be induced in the rotor windings. Finally, the second harmonic current component will be induced in the stator windings. Therefore, when a three-phase fault occurs in the grid, the stator and rotor short circuit whole currents can be expressed as

$$\boldsymbol{i}_{\mathbf{s}} = \boldsymbol{i}_{\mathbf{s}0} + \boldsymbol{\Delta i}_{\mathbf{s}} + \boldsymbol{i}'_{\mathbf{s}(1+s)} \tag{3.87}$$

$$\boldsymbol{i}_{\mathbf{r}} = \boldsymbol{i}_{\mathbf{r}0} + \boldsymbol{\Delta i}_{\mathbf{r}} + \boldsymbol{i}_{\mathbf{r}(1+s)} \tag{3.88}$$

where $\boldsymbol{i}_{\mathbf{s}0}$ and $\boldsymbol{i}_{\mathbf{r}0}$ are current increments in the normal-operation network. $\boldsymbol{\Delta i}_{\mathbf{s}}$ and $\boldsymbol{\Delta i}_{\mathbf{r}}$ are current increments in the superimposed network. $\boldsymbol{i}'_{\mathbf{s}(1+s)}$ and $\boldsymbol{i}_{\mathbf{r}(1+s)}$ are harmonic current increments caused by the transient regulatory control of the grid-side converter and rotor-side converter.

E. Simulation Verification

In order to verify the correctness of the short circuit current expressions derived here, as well as the converter regulation mechanism in the fault process, and to analyse the transient characteristics of DFIG stator and rotor currents, a 2 MW DFIG SMIB (single machine infinite bus) simulation system was established in PSCAD/EMTDC, as shown in Figure 3.6.

1. Short Circuit Currents Suppose that at $t = 2$ s, a three-phase fault occurs at the high-voltage side of the transformer, and the DFIG terminal voltage drops to about 60%. The stator and rotor three-phase short circuit currents can be calculated according to Equations (3.87) and (3.88). Taking phase A, for example, the rotor and stator current waveforms are as shown in Figure 3.18.

It can be seen that the calculated rotor and stator currents are basically consistent with the simulation results in amplitude and variation trend. The calculated values exhibit obvious transient attenuation, which is in line with the transient characteristics of rotor and stator short circuit currents at the fault instant and during the fault process. Especially in the first cycle, the calculated waveform is approximately the same as the simulation waveform. However, after 200 ms, the calculated waveform of the rotor current gradually lags behind the simulation waveform. This is because, in the calculation process, the rotor rotational speed change is not considered, i.e. the calculated rotor current frequency remains unchanged. However, in actual operation, in order to maintain the grid frequency at the same level, the DFIG rotor current frequency will increase. Considering that the grid fault duration is relatively short, the error caused by the steady-state rotational speed can be neglected.

2. DC Bus Voltage Fluctuation Characteristics The comparison between the DC capacitance power difference curve and DC bus voltage deviation curve is as shown in Figure 3.19. It can be seen that when a three-phase fault occurs in the grid, the DC capacitance power difference curve and the DC bus voltage deviation curve are basically the same in amplitude and variation trend. This is because, when a fault occurs, the power difference on the two sides of the capacitance is no longer zero, which causes considerable fluctuation in the capacitance voltage. With a big enough capacity, the grid-side converter will regulate to reduce the DC bus voltage deviation and power difference

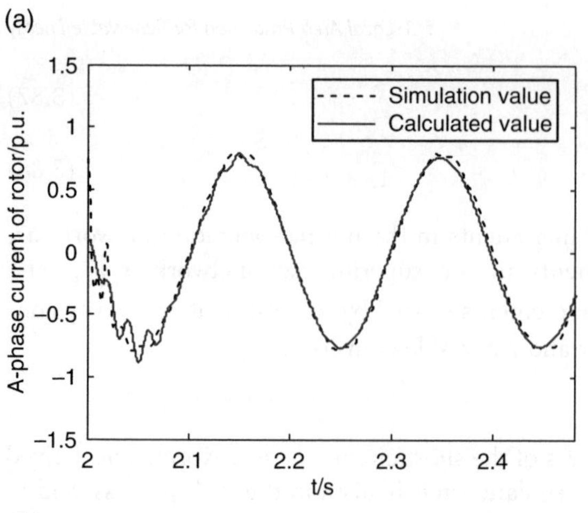

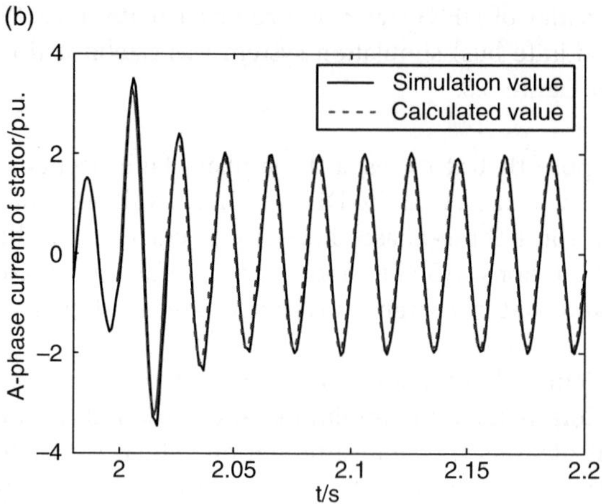

Figure 3.18 Stator and rotor short circuit currents considering the converter transient regulation. (a) Rotor phase A current. (b) Stator phase A current.

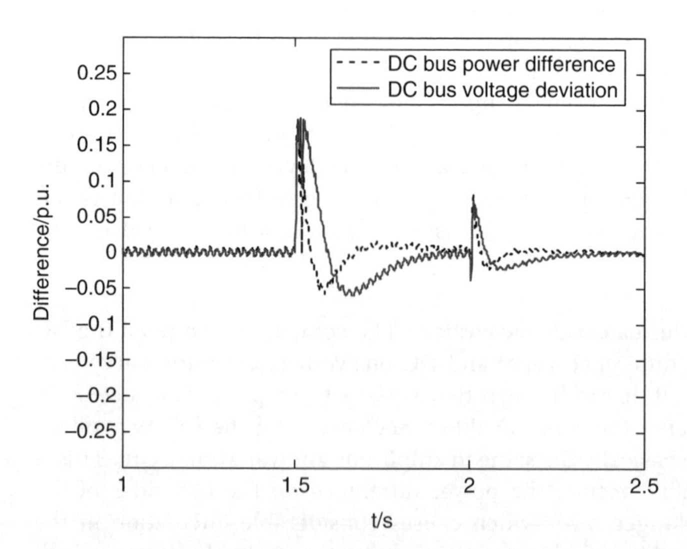

Figure 3.19 Comparison between DC bus power difference curve and voltage deviation curve in the case of grid short circuit fault.

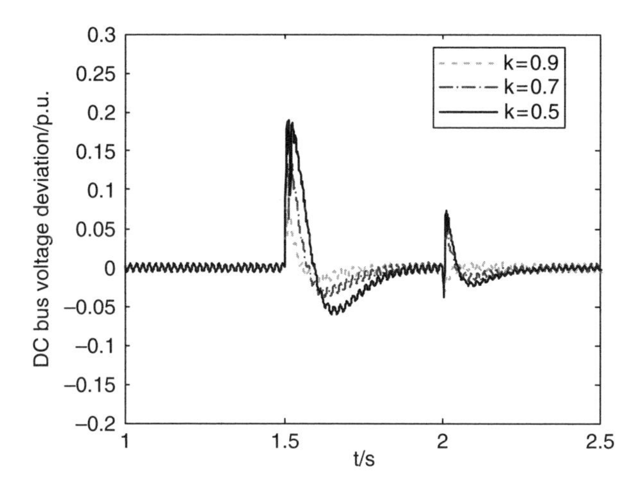

Figure 3.20 DC bus voltage deviation curves with different terminal voltage drops.

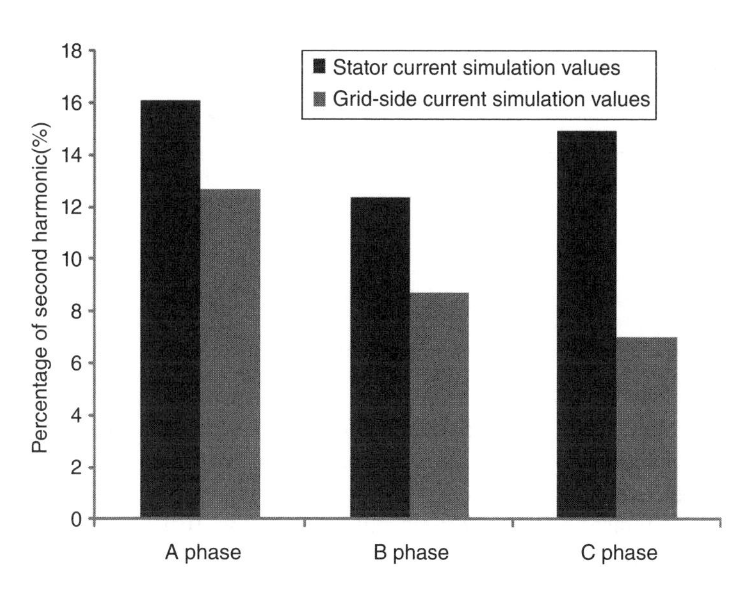

Figure 3.21 Percentage of second harmonics in the short circuit currents.

gradually, until they are stabilized to the vicinity of the reference values. Also, it can be seen from Figure 3.20 that the more serious the grid fault, i.e. the further the terminal voltage drops, the bigger the maximum value of DC bus voltage deviation, and the larger the fluctuation of DC bus voltage deviation.

3. Second Harmonic Component in the Stator Short Circuit Current To verify the second harmonic component in the stator short circuit current, the improved Fourier algorithm is used to extract the second harmonic in the short circuit current, as shown in Figure 3.21. The percentages of second harmonic

Table 3.1 Percentage of second harmonic in the stator short circuit current with different terminal voltage drops.

Terminal voltage/p.u.	Percentage of second harmonic in stator current/%		
	A	B	C
0.90	12.80	9.21	10.80
0.79	13.47	9.83	11.52
0.69	14.29	10.33	12.85
0.59	15.51	11.16	13.70
0.50	16.12	12.36	14.96

Table 3.2 Percentage of second harmonic in the grid-side short circuit current with different terminal voltage drops.

Terminal voltage/p.u.	Percentage of second harmonic in grid-side current/%		
	A	B	C
0.90	7.13	4.63	5.63
0.79	8.09	5.69	6.38
0.69	9.66	6.82	6.49
0.59	10.58	7.34	6.79
0.50	12.70	8.74	7.03

component in the stator short circuit currents and grid-side short circuit currents are shown in Tables 3.1 and 3.2, respectively. It can be seen that, when the terminal voltage drops to about 0.5 p.u., the percentage of second harmonic in the stator phase A/B/C short circuit currents is 16.12%, 12.36% and 14.96%, respectively, and that in the grid-side phase A/B/C short circuit currents are 12.70%, 8.74% and 7.03%, respectively. This demonstrates that, when a fault occurs in the grid, due to the comprehensive regulatory control of the grid-side converter and rotor-side converter, the second harmonic component will be generated in both the stator current and the grid-side current, which makes the transient characteristics of the short circuit currents even more complicated.

Also, it can be seen from Table 3.1 that, when the terminal voltage drops 10%, the percentage of second harmonic in phase A current is 12.80%, that in phase B current is 9.21% and that in phase C current is 10.80%. When the terminal voltage drops lower, for example, 50%, the percentage of second harmonic in the phase A current increases to 16.12%, that in phase B and phase C current also increases to 12.36% and 14.96%, respectively. A similar conclusion can also be made from Table 3.2. Thus, the simulation results demonstrate that the further the terminal voltage drops, the higher the percentage of second harmonic component in the short circuit current. This percentage even exceeds the setting value of the transformer second harmonic restraint (15%) in some cases [6], which might cause malfunction of the second harmonic restraint component.

3.2.3 DFIG Fault Transient Characteristics When Crowbar Protection Is Put into Operation

3.2.3.1 Short Circuit Current Calculation Model When Crowbar Resistance Remains Unchanged in the Fault Duration

A. Grid Symmetrical Short Circuit Fault

Suppose a three-phase symmetrical short circuit fault occurs in the grid at $t = t_2$; neglecting the phase jump of the DFIG terminal voltage, the pre-fault and post-fault DFIG terminal voltage vectors are

$$\boldsymbol{u}_{\mathbf{sc}} = \begin{cases} \boldsymbol{u}_{\mathbf{sc1}} = U_{\mathrm{sc0}} \mathrm{e}^{\mathrm{j}\varphi_{\mathrm{c}}}, t < t_2 \\ \boldsymbol{u}_{\mathbf{sc2}} = k_{\mathrm{c}} U_{\mathrm{sc0}} \mathrm{e}^{\mathrm{j}\varphi_{\mathrm{c}}}, t \geq t_2 \end{cases} \tag{3.89}$$

where k_{c} is the terminal voltage amplitude drop rate. U_{sc0} and φ_{c} are the amplitude and phase angle of the DFIG terminal voltage when the grid is in steady-state operation.

According to Equations (3.3) and (3.4), the rotor flux is

$$\boldsymbol{\psi}_{\mathrm{rc}} = \sigma L_{\mathrm{r}} \boldsymbol{i}_{\mathbf{rc}} + \frac{L_{\mathrm{m}}}{L_{\mathrm{s}}} \boldsymbol{\psi}_{\mathbf{sc}} \tag{3.90}$$

where $\sigma = 1 - L_{\mathrm{m}}^2/(L_{\mathrm{s}} L_{\mathrm{r}})$ is the DFIG flux leakage coefficient.

According to Equations (3.1) and (3.3), the integral first-order differential equation of the stator flux is

$$\frac{\mathrm{d}\boldsymbol{\psi}_{\mathbf{sc}}}{\mathrm{d}t} + \left(\frac{R_{\mathrm{s}}}{L_{\mathrm{s}}} + \mathrm{j}\omega_1\right)\boldsymbol{\psi}_{\mathbf{sc}} = \boldsymbol{u}_{\mathbf{sc}} + \frac{R_{\mathrm{s}} L_{\mathrm{m}}}{L_{\mathrm{s}}} \boldsymbol{i}_{\mathbf{rc}} \tag{3.91}$$

Due to the operation of the crowbar protection, the rotor voltage turns to zero instantaneously. Applying Equations (3.90) and (3.91) to the rotor voltage equation in (3.2), the stator second-order differential equation can be obtained, as shown in Equation (3.92). The stator flux eigenvalues s_1 and s_2 can be calculated by solving the second-order differential equation.

$$\frac{\mathrm{d}^2\boldsymbol{\psi}_{\mathbf{sc}}}{\mathrm{d}t^2} + \lambda_1 \frac{\mathrm{d}\boldsymbol{\psi}_{\mathbf{sc}}}{\mathrm{d}t} + \lambda_2 \boldsymbol{\psi}_{\mathbf{sc}} = \frac{\mathrm{d}\boldsymbol{u}_{\mathbf{sc}}}{\mathrm{d}t} + \frac{(R_{\mathrm{rc}} + \mathrm{j}\omega\sigma L_{\mathrm{r}})}{\sigma L_{\mathrm{r}}} \boldsymbol{u}_{\mathbf{sc}} \tag{3.92}$$

In Equation (3.92), the expressions for calculating coefficients λ_1 and λ_2 and eigenvalues s_1 and s_2 of the second-order differential equation can be found in Appendix A.

According to Equation (3.92), when a fault occurs in the grid, the stator flux is a second-order dynamic circuit without any negligence and approximation conditions, thus it could be used to reveal the variation pattern of the stator flux after a fault occurs.

Combining Equation (3.92) and the stator flux eigenvalues, it can be found that the post-fault stator flux contains a general solution and a particular solution:

$$\boldsymbol{\psi}_{\mathbf{sc}} = \boldsymbol{\psi}_{\mathbf{scf}} + \boldsymbol{\psi}_{\mathbf{scn}} = \underbrace{(R_{\mathrm{rc}} + \mathrm{j}\omega\sigma L_{\mathrm{r}})\boldsymbol{u}_{\mathbf{sc2}}/\lambda_2\sigma L_{\mathrm{r}}}_{\boldsymbol{\psi}_{\mathbf{scf}}} + \underbrace{f_{\mathrm{scn1}}(\omega_{\mathrm{r}})\mathrm{e}^{s_1 t}}_{\boldsymbol{\psi}_{\mathbf{scn1}}} + \underbrace{f_{\mathrm{scn2}}(\omega_{\mathrm{r}})\mathrm{e}^{s_2 t}}_{\boldsymbol{\psi}_{\mathbf{scn2}}} \tag{3.93}$$

$$\begin{cases} f_{\mathrm{scn1}}(\omega_{\mathrm{r}}) = -\dfrac{s_2}{(s_1 - s_2)} \dfrac{L_{\mathrm{s}}\boldsymbol{u}_{\mathbf{sc1}}}{(R_{\mathrm{s}} + \mathrm{j}\omega_1 L_{\mathrm{s}})} \\ f_{\mathrm{scn2}}(\omega_{\mathrm{r}}) = \dfrac{s_1}{(s_1 - s_2)} \dfrac{L_{\mathrm{s}}\boldsymbol{u}_{\mathbf{sc1}}}{(R_{\mathrm{s}} + \mathrm{j}\omega_1 L_{\mathrm{s}})} \end{cases} \tag{3.94}$$

where $\boldsymbol{\psi}_{\mathbf{scf}}$ is the particular solution of the stator flux differential equation and $\boldsymbol{\psi}_{\mathbf{scn}}$ is the general solution. $\boldsymbol{\psi}_{\mathbf{scf}}$ is the response of the stator flux to steady-state voltage $\boldsymbol{u}_{\mathbf{sc2}}$, i.e. the stator flux mandatory component. $\boldsymbol{\psi}_{\mathbf{scn}}$ is the stator flux transient natural component. $\boldsymbol{\psi}_{\mathbf{scn1}}$ is the slowly attenuating component, $\boldsymbol{\psi}_{\mathbf{scn2}}$ is the rapidly attenuating component.

Similarly, combining Equations (3.2)–(3.4), the second-order differential equation of the rotor flux is

$$\frac{\mathrm{d}^2\boldsymbol{\psi}_{\mathbf{rc}}}{\mathrm{d}t^2} + \lambda_1 \frac{\mathrm{d}\boldsymbol{\psi}_{\mathbf{rc}}}{\mathrm{d}t} + \lambda_2 \boldsymbol{\psi}_{\mathbf{rc}} = \boldsymbol{u}_{\mathbf{sc}} \tag{3.95}$$

By solving Equation (3.95), the rotor flux is

$$\boldsymbol{\psi}_{\mathbf{rc}} = \boldsymbol{\psi}_{\mathbf{rcf}} + \boldsymbol{\psi}_{\mathbf{rcn}} = \underbrace{\boldsymbol{u}_{\mathbf{sc2}}/\lambda_2}_{\boldsymbol{\psi}_{\mathbf{rcf}}} + \underbrace{f_{\mathrm{rcn1}}(\omega_{\mathrm{r}})\mathrm{e}^{s_1 t}}_{\boldsymbol{\psi}_{\mathbf{rcn1}}} + \underbrace{f_{\mathrm{rcn2}}(\omega_{\mathrm{r}})\mathrm{e}^{s_2 t}}_{\boldsymbol{\psi}_{\mathbf{rcn2}}} \tag{3.96}$$

$$\begin{cases} f_{\mathrm{rcn1}}(\omega_{\mathrm{r}}) = -\dfrac{s_2}{(s_1 - s_2)} \left(\sigma L_{\mathrm{r}} \boldsymbol{i}_{\mathbf{r0}} + \dfrac{L_{\mathrm{m}}\boldsymbol{u}_{\mathbf{sc1}}}{(R_{\mathrm{s}} + \mathrm{j}\omega_1 L_{\mathrm{s}})} \right) \\ f_{\mathrm{rcn2}}(\omega_{\mathrm{r}}) = -\dfrac{s_1}{(s_1 - s_2)} \left(\sigma L_{\mathrm{r}} \boldsymbol{i}_{\mathbf{r0}} + \dfrac{L_{\mathrm{m}}\boldsymbol{u}_{\mathbf{sc1}}}{(R_{\mathrm{s}} + \mathrm{j}\omega_1 L_{\mathrm{s}})} \right) \end{cases} \tag{3.97}$$

where $\boldsymbol{i}_{\mathbf{r0}}$ is the rotor current in the normal-operation network. $\boldsymbol{\psi}_{\mathbf{rcf}}$ is the rotor flux mandatory component, $\boldsymbol{\psi}_{\mathbf{rcn}}$ is the rotor flux transient natural component.

According to Equation (3.96), the real parts of the eigenvalues react to the attenuation speed of the transient component, and the imaginary parts react to the attenuation frequency of the transient component. Since the stator/rotor flux is not a simple first-order differential equation, but a second-order differential equation, its transient process is not monotonic attenuation.

According to Equations (3.3) and (3.4), the relationship between the stator current and the stator/rotor flux is

$$\boldsymbol{i}_{\mathbf{sc}} = \frac{1}{\sigma L_{\mathrm{s}}} \boldsymbol{\psi}_{\mathbf{sc}} - \frac{L_{\mathrm{m}}}{\sigma L_{\mathrm{r}} L_{\mathrm{s}}} \boldsymbol{\psi}_{\mathbf{rc}} \tag{3.98}$$

Applying the expressions of stator/rotor flux to Equation (3.98), the expression for the stator short circuit current is

$$\boldsymbol{i}_{\mathbf{sc}} = \boldsymbol{i}_{\mathbf{scf}} + \boldsymbol{i}_{\mathbf{scn1}} + \boldsymbol{i}_{\mathbf{scn2}} = \underbrace{\frac{(R_{\mathrm{rc}} + \mathrm{j}\omega\sigma L_{\mathrm{r}} - \sigma L_{\mathrm{m}})}{\lambda_2\sigma^2 L_{\mathrm{r}} L_{\mathrm{s}}}\boldsymbol{u}_{\mathbf{sc2}}}_{\boldsymbol{i}_{\mathbf{scf}}}$$
$$+ \underbrace{\left(\frac{f_{\mathrm{scn1}}(\omega_{\mathrm{r}})}{\sigma L_{\mathrm{s}}} - \frac{L_{\mathrm{m}} f_{\mathrm{rcn1}}(\omega_{\mathrm{r}})}{\sigma L_{\mathrm{r}} L_{\mathrm{s}}}\right)\mathrm{e}^{s_1 t}}_{\boldsymbol{i}_{\mathbf{scn1}}} + \underbrace{\left(\frac{f_{\mathrm{scn2}}(\omega_{\mathrm{r}})}{\sigma L_{\mathrm{s}}} - \frac{L_{\mathrm{m}} f_{\mathrm{rcn2}}(\omega_{\mathrm{r}})}{\sigma L_{\mathrm{r}} L_{\mathrm{s}}}\right)\mathrm{e}^{s_2 t}}_{\boldsymbol{i}_{\mathbf{scn2}}} \tag{3.99}$$

where $\boldsymbol{i}_{\mathbf{scf}}$ is the mandatory component of the short circuit current. $\boldsymbol{i}_{\mathbf{scn1}}$ and $\boldsymbol{i}_{\mathbf{scn2}}$ are the slowly attenuating component and rapidly attenuating component of the short circuit current, respectively.

Transforming Equation (3.99) to the three-phase stationary coordinate system, the stator short circuit currents under grid symmetrical fault in the positive sequence network are

$$\begin{bmatrix} i_{\mathrm{sca}} \\ i_{\mathrm{scb}} \\ i_{\mathrm{scc}} \end{bmatrix} = \boldsymbol{C}_{2\mathrm{r}/3\mathrm{s}} \begin{bmatrix} \mathrm{Re}\left(\boldsymbol{i}_{\mathbf{sc}}\mathrm{e}^{\mathrm{j}(\omega_1 t + \theta_1)}\right) \\ \mathrm{Im}\left(\boldsymbol{i}_{\mathbf{sc}}\mathrm{e}^{\mathrm{j}(\omega_1 t + \theta_1)}\right) \end{bmatrix} \tag{3.100}$$

According to Equation (3.100), the stator short circuit current considering crowbar protection contains three parts. The first part is the steady-state AC component which rotates at the synchronous speed. The second part is the slowly attenuating transient DC component, which corresponds to the slowly attenuating components of the flux. The third part is the rapidly attenuating transient AC component which rotates at the rotor speed, and which corresponds to the rapidly attenuating components of the flux. Also, the amplitude of the transient AC component is bigger than the amplitude of the transient DC component. Therefore, after crowbar protection operates, the stator initial short circuit current is mainly composed of a transient AC component which rotates at rotor speed frequency. Since the slip of DFIG varies between −0.3 and +0.3, the frequency of the stator short circuit current will vary between 35 and 65 Hz when the DFIG is in different operating states.

B. Grid Asymmetrical Short Circuit Fault

When an asymmetrical short circuit fault occurs in the grid, since the stator-side step-up transformer is Y/Δ connected, there is no zero sequence current, and the DFIG terminal voltage vector in the synchronous rotating coordinate system is

$$\boldsymbol{u}_{\mathbf{sc}} = \boldsymbol{u}_{\mathbf{sc+}} + \boldsymbol{u}_{\mathbf{sc-}} = U_{\mathrm{sc+}}\mathrm{e}^{\mathrm{j}\varphi_+} + U_{\mathrm{sc-}}\mathrm{e}^{\mathrm{j}\varphi_-} \tag{3.101}$$

where $\boldsymbol{u}_{\mathbf{sc+}}$ and $\boldsymbol{u}_{\mathbf{sc-}}$ are the positive sequence and negative sequence components of the DFIG terminal voltage vector.

Considering that the grid is three-phase symmetrical, the sequence components are independent of each other. The positive sequence model of a DFIG in an asymmetrical fault is the same as that in a symmetrical fault, thus the positive sequence transient flux analysis and short circuit current calculation are the same as the derivation process in a symmetrical fault, and the negative sequence component of the DFIG terminal voltage could be transformed to the dq reversed synchronous rotating coordinate system. The mathematical model of the DFIG in the new coordinate system is

$$\boldsymbol{u}_{\mathbf{sc}-} = R_{\mathrm{s}}\boldsymbol{i}_{\mathbf{sc}-} + p\boldsymbol{\psi}_{\mathbf{sc}-} - \mathrm{j}\omega_1\boldsymbol{\psi}_{\mathbf{sc}-} \tag{3.102}$$

$$\mathbf{0} = R_{\mathrm{rc}}\boldsymbol{i}_{\mathbf{rc}-} + p\boldsymbol{\psi}_{\mathbf{rc}-} - \mathrm{j}(2-s)\omega_1\boldsymbol{\psi}_{\mathbf{rc}-} \tag{3.103}$$

$$\boldsymbol{\psi}_{\mathbf{sc}-} = L_{\mathrm{s}}\boldsymbol{i}_{\mathbf{sc}-} + L_{\mathrm{m}}\boldsymbol{i}_{\mathbf{rc}-} \tag{3.104}$$

$$\boldsymbol{\psi}_{\mathbf{rc}-} = L_{\mathrm{m}}\boldsymbol{i}_{\mathbf{sc}-} + L_{\mathrm{r}}\boldsymbol{i}_{\mathbf{rc}-} \tag{3.105}$$

where $\boldsymbol{u}_{\mathbf{sc}-}$, $\boldsymbol{u}_{\mathbf{rc}-}$, $\boldsymbol{i}_{\mathbf{sc}-}$, $\boldsymbol{i}_{\mathbf{rc}-}$, $\boldsymbol{\psi}_{\mathbf{sc}-}$ and $\boldsymbol{\psi}_{\mathbf{rc}-}$ are the space vectors of the DFIG stator/rotor negative sequence voltage, current and flux in the dq reversed synchronous rotating coordinate system.

According to Equations (3.104) and (3.105), the relationship between negative sequence stator/rotor current and flux is

$$\begin{cases} \boldsymbol{i}_{\mathbf{rc}-} = -L_{\mathrm{m}}\boldsymbol{\psi}_{\mathbf{sc}-}/(\sigma L_{\mathrm{r}}L_{\mathrm{s}}) + \boldsymbol{\psi}_{\mathbf{rc}-}/(\sigma L_{\mathrm{r}}) \\ \boldsymbol{i}_{\mathbf{sc}-} = \boldsymbol{\psi}_{\mathbf{sc}-}/(\sigma L_{\mathrm{s}}) - L_{\mathrm{m}}\boldsymbol{\psi}_{\mathbf{rc}-}/(\sigma L_{\mathrm{r}}L_{\mathrm{s}}) \end{cases} \tag{3.106}$$

Applying Equations (3.106) to Equations (3.102) and (3.103), the constant-coefficient dualistic differential equation of the negative sequence stator/rotor flux is as shown in Equation (3.107). By solving Equation (3.107), the constant-coefficient dualistic differential equation eigenvalues s_3 and s_4 can be obtained.

$$\begin{cases} \dfrac{\mathrm{d}\boldsymbol{\psi}_{\mathbf{sc}-}}{\mathrm{d}t} + p_1\boldsymbol{\psi}_{\mathbf{sc}-} + p_2\boldsymbol{\psi}_{\mathbf{rc}-} = \boldsymbol{u}_{\mathbf{sc}-} \\ \dfrac{\mathrm{d}\boldsymbol{\psi}_{\mathbf{rc}-}}{\mathrm{d}t} + p_3\boldsymbol{\psi}_{\mathbf{rc}-} + p_4\boldsymbol{\psi}_{\mathbf{sc}-} = \mathbf{0} \end{cases} \tag{3.107}$$

In Equations (3.107), the expressions to calculate coefficients p_1, p_2, p_3 and p_4, and the negative sequence flux eigenvalues s_3 and s_4 can be found in Appendix B.

Combining Equations (3.107) and the negative sequence flux eigenvalues, the negative sequence stator/rotor flux are

$$\begin{aligned} \boldsymbol{\psi}_{\mathbf{sc}-} = \boldsymbol{\psi}_{\mathbf{scf}-} + \boldsymbol{\psi}_{\mathbf{scn}-} &= \underbrace{-R_{\mathrm{rc}}L_{\mathrm{m}}\boldsymbol{u}_{\mathbf{sc}-}/\sigma L_{\mathrm{r}}L_{\mathrm{s}}s_3s_4}_{\boldsymbol{\psi}_{\mathbf{scf}-}} \\ &\quad + \underbrace{f_{\mathrm{scn1}-}(\omega_{\mathrm{r}})\mathrm{e}^{s_3t}}_{\boldsymbol{\psi}_{\mathbf{scn1}-}} + \underbrace{f_{\mathrm{scn2}-}(\omega_{\mathrm{r}})\mathrm{e}^{s_4t}}_{\boldsymbol{\psi}_{\mathbf{scn2}-}} \end{aligned} \tag{3.108}$$

$$\begin{cases} f_{\mathrm{scn1}-}(\omega_{\mathrm{r}}) = \left(\dfrac{p_4}{s_3s_4} + \dfrac{s_4+p_4}{s_4(s_3-s_4)}\right)\boldsymbol{u}_{\mathbf{sc}-} \\ f_{\mathrm{scn2}-}(\omega_{\mathrm{r}}) = -\dfrac{s_4+p_4}{s_4(s_3-s_4)}\boldsymbol{u}_{\mathbf{sc}-} \end{cases} \tag{3.109}$$

where $\boldsymbol{\psi}_{\mathbf{scf}-}$ represents the mandatory components of the stator negative sequence flux. $\boldsymbol{\psi}_{\mathbf{scn}-}$ represents the transient natural components of the stator negative sequence flux. $\boldsymbol{\psi}_{\mathbf{scn1}-}$, $\boldsymbol{\psi}_{\mathbf{scn2}-}$ are the rapidly attenuating component and slowly attenuating component of the stator negative sequence flux, respectively.

$$\boldsymbol{\psi}_{\text{rc}-} = \boldsymbol{\psi}_{\text{rcf}-} + \boldsymbol{\psi}_{\text{rcn}-} = \underbrace{(-R_{\text{rc}} + \text{j}(2-s)\omega_1\sigma L_{\text{r}})\boldsymbol{u}_{\text{sc}-}/\sigma L_{\text{r}} s_3 s_4}_{\boldsymbol{\psi}_{\text{rcf}-}} + \underbrace{f_{\text{rcn1}-}(\omega_{\text{r}})\text{e}^{s_3 t}}_{\boldsymbol{\psi}_{\text{rcn1}-}} + \underbrace{f_{\text{rcn2}-}(\omega_{\text{r}})\text{e}^{s_4 t}}_{\boldsymbol{\psi}_{\text{rcn2}-}} \tag{3.110}$$

$$\begin{cases} f_{\text{rcn1}-}(\omega_{\text{r}}) = \left(\dfrac{p_3}{s_3 s_4} - \dfrac{p_3}{s_4(s_3 - s_4)}\right)\boldsymbol{u}_{\text{sc}-} \\ f_{\text{rcn2}-}(\omega_{\text{r}}) = -\dfrac{p_3}{s_4(s_3 - s_4)}\boldsymbol{u}_{\text{sc}-} \end{cases} \tag{3.111}$$

where $\boldsymbol{\psi}_{\text{rcf}-}$ represents the mandatory components of the rotor negative sequence flux. $\boldsymbol{\psi}_{\text{rcn}-}$ represents the transient natural components of the rotor negative sequence flux. $\boldsymbol{\psi}_{\text{rcn1}-}$, $\boldsymbol{\psi}_{\text{rcn2}-}$ are the rapidly attenuating component and the slowly attenuating component of the rotor negative sequence flux, respectively.

According to Equations (3.108) and (3.110), the mandatory components of the stator/rotor negative sequence flux are mainly related to the parameters and terminal negative sequence voltage drop rate of the DFIG. The amplitude and attenuation time constant of the natural components of the stator/rotor negative sequence flux are not only related to the pre-fault DFIG rotor speed and crowbar resistance, but also to the DFIG terminal negative sequence voltage.

Applying Equations (3.108)–(3.109) to Equation (3.106), the expression for the stator negative sequence short circuit current is

$$\begin{aligned} \boldsymbol{i}_{\text{sc}-} = \boldsymbol{i}_{\text{scf}-} + \boldsymbol{i}_{\text{sc1}-} + \boldsymbol{i}_{\text{sc2}-} &= \underbrace{\frac{(p_4/(\sigma L_{\text{s}}) + p_3 L_{\text{m}}/(\sigma L_{\text{r}} L_{\text{s}}))}{s_3 s_4}\boldsymbol{u}_{\text{sc}-}}_{\boldsymbol{i}_{\text{scf}-}} \\ &+ \underbrace{\frac{s_3 s_4/(\sigma L_{\text{s}}) - s_4(p_4/(\sigma L_{\text{s}}) - p_3 L_{\text{m}}/(\sigma L_{\text{r}} L_{\text{s}}))}{s_3 s_4(s_3 - s_4)}\text{e}^{s_3 t}\boldsymbol{u}_{\text{sc}-}}_{\boldsymbol{i}_{\text{sc1}-}} \\ &+ \underbrace{\frac{p_3 L_{\text{m}}/(\sigma L_{\text{r}} L_{\text{s}}) + (s_4 + p_4)/(\sigma L_{\text{s}})}{-s_4(s_3 - s_4)}\text{e}^{s_4 t}\boldsymbol{u}_{\text{sc}-}}_{\boldsymbol{i}_{\text{sc2}-}} \end{aligned} \tag{3.112}$$

where $\boldsymbol{i}_{\text{scf}-}$, $\boldsymbol{i}_{\text{sc1}-}$ and $\boldsymbol{i}_{\text{sc2}-}$ are the mandatory component, rapidly attenuating component and slowly attenuating component of the negative sequence short circuit current.

Transforming Equations (3.99) and (3.112) to the stator three-phase stationary coordinate system, the stator short circuit phase currents in the case of grid asymmetrical fault are

$$\begin{bmatrix} i_{\text{sca}} \\ i_{\text{scb}} \\ i_{\text{scc}} \end{bmatrix} = \boldsymbol{C}\begin{bmatrix} \text{Re}\left(\boldsymbol{i}_{\text{sc}+}\text{e}^{\text{j}(\omega_1 t + \theta_1)}\right) \\ \text{Im}\left(\boldsymbol{i}_{\text{sc}+}\text{e}^{\text{j}(\omega_1 t + \theta_1)}\right) \end{bmatrix} + \boldsymbol{C}\begin{bmatrix} \text{Re}\left(\boldsymbol{i}_{\text{sc}-}\text{e}^{\text{j}(-\omega_1 t + \theta_1)}\right) \\ \text{Im}\left(\boldsymbol{i}_{\text{sc}-}\text{e}^{\text{j}(-\omega_1 t + \theta_1)}\right) \end{bmatrix} \tag{3.113}$$

According to Equation (3.113), the stator short circuit current in the case of grid asymmetrical fault contains five parts: (1) the steady-state AC component which rotates at the synchronous speed; (2) the slowly attenuating transient DC component; (3) the rapidly attenuating transient AC component which rotates at the synchronous speed; (4) the rapidly attenuating transient AC component which rotates at the rotor speed; (5) the slowly attenuating transient component which rotates at the second harmonic frequency of the rotor speed. Also, the amplitude of the transient AC component is much bigger than the amplitude of the transient DC component. Therefore, when an asymmetrical fault occurs in the grid, the short circuit current characteristics of the DFIG with crowbar protection will become even more complicated.

3.2.3.2 Short Circuit Current Calculation Model when Crowbar Resistance Varies in the Fault Duration

Reference [7] introduces some types of active crowbar protection circuits in the market, the resistance of which could vary during the fault. Variation of the resistance value is mainly achieved by using PWM, providing multiple resistors and using varistors. At the initial stage after a fault occurs, the active crowbar protection will put a relatively small resistance into operation to keep the rotor voltage from exceeding the limit. Then, in the transient duration, the resistance value is increased to accelerate the attenuation of the rotor current, thus guaranteeing the safety of the converters. Applying the instantaneous values of stator/rotor flux when the resistance varies and the varied crowbar resistance to the short circuit current model, the instantaneous expressions of the DFIG stator short circuit current during the fault considering the crowbar resistance variation can be gained.

Suppose a three-phase fault occurs in the grid at $t = t_2$, when the rotor crowbar protection circuit puts a resistance of R_{cb} into operation. According to Section 3.2.2, the instantaneous values of stator/rotor flux and the instantaneous expressions for the short circuit currents at $t = t_2$ can be solved. At $t = t_3$, the crowbar resistance increases to R'_{cb}, thus $R'_{\mathrm{cr}} = R_{\mathrm{r}} + R'_{\mathrm{cb}}$. Combining Equations (3.93)–(3.97), the instantaneous values of stator/rotor flux at $t = t_3$ can be solved:

$$\boldsymbol{\psi}_{\mathbf{sc}}(t_3) = \boldsymbol{\psi}_{\mathbf{scf}}(t_3) + f_{\mathrm{scn1}}(\omega_{\mathrm{r}})\mathrm{e}^{s_1 t_3} + f_{\mathrm{scn2}}(\omega_{\mathrm{r}})\mathrm{e}^{s_2 t_3} \tag{3.114}$$

$$\boldsymbol{\psi}_{\mathbf{rc}}(t_3) = \frac{\boldsymbol{u}_{\mathbf{sc2}}(t_3)}{\lambda_2} + f_{\mathrm{rcn1}}(\omega_{\mathrm{r}})\mathrm{e}^{s_1 t_3} + f_{\mathrm{rcn2}}(\omega_{\mathrm{r}})\mathrm{e}^{s_2 t_3} \tag{3.115}$$

Applying R'_{cb} and Equation (3.114) to Equation (3.93), the stator flux second-order differential equation after the resistance changes is

$$\frac{\mathrm{d}^2\boldsymbol{\psi}_{\mathbf{sc}}}{\mathrm{d}t^2} + \lambda'_1\frac{\mathrm{d}\boldsymbol{\psi}_{\mathbf{sc}}}{\mathrm{d}t} + \lambda'_2\boldsymbol{\psi}_{\mathbf{sc}} = \frac{\mathrm{d}\boldsymbol{u}_{\mathbf{sc}}}{\mathrm{d}t} + \frac{(R'_{\mathrm{rc}} + \mathrm{j}\omega\sigma L_{\mathrm{r}})}{\sigma L_{\mathrm{r}}}\boldsymbol{u}_{\mathbf{sc}} \tag{3.116}$$

In Equation (3.116), the expressions to calculate coefficients λ'_1 and λ'_2, and eigenvalues s'_1 and s'_2 after the resistance changes can be found in Appendix C.

According to Equation (3.116) and flux eigenvalues, the analytical instantaneous expression of stator flux when $t > t_3$ can be calculated as

$$\begin{aligned}\boldsymbol{\psi}_{\mathbf{sc}}(t > t_3) = \boldsymbol{\psi}'_{\mathbf{scf}} + \boldsymbol{\psi}'_{\mathbf{scn}} &= \underbrace{(R'_{\mathrm{rc}} + \mathrm{j}\omega\sigma L_{\mathrm{r}})\boldsymbol{u}_{\mathbf{sc2}}/\lambda_2\sigma L_{\mathrm{r}}}_{\boldsymbol{\psi}'_{\mathbf{scf}}} \\ &+ \underbrace{f'_{\mathrm{scn1}}(\omega_{\mathrm{r}})\mathrm{e}^{s'_1 t}}_{\boldsymbol{\psi}'_{\mathbf{scn1}}} + \underbrace{f'_{\mathrm{scn2}}(\omega_{\mathrm{r}})\mathrm{e}^{s'_2 t}}_{\boldsymbol{\psi}'_{\mathbf{scn2}}}\end{aligned} \tag{3.117}$$

$$\begin{cases} f'_{\text{scn1}}(\omega_\text{r}) = -\dfrac{s'_2}{(s'_1 - s'_2)}\boldsymbol{\psi}_{\text{sc}}(t_3) \\ f'_{\text{scn2}}(\omega_\text{r}) = \dfrac{s'_1}{(s'_1 - s'_2)}\boldsymbol{\psi}_{\text{sc}}(t_3) \end{cases} \tag{3.118}$$

Similarly, the instantaneous expression for the rotor flux after the resistance changes can be derived:

$$\boldsymbol{\psi}_{\text{rc}}(t > t_3) = \frac{\boldsymbol{u}_{\text{sc2}}}{\lambda'_2} + f'_{\text{rcn1}}(\omega_\text{r})e^{s'_1 t} + f'_{\text{rcn2}}(\omega_\text{r})e^{s'_2 t} \tag{3.119}$$

$$\begin{cases} f'_{\text{rcn1}}(\omega_\text{r}) = -\dfrac{s'_2}{(s'_1 - s'_2)}\boldsymbol{\psi}_{\text{rc}}(t_3) \\ f'_{\text{rcn2}}(\omega_\text{r}) = \dfrac{s'_1}{(s'_1 - s'_2)}\boldsymbol{\psi}_{\text{rc}}(t_3) \end{cases} \tag{3.120}$$

Applying the instantaneous expressions of stator/rotor flux to Equation (3.98), the instantaneous expression for the stator short circuit current when $t > t_3$ is

$$\boldsymbol{i}_{\text{sc}} = \frac{1}{\sigma L_\text{s}}\boldsymbol{\psi}_{\text{sc}}(t > t_3) - \frac{L_\text{m}}{\sigma L_\text{r} L_\text{s}}\boldsymbol{\psi}_{\text{rc}}(t > t_3) \tag{3.121}$$

Similarly, the stator negative sequence short circuit current in the case of grid asymmetrical fault when the crowbar resistance varies during the fault can be derived.

3.2.3.3 Analysis of the Short Circuit Current Calculation Model

A. *Influence of Crowbar Resistance on the Short Circuit Current Calculation Model*

According to Equations (3.93) and (3.108), the natural component of the stator flux has to do with the crowbar resistance. Compared with stator/rotor resistance, the crowbar resistance should be at least 30 times the rotor resistance in value to suppress the stator fault peak current. Therefore, it is necessary to study the variation pattern of the eigenvalues of the stator flux natural component as the crowbar resistance changes, which could then reflect the variation pattern of the stator flux and the stator short circuit current.

1. Positive Sequence Flux Eigenvalues Figure 3.22 shows the variation curves of eigenvalue s_1 and s_2 as the crowbar resistance changes. As the crowbar resistance increases, the absolute value of the real part of s_1 decreases until close to the imaginary axis, while the absolute value of the imaginary part of s_1 remains close to the synchronous speed. This means that, when the crowbar resistance increases, the stator-side damping decreases and the attenuation speed slows down. Thus, $\boldsymbol{\psi}_{\text{scn1}}$ in Equation (3.93) is a slowly attenuating transient DC component in the three-phase stationary coordinate system. At the same time, as the crowbar resistance increases, the absolute value of the real part of s_2 increases to farther away from the imaginary axis, while the absolute value of the imaginary part of s_2 remains close to the slip speed. This means that, when the crowbar resistance increases, the rotor-side damping increases and the attenuation speed rises. Thus, $\boldsymbol{\psi}_{\text{scn2}}$ in Equation (3.93)

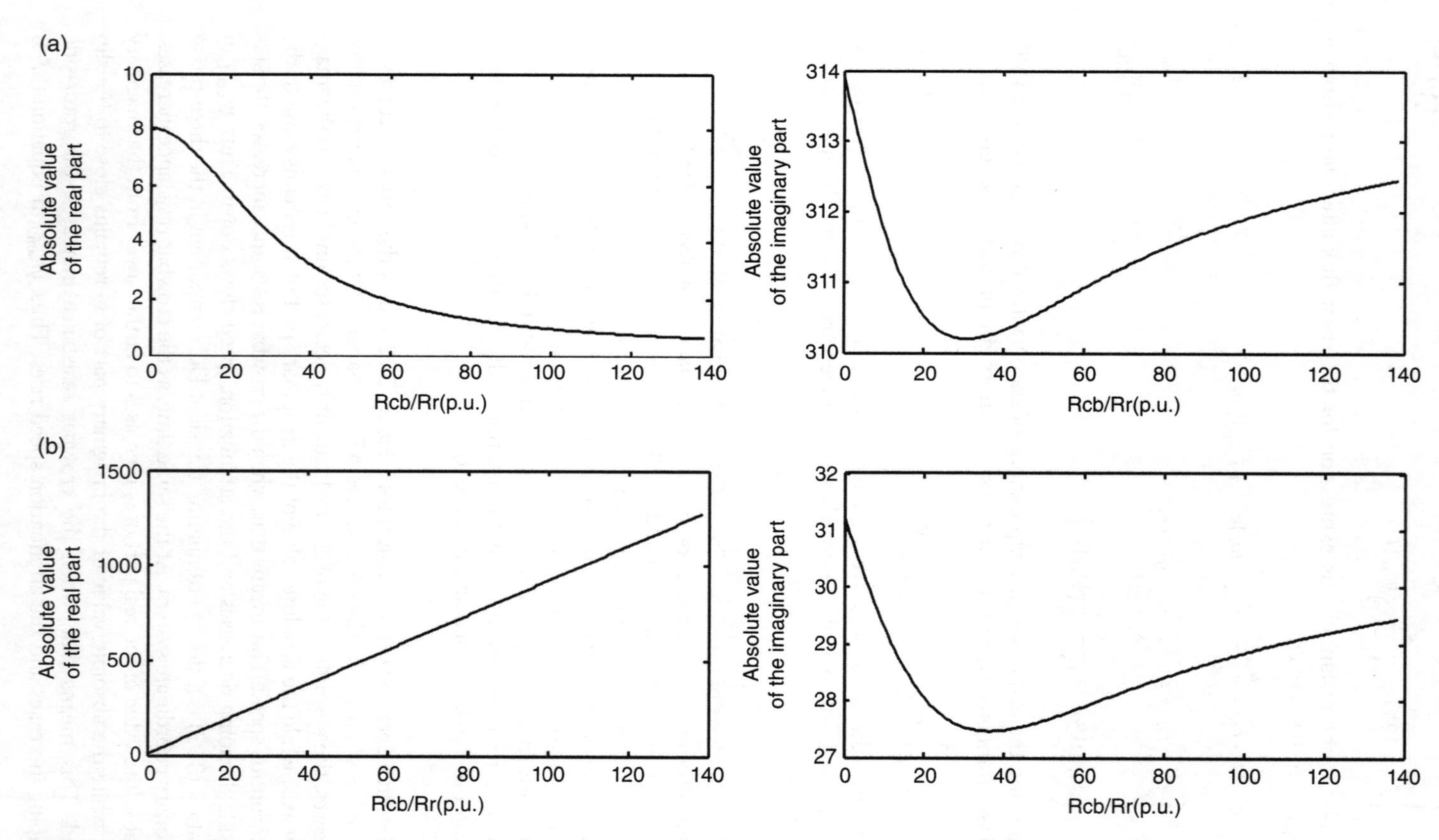

Figure 3.22 Variation of positive sequence flux eigenvalues as the crowbar resistance changes. (a) Variation of s_1. (b) Variation of s_2.

is a rapidly attenuating transient AC component in the three-phase stationary coordinate system, which rotates at the rotor speed.

2. Negative Sequence Flux Eigenvalues The variation curves of negative sequence flux eigenvalues as the crowbar resistance changes are shown in Figure 3.23.

It can be seen that, as the crowbar resistance increases, the absolute value of the real part of s_3 increases, while the absolute value of the imaginary part of s_3 increases first and then decreases, which corresponds to a rapidly attenuating transient component rotating at synchronous speed in the three-phase stationary coordinate system. At the same time, as the crowbar resistance increases, the absolute value of the real part of s_4 (being small in value) decreases slowly, while the absolute value of the imaginary part of s_4 decreases first and then increases, which corresponds to the slowly attenuating transient component rotating at the second harmonic frequency of the rotor speed in the three-phase stationary coordinate system.

B. Influence of Rotor Speed on the Short Circuit Current Calculation Model
According to Equations (3.93)–(3.97) and Equations (3.108)–(3.111), the coefficients of the stator flux components are functions of the DFIG rotor speed ω_r. Therefore, the pre-fault rotor speed ω_r will affect the amplitude of the stator flux, thus affecting the characteristics of the stator short circuit current. The variation curves of the coefficients of different flux components as the DFIG rotor speed (per unit value) changes are shown in Figure 3.24.

1. Positive Sequence Flux Components It can be seen from Figure 3.24(a) that when the rotor speed is close to the synchronous speed 1.0, the mandatory component of the stator flux varies significantly as rotor speed changes. However, when the DFIG is in a sub-synchronous or over-synchronous operating state, the mandatory component of the stator flux remains almost unchanged. As the DFIG turns from a sub-synchronous operating state to an over-synchronous operating state, the rapidly attenuating transient component of the stator flux increases. When the rotor speed is close to the synchronous speed, the slowly attenuating transient component of the stator flux is the minimum. As the rotor speed deviates further from the synchronous speed, the slowly attenuating transient component of the stator flux increases gradually. This demonstrates that different components of the positive sequence stator flux are significantly affected by the pre-fault DFIG rotor speed. Therefore, the stator short circuit current will manifest different characteristics when the DFIG is in different operating states.

2. Negative Sequence Flux Components It can be seen from Figure 3.24(b) that when the pre-fault rotor speed varies between −0.7 and −1.3, the mandatory component and the rapidly attenuating component of the stator negative sequence flux decrease constantly. The slowly attenuating component remains almost unchanged when the DFIG is in the over-synchronous operating state, but decreases constantly when the DFIG turns from the sub-synchronous operating state to the synchronous operating state. This demonstrates that the amplitude of the stator negative sequence flux is also affected by the pre-fault rotor speed.

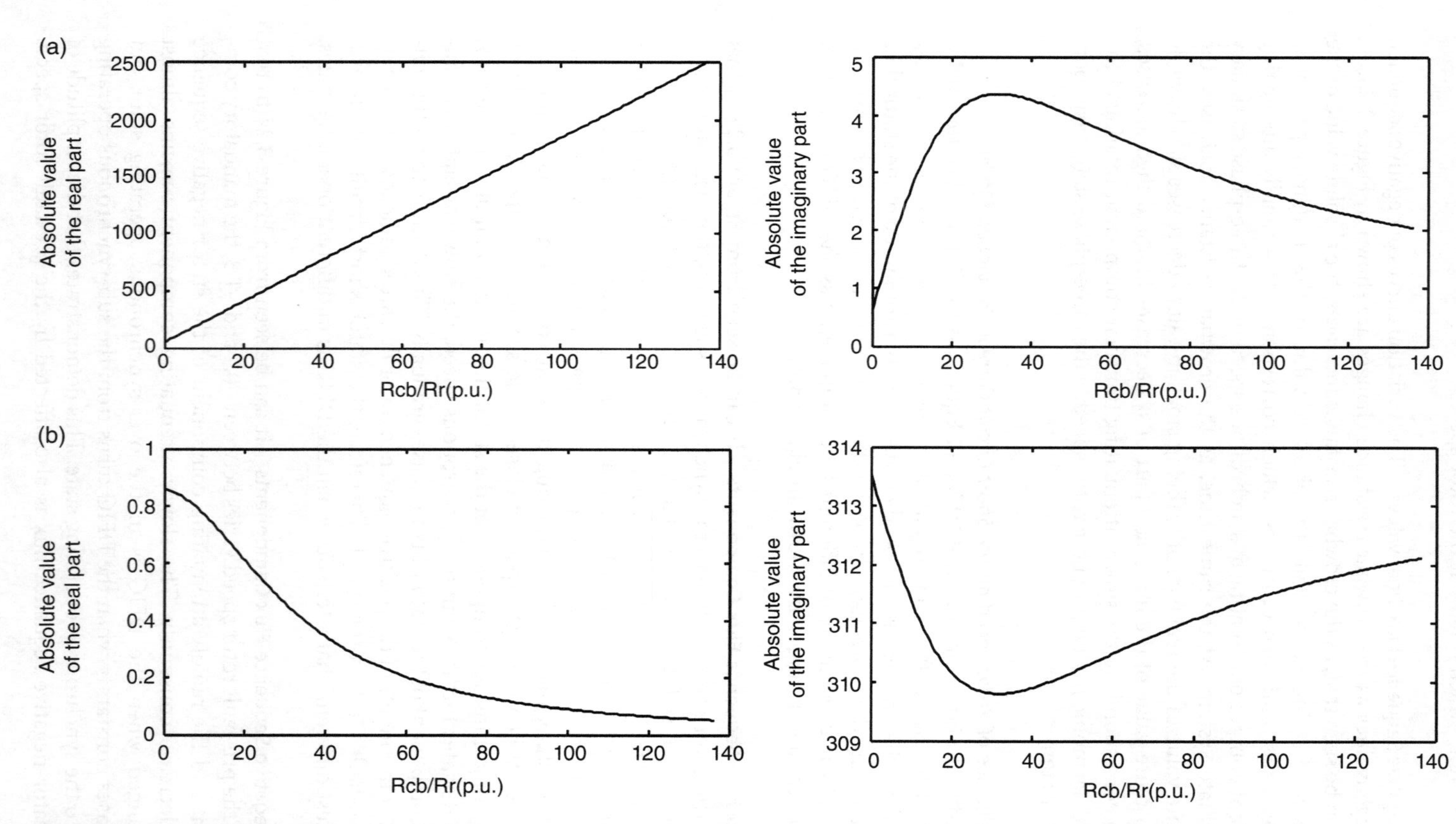

Figure 3.23 Variation of negative sequence flux eigenvalues as the crowbar resistance changes. (a) Variation of s_3. (b) Variation of s_4.

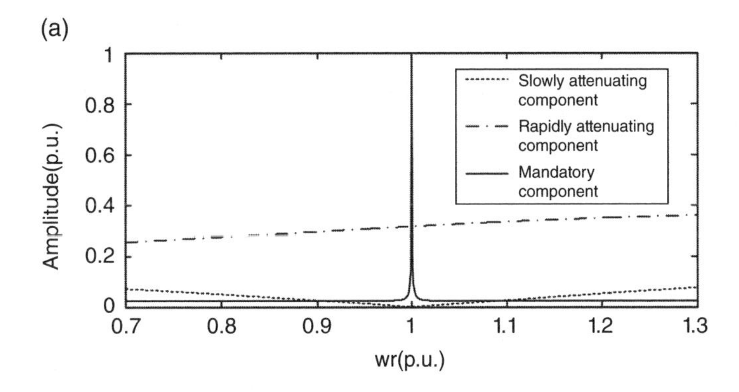

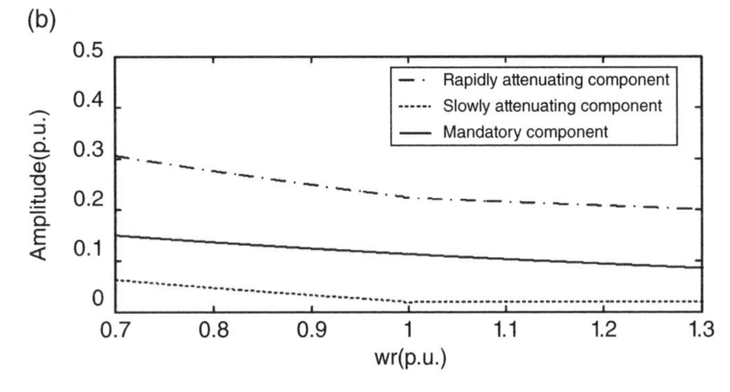

Figure 3.24 Variation of different components of the stator flux as the DFIG rotor speed changes. (a) Positive. (b) Negative.

Then the stator short circuit current characteristics are jointly affected by the stator negative sequence flux and the stator positive sequence flux.

As the capacity of the power grid with wind power access constantly increases, grid codes have posed certain requirements on the wind farm active and reactive power regulatory control, voltage and frequency operational constraints, wind power behaviour during grid disturbance, etc. This facilitates the application of wind power technology in the grid and improves the stability and safety of a large-capacity system with wind power access. Grid codes of different countries are different from each other due to various power grid structures and protection system response times. When the system voltage drops below a certain percentage value of the rated voltage, large-scale wind farms need to be connected to the grid continuously for a certain length of time, and certain requirements are posed on the reactive power output in the fault duration. Some countries (e.g. Germany and Ireland) require that the wind farm inject reactive power within 20 ms after a fault occurs [8]. Crowbar access here cannot meet such a requirement, so other means such as reactive power compensation equipment will be needed for realizing large-scale application of wind power. However, other countries (e.g. Spain) have no requirement on the reactive power output in the first 100 ms after a fault occurs, and require that the wind farm provide dynamic reactive power after 100 ms, in response to the grid voltage drop. Here, the crowbar is accessed until 100 ms after the fault occurs, so that the wind generator does not drop off the grid. Then the crowbar withdraws and the wind generator restores the output of its reactive power. Thus, crowbar access here could meet the grid code requirements of these countries on LVRT and reactive power output. Therefore, the proposed method has practical value in improving wind farm short circuit calculation under modern grid codes.

3.2.3.4 Simulation Verification

A. Simulation System

In order to verify the correctness of the DFIG short circuit current expressions derived here, a DFIG simulation system was established in RTDS, which is shown in Figure 3.25. The main parameters of the DFIG are shown in Table 3.3. The RTDS simulation platform could simulate the transient characteristics of a real DFIG accurately and provide a test environment for a real crowbar protection circuit. Compared with other simulation platforms, RTDS has higher practical value.

B. Crowbar Resistance Remains Unchanged in the Fault Duration

1. Stator Short Circuit Current in the Case of Grid Symmetrical Fault Suppose that at $t = 0.1$ s, a three-phase short circuit fault occurs at the high-voltage side of the transformer, the rotor current increases and the crowbar protection operates. When the crowbar resistance is fixed at 30 times the rotor resistance, the current waveforms of phase A are as shown in Figure 3.26(a). Figure 3.26(b) shows the stator short circuit current of phase A when various levels of crowbar resistance are put into operation and it remains the same during the fault. The variation of phase A stator short circuit current as the pre-fault rotor speed changes is shown in Figure 3.26(c).

It can be seen from Figure 3.26(a) that the calculated stator short circuit phase currents are basically consistent with the simulation results. The calculated values exhibit obvious transient attenuation, which could reflect the transient characteristics of stator short circuit currents after the crowbar protection operates. Especially in the first three cycles, the calculated waveform is approximately the same as the simulation waveform. However, after the first three cycles, the calculated waveform deviates slightly from the simulation waveform. This is because, in the calculation process, the steady-state rotor speed is used without considering the change of rotor speed in the fault transient. However, since the fault transient is relatively short, the error caused by using the steady-state rotor speed can be neglected.

According to the analysis in Section 2.2.3.3(A), the crowbar resistance will affect the natural attenuating component of the stator flux, thus affecting the stator short circuit current characteristics. It can be seen from Figure 3.26(b) that, when the crowbar resistance is $10R_r$ and remains the same during the fault, the stator short circuit current peak value is 5.63 p.u., and the time it takes to reach the peak is the longest – 0.009 s. As crowbar resistances of larger value are put into operation and remain the same during the fault, the peak value

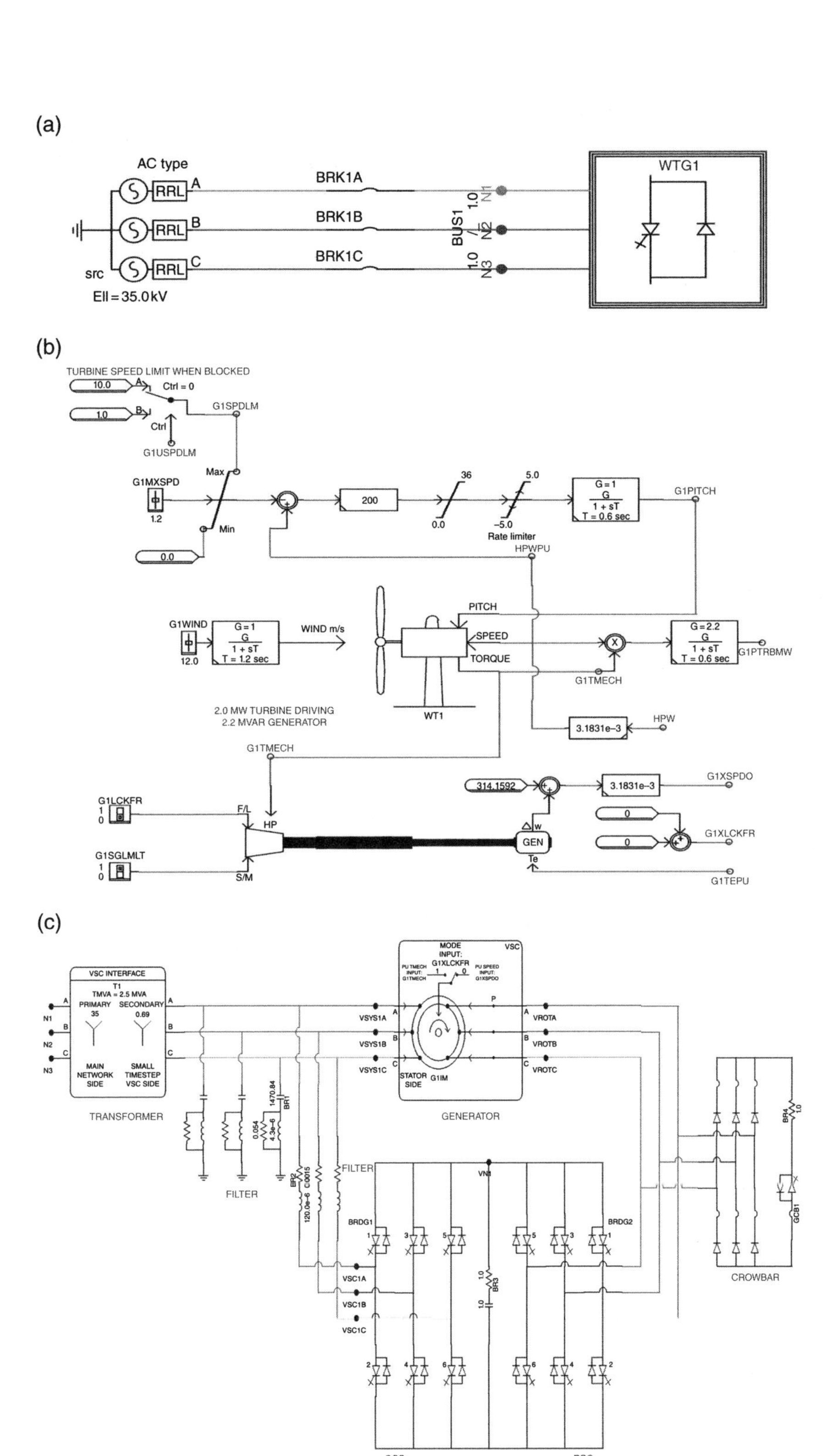

Figure 3.25 RTDS models of DFIG simulation system. (a) RTDS model for the simulation system. (b) RTDS model for a wind turbine DFIG. (c) RTDS model for the electrical part of a DFIG.

Table 3.3 The main parameters of the DFIG.

Parameter name	Value	Parameter name	Value
Rated power	2 MW	R_r	0.00607 p.u.
Stator rated voltage	690 V	L_{ls}	0.102 p.u.
Rated frequency	50 Hz	L_{lr}	0.11 p.u.
R_s	0.0054 p.u.	L_m	4.362 p.u.

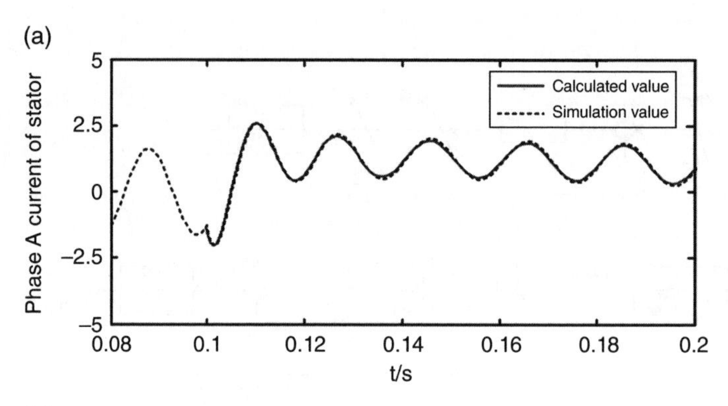

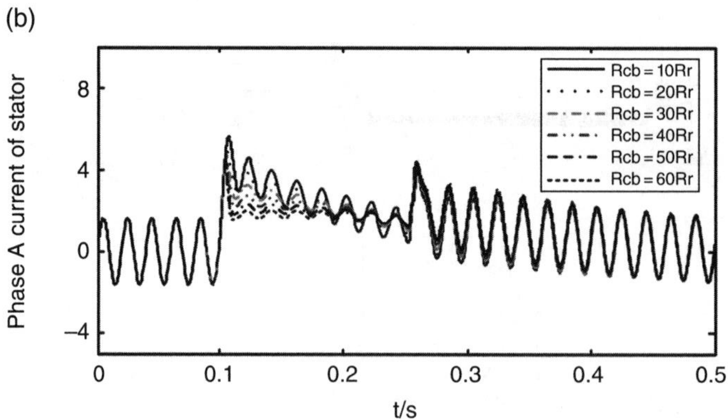

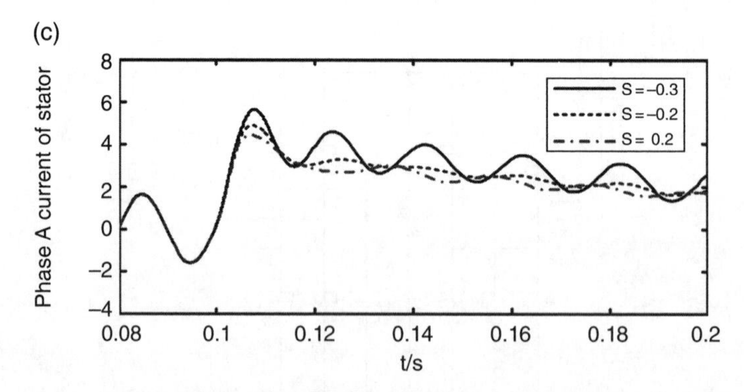

Figure 3.26 Stator short circuit phase currents in the case of a three-phase symmetrical fault. (a) Short circuit current of phase A under a symmetrical fault. (b) Variation of stator short circuit current as the crowbar resistance changes under a symmetrical fault. (c) Influence of pre-fault rotor speed on the stator short circuit current under a symmetrical fault.

and the peak time of the stator short circuit current both become smaller, until the crowbar resistance exceeds $40R_r$, the stator short circuit current peak value remains at 4.816 p.u., and the peak time also remains basically unchanged. This is because the initial stator current is mainly composed of the rapidly attenuating AC component rotating at the rotor speed. As the crowbar resistance increases, the absolute value of the real part of the rapidly attenuating AC component also increases, thus the attenuating speed and attenuating degree of the stator short circuit current both increase.

According to the analysis in Section 2.2.3.3(B), when a symmetrical fault occurs in the grid, the coefficients of different components of the stator flux have to do with the pre-fault DFIG rotor speed. By affecting the amplitude characteristics of different components of the stator flux, the pre-fault rotor speed then affects the stator short circuit current. It can be seen from Figure 3.26(c) that, as the rotor speed increases, the frequency and peak value of stator short circuit current both increase. This is because, as the rotor speed increases, the amplitude of the rapidly attenuating component also increases. Also, the rapidly attenuating component rotates at the rotor speed, thus the frequency of the stator short circuit current also increases with the increase of the rotor speed.

2. Stator Short Circuit Current in the Case of Grid Asymmetrical Fault Suppose that at $t = 2$ s, an asymmetrical short circuit fault occurs at the high-voltage side of the transformer. Taking a phase-C-to-ground fault and a phase BC-to-ground fault, for example, the correctness of the stator short circuit current expressions derived here is verified. The current waveforms are shown in Figure 3.27, and the calculated stator short circuit phase currents are basically consistent with the simulation results. In the first few cycles after the fault, the fault phase current increases significantly and then attenuates rapidly, while the non-fault phase current increases rapidly to the steady-state value. This is in line with the transient characteristics of stator short circuit currents after an asymmetrical fault occurs. Thus, the analytical expressions of stator short circuit currents in the case of a grid asymmetrical fault derived here are proved correct.

The crowbar resistance will affect the transient natural component of the stator negative sequence flux, thus affecting the stator short circuit current characteristics in the case of asymmetrical fault (which is jointly affected by the transient natural components of the stator positive sequence flux and the stator negative sequence flux). When a phase-C-to-ground fault occurs in the grid, the variation curves of the stator short circuit currents as various levels of crowbar resistance are put into operation and remain the same during the fault are shown in Figures 3.28(a) and 3.28 (b). It can be seen that the influence of crowbar resistance on the fault phase stator current and non fault phase stator current is different. Since the non-fault phase contains fewer transient attenuating components, it takes less time for the non-fault phase current to reach the peak value, and the current transits to the steady-state value quickly. However, for the fault phase current, as crowbar resistances of larger value are put into operation and remain the same during the fault, the peak value becomes larger, and the peak time remains the same. The characteristics of stator short circuit current in the case of an asymmetrical fault are different from those in the case of a

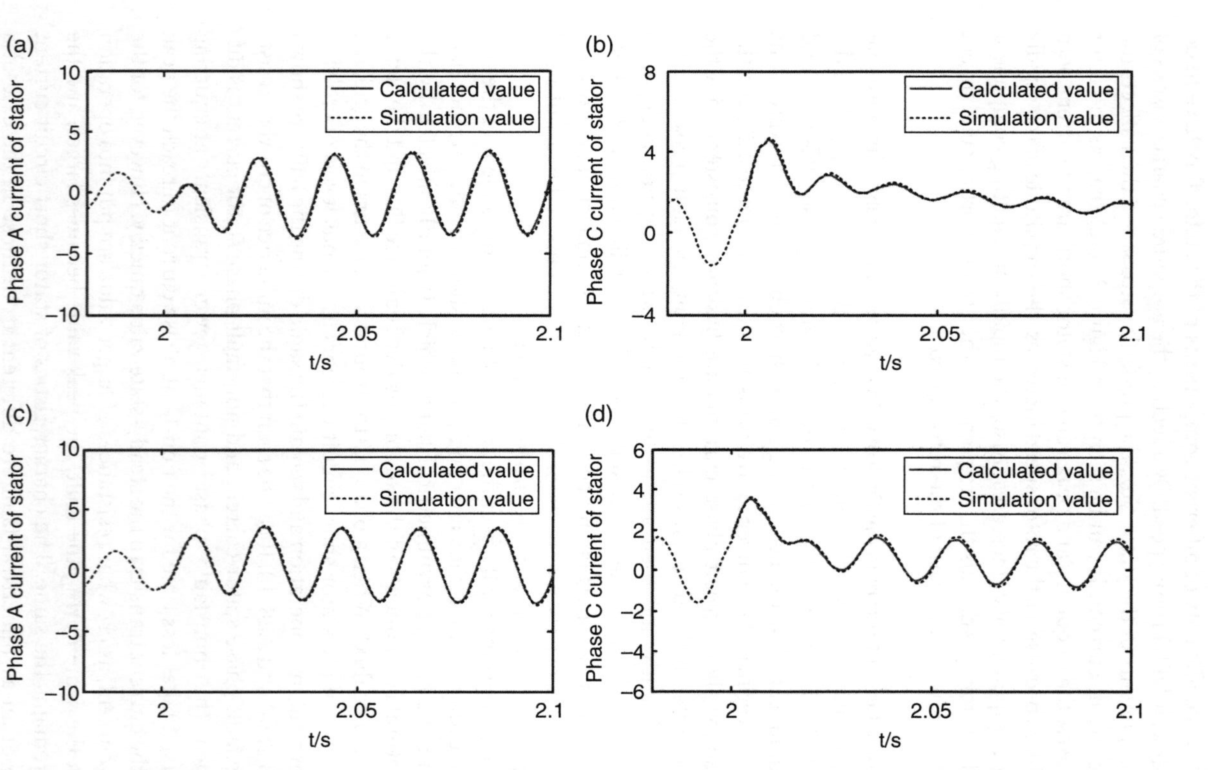

Figure 3.27 Stator short circuit phase currents in the case of a grid asymmetrical fault. (a) Short circuit current of phase A under a phase-C-to-ground fault. (b) Short circuit current of phase C under a phase-C-to-ground fault. (c) Short circuit current of phase A under a phase-BC-to-ground fault. (d) Short circuit current of phase C under a phase-BC-to-ground fault.

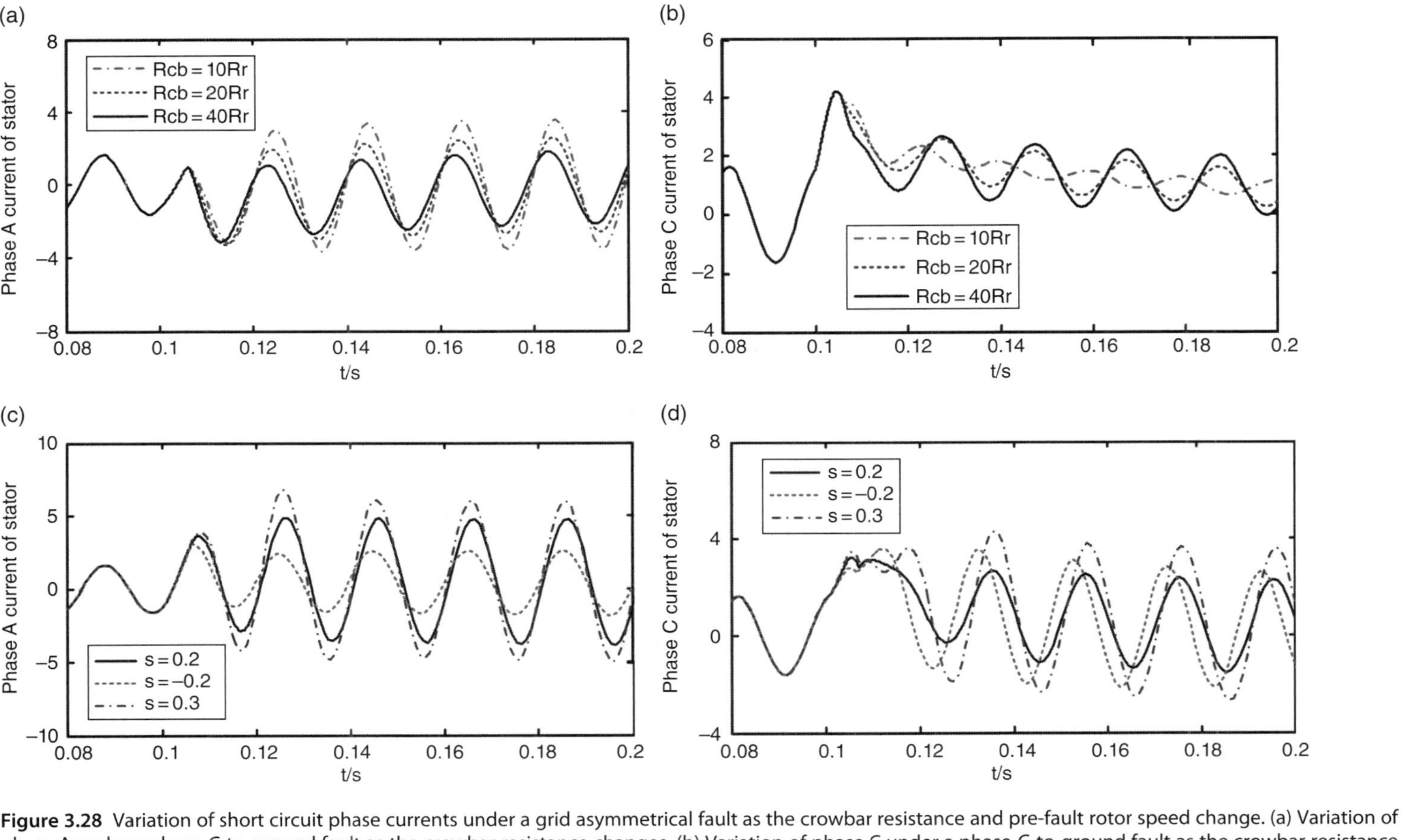

Figure 3.28 Variation of short circuit phase currents under a grid asymmetrical fault as the crowbar resistance and pre-fault rotor speed change. (a) Variation of phase A under a phase-C-to-ground fault as the crowbar resistance changes. (b) Variation of phase C under a phase-C-to-ground fault as the crowbar resistance changes. (c) Variation of phase A under a phase-BC-to-ground fault as the pre-fault rotor speed changes. (d) Variation of phase C under a phase-BC-to-ground fault as the pre-fault rotor speed changes.

symmetrical fault. This is mainly because, in the case of the asymmetrical fault, the stator short circuit current contains five components, including the rapidly attenuating transient component which rotates at the synchronous speed and the slowly attenuating transient component which rotates at the second harmonic frequency. The values of these two components have to do with the crowbar resistance and the terminal negative sequence voltage. As the crowbar resistance increases, the terminal negative sequence voltage increases, thus the stator short circuit current peak value increases. Also, with the same crowbar resistance, the attenuating speed of the asymmetrical fault current is smaller than that of the symmetrical fault current due to the influence of the negative sequence voltage.

The components of the stator negative sequence flux have to do with the pre-fault DFIG rotor speed. The stator short circuit phase currents in the case of phase-B-to-C fault when the DFIG is at different rotor speeds are shown in Figures 3.28(c) and 3.28 (d). As the rotor speed increases, the frequency of the stator short circuit current increases, due to the attenuating AC components rotating at the rotor speed and at the second harmonic frequency of the rotor speed. Meanwhile, as the rotor speed increases, the non-fault phase current peak value decreases; the fault phase current peak value first decreases and then increases again. This is mainly because, as the rotor speed increases, the amplitude of the positive sequence rapidly attenuating AC component increases, while the amplitude of the negative sequence mandatory component and amplitude of the negative sequence rapidly attenuating component both decrease. The amplitude of the negative sequence slowly attenuating component decreases in the sub-synchronous state and remains the same in the over-synchronous state. Therefore, after the positive and negative sequence components are added up, the stator short circuit current may increase or decrease, depending on the variation of the positive and negative sequence attenuating components with the rotor speed.

Based on the above analysis, when an asymmetrical fault occurs, the stator positive sequence components and negative sequence components of the short circuit current are both closely related to the pre-fault DFIG rotor speed. The coupling between the stator short circuit currents superimposed with corresponding components and the rotor speed is different from the simple relationship between the positive/negative sequence short circuit current and the rotor speed. Therefore, the influence of pre-fault rotor speed on the stator short circuit current in the case of an asymmetrical fault is different from that in the case of a symmetrical fault.

C. *Crowbar Resistance Variation in the Grid Fault Duration*

Suppose that a symmetrical short circuit fault occurs at the high-voltage side of the transformer at $t = 2$ s. The crowbar protection operates and puts a $20R_r$ resistance into operation. Then, at $t = 2.02$ s, the crowbar resistance increases to $30R_r$. Shown in Figure 3.29(a) is a comparison between the simulation values of phase A stator current when the crowbar resistance remains the same and changes during the fault. It can be seen that, at the initial stage after the fault occurs, the crowbar resistance is $20R_r$ in both cases, and the simulation curves

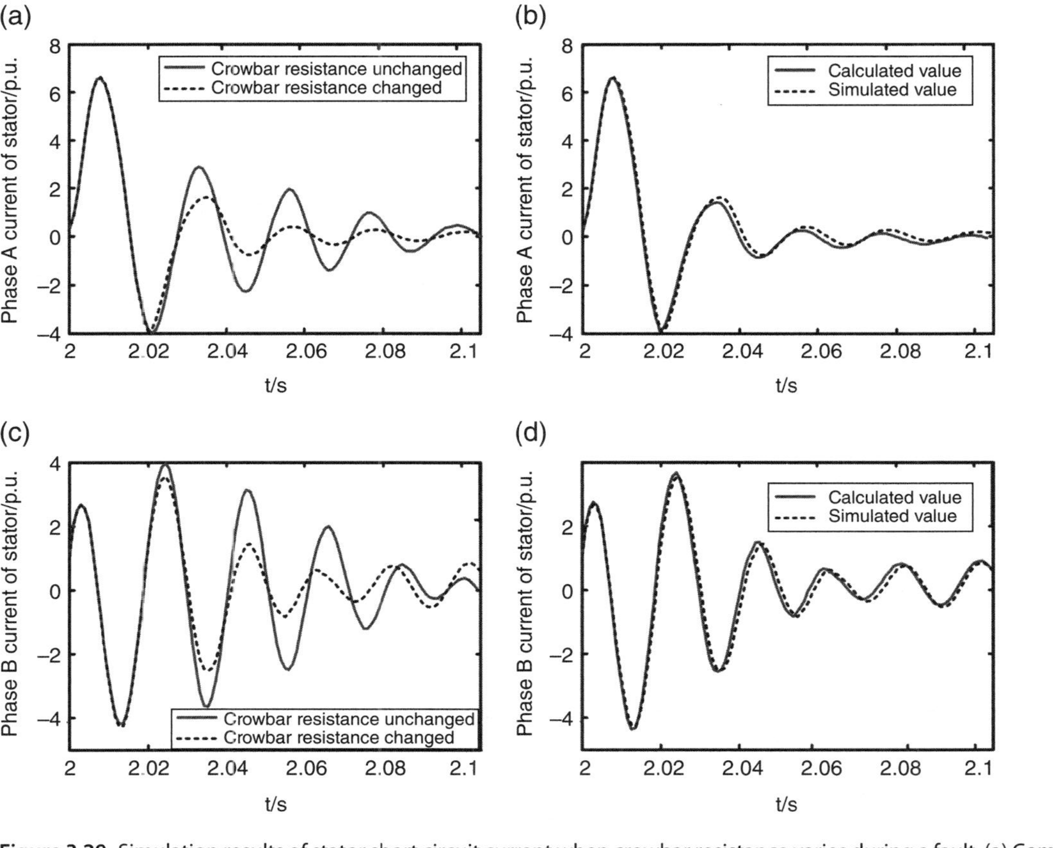

Figure 3.29 Simulation results of stator short circuit current when crowbar resistance varies during a fault. (a) Comparison of resistance changing and unchanging during a symmetrical fault. (b) Comparison of calculation and simulation when resistance changes during a symmetrical fault. (c) Comparison of resistance changing and unchanging during an asymmetrical fault. (d) Comparison of calculation and simulation when resistance changes during an asymmetrical fault.

are the same. At $t = 2.02$ s, for the case of the crowbar resistance increasing to $30R_r$, the corresponding short circuit current attenuates faster with smaller amplitude than the case of the crowbar resistance remaining the same. Shown in Figure 3.29(b) is a comparison between the calculated values and simulation values of phase A stator current when the crowbar resistance changes during the fault. It can be seen that the calculation curve is basically consistent with the simulation curve.

Suppose a phase-BC-to-ground fault occurs at $t = 2$ s. The crowbar protection operates and the resistance put into operation is the same as for the symmetrical fault case. During the fault, the comparison between the simulation values of phase B stator current when the crowbar resistance remains the same and changes are shown in Figure 3.29(c). In the case of an asymmetrical fault, due to the influence of negative sequence voltage, the short circuit current attenuation amplitude when the crowbar resistance increases is smaller than that of the symmetrical fault. Thus, in order to guarantee safe operation of the converters, the time for which the crowbar is put into operation is usually longer in asymmetrical fault cases than in symmetrical fault cases. Figure 3.29(d) shows the comparison between the calculated values and the simulation values of phase B stator current when the crowbar resistance changes during the phase-BC-to-ground fault. The two groups of curves are both consistent with each other. Therefore, the proposed DFIG short circuit current calculation model is applicable to industrial crowbar protection circuits whose resistance varies during the fault.

3.3 Local Area Protection for Centralized Renewable Energy

In China, renewable energy resources are relatively far from centralized power loads, thus it is suitable to establish centralized renewable energy plant for power generation. Affected by the natural characteristics of renewable energy, the electronic devices and the control strategy, the fault transient characteristics of renewable power generation systems and traditional power generation systems are very different, thus new problems and requirements have arisen concerning the reliable operation of traditional protection. Due to the power converter used and the characteristic of variable speed, constant frequency (VSCF), the DFIG has become the mainstream model for wind generators. In this section we take a wind farm based on a DFIG as an example; the problems of traditional protection applied in renewable energy plant are analysed, and new protection schemes for renewable energy plant are put forward with the collector line and outgoing transmission line as the object of research. In Section 3.3.1, the wiring and protection configuration of wind farms are introduced. On this basis, Section 3.3.2 proposes a novel adaptive distance protection scheme for a collector line based on a voltage drop equation. The protection scheme is easily adaptable to different operating modes of the plant, is immune to fault resistance and is unaffected by the weak feed characteristics of the collector system. In Section 3.3.3 we propose a current differential protection scheme for the outgoing transmission line based on fault steady-state components, which is not affected by fault resistance or load current, and is highly reliable and sensitive in fault identification.

3.3.1 Connection Form of a Wind Farm and its Protection Configuration

Currently, a wind farm includes the following parts: the wind power generation units that are responsible for wind power generation, the collector lines for power transmission and a step-up substation to raise the output voltage to higher levels such as 110 kV or 220 kV. A wind power generation unit mainly consists of the wind turbine, which outputs mechanical power, the generator, which converts mechanical power into electricity, the inverter inside of the unit that is responsible for frequency regulating and the corresponding internal step-up transformer. For current wind farms, the output terminal voltage of the mainstream wind power generator is commonly 690 V, and the voltage level of the collector lines is generally set to 35 kV. Therefore, each unit is equipped with an internal step-up transformer, usually a box transformer with a ratio of 0.69/35 kV. According to the characteristics of the wind power generator, a relay protection scheme such as low-voltage protection for a low-voltage fault or over-current protection in view of a likely over-current are usually configured.

For the box transformer of the wind power generator, a circuit breaker is used as protection on the low-voltage side. When a fault occurs in the fan generator or in the box transformer, a low-voltage-side circuit breaker can trip in time, ensuring the reliability and stability of the power system operation. At the same time, a fuse is configured to protect the high-voltage side. However, due to the required action condition of the fuse, it may not operate reliably under all operational modes. Considering the economics, there is generally no other special protection equipment on the high-voltage side. In this case, a fault on the high-voltage side can be cleared through the circuit breaker in the line protection, which is installed on the 35 kV collector line.

The main wiring and relay protection configuration diagram of a wind farm are shown in Figure 3.30. As can be seen, on the high-voltage sides of the box transformers, 3–8 wind generators are paralleled (according to the principle of proximity), and then via a 35 kV collector line they are transmitted to the step-up substation. In the step-up substation, the collected power is boosted to 110 kV (or 220 kV) and then connected to the power system. For extra large wind farms, the voltage level could be boosted to 500 kV.

Due to the weak feed characteristics of the wind generator, the proportion of positive and negative sequence components is much smaller than the proportion of zero sequence components in the collector system side short circuit current. Thus, traditional distance protection may not operate correctly. Since the lengths of the collector lines differ from one another, it is not easy for the setting values of line distance protection to cooperate with each other, thus poor selectivity may result. Meanwhile, the wind farm collector lines are grounded via resistance or an arc suppression coil. When a fault occurs, the transition resistance will greatly affect the operational performance of distance protection. Based on the above analysis, it is not easy for wind farm collector line protection to clear a fault quickly and accurately.

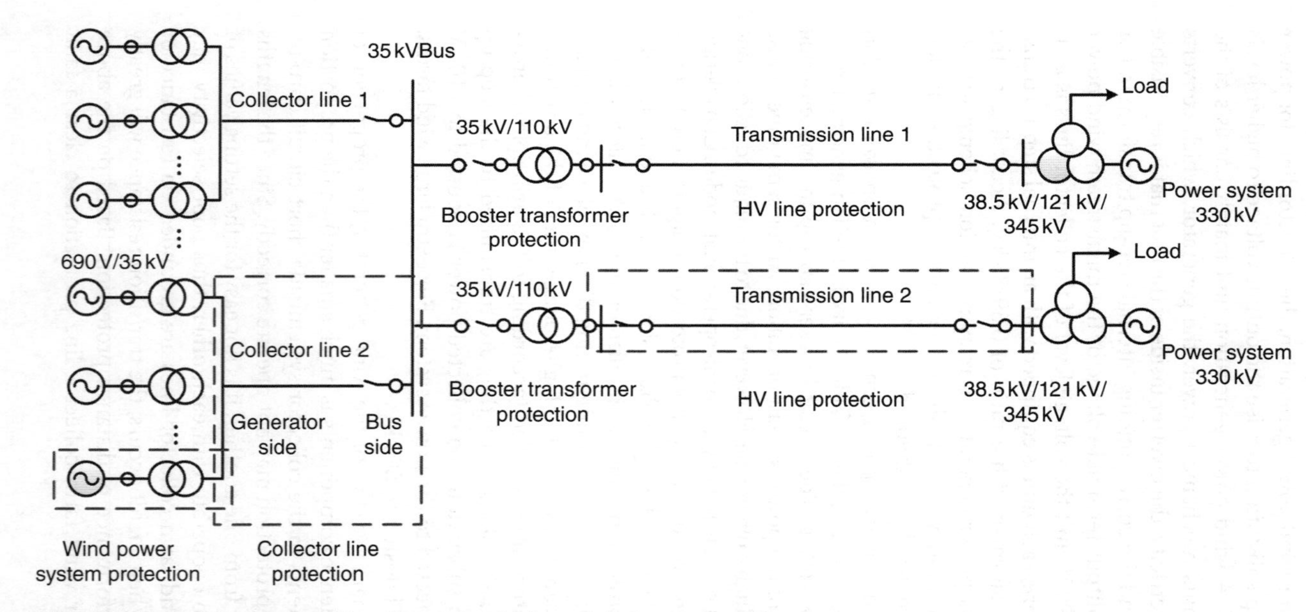

Figure 3.30 Wind farm main wiring diagram and relay protection configuration.

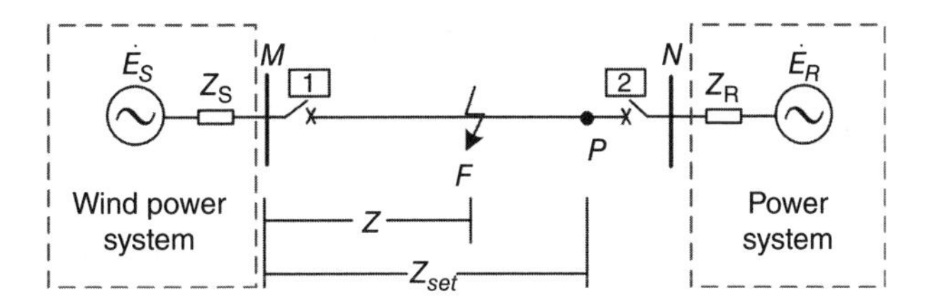

Figure 3.31 Adaptive distance protection analysis model.

3.3.2 Adaptive Distance Protection Scheme for Wind Farm Collector Lines

3.3.2.1 Adaptive Distance Protection Configuration of Wind Farm Collector Lines

When a fault occurs on the collector line of the wind farm shown in Figure 3.30, the distance protection analysis model shown in Figure 3.31 can be established. In Figure 3.31, M is the collector system side, the equivalent system voltage being $\dot{E}_S$; N is the grid side, the equivalent system voltage being $\dot{E}_R$. The equivalent system impedances of the two sides are Z_S and Z_R, respectively. Take the protection on the collector system side as an example, the protection setting range is MP, the setting value is Z_{set}. F is the fault point and Z is the positive sequence impedance of the line length from the fault point to the relaying point.

When a fault occurs on the collector line, the measured voltage $\dot{U}_{\text{m}}$ and the measured current $\dot{I}_{\text{m}}$ at the relaying point meet the following equation:

$$\dot{U}_{\text{m}} = \dot{I}_{\text{m}} Z + \dot{U}_{\text{f}} \tag{3.122}$$

where $\dot{U}_{\text{f}}$ is the voltage at the fault point.

Divide both sides of Equation (3.122) by $\dot{I}_{\text{m}}$, so that the measured impedance at the relaying point Z_{m} can be obtained:

$$Z_{\text{m}} = Z + \frac{\dot{U}_{\text{f}}}{\dot{I}_{\text{m}}} = Z + \frac{\dot{I}_{\text{f}} R_g}{\dot{I}_{\text{m}}} \tag{3.123}$$

where $\dot{I}_{\text{f}}$ is the short circuit current and R_g is the transition resistance. It can be seen that the measured impedance Z_{m} reflects not only the real fault impedance Z, but also a supplementary impedance $\dot{I}_{\text{f}} R_g / \dot{I}_{\text{m}}$.

As can be seen from Figure 3.31, when a fault occurs on the line, the voltage drop from the relaying point to the fault point equals the voltage drop on Z, which is not related to the transition resistance. Referring to Section 2.3 of Chapter 2 for the analysis of local line protection, an adaptive distance protection criterion can be constructed for wind farm collector lines as

$$\begin{cases} |Z_{\text{m}}| \le k_{\text{a}} |Z_{\text{set}}| & \textit{internal fault} \\ |Z_{\text{m}}| > k_{\text{a}} |Z_{\text{set}}| & \textit{external fault} \end{cases} \tag{3.124}$$

where Z_{m} is the measured impedance. Z_{set} is the setting value of the relay. k_{a} is the adaptive setting coefficient, which equals $\frac{\cos(90^\circ - \varphi_{line} - \psi)}{\sin(\varphi_{ui} + \psi)}$. φ_{line} is the impedance angle of the protected line. φ_{ui} is the angular difference between the measured voltage and current. ψ is the angular difference between the measured current $\dot{I}_{\text{m}}$ and the voltage at the fault point $\dot{U}_{\text{f}}$.

For a single-phase grounding fault,

$$\dot{I}_{\text{f}} = 3\dot{I}_{\text{f0}} = 3\dot{I}_{\text{f2}} \tag{3.125}$$

Thus, according to Equation (3.125), the deviation angle ψ of the measured current $\dot{I}_{\text{m}}$ from the fault point current can be calculated as

$$\psi = \arg\left(\frac{\dot{I}_{\text{m}}}{R_g\dot{I}_{\text{f}}}\right) = \arg\left(\frac{\dot{I}_{\text{m}}}{\dot{I}_{\text{f0}}}\right) = \arg\left(\frac{\dot{I}_{\text{m}}}{\dot{I}_{\text{f2}}}\right) \tag{3.126}$$

where $\dot{I}_{\text{f2}}$ and $\dot{I}_{\text{f0}}$ represent the negative sequence current and zero sequence current at the fault point.

Since the wind farm collector system is a weak feed system and the system zero sequence impedance angle has nothing to do with the load, the zero sequence current distribution coefficient can be taken as a real number, i.e. it can be taken that the zero sequence current at the relaying point is approximately the same as the zero sequence current at the fault point in phase angle. Thus, the deviation angle in Equation (3.126) can be expressed as

$$\psi = \arg\left(\frac{\dot{I}_{\text{m}}}{\dot{I}_{\text{m0}}}\right) \tag{3.127}$$

According to Equation (3.127), the deviation angle ψ is only related to the zero sequence current and the measured current at the relaying point. By calculating ψ, the adaptive distance protection setting value $k_{\text{a}}|Z_{\text{set}}|$ can be obtained from Equation (3.124), and then, according to the protection criterion, the fault location can be identified.

3.3.2.2 Operational Characteristics Analysis

According to the above analysis, the proposed adaptive distance protection scheme for wind farm collector lines has the following characteristics:

1) Adaptability. The adaptive setting coefficient k_{a} could adjust adaptively according to the measured values of collector line electrical variables, and has nothing to do with the back side system of the line. Thus, the proposed scheme is easily adaptable to the variation in system operating mode.
2) Unaffected by transition resistance. The protection criterion is formed according to the voltage drop from the relaying point to the fault point, which is not related to the transition resistance. Therefore, the protection scheme has good immunity to transition resistance.
3) Applicable to a weak feed system. A wind generator manifests weak feed characteristics. When a grounding fault occurs on the collector line, the zero sequence current is the main component of the collector system side fault current. Thus, the proposed scheme is applicable to wind farm collector line protection.

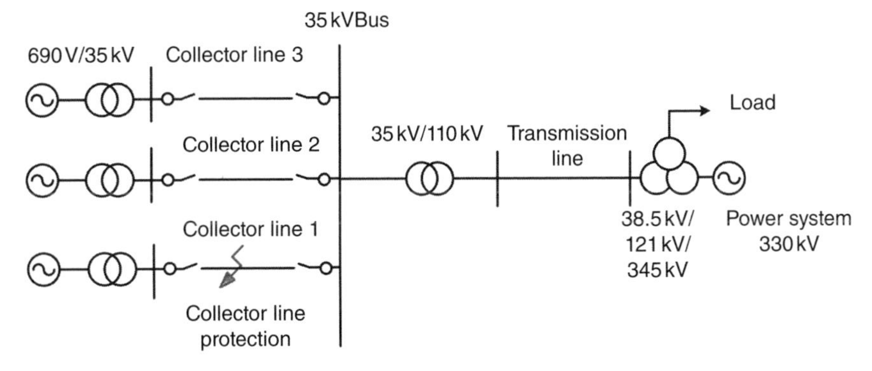

Figure 3.32 Wind farm simulation model.

3.3.2.3 Simulation Verification

To verify the correctness and effectiveness of the adaptive distance protection scheme proposed here, a doubly-fed wind farm simulation system was established in RTDS for simulation tests, shown in Figure 3.32. The detailed parameters are system positive sequence impedance $Z_{S1} = 4.264 + j85.15\ \Omega$, zero sequence impedance $Z_{S0} = 0.6 + j9.91\ \Omega$, system voltage $E_S = 330\angle\delta°$ kV, short-circuit capacity 1000 MVA; the 110 kV sending line: positive sequence impedance $Z_{L1} = 0.131 + j0.401\ \Omega$/km, zero sequence impedance $Z_{L0} = 0.328 + j0.197\ \Omega$/km, line length $L = 60$ km; the main transformer: rated capacity 100 MVA, rated voltage 38.5/110 kV, short-circuit voltage $U_d = 10.5\%$; the box transformer: rated capacity 20 MVA, rated voltage 690 V, short-circuit voltage $U_d = 6.5\%$; the 35 kV collector lines: $Z_1 = 0.020 + j0.894\ \Omega$/km, $Z_0 = 0.114 + j2.288\ \Omega$/km, line length $L_1 = 20$ km, $L_2 = 14$ km, $L_3 = 8$ km. Each collector line is connected with 10 doubly-fed wind generators (DFIG), which are simplified as one wind generator in Figure 3.32. The main parameters of the DFIG are rated power 2 MW, rated frequency 50 Hz, stator rated voltage 690 V, stator resistance 0.0054 p.u., rotor resistance 0.00607 p.u., stator leakage inductance 0.102 p.u., rotor leakage inductance 0.11 p.u., mutual inductance 4.362 p.u.

A. *Different Transition Resistances*

Suppose at $t = 0.5$ s, three types of fault occur on the wind farm collector line respectively: (1) phase-A-to-ground fault at 25% of line L_1, and none of the crowbar protection of wind generators connected to L_1 is put into operation; (2) phase-A-to-ground fault at 55% of line L_1, and some of the crowbar protection of wind generators connected to L_1 is put into operation; (3) phase-A-to-ground fault at 75% of line L_1, and all of the crowbar protection of wind generators connected to L_1 is put into operation. The operational curves of adaptive distance protection when the collector system is in the above three operating conditions are shown in Figures 3.33–3.35. It can be seen that, in any operating conditions, when a fault occurs on line L_1 via transition resistance, the measured values decrease rapidly, while the setting values adjust adaptively according to the measured information. Since the setting values remain smaller than the measured values, the adaptive

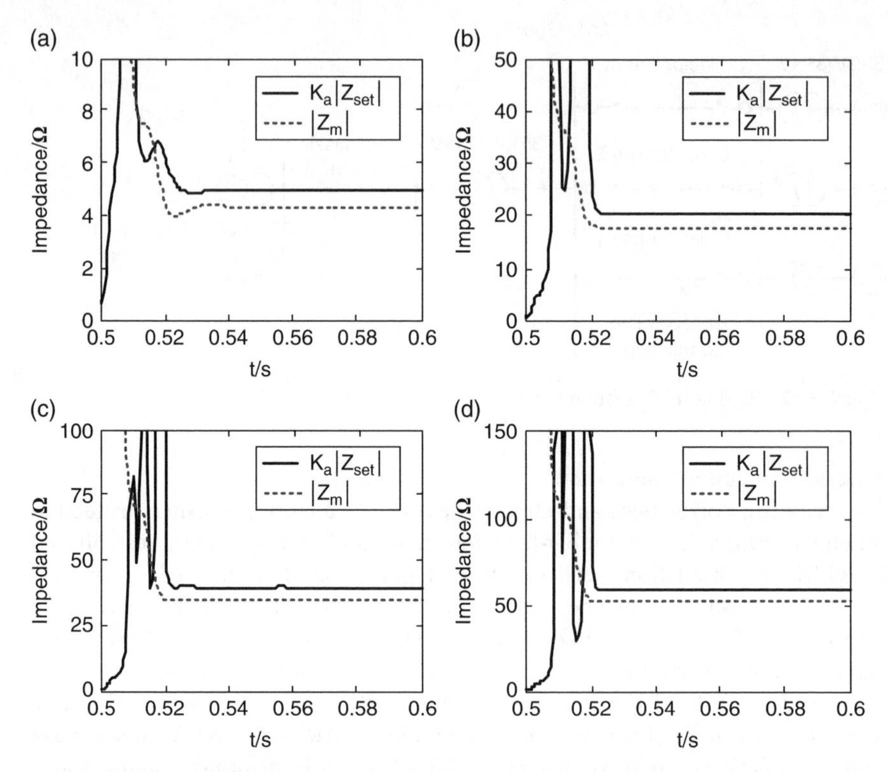

Figure 3.33 Operational curve of the adaptive distance protection on the collector system side when a phase-A-to-ground fault occurs at 25% of L_1 via different transition resistances. (a) 0 Ω. (b) 10 Ω. (c) 20 Ω. (d) 30 Ω.

distance protection can operate correctly in all cases. Thus, the proposed scheme is proved to be easily adaptable to the variation of wind generator operating mode, and strongly immune to transition resistance.

B. Different Fault Locations

When a phase-A-to-ground fault occurs at different locations on L_1 via different transition resistances, the setting values and measured values of the adaptive distance protection on the collector system side and on the system side are shown in Tables 3.4 and 3.5, respectively. In the tables, ' + ' means that the adaptive distance protection operates, and '–' means that the adaptive distance protection does not operate.

It can be seen from Tables 3.4 and 3.5 that when a fault occurs within the protection range, the setting values are bigger than the measured values, and the protection operates correctly. When a fault occurs outside the protection range, the setting values are smaller than the measured values, and the protection does not operate. Also, when a fault occurs via different transition resistances, the adaptive distance protection on the collector system side and on the system side both could operate correctly.

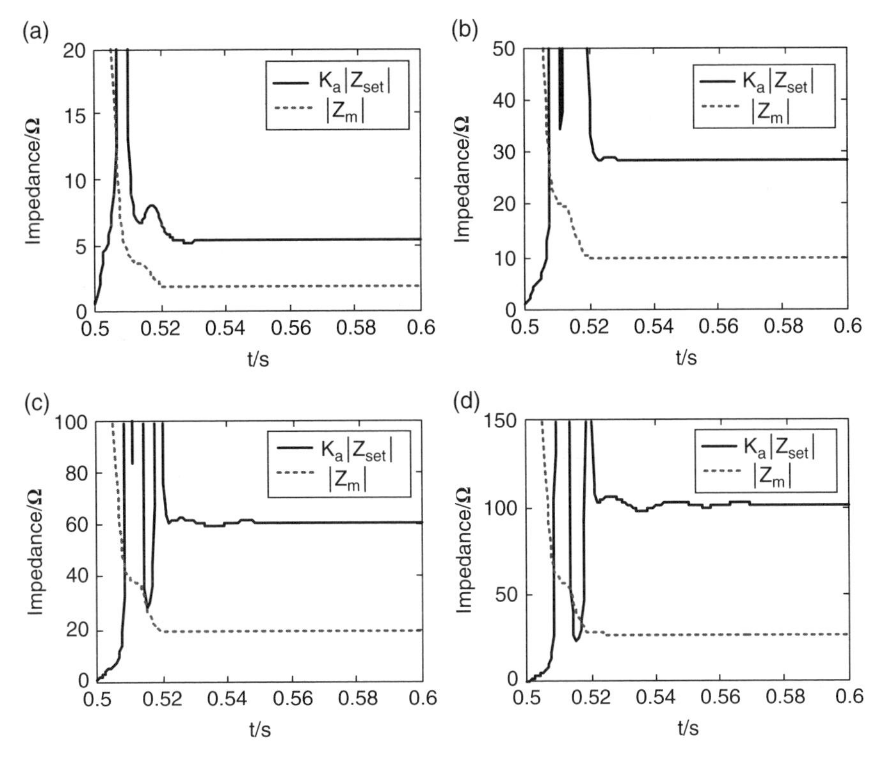

Figure 3.34 Operational curve of the adaptive distance protection on the collector system side when a phase-A-to-ground fault occurs at 55% of L_1 via different transition resistances. (a) 0 Ω. (b) 10 Ω. (c) 20 Ω. (d) 30 Ω.

C. *Adjacent Line Fault*

When a fault occurs on an adjacent collector line, the crowbar protection of wind generators connected to this collector line may operate. In this case, the short circuit current from wind generators connected to this collector line may cause the line protection of this line to malfunction, and the malfunction of collector lines with different line lengths is different. If the fault occurs on line L_2 near the 35 kV bus, this will cause the crowbar protection of wind generators connected to L_1 to operate in three possible ways: none of the crowbar protection put into operation; part of the crowbar protection put into operation; all of the crowbar protection put into operation. The operational curves of adaptive distance protection in the above three cases are shown in Figures 3.36–3.38. It can be seen that, after a fault occurs on L_2, no matter what operating condition the wind generators connected to L_1 are in, the adaptive distance protection does not operate, unaffected by the line length.

D. *Comparison with Traditional Distance Protection*

To analyse the immunity of traditional distance protection to transition resistance, suppose a phase-A-to-ground fault occurs at different locations on collector

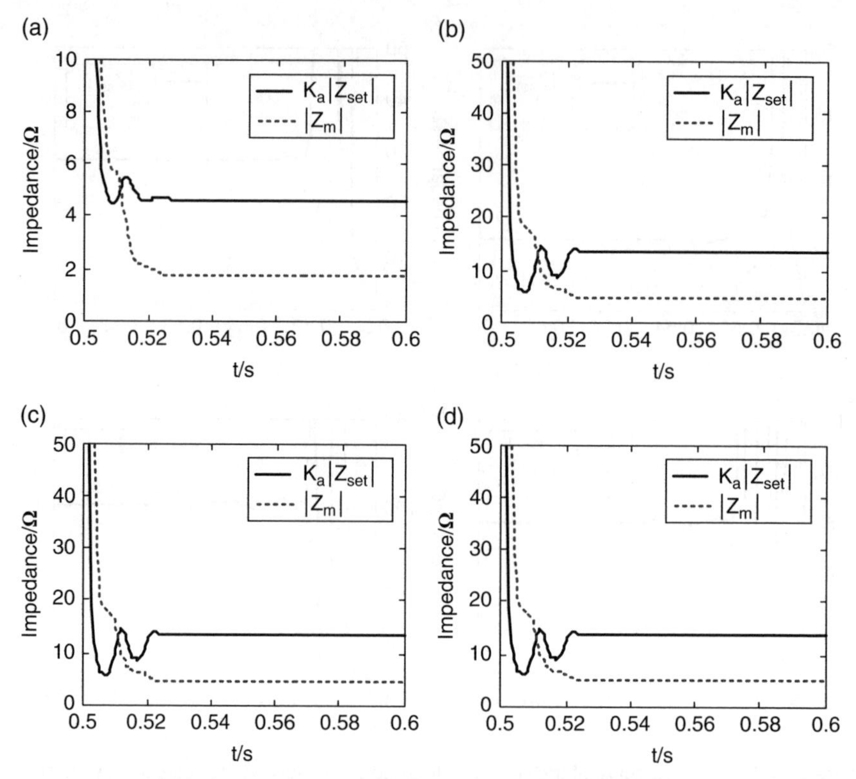

Figure 3.35 Operational curve of the adaptive distance protection on the collector system side when a phase-A-to-ground fault occurs at 75% of L_1 via different transition resistances. (a) 0 Ω. (b) 10 Ω. (c) 20 Ω. (d) 30 Ω.

line L_1 via different transition resistances. The operational characteristics of traditional distance protection apply the directional circle characteristics, and the setting range covers 80% of the whole line length. When the measured impedance falls inside the operating circle with the setting value Z_{set}(Z_{set} = 1.1812 + j5.5597 Ω) as diameter, distance protection will operate. When the measured impedance falls outside the operating circle, distance protection will not operate.

If a phase-A-to-ground fault occurs at 4 km from the relaying point on collector line L_1 via different transition resistances, the operational curves of traditional distance protection will be as shown in Figure 3.39. It can be seen that, when the transition resistance is 5 Ω, the measured value falls inside the operational area and traditional distance protection operates correctly. When the transition resistance is bigger than 10 Ω, the measured value falls outside the operational area, and traditional distance protection refuses to operate.

When a phase-A-to-ground fault occurs at different locations on L_1 via 30 Ω transition resistance, the variation trajectory of the measured impedance at the relaying point of the collector system side is as shown in Figure 3.40. It can be seen that, even when the fault location is at 10% of L_1, the measured impedance

Table 3.4 Simulation results of adaptive distance protection when a phase-A-to-ground fault occurs at different locations on L_1 via different transition resistances.

Transition resistance	Fault location	$k_a\|Z_{set}\|$	$\|Z_m\|$	Operational status
0 Ω	15%	6.8014 Ω	1.2538 Ω	+
	20%	6.0223 Ω	1.4866 Ω	+
	25%	5.6056 Ω	1.7340 Ω	+
	75%	4.9869 Ω	4.6766 Ω	+
	*80%	5.0409 Ω	5.0529 Ω	–
	85%	5.1233 Ω	5.4732 Ω	–
15 Ω	15%	43.2707 Ω	17.3421 Ω	+
	20%	48.7900 Ω	15.2218 Ω	+
	25%	60.8661 Ω	13.5285 Ω	+
	75%	30.2294 Ω	29.0758 Ω	+
	*80%	30.8418 Ω	32.7110 Ω	–
	85%	31.6569 Ω	37.3883 Ω	–
30 Ω	15%	86.5958 Ω	30.0116 Ω	+
	20%	77.5041 Ω	31.9377 Ω	+
	25%	71.1944 Ω	34.1173 Ω	+
	75%	58.0771 Ω	57.3266 Ω	+
	*80%	57.5276 Ω	64.6108 Ω	–
	85%	56.7990 Ω	74.0169 Ω	–

Note: 80% is the end of the adaptive distance protection setting range.

still falls outside the operating circle, and traditional distance protection refuses to operate.

According to the above simulation analysis, traditional distance protection has poor immunity to transition resistance, its operational performance being greatly affected by transition resistance. On the other hand, according to Figures 3.33–3.35, Tables 3.4 and 3.5, the proposed protection scheme is strongly immune to transition resistance.

3.3.3 Differential Protection Scheme for Wind Farm Outgoing Transmission Line

3.3.3.1 Fault Equivalent Model

When a short circuit fault occurs in the power grid connected with a doubly-fed wind farm, the rotor-side converter of the DFIG will automatically lock in 3–5 ms and the rotor-side crowbar protection will be put into operation. At this point, protection components on the 110 kV outgoing transmission line have not yet operated. Therefore, in the fault duration, the DFIG is equivalent to an asynchronous generator for fault analysis and protection calculation. The frequency of the

Table 3.5 Simulation results of the system side when a phase-A-to-ground fault occurs at different locations on L_1 via different transition resistances.

Transition resistance	Fault location	$k_a\|Z_{set}\|$	$\|Z_m\|$	Operational status
0 Ω	15%	7.5309 Ω	1.4015 Ω	+
	20%	6.5902 Ω	1.6397 Ω	+
	25%	6.0840 Ω	1.8956 Ω	+
	75%	5.7890 Ω	5.6451 Ω	+
	*80%	6.0844 Ω	6.4881 Ω	–
	85%	6.5428 Ω	7.8058 Ω	–
15 Ω	15%	87.6266 Ω	14.7498 Ω	+
	20%	66.6868 Ω	15.7065 Ω	+
	25%	55.1868 Ω	16.7853 Ω	+
	75%	35.0812 Ω	27.9854 Ω	+
	*80%	33.5438 Ω	35.7723 Ω	–
	85%	31.5469 Ω	49.5361 Ω	–
30 Ω	15%	105.9732 Ω	29.2081 Ω	+
	20%	120.5265 Ω	31.0505 Ω	+
	25%	112.0237 Ω	33.1314 Ω	+
	75%	60.4807 Ω	54.9488 Ω	+
	*80%	56.8389 Ω	61.6588 Ω	–
	85%	53.1770 Ω	70.2258 Ω	–

Note: 80% is the end of the adaptive distance protection setting range.

DFIG fault current is the product of the pre-fault rotational speed and the power frequency 50 Hz. Usually the rotational speed of the DFIG is 0.7–1.3 p.u., thus when a crowbar is put into operation, the frequency of the DFIG fault current is 35–65 Hz. Traditional transmission line pilot protection is based on the frequency of currents on the two line ends, thus the reliability is not high. In view of this problem, this section introduces a current differential protection strategy based on fault steady-state components, which could improve the sensitivity of the wind farm outgoing transmission line.

Taking line MN, for example, the model of the normal-operation network is as shown in Figure 3.41, where Z_{mn} is the equivalent impedance; y_m and y_p are the equivalent admittance; $\dot{I}_m$ is the measured current at bus M; $\dot{I}_n$ is the measured current at bus N.

When a fault occurs on line mp, the model of fault steady-state network is as shown in Figure 3.42, where α represents the percentage of the distance from the fault point to the bus M in the whole line length ($0 \le \alpha \le 1$). $\dot{U}_f$ and $\dot{I}_f$ are the voltage and current at the fault point. $\dot{I}_{mf}$ and $\dot{I}_{nf}$ are the currents at the two line ends. $\dot{U}_{mf}$ and $\dot{U}_{nf}$ are the voltages at the two line ends.

According to electrical circuit theory, the node voltage equation can be written as

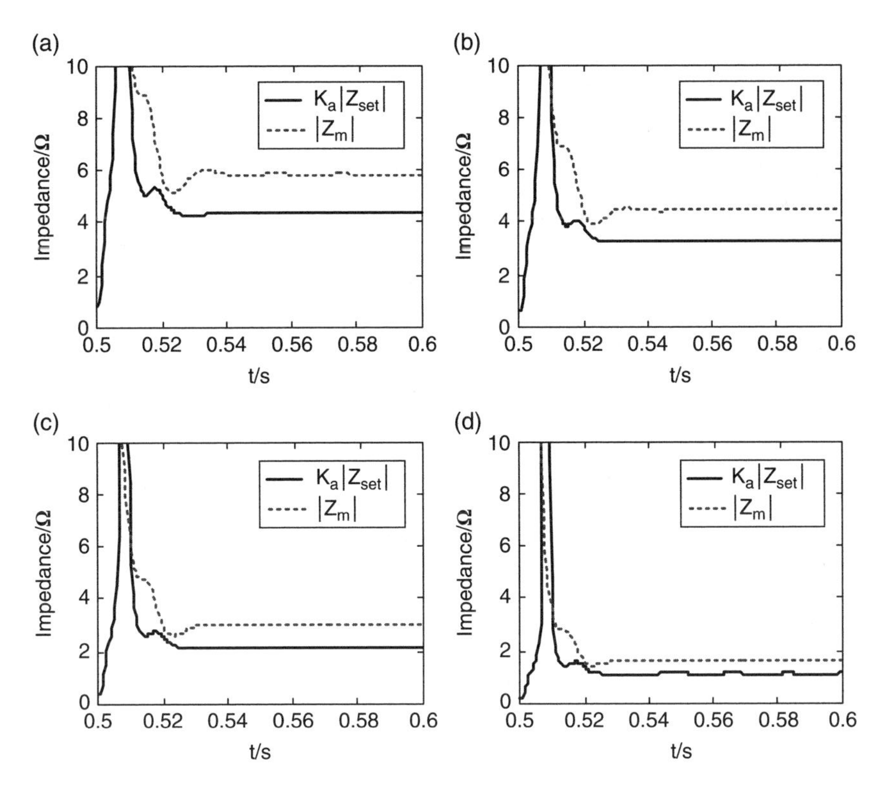

Figure 3.36 Operational curve of the adaptive distance protection on the collector system side when none of the crowbar protection of the wind turbines connected to line L_1 is put into operation. (a) $L_1 = 20$ km. (b) $L_1 = 15$ km. (c) $L_1 = 10$ km. (d) $L_1 = 5$ km.

$$\begin{cases} \left(y_{\mathrm{m}} + \dfrac{1}{\alpha Z_{\mathrm{mn}}}\right)\dot{U}_{\mathrm{mf}} - \dfrac{1}{\alpha Z_{\mathrm{mn}}}\dot{U}_{\mathrm{f}} = \dot{I}_{\mathrm{mf}} \\ \left(y_{\mathrm{n}} + \dfrac{1}{(1-\alpha)Z_{\mathrm{mn}}}\right)\dot{U}_{\mathrm{nf}} - \dfrac{1}{(1-\alpha)Z_{\mathrm{mn}}}\dot{U}_{\mathrm{f}} = \dot{I}_{\mathrm{nf}} \\ -\dfrac{1}{\alpha Z_{\mathrm{mn}}}\dot{U}_{\mathrm{mf}} - \dfrac{1}{(1-\alpha)Z_{\mathrm{mn}}}\dot{U}_{\mathrm{nf}} + \left(\dfrac{1}{\alpha Z_{\mathrm{mn}}} + \dfrac{1}{(1-\alpha)Z_{\mathrm{mn}}}\right)\dot{U}_{\mathrm{f}} = \dot{I}_{\mathrm{f}} \end{cases} \tag{3.128}$$

By eliminating the fault point voltage $\dot{U}_{\mathrm{f}}$, Equations (3.128) can be rewritten as

$$\begin{bmatrix} y_{\mathrm{m}} + \dfrac{1}{Z_{\mathrm{mn}}} & -\dfrac{1}{Z_{\mathrm{mn}}} \\ -\dfrac{1}{Z_{\mathrm{mn}}} & y_{\mathrm{n}} + \dfrac{1}{Z_{\mathrm{mn}}} \end{bmatrix} \begin{bmatrix} \dot{U}_{\mathrm{mf}} \\ \dot{U}_{\mathrm{nf}} \end{bmatrix} = \begin{bmatrix} \dot{I}_{\mathrm{mf}} + (1-\alpha)\dot{I}_{\mathrm{f}} \\ \dot{I}_{\mathrm{nf}} + \alpha\dot{I}_{\mathrm{f}} \end{bmatrix} \tag{3.129}$$

According to Equation (3.129), the equivalent model of the fault supplementary network (line MN fault) is as shown in Figure 3.43.

Comparison with Figure 3.42 shows that, in the equivalent model of the fault supplementary network, the fault point current is distributed to the two line ends,

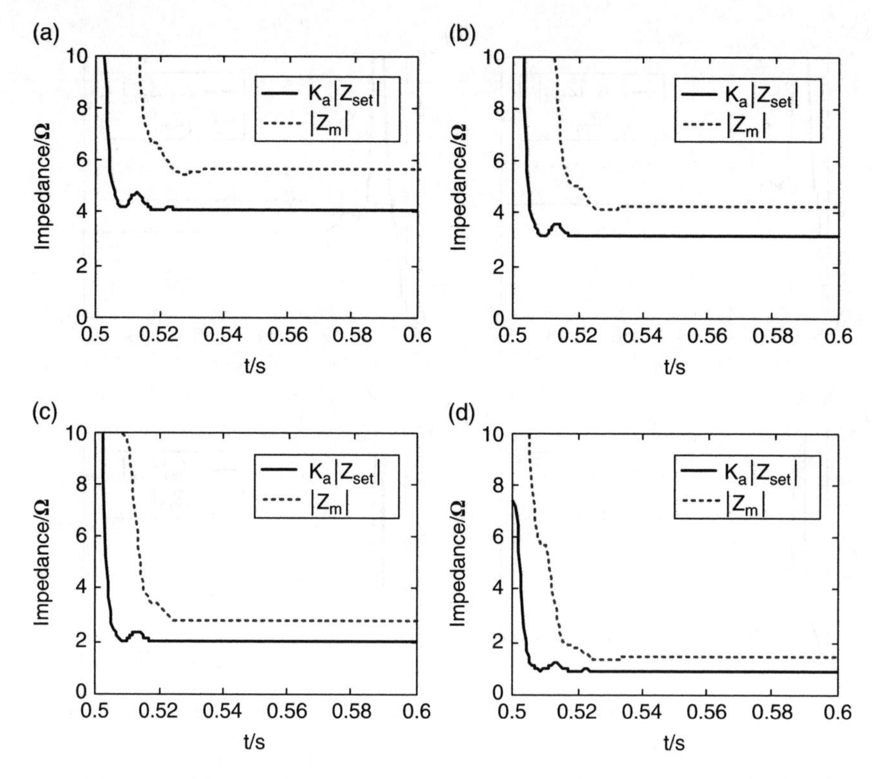

Figure 3.37 Operational curve of the adaptive distance protection on the collector system side when the crowbar protection of some of the wind turbines connected to line L_1 is put into operation. (a) $L_1 = 20$ km. (b) $L_1 = 15$ km. (c) $L_1 = 10$ km. (d) $L_1 = 5$ km.

the value of current related to the fault location. The system node number and topology remain unchanged after the fault occurs.

3.3.3.2 Fault Criteria

In the equivalent model of the fault supplementary network shown in Figure 3.43, the injection current only exists at the bus nodes of line MN, i.e. in the node injection current column vector $\mathbf{J_s}$, only the elements corresponding to bus M and bus N are not 0, and the other elements are all 0,

$$\mathbf{J_S} = \begin{bmatrix} 0 & \cdots & (1-\alpha)\dot{I}_f & \cdots & \alpha\dot{I}_f & \cdots & 0 \end{bmatrix}^T \tag{3.130}$$

where T represents the transposition of the vector.

According to the knowledge of electrical networks, the current generated on each line by the node injection current column vector $\mathbf{J_f}$ is

$$\mathbf{I} = \mathbf{Y_b A^T Y^{-1} J_s} = \mathbf{C J_s} \tag{3.131}$$

where $\mathbf{I}$ is the column vector of branch currents. $\mathbf{Y}$ is the node admittance matrix. $\mathbf{Y_b}$ is the branch admittance matrix. $\mathbf{A}$ is the node-branch incidence matrix. $\mathbf{C}$ is the correlation coefficient matrix, the elements of which take values between [−1,1].

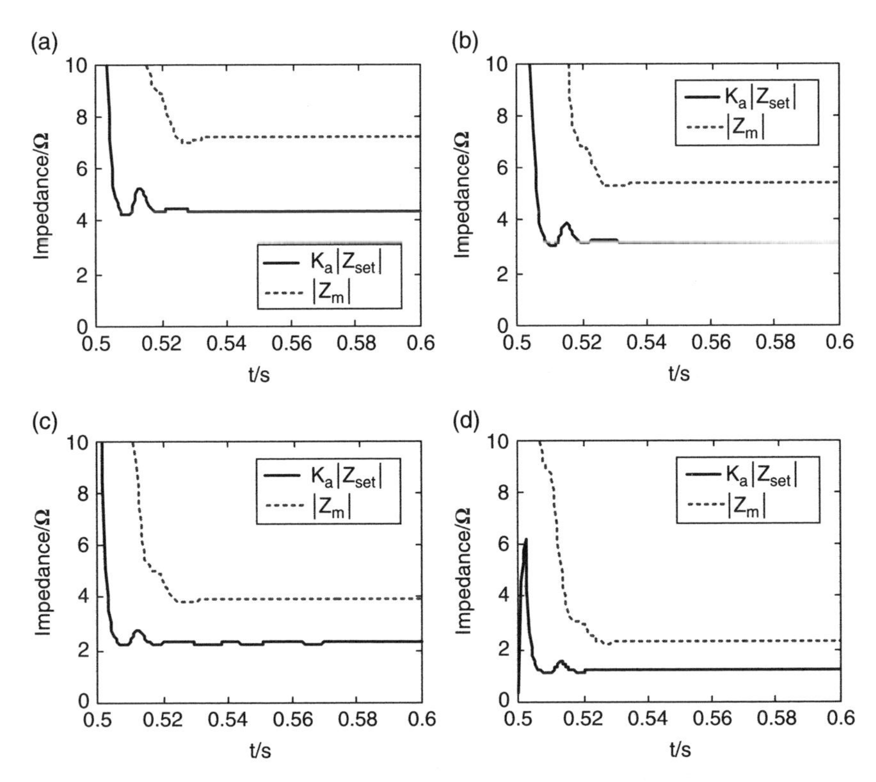

Figure 3.38 Operational curve of the adaptive distance protection on the collector system side when the crowbar protection of all the wind turbines connected to line L_1 is put into operation. (a) $L_1 = 20$ km. (b) $L_1 = 15$ km. (c) $L_1 = 10$ km. (d) $L_1 = 5$ km.

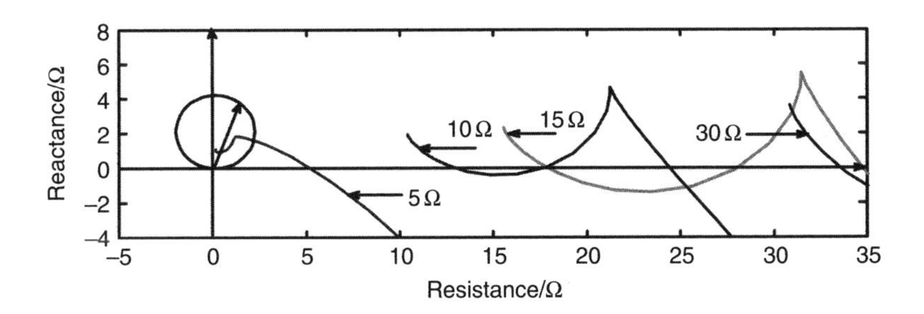

Figure 3.39 Operational curves of traditional distance protection when a phase-A-to-ground fault occurs at 4 km from the relaying point via different transition resistances.

According to Figure 3.43, when a fault occurs on line MN, the injection currents to two line ends in the fault supplementary network are, respectively

$$\begin{cases} \dot{I}_{\mathrm{mf}} = C_{\mathrm{MN,M}} * (1-\alpha)\dot{I}_{\mathrm{f}} + C_{\mathrm{MN,N}} * \alpha\dot{I}_{\mathrm{f}} - (1-\alpha)\dot{I}_{\mathrm{f}} \\ \dot{I}_{\mathrm{nf}} = -C_{\mathrm{MN,N}} * (1-\alpha)\dot{I}_{\mathrm{f}} - C_{\mathrm{MN,N}} * \alpha\dot{I}_{\mathrm{f}} - \alpha\dot{I}_{\mathrm{f}} \end{cases} \tag{3.132}$$

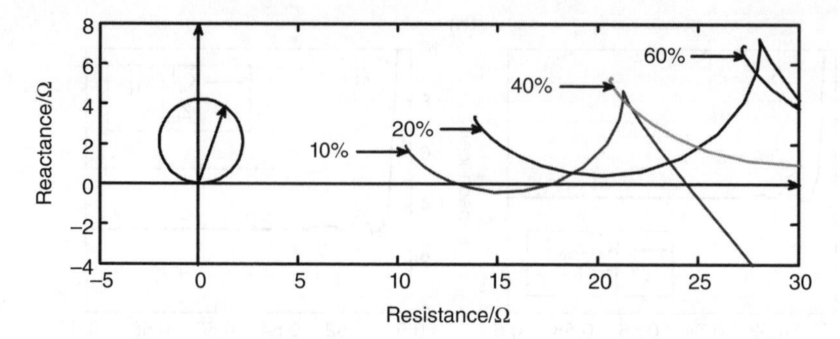

Figure 3.40 Operational curves of traditional distance protection when a phase-A-to-ground fault occurs via 30 Ω transition resistance at different locations.

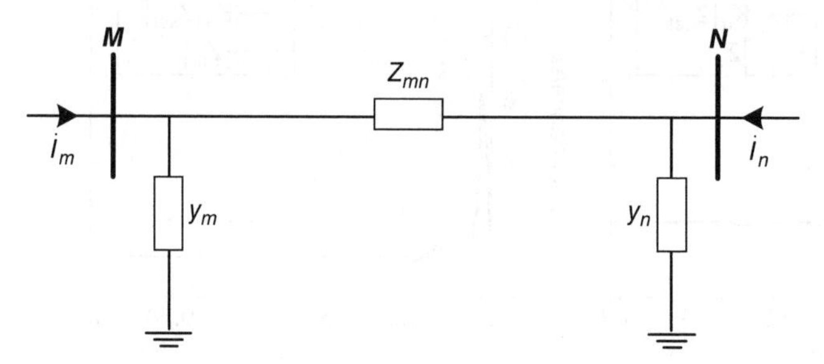

Figure 3.41 Model of network in normal operating state.

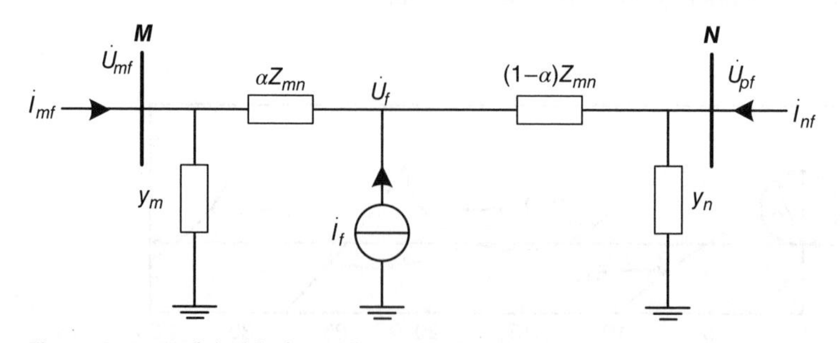

Figure 3.42 Model of fault steady-state network.

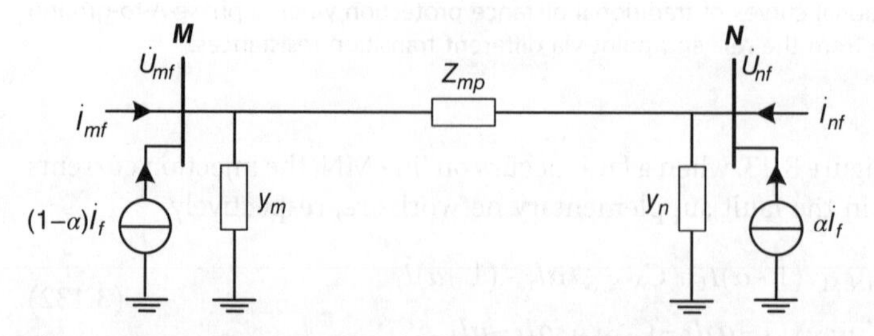

Figure 3.43 Equivalent model of fault supplementary network.

where $C_{MN,M}$ and $C_{MN,N}$ are the Mth and Nth elements in the row corresponding to line MN in the branch correlation factor matrix.

The phasor sum modulus value and phasor difference modulus value of the currents at the two line ends $\dot{I}_{mf}$ and $\dot{I}_{nf}$ are, respectively

$$\begin{cases} \left|\dot{I}_{mf} + \dot{I}_{nf}\right| = \left|\dot{I}_{f}\right| \\ \left|\dot{I}_{mf} - \dot{I}_{nf}\right| = \left|2*(C_{MN,M}*(1-\alpha) + C_{MN,N}*\alpha) + (2\alpha - 1)\right|\left|\dot{I}_{f}\right| \end{cases} \tag{3.133}$$

When there is no fault on line MN, the currents at the two line ends are

$$\begin{cases} \dot{I}_{mf} = C_{MN,M}*(1-\alpha)\dot{I}_{f} + C_{MN,N}*\alpha\dot{I}_{f} \\ \dot{I}_{nf} = -C_{MN,N}*(1-\alpha)\dot{I}_{f} - C_{MN,N}*\alpha\dot{I}_{f} \end{cases} \tag{3.134}$$

The phasor sum modulus value and phasor difference modulus value of the currents at the two line ends $\dot{I}_{mf}$ and $\dot{I}_{nf}$ are, respectively

$$\begin{cases} \left|\dot{I}_{mf} + \dot{I}_{nf}\right| = 0 \\ \left|\dot{I}_{mf} - \dot{I}_{nf}\right| = \left|2*(C_{MN,M}*(1-\alpha) + C_{MN,N}*\alpha)\right|\left|\dot{I}_{f}\right| \end{cases} \tag{3.135}$$

According to the phasor sum modulus value and phasor difference modulus value of $\dot{I}_{mf}$ and $\dot{I}_{nf}$, the fault identification criteria can be formed:

$$\left|\dot{I}_{Nf} + \dot{I}_{Mf}\right| > K\left|\dot{I}_{Nf} - \dot{I}_{Mf}\right| \tag{3.136}$$

Based on the above analysis, the proposed scheme has the following characteristics:

1) Unaffected by the fault type. No matter what type of fault occurs, the equivalent model of the fault steady-state network shown in Figure 3.42 is valid. Therefore, this scheme could correctly identify all types of fault.
2) Unaffected by the fault location. No matter where the fault location is, in the case of an in-zone fault, the fault identification criteria in Inequality (3.136) are satisfied; in the case of an out-of-zone fault, the fault identification criteria in Inequality (3.136) are not satisfied.

3.3.3.3 Scheme Verification

A simulation model of a doubly-fed wind farm connected to a power grid was built in PSCAD, as shown in Figure 3.44. Taking one of the collector lines as an example, the corresponding scheme is verified. Six DFIG wind turbines are connected in parallel on the line, the rated capacity of each DFIG being 1.5 MW, thus rated capacity of the collector line is 9 MW. The voltage level is 35 kV.

When $t = 1.5$ s, an in-zone fault occurs at different locations on the 110 kV outgoing transmission line, and the operational status of line protection is as shown in Figures 3.45–3.47. It can be seen that, whether it is a three-phase fault or a phase-to-phase fault, both traditional current differential protection [9] and the proposed protection scheme could operate reliably without malfunction.

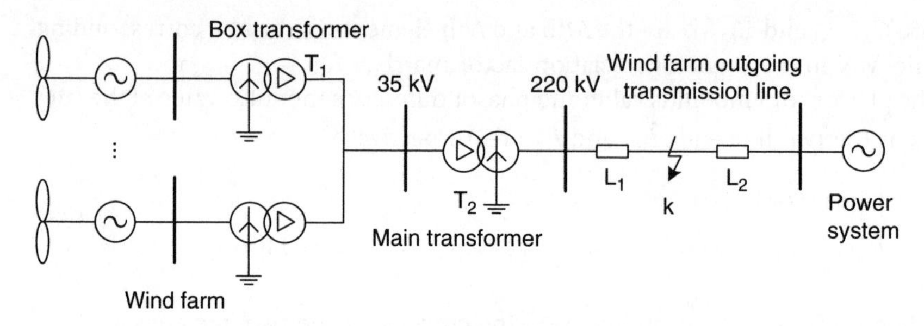

Figure 3.44 Equivalent diagram of wind farm outgoing transmission line.

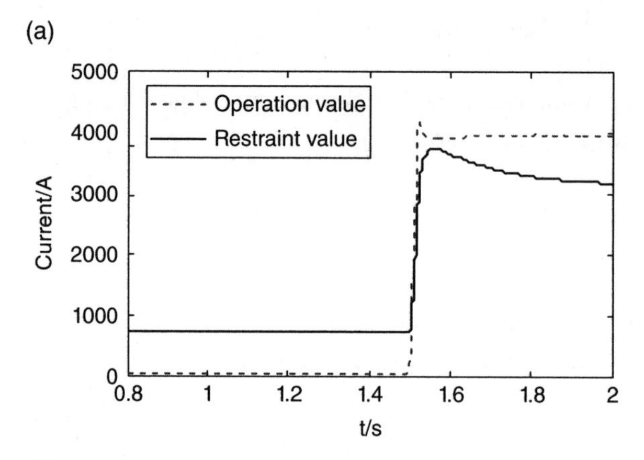

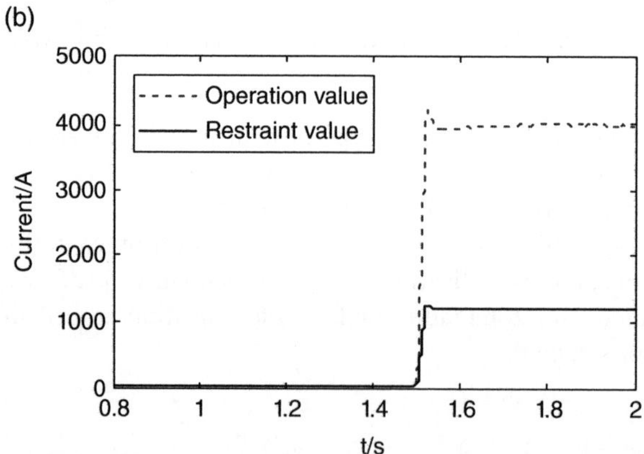

Figure 3.45 Protection operational status when a three-phase fault occurs at 50% of the protection line. (a) Traditional current pilot differential protection. (b) Proposed current pilot differential protection.

Meanwhile, the sensitivity of the proposed scheme is obviously higher than that of traditional pilot differential protection.

When an out-of-zone fault occurs on the 110 kV outgoing transmission line, as shown in Figure 3.47, neither traditional current differential protection nor the proposed protection scheme operates. However, affected by the fault injection

Figure 3.46 Protection operational status when a two-phase fault occurs at 50% of the protection line. (a) Traditional current pilot differential protection. (b) Proposed current pilot differential protection.

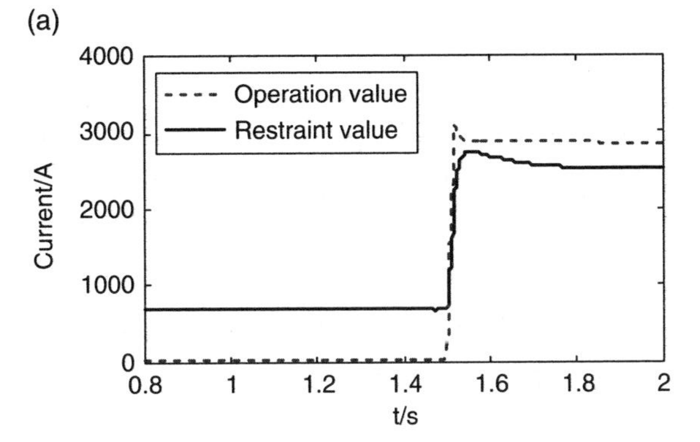

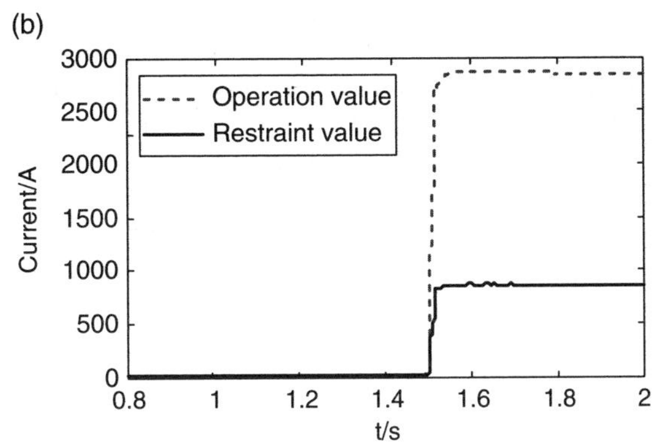

Figure 3.47 Protection operational status when a two-phase fault occurs at 90% of the protection line. (a) Traditional current pilot differential protection. (b) Proposed current pilot differential protection.

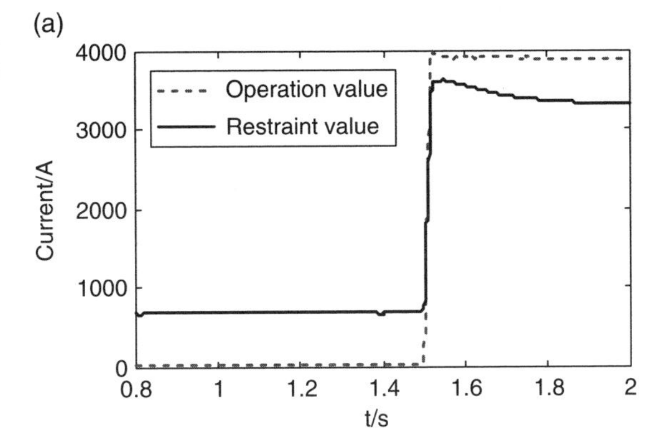

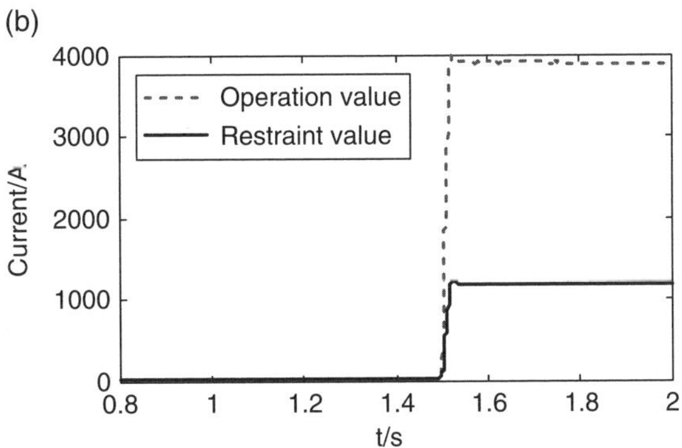

current of a doubly-fed wind farm, the reliability of traditional current differential protection is poor. The proposed scheme is based on fault steady-state components which are not affected by the transient characteristics of a doubly-fed wind farm, thus the reliability is higher.

3.4 Local Area Protection for Distributed Renewable Energy

As an effective supplement to centralized grid-connected power generation, distributed grid-connected renewable power generation technology [10] and its integration system (microgrid) are maturing. With the development of large-scale distributed generation (DG), the distribution system is gradually developing into a two-terminal or multi-terminal active network from a single power supply radial network, and this has a large influence on the traditional distribution network protection system. On the one hand, the infeeding current and the outflowing current of DG can cause the malfunction or miss-trip of the corresponding distribution line protection. On the other hand, the operation of the protective devices may lead to isolated islands. In view of this, Section 3.4.1 introduces an adaptive protection approach for distribution networks containing distributed generation [11], which can eliminate the influence of DG on the branch current and increase the protection range of the main protection and backup protection. In Section 3.4.2, two islanding detection methods based on voltage harmonic distortion positive feedback [12] and negative sequence voltage contribution factor [13] are introduced. The methods are capable of fast and accurate islanding detection, effective in open-phase operation and immune to false islanding.

3.4.1 Adaptive Protection Approach for a Distribution Network Containing Distributed Generation

3.4.1.1 Impact of DG on Protection Coordination

In spite of the positive impacts of DG on system design and operation, it changes the original steady-state and fault current directions and values. The severity of these changes is based on the DG's location, capacity and number in distribution systems. As pointed out in [14], the fault contribution from a single small DG unit may not be large, but the aggregate contributions of a few larger units can alter the short-circuit levels enough to cause protective devices to malfunction.

A typical case is shown in Figure 3.48. An industrial power network is fed through source G and protected by R1, R2 and R3. Each protective device is assigned a primary function to clear faults in a specific zone and a secondary function to clear faults in the adjacent or downstream zones to the extent within the range that the device permits. In this situation, the next upstream device, or device combination, must operate to provide backup protection. When two devices operate properly in this primary/secondary mode for any system fault, they are said to be coordinated. Proper coordination is achieved by this discrimination between successive devices. Good practice dictates that when a fault F1 occurs, the time of operation of relay R2 should be made longer than the time of operation of R1 at least by a time interval called the 'coordination time interval'. As clearly shown in Figure 3.48, R2 will back up R1.

It is clear that protection for distribution systems with DG cannot be achieved with the same philosophies that have been used to protect traditional distribution systems. At the very least, a system designed to protect distribution systems with DG should take the following into consideration.

1) Bidirectional current flow – If DG1 and DG2 connect to the system, as shown in Figure 3.49. For F2, selectivity requires that R2 operates before R1, and for F1, R1 should operate before R2. Therefore, the system should be facilitated with two directional protection devices at each line to ensure the correct fault isolation, as shown in Figure 3.50.

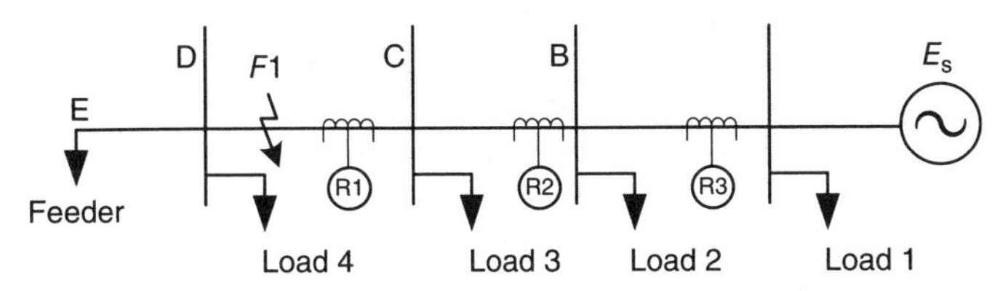

Figure 3.48 An industrial power network fed through source G and protected by R1, R2 and R3.

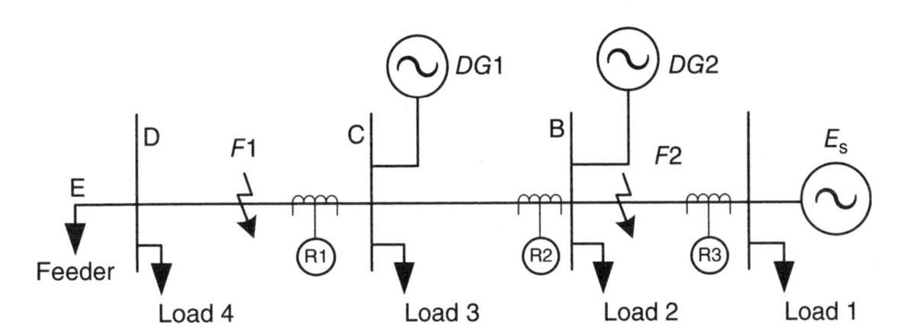

Figure 3.49 An industrial power network fed through source G, DG1 and DG2, and protected by R1, R2 and R3.

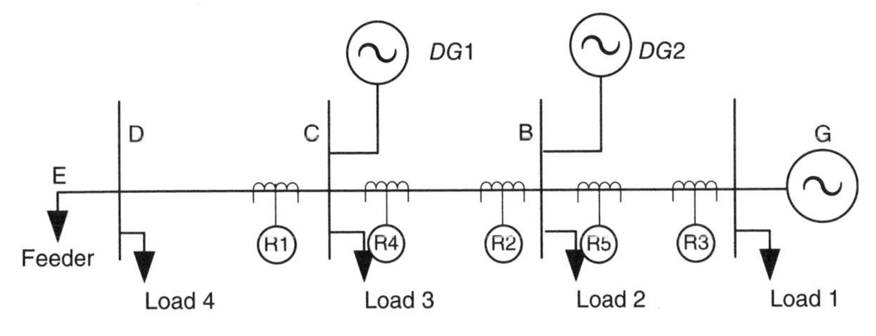

Figure 3.50 An industrial power network with directional protection devices to ensure the correct fault isolation.

2) Changing system conditions – For F1, the fault current now has two components, one coming from the supply and the other from DG1 and DG2. Under the condition of changing of capacities of DG units, the maximum and minimum fault currents will change. This will require R1 and R2 to be coordinated at changing current conditions.

In addition, due to the size and placement of the DG units, some potential cases are not well coordinated in an actual distribution system [15]. The only way to withhold coordination in the presence of DG penetration is to throw off all DG units when a fault occurs. However, this would cause a DG to be disconnected for temporary faults.

3.4.1.2 Adaptive Protection Scheme

A. Adaptive Primary Protection Setting for Feeders with DG

For traditional distribution systems, the substation is the only source of power. In addition, due to the substations being away from large generation units, the initial high 'subtransient component' does not exist in the fault current transients. Assuming that the faulted phase can be determined previously, the fault current can be approximated by its steady-state value. In this way, the feeder can be represented by a steady-state model, in which the substation is represented as a voltage source behind a Thévenin impedance. If there are conventional DG units on the feeder, the above feeder model can be obtained easily by employing the simple Thévenin equivalent models for these generators. However, for other kinds of DG that respond considerably after a fault occurs, the same technique cannot be applied. Therefore, a new scheme is needed to incorporate all DG units into the fault analysis.

As shown in Figure 3.50, DG1 can be represented as an injected phase current i_1, the lines CD and DE are represented by their series impedance z_{CE}, and the Load 3 can be represented by its equivalent impedances z_{d3}. The corresponding equivalent circuit of these devices can then be represented by a Norton equivalent model, as shown in Figure 3.51, which can then be easily transformed to a Thévenin equivalent model, as shown in Figure 3.52. The voltage u_1 behind the Thévenin impedance is calculated as

$$u_1 = i_1 \frac{z_{CE} z_{d3}}{z_{CE} + z_{d3}} \tag{3.137}$$

Thus, the adaptive primary protection setting for R4 can be formulated as

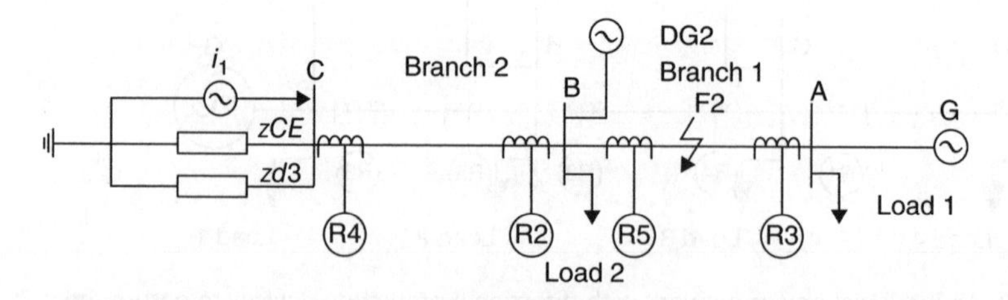

Figure 3.51 Norton equivalent model behind R4.

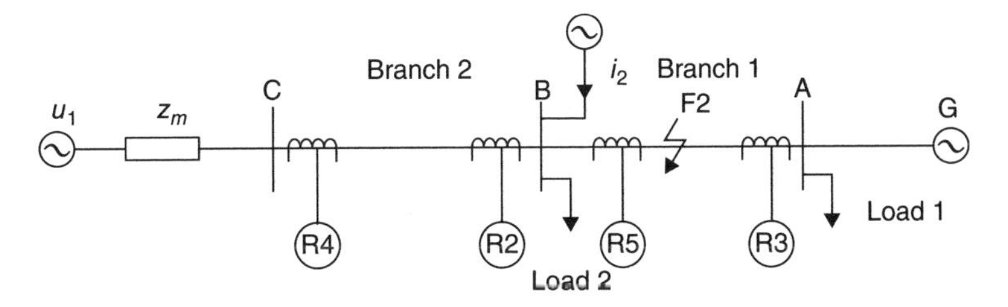

Figure 3.52 Thévenin equivalent model behind R4.

$$i_{\mathrm{ps4}} = \frac{k_{\mathrm{k}} k_{\mathrm{d}} u_1}{z_{\mathrm{m}} + z_{\mathrm{l}}} \tag{3.138}$$

$$z_{\mathrm{m}} = \frac{z_{\mathrm{CE}} z_{\mathrm{d3}}}{z_{\mathrm{CE}} + z_{\mathrm{d3}}} \tag{3.139}$$

where i_{ps4} is the adaptive primary protection setting for R4. z_{m} is the integrated impedance of the source side. z_{l} is the protected line impedance. k_{d} is the coefficient of fault type and can be previously determined by software used by utilities. k_{k} is the coefficient of reliability.

A similar procedure can also be applied to R5 for its adaptive primary protection setting i_{ps5}.

B. Adaptive Backup Protection Setting for Feeders with DG

In the event of a switching device failure, the next upstream device, or device combination, must operate to provide backup protection. A typical case is shown in Figure 3.52. For F2, if the measured current of R5 is still above the threshold after a certain time delay, measurements of the current (including the injected current of DG2, i_2 and switching device status at bus B) will be sent to R4 to provide backup protection. However, since DG2 may respond considerably after a fault occurs, measured currents of R4 and R5 will change dramatically. Therefore, it is recommended to eliminate the effect of DG2 on these measurements.

The nodal equations of the distribution systems can be written as

$$\mathbf{I_N} = \mathbf{Y_N U_N} \tag{3.140}$$

$$\mathbf{Y_N} = \mathbf{A Y A^T} \tag{3.141}$$

where $\mathbf{Y_N}$ is the node admittance matrix. $\mathbf{U_N}$ is the voltage at each node. $\mathbf{I_N}$ is the current injected at each node. $\mathbf{A}$ is the node correlation matrix. $\mathbf{Y}$ is the branch admittance matrix.

The branch current can be obtained by multiplying the voltage difference between two ends of a branch with that branch admittance. Therefore, the relationship between the branch current $\mathbf{I_B}$ and the node voltage $\mathbf{U_N}$ can be expressed as

$$\mathbf{I_B} = \mathbf{Y A^T U_N} \tag{3.142}$$

From Equations (3.140)–(3.142), the relationship between the branch current $\mathbf{I_B}$ and the injected current $\mathbf{I_N}$ can be deduced as

$$\mathbf{I_B} = \mathbf{Y}\mathbf{A}^{\mathbf{T}}\mathbf{Y}_{\mathbf{N}}^{-1}\mathbf{I_N} \tag{3.143}$$

where $\mathbf{C}(\lambda) = \mathbf{Y}\mathbf{A}^{\mathbf{T}}\mathbf{Y}_{\mathbf{N}}^{-1}$ is defined as the system correlation coefficient matrix.

The current of the kth branch caused by the injected current of DG2 is calculated as

$$i_{k,\mathrm{B}} = \lambda_{k2} i_2 \tag{3.144}$$

where $i_{k,\mathrm{B}}$ is the current of the kth branch caused by the injected current of DG2. λ_{k2} is the element of the system correlation coefficient matrix $\mathbf{C}(\lambda)$.

The impact of DG2 on the current of the kth branch can be eliminated by

$$i_{k,\mathrm{d}} = i_{k,\mathrm{m}} - i_{k,\mathrm{B}} \tag{3.145}$$

where $i_{k,\mathrm{m}}$ is the measured current of the kth branch. $i_{k,\mathrm{d}}$ is the current of the kth branch without the impact of DG2.

A similar procedure is then applied to the branches connected with R4 and R5. In this situation, the adaptive backup protection setting can be formulated as

$$i_{\mathrm{bs4}} = k_{\mathrm{k}}' i_{\mathrm{ps5}} / k_{\mathrm{b}} \tag{3.146}$$

$$k_{\mathrm{b}} = i_{1,\mathrm{d}} / i_{2,\mathrm{d}} \tag{3.147}$$

where i_{bs4} is the adaptive backup protection setting for R4. i_{ps5} is the primary protection setting for R5. k_{b} is the branch coefficient. k_{k}' is the coefficient of reliability. $i_{1,\mathrm{d}}$ is the current of branch 1 without the impact of DG2. $i_{2,\mathrm{d}}$ is the current of branch 2 without the impact of DG2.

3.4.1.3 Test Systems and Results

A. Test Systems

To test the performance of the proposed adaptive protection method, a practical 10 kV distributed system in the Tianjin power network is used in this study, as shown in Figure 3.53.

The base capacity is 500 MVA, and the base voltage is 10.5 kV. Branches AB, BC and AF are all overhead lines. The parameters of these lines are $r_1 = 0.27\ \Omega/\mathrm{km}$, $x_1 = 0.347\ \Omega/\mathrm{km}$. Branches CD, DE and FG are all underground cables. The parameters of these lines are $r_1 = 0.259\ \Omega/\mathrm{km}$, $x_1 = 0.093\ \Omega/\mathrm{km}$. Each load has the nominal capacity 6 MVA and nominal power factor 0.85. DG1 and DG2 with

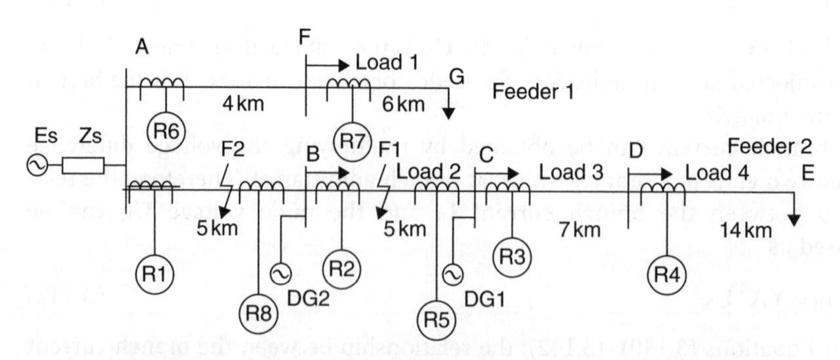

Figure 3.53 A practical 10 kV distribution system in the Tianjin power network.

P-Q control schemes are connected to bus C and bus E, respectively. Their nominal capacities are both 10 MVA. Additionally, any lines that will experience bidirectional current flow will need two protection devices at each line, whereas any lines that will not experience bidirectional current flow will only need one protection device instead of two, thus reducing the cost.

The test system is simulated by using PSCAD/EMTDC. In the following analysis, I_{ps5} and I_{pm5} represent the adaptive primary protection setting and the measured current of R5, respectively. I_{ps8} and I_{pm8} represent the adaptive primary protection setting and the measured current of R8, respectively. I_{bs5} and I_{bm5} represent the adaptive backup protection setting and the measured current of R5 without the impact of DG2, respectively.

B. Responses of Adaptive Primary Protection System

For a conventional source in the distribution system, the fault current can be approximated by its steady-state value. But for DG, that response is considerably after a fault occurs, so the same technique cannot be applied. Therefore, the proposed adaptive protection scheme mainly focuses on lines fed through DG units. A typical case is shown in Figure 3.54. A three-phase fault was applied to the middle of the line between buses B and C at 0.30 seconds. The fault is cleared after 0.5 seconds. When the fault occurs, the fault current flowing through R5 is fed by DG1. The Thévenin equivalent model behind R5 can be obtained by using Equations (3.137) and (3.139), and I_{ps5} can be calculated by using Equation (3.138).

Figure 3.54 shows the curves of I_{ps5} and I_{pm5}. It can be found that after the fault occurs, I_{pm5} responds considerably and becomes larger than I_{ps5}. Thus, the primary protection issues a signal to the switching device to trip the fault.

Additionally, a case of phase-to-phase fault was simulated in the middle of Section BC at 0.3 seconds and was cleared after 0.5 seconds. I_{pm5} and I_{ps5} are shown in Figure 3.55. It can be seen that I_{pm5} is larger than I_{ps5} after the fault occurs, which makes the relay trip the fault accurately.

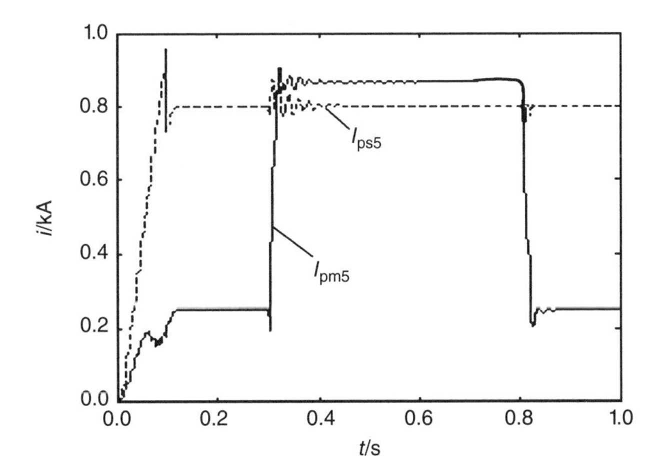

Figure 3.54 The adaptive primary protection setting and the measured current of R5 when a three-phase fault occurs in the middle of Section BC.

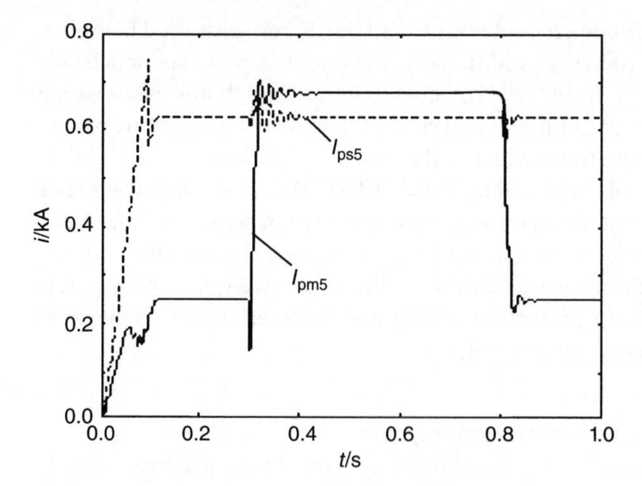

Figure 3.55 The adaptive primary protection setting and the measured current of R5 when a phase-to-phase fault occurs in the middle of Section BC.

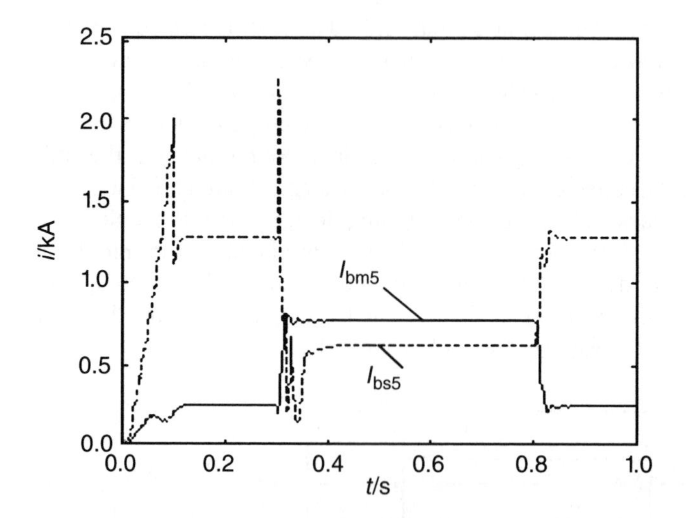

Figure 3.56 The adaptive backup protection setting and the measured current of R5 when a three-phase fault occurs on Section AB close to bus B.

C. *Responses of Adaptive Backup Protection System*

If a fault occurs on Section AB close to bus B, and the measured current of R8 is still above the threshold after a certain time delay, then R5 must operate to provide adaptive backup protection. In this situation, the impact of DG2 on the branch currents is first eliminated by using Equations (3.140)–(3.145). Then, I_{bs5} is calculated by using Equations (3.146) and (3.147).

Figure 3.56 shows I_{bs5} and I_{bm5} in the case of a three-phase fault, whereas Figure 3.57 shows I_{bs5} and I_{bm5} in the case of a phase-to-phase fault. It can be found that no matter what types of fault occur, I_{bm5} is greater than I_{bs5} in both

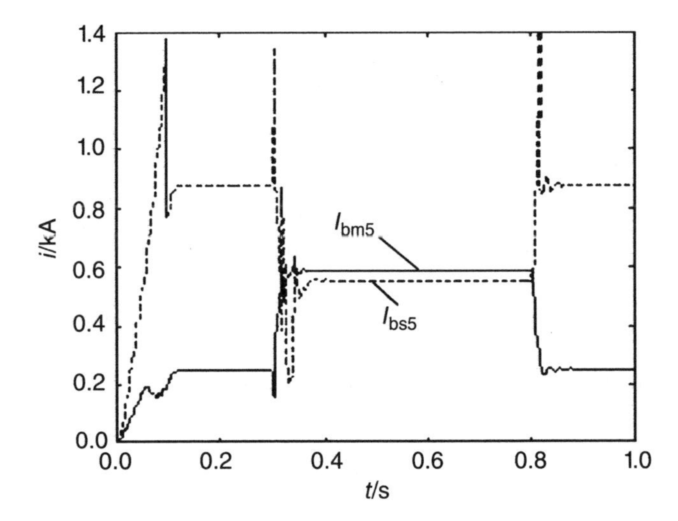

Figure 3.57 The adaptive backup protection setting and the measured current of R5 when a phase-to-phase fault occurs on Section AB close to bus B.

figures. Thus, in each case, the adaptive backup protection at R5 will issue a signal to the switching device to clear the fault on the adjacent Section AB.

A comprehensive fault analysis has been performed for different locations on the system, different fault types (balanced and unbalanced) and different capacities of DG1. It can be concluded that the extents of the adaptive primary and backup protection are about 80% and 140% of the protected line, respectively. However, the extents of the traditional current primary and backup protection are less than 20% and less than 80% of the protected line, respectively. Furthermore, the proposed scheme is not affected by the fault types and capacities of DG1. Thus, the protection performance has been greatly improved.

3.4.2 Islanding Detection Method

Islanding detection is a critical technique in flexible DG (distributed generation) control. Islanding refers to the breaking away of a grid-connected DG unit from the grid, thus forming a micro-self-supplying system with local loads [16]. Due to the difference in controlling methods for DG under grid-connected mode and islanded mode, a smooth switch between different controlling methods can be achieved only when fast and accurate detection of islanding is available. So islanding detection is of great practical value.

Currently, islanding detection methods can be sorted into three major groups: passive, active and switch state methods. The passive methods include those of using over/under-voltage, over/under-frequency, phase mutation, voltage harmonic, etc. as the detection signal, and these are easy to implement. However, recognition failure may occur when the voltage at the PCC falls into the non-detection zone (NDZ) after islanding. The active methods include active frequency offset, sliding mode frequency offset, active frequency offset with positive feedback and automatic phase shift. For this group of methods, islanding is detected when the voltage or frequency at the PCC exceeds the threshold range. This can be caused by different disturbance inputs which can break the balance of the DG in islanded mode. They have the advantage of high detection precision and small NDZ, but they are often destructive because they lower the power quality of the inverter output. The third group of methods, which works on the basis of utilizing communication means such as power line carrier communication (PLCC) and supervisory control and data acquisition (SCADA), despite its small NDZ, has not been widely applied in DG due to its high cost and complicated design.

Above all, islanding detection methods should ensure the correct judgement when islanding occurs in a micro network and should not malfunction when there is no islanding. Good islanding detection methods should avoid damage to stable operation and power quality of the micro network, and reduce the detection blind area as much as possible.

3.4.2.1 Application of Voltage Harmonic Distortion Positive Feedback for Islanding Detection

A. Islanding Detection Principle of DG

As shown in Figure 3.58, sinusoidal pulse width modulation (SPWM) is applied to the voltage source inverter. U_{dc} is the equivalent voltage source on the DC side and C_{dc} is the capacitor on the DC side. In the LCL-filter, L_1 is the inductance on the inverter side, C is the filter capacitor and L_2 is the inductance on the grid side. The equivalent resistance of the filter capacitor C is R_c and Z represents the local load. The schematic diagram of the controller is the dash-dotted-line block shown in Figure 3.58. By calculating the real-time output power of DG with the load voltage and the inverter output current acquired by sensors, the average

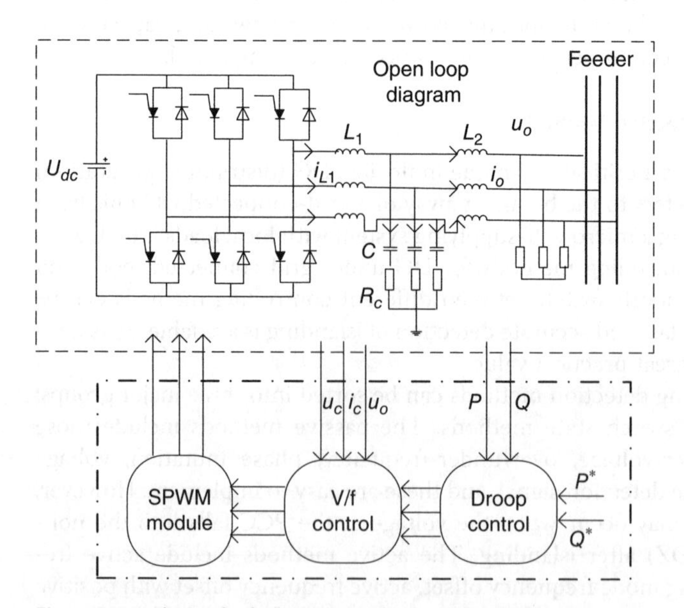

Figure 3.58 The multiple-loop control schematic of the DG.

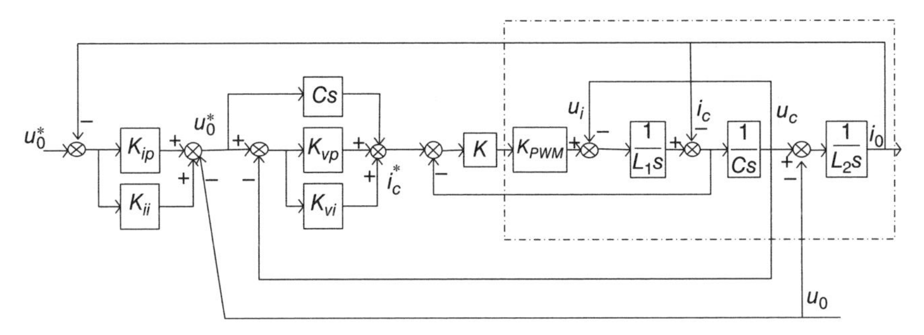

Figure 3.59 Schematic diagram of the multi-loop control system.

power is obtained after passing through a low-pass filter, which is then compared with a reference value. The outcome of the former process going through the PQ controller can produce the input signals needed for the voltage/frequency (v/f) control module. Finally, with the outcome of the v/f module, the desired sinusoidal modulation wave is obtained, which is further applied to the SPWM module to modulate the output voltage.

According to the mathematical model under the dq coordinate system, a multiple-loop feedback control system is established as shown in Figure 3.59, which consists of an inner loop capacitor current controller, a middle loop capacitor voltage controller and an outer loop output voltage controller.

The inner loop is designed to regulate the capacitor current by utilizing a proportional controller which can increase the dynamic response speed of the system, realizing dynamic following of the capacitor current. In accordance with this objective, the current loop is designed to cover the maximum bandwidth possible. On the other hand, the ratio of the capacitor current to the inverter output current and the amplitude of the capacitor current itself should be kept at a low value.

The middle loop is designed to regulate the capacitor voltage applying a proportional-integral (PI) controller, which can inhibit the impact of the grid voltage fluctuations and load imbalances on the inverter output power.

The outer loop is designed for controlling the inverter output voltage by applying a PI controller. It can regulate the active and reactive power of the DG, improve its power factor and power quality, as well as keeping the steady-state error of the load voltage at zero.

When a DG unit operates under grid-connected mode, the voltage and frequency at the point of common coupling (PCC) are determined by the main grid. The inverter, under this grid-tied mode, applies a constant power control strategy. The principle of the control system is shown in Figure 3.60.

In Figure 3.60, $G_i(s)$ is the follow-up used by the current controller to produce the current output. K_{pwm} is the gain of the inverter output voltage. L and R are the equivalent serial conductance and resistance, respectively. $H(U)$ is a function of the voltage total harmonic distortion (THD), which refers to the root mean square (rms) value of the harmonic parts in a periodical AC wave divided by that of the fundamental part, i.e.

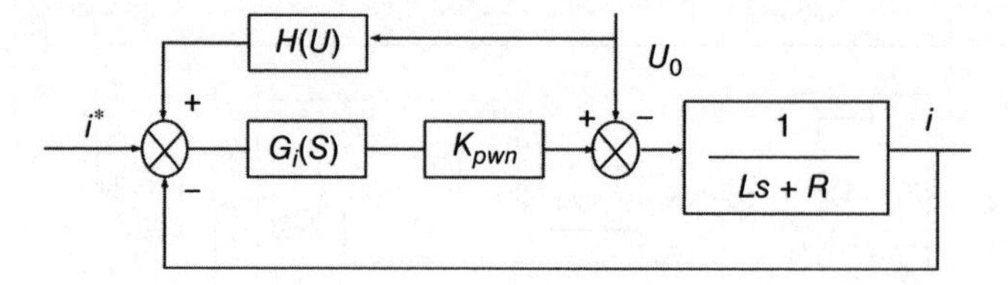

Figure 3.60 Block diagram of grid-tied control for inverter.

$$H = \frac{\sqrt{\sum_{k=2}^{\infty} U_k^2}}{U_1} = \sqrt{\sum_{k=2}^{\infty}\left(\frac{U_k}{U_1}\right)^2} \times 100\% \tag{3.148}$$

where U_k is the amplitude of the kth voltage harmonic, and U_1 is that of the voltage fundamental.

As seen in Figure 3.60, the inverter output voltage, after being extracted by the THD circuit, becomes

$$\begin{bmatrix} U_{Ha} \\ U_{Hb} \\ U_{Hc} \end{bmatrix} = \begin{bmatrix} KHU_m \cos(\omega t) \\ KHU_m \cos(\omega t - 120°) \\ KHU_m \cos(\omega t + 120°) \end{bmatrix} \tag{3.149}$$

where U_m is the amplitude of the grid voltage, while K is the positive feedback coefficient.

Assuming that the current output of the inverter lags behind the voltage by a phase angle of θ_i, the resulting reference value of the current at $t = 0$ can be shown as

$$\begin{bmatrix} i_a^* \\ i_b^* \\ i_c^* \end{bmatrix} = \begin{bmatrix} I_m \cos(\omega t + \theta_i) \\ I_m \cos(\omega t + \theta_i - 120°) \\ I_m \cos(\omega t + \theta_i + 120°) \end{bmatrix} \tag{3.150}$$

where I_m is the amplitude of the current output.

By applying the Park transformation to Equations (3.149) and (3.150), the voltage and current components under the two-phase rotating coordinate system are composed as

$$\begin{bmatrix} U_{Hd} \\ U_{Hq} \end{bmatrix} = \begin{bmatrix} KHU_m \\ 0 \end{bmatrix} \tag{3.151}$$

$$\begin{bmatrix} i_d^* \\ i_q^* \end{bmatrix} = \begin{bmatrix} I_m \cos(\theta_i - \alpha_0) \\ -I_m \sin(\theta_i - \alpha_0) \end{bmatrix} \tag{3.152}$$

where α_0 is the initial value of the phase angle in the rotating coordinate system.

The current output of the DG in grid-connected mode can be calculated by summing Equations (3.151) and (3.152):

$$\begin{bmatrix} i_d \\ i_q \end{bmatrix} = \begin{bmatrix} i_d^* + U_{Hd} \\ i_q^* + U_{Hq} \end{bmatrix} = \begin{bmatrix} I_m \cos(\theta_i - \alpha_0) + KHU_m \\ -I_m \sin(\theta_i - \alpha_0) \end{bmatrix} \tag{3.153}$$

The power factor of the inverter output is kept at 1.0 by the constant power control applied when DG operates in grid-connected mode, i.e. the phase angle of the voltage output and the current output remains the same, or $\theta_i = \alpha_0$. By applying a Park inverse transformation to Equation (3.153), phase components of the currents can be obtained as

$$\begin{bmatrix} i_a \\ i_b \\ i_c \end{bmatrix} = \begin{bmatrix} (I_m + KHU_m)\cos(\omega t) \\ (I_m + KHU_m)\cos(\omega t - 120^\circ) \\ (I_m + KHU_m)\cos(\omega t + 120^\circ) \end{bmatrix} \tag{3.154}$$

Equation (3.154) shows that the inverter output current frequency is not affected by the introduction of the voltage THD positive feedback circuit, and phases A, B and C remain 120° apart, which is the ideal condition. This occurs because when DG is grid-connected, the voltage and frequency are determined by the main grid. According to the state standard GB/T 14549-93, the THD of the grid voltage under the level of 0.38 kV should not exceed 5%. So when a DG unit operates in grid-connected mode, the voltage THD coefficient KHU_m is quite small, the positive feedback circuit hardly affecting the inverter output current; thus destruction of the power quality can be avoided.

When the circuit breaker at the PCC breaks, DG is separated from the main grid to supply only local loads. The control system for DG in this case is as shown in Figure 3.60, where the voltage is supported by the DG alone, and the control strategy for the inverter remains the same, i.e. constant power control. For a three-phase balanced load, the voltage at the PCC after islanding is

$$\begin{bmatrix} u_{oa} \\ u_{ob} \\ u_{oc} \end{bmatrix} = \begin{bmatrix} Z_a i_a \\ Z_b i_b \\ Z_c i_c \end{bmatrix} = \begin{bmatrix} |Z|(I_m + KHU_m)\cos(\omega t + \varphi_z) \\ |Z|(I_m + KHU_m)\cos(\omega t - 120^\circ + \varphi_z) \\ |Z|(I_m + KHU_m)\cos(\omega t + 120^\circ + \varphi_z) \end{bmatrix} \tag{3.155}$$

where $Z(s) = RLs/(RLCs^2 + Ls + R)$, $|Z| = 1/\sqrt{\left(\frac{1}{R}\right)^2 + \left(\varpi C - \frac{1}{\omega L}\right)^2}$, $\angle\varphi_z = \arctan(R(1-\omega^2 LC)/(\omega L))$.

As seen in Figure 3.60, when the DG unit is grid-connected, the voltage is determined by the main grid, so the THD positive feedback circuit has little impact on it. But after islanding occurs, as shown in Figure 3.61, without any voltage regulation measure by the DG, the positive feedback circuit will magnify the voltage THD, which can be used as a characteristic for detecting an islanding event.

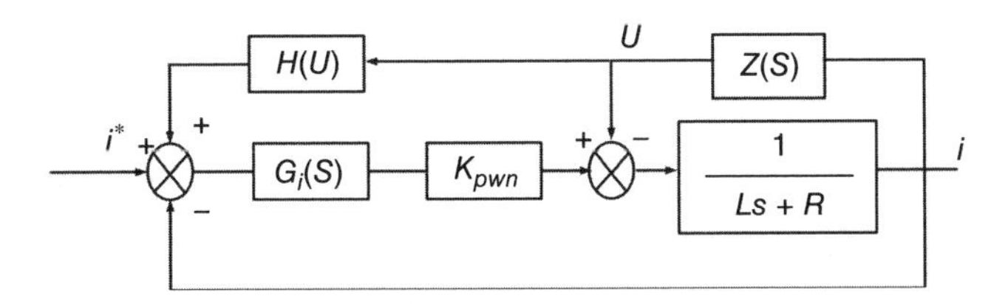

Figure 3.61 Block diagram of inverter control after islanding.

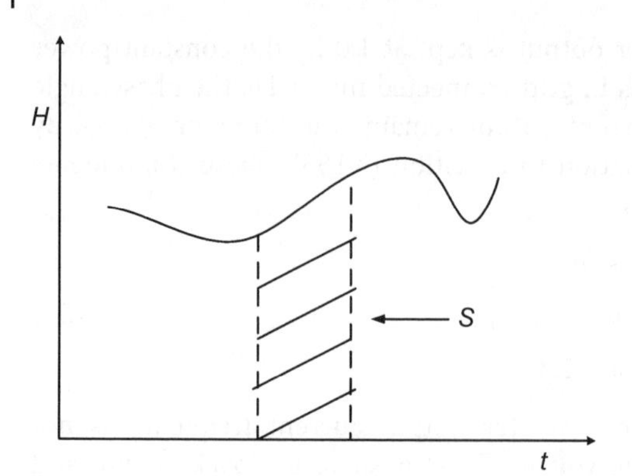

Figure 3.62 Schematic diagram of integral sum.

It should be noted that even in islanded mode, the THD must not be too big or it will not meet the requirements of the state standard. Therefore, the following method is proposed for islanding detection. First, a sliding data window is used to calculate the area S bounded between the THD curve (H-curve) and the time axis, as shown in Figure 3.62. To avoid false detection caused by a sudden disturbance in the grid which can lead to transitory surge in the voltage THD, the chosen window of THD data (H-value) is set to cover a time of 20 ms here. So, islanding can be determined when S exceeds the preset minimum threshold and remains within the maximum threshold for the next 20 ms. Furthermore, this method is not restricted by the type or capacity of the DG, since it is based on the control of the grid-tied inverter. The flowchart of islanding detection is shown in Figure 3.63.

B. Simulation Verification

To verify the effectiveness of the proposed method, simulation tests under the worst case scenario defined by the IEEE Std.1547 were carried out using PSCAD. The type and capacity of the load provided by the standard is close to that of the real load, thus faithfully reflecting the operating condition of the real system. According to the circuit diagram shown in Figure 3.64, the main circuit is established with element parameters as follows: grid line voltage 380 V, frequency 50 Hz (with high harmonics of small amplitude); rated reference power of the inverter 50 kW, DC bus voltage 800 V, LC filter ($L_f = 0.6$ mH, $C_f = 1500$ μF, equivalent resistance $R = 0.01\ \Omega$); rated power of the load 50 kW ($R = 2.904\ \Omega$, $L = 3.698$ mH, $C = 2740.2$ μF, $Q_f = 2.5$).

The length of time of the data window is 20 ms with sampling intervals of 0.1 ms, so that 200 points are included in a data window. According to the state standard, the THD of a grid voltage of 0.38 kV should not exceed 5%, so the maximum threshold should be $\varepsilon_1 = 5\% \times 200 = 10$; then, in view of a small margin for the specification of ε_2, it is set as $\varepsilon_2 = 0.3$.

It should be noted that the aim here is to study islanding detection techniques. Therefore, the switching of control mode for the inverter after a successful detection is not discussed.

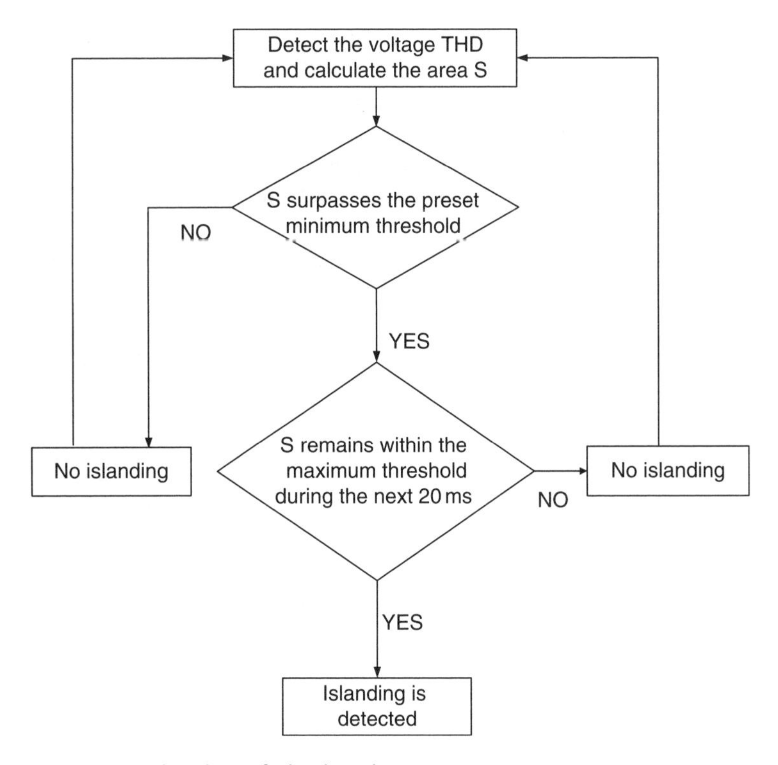

Figure 3.63 Flowchart of islanding detection.

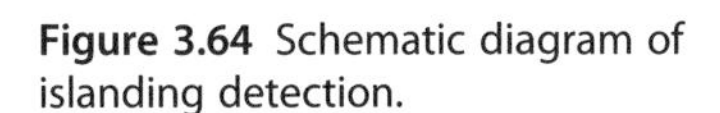

Figure 3.64 Schematic diagram of islanding detection.

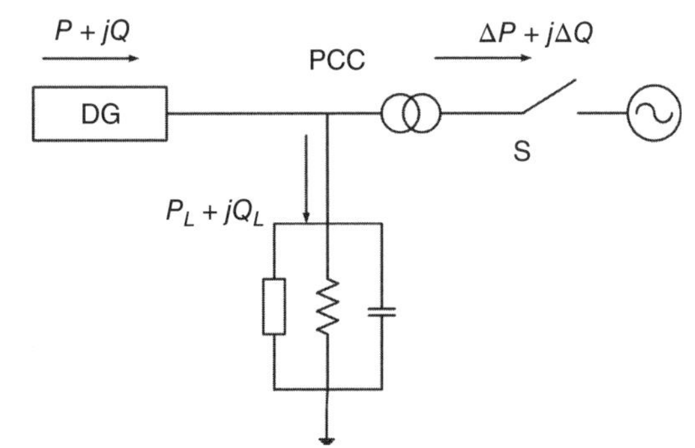

Figure 3.65 shows the simulation results. In the beginning, the DG unit is grid-connected, and the three-phase circuit breaker is disconnected at $t = 0.5$ s. The inverter applies the constant power control strategy.

The waveform of the grid voltage is shown in Figure 3.65(a). DG breaks from the main grid at $t = 0.5$ s and enters islanded mode, supplying power to local loads. When $t > 0.5$ s, voltage THD increases greatly due to the positive feedback circuit, but still remains within the maximum allowed value of 5%, maintaining a good voltage quality, as seen in Figure 3.65(b). At the same time, with the THD increasing, the area bounded between the H-curve and the time axis S is also rising. S surpasses the minimum threshold at $t = 0.54$ s, while remaining within the maximum threshold during the next 20 ms (see Figure 3.65(c)). Thus, the islanding at $t = 0.5$ s is successfully detected using the proposed method. Figure 3.65(d) provides the waveform of the inverter output current.

The output current is scarcely affected by the positive feedback circuit when the DG unit is grid-connected; if islanding is detected, the inverter will be pulled out of service. Therefore, the introduction of the positive feedback circuit does not interfere with the normal operation of the DG unit under either grid-connected or islanded mode. The voltage and frequency of the DG unit during the detection process are shown in Figure 3.65(e) and Figure 3.65(f), respectively. They both stay within the normal operational range given by the state standard GB/T 14549-93. So, the proposed detection method is proved to be non-destructive and NDZ-free.

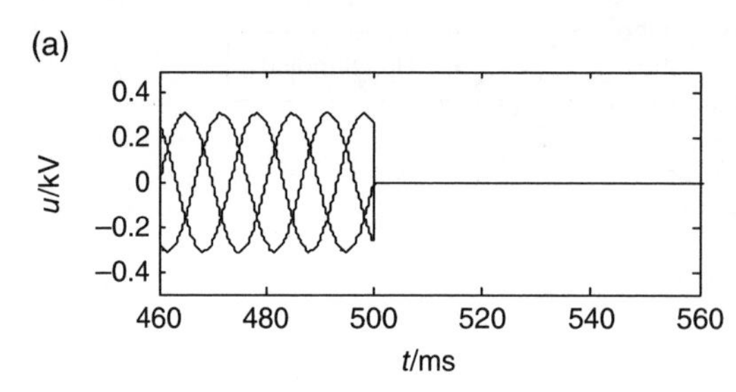

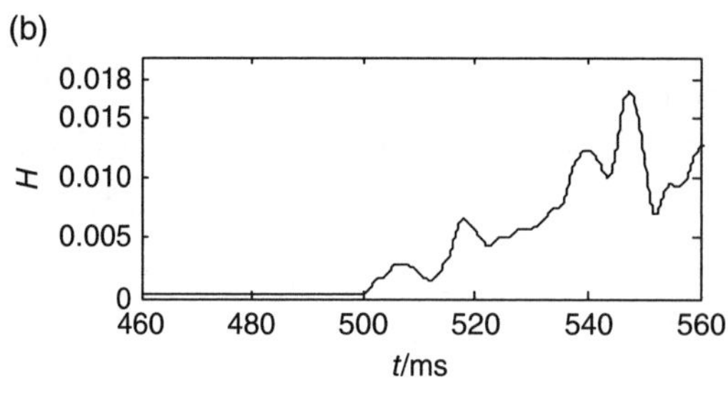

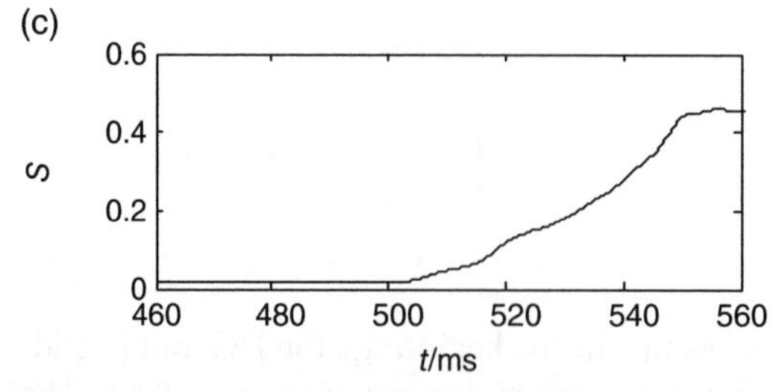

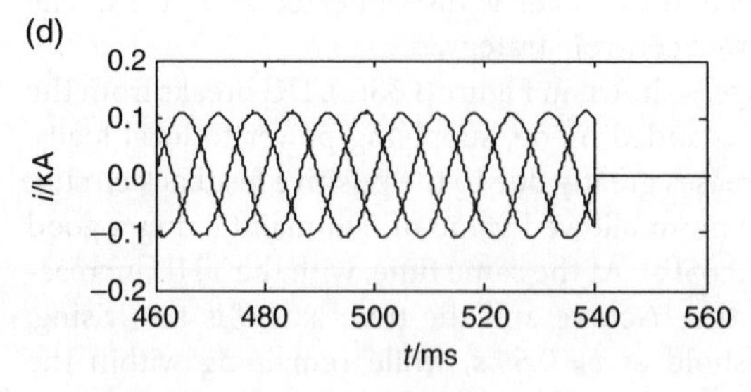

Figure 3.65 Simulation results of islanding detection. (a) Power grid voltage. (b) *H*-curve. (c) Area between *H*-curve and the time axis. (d) Inverter output current. (e) Microgrid frequency. (f) Amplitude of the microgrid voltage.

Figure 3.65 (Continued)

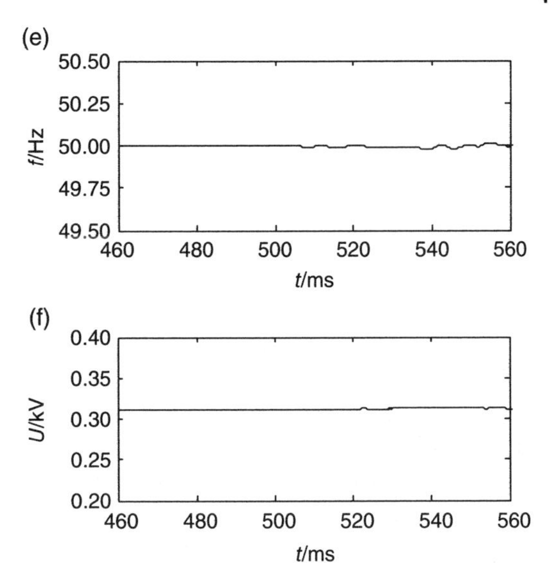

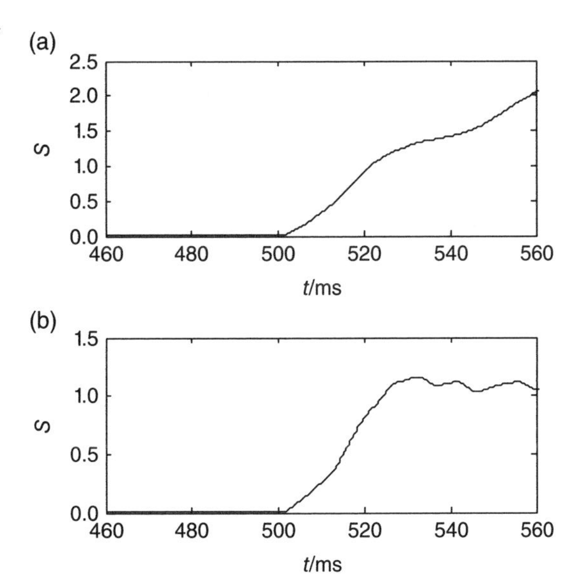

Figure 3.66 Simulation results of islanding detection for open phase. (a) Single-phase open. (b) Two-phase open.

On the other hand, disconnection of the circuit breaker at the PCC is not confined only to the case of three-phase open. The other two cases, i.e. single-phase open and two-phase open are also studied here. The area S in these two cases is shown in Figure 3.66(a) and Figure 3.66(b), respectively. When $t > 0.5$ s, voltage THD increases greatly due to the positive feedback circuit, but it still remains within the maximum allowed value of 5%, maintaining a good voltage quality. It can be seen that as soon as the single-phase/two-phase open state occurs at

$t = 0.5$ s, S is rising, and surpasses the minimum threshold quickly, while remaining within the maximum threshold in the following 20 ms, similar to the previously studied case of three-phase open. So the proposed method proves to be capable of detecting islanding with an open-phase operation.

C. Prevention of False Islanding

Currently, for wind power generation systems, continuity in connection with the main grid is required even when there is a fault such as a voltage drop in the grid, and disconnection is allowed only when major faults occur. This means that the wind power generation system needs solid low-voltage, ride-through (LVRT) capability. The wind turbines are required to continue in operation for 0.625 s when the terminal voltage suddenly drops to 15% of its rated value. This poses a new challenge to islanding detection techniques. For conventional passive detection methods, false detection can occur when a DG unit is experiencing LVRT, but by using the proposed method, this problem can be effectively avoided.

As shown in Figure 3.67(a), the voltage of the DG unit suddenly drops to 15% of its rated value at $t = 0.5$ s. Because the DG unit remains connected to the main grid, its voltage is determined by the grid, and its frequency remains consistent with the grid frequency. This is actually a false islanding event. Now, the THD of the DG voltage surpasses the maximum threshold 5% quickly, as seen in

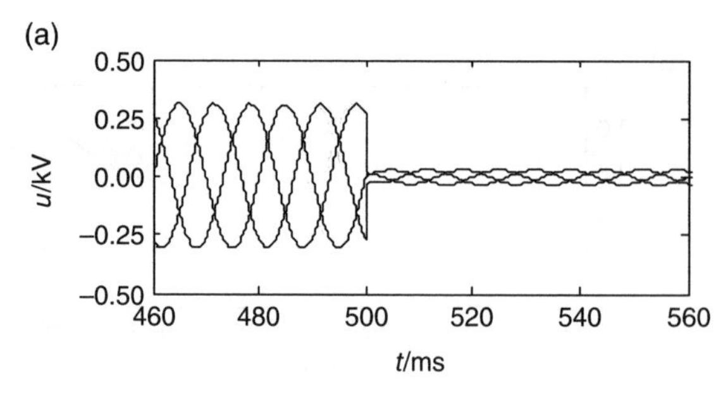

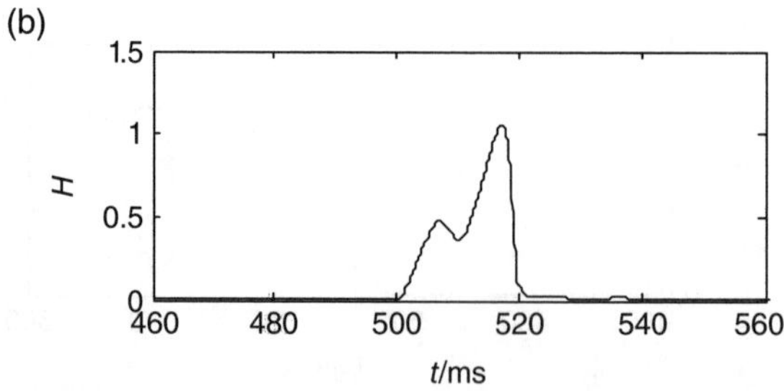

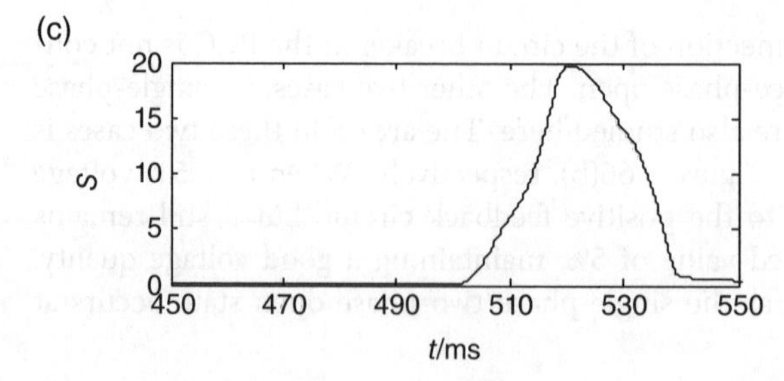

Figure 3.67 Simulation results under fault islanding. (a) power grid voltage. (b) *H*-curve. (c) Area between *H*-curve and the time axis.

Figure 3.67(b). At the same time, the area S surpasses the maximum threshold within 20 ms of it surpassing the minimum value, as shown in Figure 3.67(c). So, based on the proposed detection method, no islanding will be determined. In this way, the detection method proves to be immune to false islanding events.

3.4.2.2 Islanding Detection Method Based on Negative Sequence Voltage Contribution Factor

A. Basic Theory

The islanding detection method is shown in Figure 3.68. The DG unit, local equivalent impedance, distributed equivalent source and equivalent impedance of the source connect with the PCC. At the same time, a negative sequence voltage source with low voltage is added to the PPC through a constant impedance. This circuit is used to detect the change of negative sequence voltage when the microgrid operates in both grid-connected and islanded modes. U_G is the terminal voltage of the negative sequence voltage source, and U_2 is the measured voltage in the PCC.

Decomposing the islanding detection network in Figure 3.68 could result in the negative sequence network shown in Figure 3.69. Generally speaking, the equivalent negative impedance of detection network is small. Correspondingly, the equivalent negative impedance $Z_{DG}^{(2)}$ of the microgrid is large. When the microgrid operates in grid-connected mode, the equivalent negative impedance $Z_2^{(2)}$ in the PCC is a parallel connection of equivalent negative impedance of the detection network and the equivalent negative impedance of the microgrid, $Z_2^{(2)} = Z_S^{(2)}//Z_{DG}^{(2)}$. However, when the microgrid operates in islanded mode, as

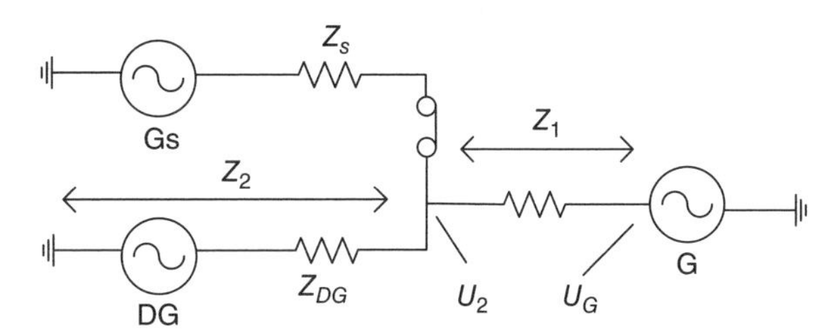

Figure 3.68 Schematic diagram of islanding detection.

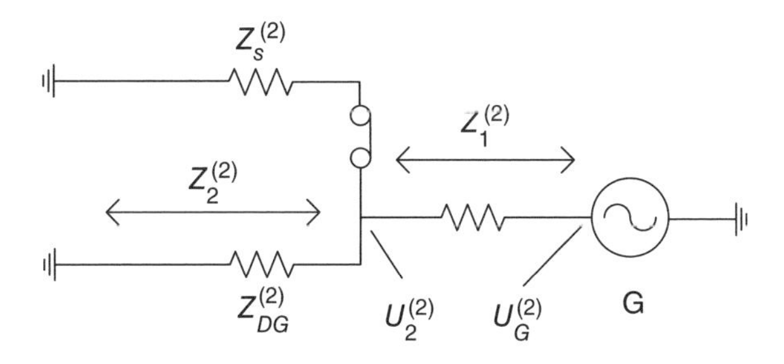

Figure 3.69 Negative sequence network of a grid-tied microgrid.

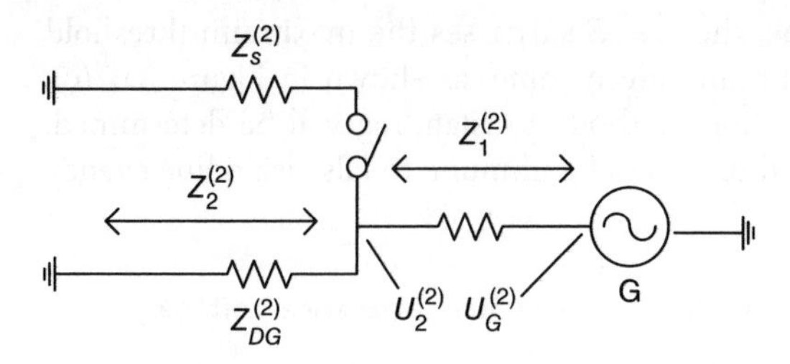

Figure 3.70 Negative sequence network of an islanding microgrid.

shown in Figure 3.69, the equivalent negative impedance in the PCC is equivalent to the negative impedance of the microgrid, $Z_2^{(2)} = Z_{DG}^{(2)}$.

Figures 3.69 and 3.70 show that voltage caused by the negative sequence voltage source will be divided by the constant negative impedance $Z_1^{(2)}$ and the equivalent negative impedance $Z_2^{(2)}$ in the PCC:

$$k = \frac{U_2^{(2)}}{U_G^{(2)}} = \frac{Z_2^{(2)}}{Z_1^{(2)} + Z_2^{(2)}} = \frac{1}{Z_1^{(2)}/Z_2^{(2)} + 1} \tag{3.156}$$

where k represents the negative sequence voltage distribution factor, $U_2^{(2)}$ represents the negative measured voltage, $U_G^{(2)}$ represents the terminal sequence voltage of the negative sequence voltage source G. $Z_1^{(2)}$ represents the constant negative impedance in Equation (3.156), so the negative sequence voltage contribution factor k is only related to $Z_2^{(2)}$. According to the equation for $Z_2^{(2)}$, when the microgrid operates in grid-connected and islanded modes, it is inferred that when the microgrid operates in grid-connected mode, k is very small, but when the microgrid operates in islanded mode, k become larger. So, islanding could be detected by detecting a change in k.

It is noted that $Z_{DG}^{(2)}$ may change when load is shed from or added in the microgrid. When the microgrid operates in grid-connected mode, $Z_2^{(2)}$ is far less than $Z_{DG}^{(2)}$, because $Z_S^{(2)}$ is far less than $Z_{DG}^{(2)}$, if partial load is shed or added, so that $Z_{DG}^{(2)}$ hardly affects k. When the microgrid operates in islanded mode, adding partial load leads to decreases of $Z_2^{(2)}$, taking $Z_2^{(2)} = Z_{DG}^{(2)}$ into account. But $Z_2^{(2)}$ remains considerable at the moment, and differentiation of k is obvious. $Z_2^{(2)}$ will increase with partial load shedding, when the microgrid operates in islanded mode, so the change in k is obvious, compared with $Z_2^{(2)}$ in a microgrid operating in grid-connected mode. The negative sequence voltage contribution factor is used for islanding detection when load is shed or added in a microgrid.

B. Simulation Results

Islanding detection in the case of IEEE Std.1547 is simulated by PSCAD, as shown in Figure 3.68. A negative sequence voltage source is added in the PCC. Parameters are as follows: distribution network line voltage is 380 V, frequency is 50 Hz, reference active power of the inverter is 50 kW, DC bus voltage is 800 V, LC filter ($L_f = 0.6$ mH, $C_f = 1500$ μF, $R = 0.01$ Ω), rated power of load is

50 kW, ($R = 2.904\,\Omega$, $L = 3.698$ mH, $C = 2740.2\,\mu$F, $Q_f = 2.5$), line voltage of the negative sequence voltage source is 3.8 V, series impedance $Z_1 = 1\,\Omega$.

At the beginning of the simulation, the microgrid operates in grid-connected mode, the switch in the PCC turns off at 0.5 s when the microgrid operates in islanded mode, as shown in Figure 3.72. In order to prevent damage caused by adding too large a source to the microgrid voltage, rms voltage of the added source is set to 1% of the microgrid voltage, as shown in Figure 3.71(a).

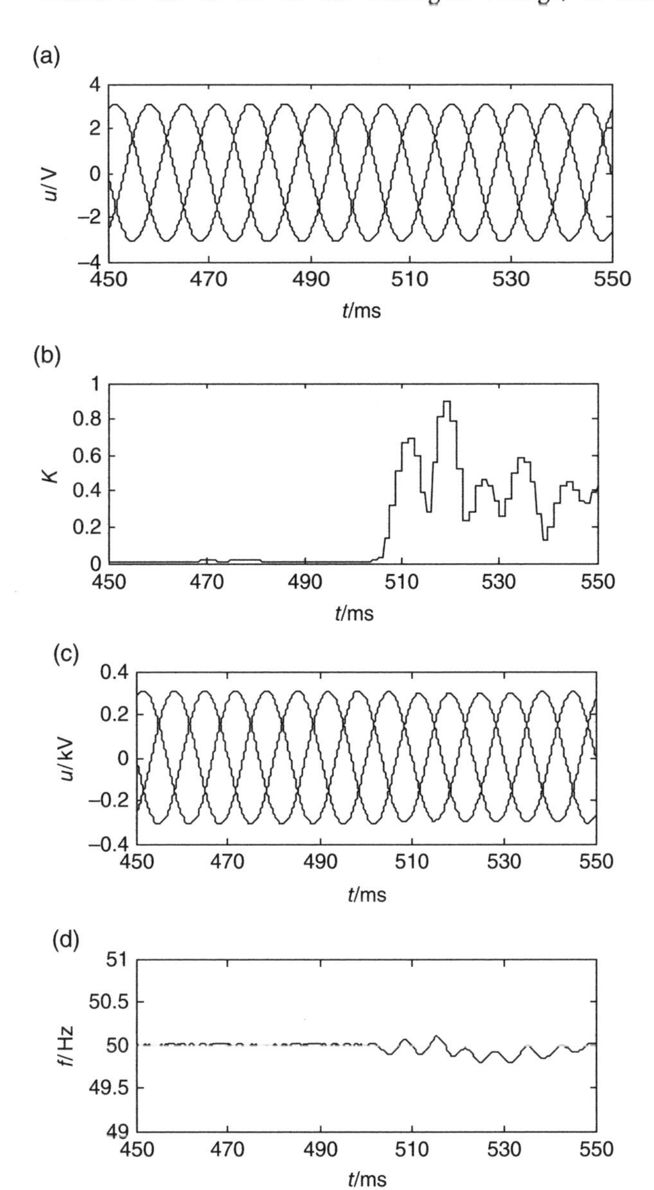

Figure 3.71 Simulation results of islanding detection. (a) Phase voltage of negative sequence voltage source. (b) Negative sequence voltage contribution factor *k*. (c) Microgrid voltage. (d) Microgrid frequency.

The negative sequence voltage contribution factor k curve is shown in Figure 3.71 (b). When the microgrid operates in grid-connected mode, the equivalent negative impedance $Z_2^{(2)}$ in the PCC is extremely small. So the ratio of negative measured voltage $U_2^{(2)}$ to added negative voltage source $U_G^{(2)}$ is extremely small too, and k is close to zero. The microgrid operates in islanded mode at 0.5 s. As the equivalent negative impedance $Z_2^{(2)}$ in the PCC increases, the ratio of $U_2^{(2)}$ to $U_G^{(2)}$ increases too. Therefore, the k curve is higher than before. Furthermore, the k curve remains very high after 20 ms, and the features of the curve could be used for islanding detection. Microgrid rms voltage and frequency are shown in Figure 3.71(c) and Figure 3.71(d), which are both in the range of national standards GB/T 14549-93. Also, in islanding detection, power quality has not been damaged, and islanding detection without devastation and without a non-detection zone is realized.

Furthermore, there are two kinds of situation where single-phase and two-phase breakers in the PCC are disconnected, where negative sequence voltage contribution factor k curves are shown in Figure 3.72(a) and Figure 3.72(b). k curves rise when single-phase and two-phase breakers are disconnected, which is similar to the situation of disconnection of a three-phase breaker, and clearly remain high after 20 ms. Therefore, this method is effective when single-phase and two-phase breakers are disconnected.

There are some requirements for wind power systems that wind generators should connect with the grid when the voltage dips because of a power fault. Only when there is a serious fault will wind generators be allowed to disconnect from

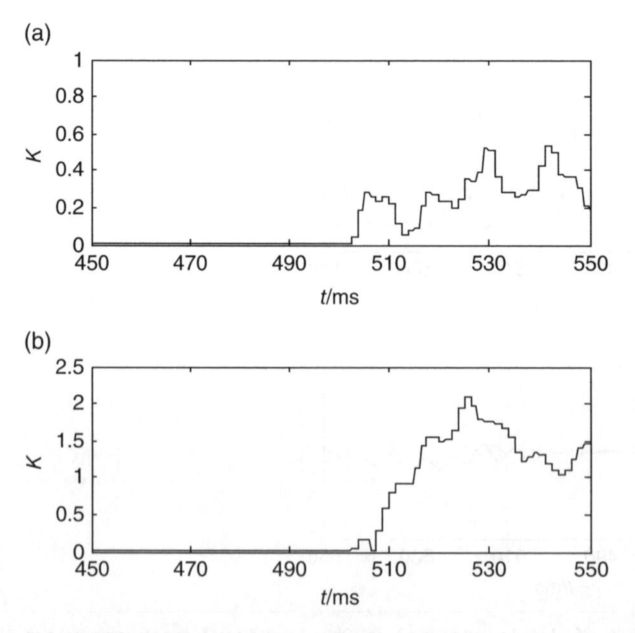

Figure 3.72 Schematic diagram of the simulation results of islanding detection for open phase. (a) Single-phase breaker open. (b) Two-phase breaker open.

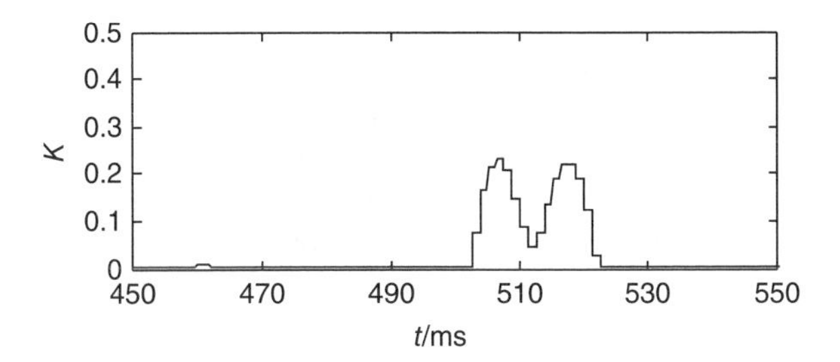

Figure 3.73 Simulation results under islanding fault.

the grid. So, wind generators must possess low-voltage, ride-through (LVRT) capability. In particular, when the terminal voltage of a wind generator is decreased to 15% of the rated voltage, they should continue to operate for 0.625 s, which presents new challenges for islanding detection technology. Traditional negative detection is invalid in LVRT. However, the islanding detection method which is based on the negative sequence voltage contribution factor is able to overcome the challenges. Figure 3.73 shows the k curve when the terminal voltage of the wind generator is suddenly decreased to 15% of the rated voltage. Because the equivalent negative impedance $Z_2^{(2)}$ in the PCC has not changed, k almost remains zero after 20 ms. So the trip on false islanding can be avoided.

3.5 Summary

In order to solve the problems of the protection of renewable energy integrated in the grid, this chapter analyses in depth the fault characteristics of renewable energy sources, centralized renewable energy protection and distributed renewable energy protection.

The methods for analysing the fault characteristics of renewable energy are based on whether a crowbar is put into operation or not. The accurate expressions for the DFIG short circuit currents considering the grid-side and rotor-side converter control are derived. Experimental results demonstrate that, in both asymmetrical short circuit fault and symmetrical short circuit fault, the calculated values of fault transient current are consistent with the experimental values. Thus, the fault analysis method could correctly reflect the fault transient characteristics of renewable energy sources.

The novel adaptive distance protection scheme for the collector line is very immune to fault resistance and is not affected by the weak feed characteristics of renewable energy and faults on the adjacent line. Thus, the selectivity problem caused by different collector line lengths can be solved. The current pilot differential protection scheme for an outgoing transmission line is not affected by the fault type with high sensitivity and reliability and could solve the problem that the traditional pilot protection is easily affected by load current.

Adaptive current protection with distributed power sources is not affected by the integration of distributed power sources and the fault type, and it could operate reliably in both symmetrical and asymmetrical faults. Also, the protection range of primary protection and backup protection can be effectively expanded, with simple calculation and easy setting. The islanding detection method is non-destructive without a dead zone and could judge correctly in non-full-phase disconnection and LVRT cases. It will not misjudge in the case of false islanding. It can be used for islanding detection not only in the case of three-phase disconnection, but also in the case of single-phase disconnection and two-phase disconnection.

References

1 Ma, J., Kang, S. and Thorp, J. S. (2016) Research on transient characteristics of doubly fed wind power generator under symmetrical short-circuit fault considering grid-side converter control, *Electric Power Components & Systems*, **44** (12), 1396–1407.

2 Djeghloud, H., Bentounsi, A. and Benalla, H. (2011) Sub- and super-synchronous wind turbine doubly fed induction generator system implemented as an active power filter, *International Journal of Power Electronics*, **3** (2), 189–212.

3 Mendes, V. F., De Sousa, C. V., Silva, S. R. and Cezar Rabelo, B. (2011) Modeling and ride-through control of doubly fed induction generators during symmetrical voltage sags, *IEEE Transactions on Energy Conversion*, **26** (4), 1161–1171.

4 Harnefors, L. and Nee, H. P. (1998) Model-based current control of AC machines using the internal model control method, *IEEE Transactions on Industry Applications*, **34** (1), 133–141.

5 Kim, J. S. and Sul, S. K. (1996) A novel voltage modulation technique of the space vector PWM, *Transactions-Korean Institute of Electrical Engineers*, **116** (8), 820–825.

6 Shin, M. C., Park, C. W. and Kim, J. H. (2003) Fuzzy logic-based relaying for large power transformer protection, *IEEE Transactions on Power Delivery*, **18** (3), 718–724.

7 Abad, G., *et al.* (2011) *Doubly fed induction machine: modeling and control for wind energy generation*, John Wiley & Sons, Chichester.

8 Tsili, M. and Papathanassiou, S. (2009) A review of grid code technical requirements for wind farms, *IET Renewable Power Generation*, **3** (3), 308–332.

9 Ziegler, G. (2005) *Numerical Differential Protection: Principles and Applications*, Wiley-VCH, Weinheim.

10 Willis, H. L. and Scott, W. G. (2000) *Distributed Power Generation Planning and Evaluation*, Marcel Dekker, New York.

11 Ma, J., Wang, X., Zhang, Y. G. *et al.* (2012) A novel adaptive current protection scheme for distribution systems with distributed generation, *International Journal of Electrical Power and Energy Systems*, **43** (1), 1460–1466.

12 Ma, J., Mi, C. and Wang, Z. (2012) Application of negative sequence voltage contribution factor to detect islanding, *Power and Energy Engineering Conference (APPEEC), 2012 Asia-Pacific, IEEE*, **29**, 1–4.

13 Ma, J., Mi, C., Zheng, S. X. *et al.* (2013) Application of voltage harmonic distortion positive feedback for islanding detection, *Electric Power Components & Systems*, **41** (6), 641–652.
14 Barker, P. and DeMello, R.W. (2000) Determining the impact of DG on power systems, radial distribution, in *Proc. IEEE Power Eng. Soc. Summer Meeting*, pp. 1645–1656.
15 Blackburn, J. L. (1998) *Protective Relaying Principles and Applications*, Marcel Dekker, New York.
16 Draft Guide for Design, Operation, and Integration of Distributed Resource Island Systems with Electric Power Systems, in *IEEE Standard*, P1547.4, 2008.

4

Topology Analysis

4.1 Introduction

Power system topology analysis is mainly focused on the classification of buses and electrical islands according to the switch status [1,2]. It is able to provide network topology parameters for power system analysis such as flow calculation, state estimation and fault analysis, etc. Topology analysis plays a crucial role in the energy management systems (EMS) and the distribution management systems (DMS). All the processes and methods in EMS and DMS applications are highly dependent on the outcome of the system topology processor [3,4]. Meanwhile, it is the premise and foundation of substation area protection and wide area protection. With the construction and development of smart grids, it is urgent to fundamentally improve the adaptability of power grid topology to non-scheduled changes in system structure. This requires that when the network structure changes, topology analysis can update quickly to adapt to the current operating mode. However, traditional methods [5] are mainly used for the topology analysis of static networks, and fast tracking in the case of dynamic topology is difficult. In addition, for loop networks, topology analysis is slow and the efficiency is not high.

In view of these problems, this chapter introduces a substation area protection topology analysis scheme and a wide area protection topology analysis scheme. Both of these schemes could improve the efficiency of static topology analysis and could realize fast updating of network topology through local modification. On this basis, concerning problems such as information damage or deviation caused by a device or network, a new method for identifying topological faults is introduced, which could realize wide area topology identification in the case of a single topological error, multiple topological errors and multiple undesirable data.

4.2 Topology Analysis for the Inner Substation

This section introduces a novel plant-station topology analysis method based on the main electrical wiring characteristics. First, the nodes that represent the circuit breakers are numbered according to graph theory [6,7]. On this basis, combining the main wiring characteristics and the switch status information, the

Hierarchical Protection for Smart Grids, First Edition. Jing Ma and Zengping Wang.

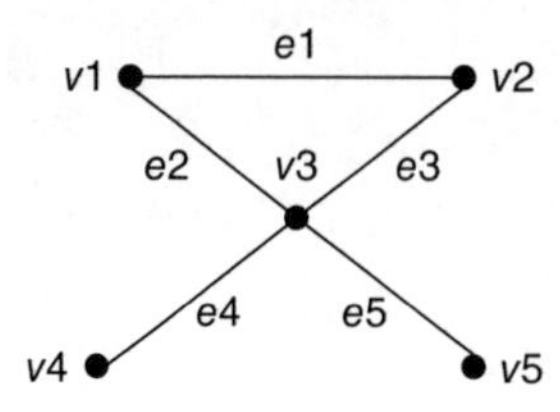

Figure 4.1 Simple connected graph.

plant-station topology is analysed using the topological base theorem. When the switch status changes, the affected part of the topology is modified, using the radius search method, so that re-forming of the topology is avoided and a fast update of the topology is guaranteed.

4.2.1 Characteristic Analysis of the Main Electrical Connection

According to graph theory [8,9], a graph can be expressed as $\boldsymbol{G} = \langle \boldsymbol{V}, \boldsymbol{E} \rangle$, where $\boldsymbol{V}$ is the vertex set and $\boldsymbol{E}$ is the set of effective pairs formed by the elements in $\boldsymbol{V}$, also called the side set. As shown in Figure 4.1, the vertex set is

$$V = \{v_1, v_2, v_3, v_4, v_5\} \tag{4.1}$$

The side set is

$$E = \{e_1, e_2, e_3, e_4, e_5\} = \{(v_1, v_2), (v_1, v_3), (v_2, v_3), (v_3, v_4), (v_3, v_5)\} \tag{4.2}$$

It can be seen from Equation (4.2) that if there is only one side between two vertices, then the two vertices can characterize the side. Similarly, if the circuit breaker is equated to a side, and the bus nodes or nodes connecting the circuit breakers are equated to vertices, then, according to graph theory, the circuit breaker could be expressed by its terminal nodes.

Before performing topology analysis, the nodes and circuit breakers need to be numbered. It should be noted that a circuit breaker together with its two disconnecting switches counts as one equivalent switch. The bus nodes are numbered as 'B_i' ($i = 1, 2, 3, \ldots$), the nodes connecting circuit breakers are numbered as 'J_i' ($i = 1, 2, 3, \ldots$, virtual node), and the circuit breakers are expressed by the terminal nodes. The main types of electrical wiring are shown in Figure 4.2.

In Figure 4.2, the plant-station main electrical wiring types can be divided into three categories according to the structure:

1) Chain structure. This category includes single-bus segmented wiring, double-bus wiring and double-bus segmented wiring. There are only bus nodes, tie breakers or section breakers in the station. As long as all the breakers in the station are switched on, the buses in the station all belong to the same electrical island.
2) Loop structure, i.e. angular wiring. There is no bus node in the station, there are only virtual nodes. When the number of breakers switched on is not less than the total number of breakers in the station minus 1, the whole station belongs to one electrical island.
3) Network structure. There are both bus nodes and virtual nodes in the station, including 3/2 wiring and 4/3 wiring. For 3/2 wiring, when a certain series of breakers are all switched on, and at least two breakers in other series are switched on, the whole station belongs to one electrical island. 4/3 wiring can be analysed similarly.

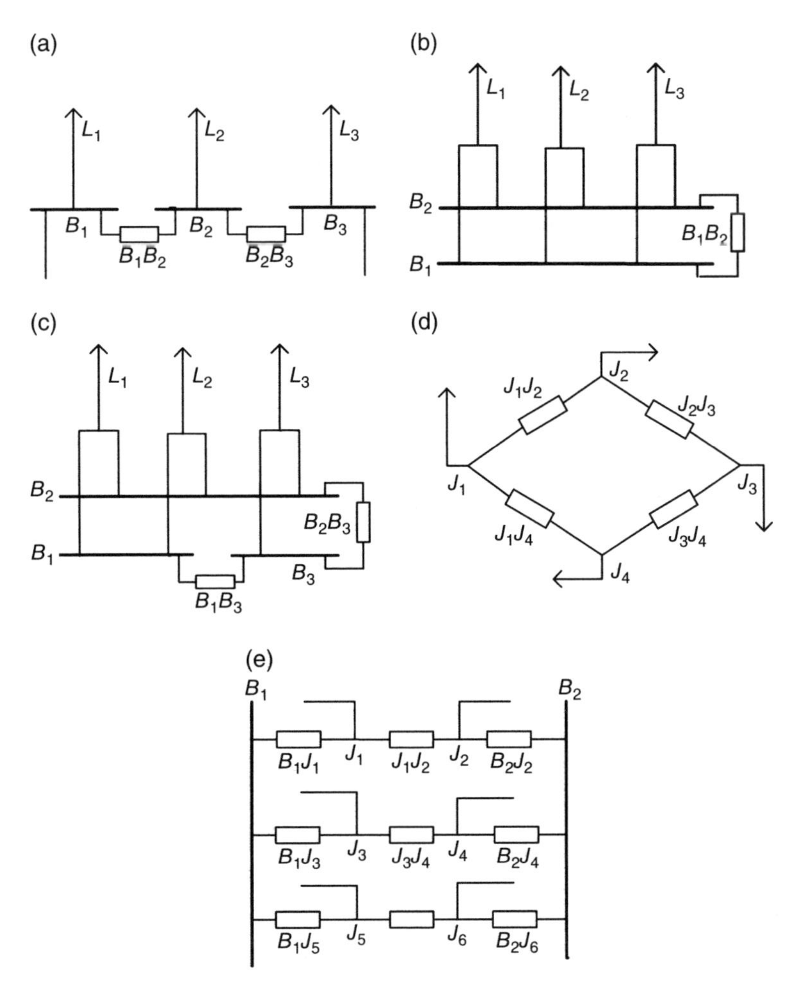

Figure 4.2 Different types of main electrical wiring. (a) Single-bus segmented wiring. (b) Double-bus wiring. (c) Double-bus segmented wiring. (d) Angular wiring. (e) 3/2 wiring.

4.2.2 Topology Analysis Method Based on Main Electrical Wiring Characteristics

4.2.2.1 Plant-Station Topology Analysis Process

According to the main electrical wiring characteristics, the plant-station electrical islands are first analysed in the following steps:

Step 1: The nodes and circuit breakers in the station are named and the nodes are entered into the corresponding plant-station set $\boldsymbol{Z}_i$ ($i = 1, 2, \ldots, n$; where n is the number of plant stations in the analysed system).

Step 2: Read the switch status information and decide whether the current station belongs to the same electrical island according to the type of its main electrical wiring. If yes, a plant-station electrical island is formed. Otherwise, further analysis needs to be conducted using the topological base theory.

Take the plant station with single-bus segmented wiring as an example for analysing, which is numbered as the first plant station in the system. First, the nodes and circuit breakers are numbered in the plant station, which can be seen in Figure 4.2(a). Then, the nodes B_1, B_2 and B_3 are entered into the corresponding plant-station set $\boldsymbol{Z}_1$. Finally, the switch status information of the breakers B_1B_2 and B_2B_3 is read. If the breakers B_1B_2 and B_2B_3 are all switched on, the nodes B_1, B_2 and B_3 are in the same electrical island. Otherwise, the topology of this plant station needs further determination.

4.2.2.2 Plant-Station Topology Analysis Based on the Topological Base

Suppose $\boldsymbol{X}$ is a non-empty set, and $2^{\boldsymbol{X}}$ is the power set of $\boldsymbol{X}$, which is composed of all the subsets of $\boldsymbol{X}$ (including the empty set $\emptyset$ and $\boldsymbol{X}$ itself). The subsets of $2^{\boldsymbol{X}}$ are called the subset family of $\boldsymbol{X}$: $\boldsymbol{\beta}$, and $\boldsymbol{\beta}$ is called a topology of $\boldsymbol{X}$. If the following conditions are met:

1) Both $\boldsymbol{X}$ and $\emptyset$ belong to $\boldsymbol{\beta}$
2) The union of any elements in $\boldsymbol{\beta}$ still belongs to $\boldsymbol{\beta}$
3) The intersection of any elements in $\boldsymbol{\beta}$ still belongs to $\boldsymbol{\beta}$

then $\boldsymbol{X}$ and topology $\boldsymbol{\beta}$ together are called a topology space, denoted $(\boldsymbol{X}, \boldsymbol{\beta})$.

Define a new subset family $\bar{\boldsymbol{\beta}}$:

$$\begin{aligned}\bar{\boldsymbol{\beta}} &= \{\boldsymbol{U} \subset \boldsymbol{X} | \boldsymbol{U} \text{ is the union of some elements in } \boldsymbol{\beta}\} \\ &= \{\boldsymbol{U} \subset \boldsymbol{X} | \forall x \in \boldsymbol{U}, \text{ there exists } \boldsymbol{B} \in \boldsymbol{\beta}, \text{ which makes } x \in \boldsymbol{B} \subset \boldsymbol{U}\}\end{aligned} \tag{4.3}$$

$\bar{\boldsymbol{\beta}}$ is called a subset family generated by $\boldsymbol{\beta}$. Obviously, $\boldsymbol{\beta} \subset \bar{\boldsymbol{\beta}}$ and $\emptyset \in \bar{\boldsymbol{\beta}}$. If $\bar{\boldsymbol{\beta}}$ is a topology of $\boldsymbol{X}$, then the subset family of $\boldsymbol{X}$: $\boldsymbol{\beta}$ is called a topological base of $\boldsymbol{X}$.

The necessary and sufficient conditions for $\boldsymbol{B}$ to be a topological base of $\boldsymbol{X}$ are

1) $\bigcup_{\boldsymbol{B} \in \boldsymbol{\beta}} \boldsymbol{B} = \boldsymbol{X}$
2) If $\boldsymbol{B}_1, \boldsymbol{B}_2 \in \boldsymbol{\beta}$, then $\boldsymbol{B}_1 \cap \boldsymbol{B}_2 \in \bar{\boldsymbol{\beta}}$ (i.e. $\forall x \in \boldsymbol{B}_1 \cap \boldsymbol{B}_2$, there exists $\boldsymbol{B} \in \boldsymbol{\beta}$, which makes $x \in \boldsymbol{B} \subset \boldsymbol{B}_1 \cap \boldsymbol{B}_2$).

Here, $\boldsymbol{X}$ is the plant-station node set $\boldsymbol{Z}_i$, and $\boldsymbol{\beta}$ is the ultimate electrical islands formed. Meanwhile, it can be seen from Equation (4.3) that intersected bases must be bases of the same electrical island. Therefore, the intersected bases are merged to form new topological bases, and the intersection points of the electrical islands are added to the intersection points set $\boldsymbol{L}_D$. In this way, when none of the bases are intersected, analysis of the electrical islands will end.

If there is still any node in the station that falls outside the electrical islands finally formed, then the leftover node alone forms an electrical island.

4.2.2.3 Plant-Station Topology Analysis after Switch Status Change Based on the Radius Search Method

The switch status change involves two cases, i.e. switch 'on' (the circuit breaker is switched on from switched-off status) and switch 'off' (the circuit breaker is switched off from switched-on status).

A. Switch 'Off'

Switching off the circuit breaker will cause the main electrical wiring to disconnect, thus the electrical island will split. After the split, the new electrical island starts from the switch-off points and ends in the boundary points of the connected graph (the boundary points do not belong to the intersection points set $\boldsymbol{L_D}$). The switch-off points are the terminal points of the circuit breaker that is switched off; while the search for the boundary points is related to the radius of the graph. According to graph theory [3], suppose v and u are any two vertices in graph $\boldsymbol{G}$, the eccentricity of v refers to the distance between v and the vertex farthest from v, denoted $e_G(v)$, i.e. $e_G(v) = \max_{u\in V(G)}\{d(u,v)\}$. The radius of graph $\boldsymbol{G}$ is the minimum eccentricity of all the vertices, denoted $r(\boldsymbol{G})$, i.e.

$$r(G) = \min_{v\in V}\{e(v)\} = \min_{v\in V}\max_{u\in V(G)}\{d(u,v)\} \tag{4.4}$$

For a certain electrical wiring, if the connected graph radius before the switch status change is *rad*, then after the switch 'off', it requires a search depth of $n = r(\boldsymbol{G}) - 1$ for relevant circuit breakers to find the nodes that do not belong to $\boldsymbol{L_D}$, so that the original electrical island can be split into new electrical islands. The newly formed electrical islands are numbered according to the number of nodes in the island, i.e. the island with the most nodes is numbered the same as the original island, and the island with the fewest nodes is numbered the last.

Take hexagonal wiring and 3/2 breaker wiring as examples for analysis. First, for hexagonal wiring, if, before the switch status change, there are multiple electrical islands in the plant-station, then the maximum radius of the connected graph is 2, as shown in Figure 4.3(a). The intersection points set $\boldsymbol{L_D} = \{J_3, J_4, J_5\}$. When a certain breaker is switched off, according to the above analysis, only those breakers firstly (directly) related to the terminal nodes of the switched-off breaker need to be analysed. For example, if breaker J_3J_4 is switched off, since $J_3, J_4 \in \boldsymbol{L_D}$, breakers J_2J_3 and J_4J_5 which are connected to J_3J_4 are searched. Since $J_5 \in \boldsymbol{L_D}$ and $J_2 \notin \boldsymbol{L_D}$, the search ends and two new electrical islands result, i.e. $\{J_2, J_3\}$ and $\{J_4, J_5, J_6\}$. Similar to the angular wiring, the double-bus wiring, double-bus segmented wiring, and single-bus segmented wiring, or more generally, for main electrical wiring with less than 6 buses, only the next level of switched-on breakers need to be searched for determination.

For 3/2 breaker wiring, in the two cases shown in Figure 4.3(b), the maximum radius is 2. Therefore, only the next level of switched-on breakers need to be analysed. Similarly, for 4/3 breaker wiring, the next two levels of breakers need to be searched for determination.

(a)

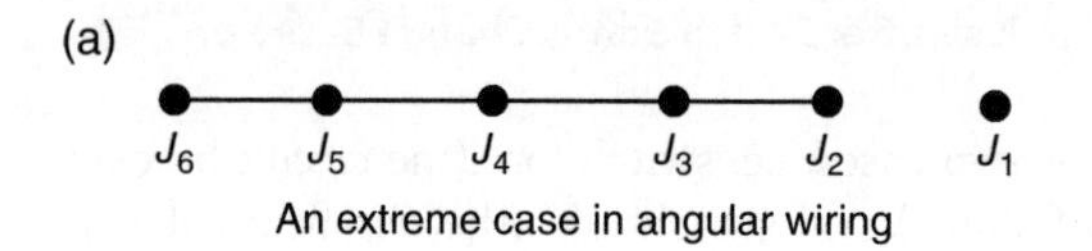

(b)

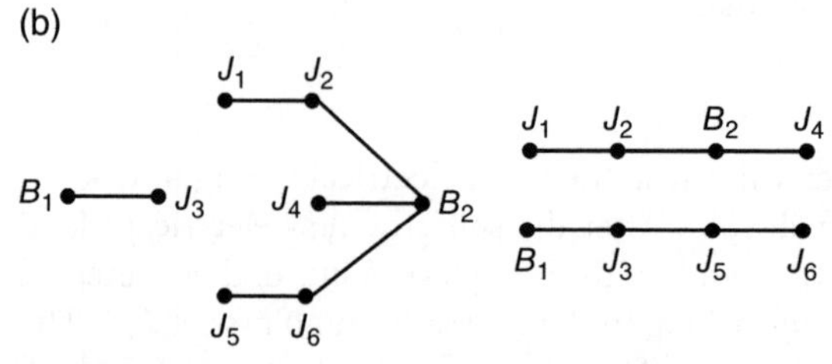

Figure 4.3 Switch 'off' analysis.

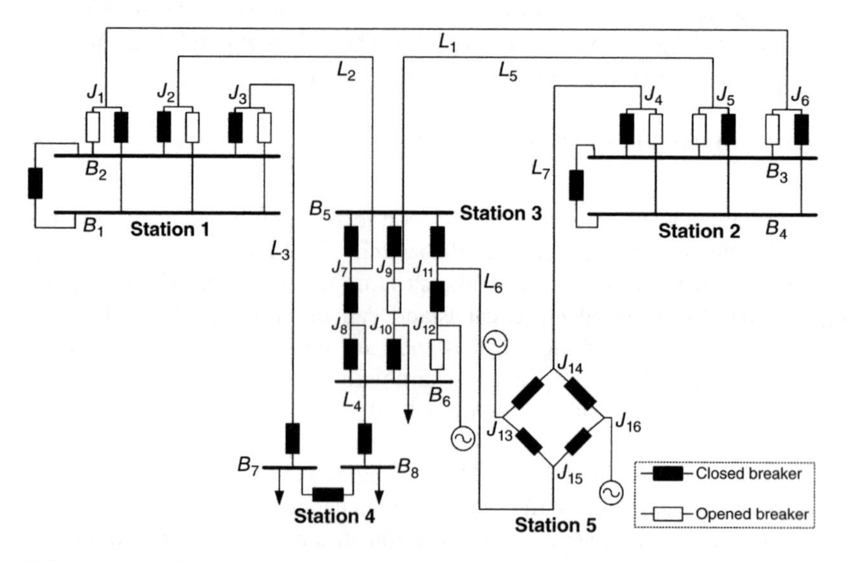

Figure 4.4 Simple network system example.

B. Switch 'On'

1) If the terminal nodes of the switched-on breaker are in the same electrical island, then the island will not change.
2) If the terminal nodes of the switched-on breaker are in different electrical islands, then the island with the smaller number is merged into the island with the bigger number, and the islands with even smaller numbers are renumbered accordingly.

4.2.3 Scheme Verification

Electrical island analysis is carried out on the system shown in Figure 4.4. First, the nodes and circuit breakers are numbered (in Figure 4.4 only the numbers of the nodes are displayed).

4.2.3.1 Plant-Station Topology Analysis of the Original System

According to Figure 4.4, the nodes in the plant-stations are attributed to separate plant-station sets, and the switch status information is acquired, shown in Table 4.1 and Table 4.2.

According to the initialization information, the nodes in Stations 1, 4 and 5 are in the same electrical island, but the nodes in Stations 2 and 3 need further determination. For Station 2, when the only tie breaker is switched off, the two buses will be in two separate electrical islands. For Station 3, the topological bases are formed according to the switch status information shown in Table 4.3. The final electrical islands of the original system can be obtained according to Section 4.2.2 part A, and these are shown in Table 4.4.

Table 4.1 Nodes in stations.

Plant-station set	Nodes in the station
Z_1	B_1, B_2
Z_2	B_3, B_4
Z_3	B_5, B_6, J_7, J_8, J_9, J_{10}, J_{11}, J_{12}
Z_4	B_7, B_8
Z_5	J_{13}, J_{14}, J_{15}, J_{16}

Table 4.2 Circuit breakers and switch status.

Plant station	Circuit breakers	Switch status
Station 1	B_1B_2	1
Station 2	B_3B_4	0
Station 3	B_5J_7	1
	J_7J_8	1
	B_6J_8	0
	B_5J_9	0
	J_9J_{10}	1
	B_6J_{10}	1
	B_5J_{11}	1
	$J_{11}J_{12}$	1
	B_6J_{12}	0
Station 4	B_7B_8	1
Station 5	$J_{13}J_{14}$	1
	$J_{13}J_{15}$	1
	$J_{14}J_{16}$	1
	$J_{15}J_{16}$	1

Table 4.3 Topological bases of Station 3.

Topological bases	Nodes in the island
TPJ3.1	B_5, J_7
TPJ3.2	J_7, J_8
TPJ3.3	J_9, J_{10}
TPJ3.4	J_6, J_{10}
TPJ3.5	B_5, B_{11}
TPJ3.6	B_{11}, B_{12}

Table 4.4 Ultimate electrical islands of the system.

Plant station	Electrical island/Intersected nodes set	Nodes
Station 1	Island 1.1	B_1, B_2
Station 2	Island 2.1	B_3
	Island 2.2	B_4
Station 3	Island 3.1	$B_5, J_7, J_8, J_{11}, J_{12}$
	Island 3.2	B_6, J_9, J_{10}
	L_D3	B_7, B_5, J_{11}, J_{10}
Station 4	Island 4.1	B_7, B_8
Station 5	Island 5.1	$J_{13}, J_{14}, J_{15}, J_{16}$

4.2.3.2 Plant-Station Topology Analysis after the Switch Status Changes

In Figure 4.4, when breaker B_5J_7 in Station 3 is switched off, since the terminal nodes of the switched-off breaker are both intersection points, according to the radius search method, the switched-on breakers that are connected to B_5J_7 need to be analysed. As shown in Table 4.5(a), Island 3.1 is split into two islands. When breaker J_7J_8 in Station 3 is switched off, since the switched-off points are boundary points, according to the radius search method, Island 3.1 is split into two islands, shown in Table 4.5(b). When breaker B_6J_8 in Station 3 is switched on, since B_6 and J_8 are not in the same electrical island, Islands 3.1 and 3.2 are merged into one island (see Table 4.5(c)).

To further verify the advantage of the proposed method, topology analysis of Station 4 in the complex system in Figure 4.5 is carried out.

A. *Plant-Station Topology Analysis of the Original System*

First, the nodes and circuit breakers in the power grid are numbered, as shown in Figure 4.5. Then the nodes in Station 4 are attributed to the station set and the switch status information is read, as shown in Table 4.6.

Table 4.5 Electrical islands in Station 3 after switch status change.

(a) Plant-station topology after breaker B_5J_7 switch 'off'

Plant station	Electrical island	Nodes
Station3	Island3.1	B_5, J_{11}, J_{12}
	Island3.2	B_6, J_9, J_{10}
	Island3.3	J_7, J_8

(b) Plant-station topology after breaker J_7J_8 switch 'off'

Plant station	Electrical island	Nodes
Station3	Island3.1	B_5, J_7, J_{11}, J_{12}
	Island3.2	B_6, J_9, J_{10}
	Island3.3	J_8

(c) Plant-station topology after breaker B_6J_8 switch 'on'

Plant station	Electrical island	Nodes
Station3	Island3.1	B_5, B_6, J_7, J_8, J_9, J_{10}, J_{11}, J_{12}

According to the switch status information, the topological bases of Station 4 are formed, as shown in Table 4.7. Then the ultimate electrical islands can be formed; see Table 4.8.

B. Plant-Station Topology Analysis After the Switch Status Changes

Suppose circuit breakers B_8J_{19}, B_8J_{20} and B_8J_{25} in Station 4 are switched on at the same time. Since B_8 is not in the same electrical island as J_{19}, J_{20} and J_{25}, Islands 4.1 and 4.2 are merged into one island, see Table 4.9(a). If B_7J_{25} breaks off after B_8J_{19}, B_8J_{20} and B_8J_{25} are switched on, then, according to the radius search method, Island 4.1 splits into two islands, with the break-off point as the boundary point, shown in Table 4.9(b).

4.2.3.3 Comparisons with Other Existing Methods

The proposed electrical island analysis method is applicable to different types of main electrical wiring. For plant-stations containing multiple electrical islands, the 'and' and 'or' operations are needed in the analysis. Suppose the number of switched-on breakers is z, then the total calculation amount is $z(z+1)/2-1$. Take 3/2 breaker wiring which contains a relatively large number of nodes, for example. The calculation amount of the traditional matrix multiplication method is $n^2(2n-1)(n-1)$; that of the matrix square method is $n^2(2n-1)\log_2(n-1)$; that of the method proposed in reference [5] is $24\,m^3 - 39\,m^2 + 20\,m - 1$ (n is the number of nodes, m is the number of breaker series, $n = 2(m+1)$ and $z < 3\,m$). It can be seen that the method proposed here has obvious advantages over the other

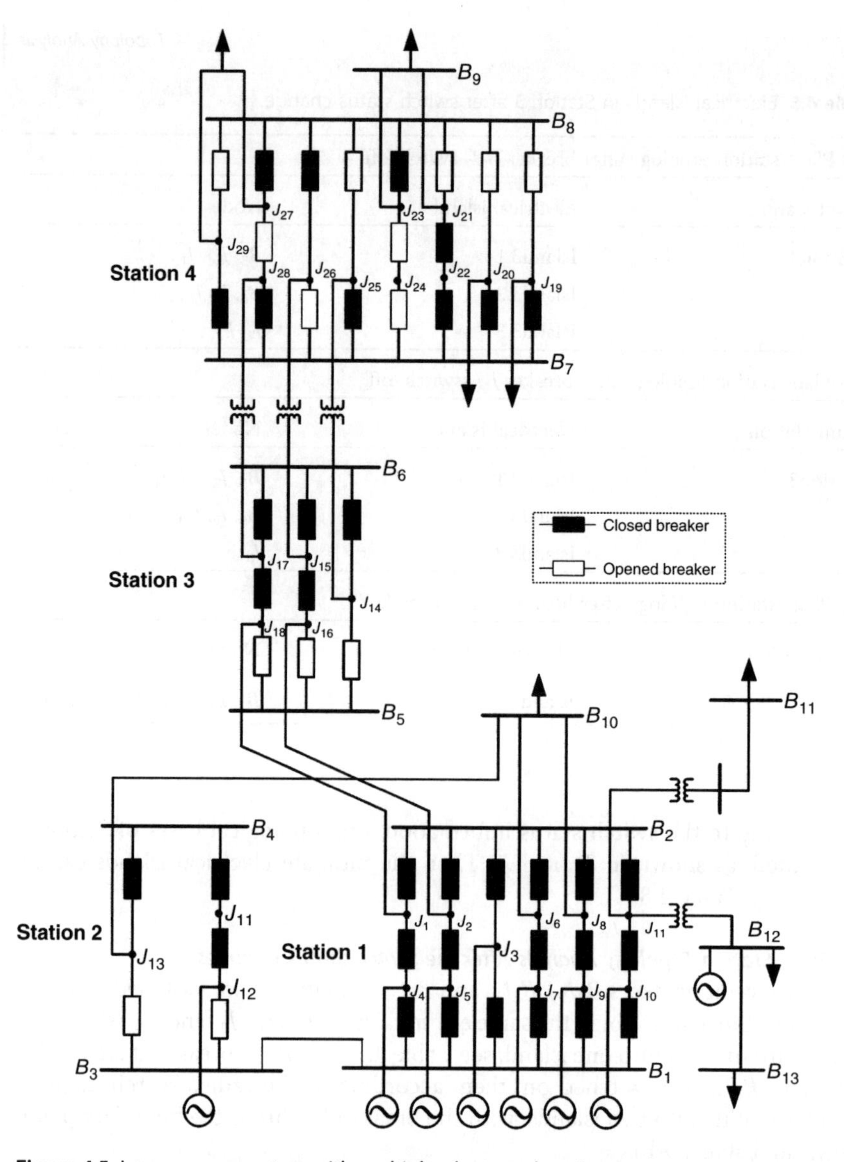

Figure 4.5 Large power system with multiple plants and stations.

Table 4.6 Circuit breakers and switch status in Station 4.

Circuit breakers	Switch status	Circuit breakers	Switch status	Circuit breakers	Switch status
B_7J_{19}	1	B_7J_{29}	1	B_8J_{25}	0
B_7J_{20}	1	$J_{21}J_{22}$	1	B_8J_{26}	1
B_7J_{22}	1	$J_{27}J_{28}$	0	B_8J_{27}	1
B_7J_{24}	0	B_8J_{19}	0	B_8J_{29}	0
B_7J_{25}	1	B_8J_{20}	0	$J_{23}J_{24}$	0
B_7J_{26}	0	B_8J_{21}	0		
B_7J_{28}	1	B_8J_{23}	1		

Table 4.7 Topological bases of Station 4.

Topological bases	Nodes in the island	Topological bases	Nodes in the island	Topological bases	Nodes in the island
TPJ4.1	B_7, J_{19}	TPJ4.5	B_7, J_{28}	TPJ4.9	B_8, J_{27}
TPJ4.2	B_7, J_{20}	TPJ4.6	B_7, J_{29}	TPJ4.10	J_{21}, J_{22}
TPJ4.3	B_7, J_{22}	TPJ4.7	B_8, J_{23}		
TPJ4.4	B_7, J_{25}	TPJ4.8	B_8, J_{26}		

Table 4.8 Ultimate electrical islands of Station 4.

Electrical island/Intersected nodes set	Nodes
Island 4.1	$B_7, J_{19}, J_{20}, J_{21}, J_{22}, J_{25}, J_{28}, J_{29}$
Island 4.2	$B_8, J_{23}, J_{26}, J_{27}$
Island 4.3	J_{24}
L_D4	B_7, B_8, J_{22}

Table 4.9 Electrical islands in Station 4 after switch status change.

(a) Plant-station topology after breaker B_8J_{19}, B_8J_{20}, B_8J_{25} switch 'on'	
Electrical island	Nodes
Island 4.1	$B_7, B_8, J_{19}, J_{20}, J_{21}, J_{22}, J_{23}, J_{24}, J_{25} J_{26}, J_{27}, J_{28}, J_{29}$
Island 4.2	J_{24}
(b) Plant-station topology after breaker B_7J_{25} switch 'off'	
Electrical island	Nodes
Island 4.1	$B_7, B_8, J_{19}, J_{20}, J_{21}, J_{22}, J_{23}, J_{24}, J_{25} J_{26}, J_{27}, J_{28}, J_{29}$
Island 4.2	J_{24}
Island 4.3	J_{25}

methods in calculation effort. With the proposed topology analysis method, fast tracking of the switch status can be realized.

On a notebook with 2 GHz basic frequency, 2 GB memory and Intel® Core™ 2 Duo CPU processor, programming was carried out in Microsoft Visual C++6.0 to compare the computational time of different topology analysis methods

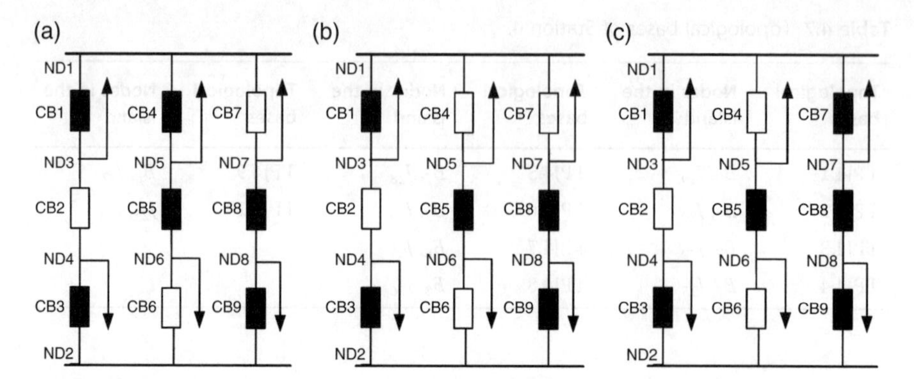

Figure 4.6 Plant-station topology analysis example. (a) Initial topology. (b) Switch 'off'. (c) Switch 'on'.

Table 4.10 Comparison between computational time of different analysis methods.

Methods	Computational time (μs)			
	Initial topology	Switch 'off'	Switch 'on'	Total computational time
Self-multiplication	18.5	18.5	30.8	67.8
Square method	17.97	18.0	22.1	58.07
Row accumulation method	11.3	10.3	9.75	31.35
Row scanning method	10.3	9.75	8.73	28.78
Reference [4]	4.11	4.11	4.11	12.33
Proposed method	3.08	2.05	2.57	7.7

(shown in Figure 4.6). The results are shown in Table 4.10. It can be seen that the proposed method is far superior to traditional methods in computation speed, especially in topology update speed.

4.3 Topology Analysis for Inter-substation

This section introduces a novel power network analysis method based on the incidence matrix notation method and the loop matrix. First, when the system is in normal operating mode, the incidence matrix notation method is used to carry out static topology analysis of the power network. Then, when the network topology changes, the property of the changing branch is determined with the help of the loop matrix. On this basis, the local network topology and the loop matrix are updated by means of the broken circle method and the radius search method.

4.3.1 Static Topology Analysis for a Power Network

4.3.1.1 The Incidence Matrix and BFS

A. Incidence Matrix

Suppose the vertex set and side set of graph $\boldsymbol{G} = (\boldsymbol{V},\boldsymbol{E})$ are $\boldsymbol{V} = \{v_1, v_2, v_3, \ldots, v_p\}$ and $\boldsymbol{E} = \{e_1, e_2, e_3, \ldots, e_q\}$, respectively, where b_{ij} is the association value between vertex v_i and side e_j, then $\mathbf{B} = (b_{ij})_{p\times q}$ is the incidence matrix of graph $\boldsymbol{G}$ [4].

For the simple graph shown in Figure 4.7, the incidence matrix **B** can be expressed as

$$
\mathbf{B} = \begin{array}{c} \\ v_1 \\ v_2 \\ v_3 \\ v_4 \\ v_5 \\ v_6 \end{array}
\begin{array}{c}
\begin{array}{cccccccc} e_1 & e_2 & e_3 & e_4 & e_5 & e_6 & e_7 & e_8 \end{array} \\
\begin{bmatrix}
0 & 1 & 1 & 0 & 0 & 0 & 0 & 1 \\
1 & 0 & 0 & 1 & 0 & 0 & 0 & 1 \\
0 & 1 & 0 & 0 & 1 & 0 & 1 & 0 \\
1 & 0 & 0 & 0 & 0 & 1 & 1 & 0 \\
0 & 0 & 0 & 1 & 0 & 1 & 0 & 0 \\
0 & 0 & 1 & 0 & 1 & 0 & 0 & 0
\end{bmatrix}
\end{array}
\tag{4.5}
$$

It can be seen from Equation (4.5) that the incidence matrix **B** has the following characteristics:

- Each column in **B** corresponds to one side, and the rows in each column with element 1 correspond to the terminal nodes of the side. The sum of the elements in each column is 2. This is because each side is related to two vertices.
- Each row in **B** corresponds to one vertex, and the columns in each row with element 1 correspond to the sides related to the vertex. Therefore, the sum of the elements in each row represents the number of sides to which this vertex is related.
- If all the elements in a row are 0, then the vertex corresponding to this row is an isolated node.

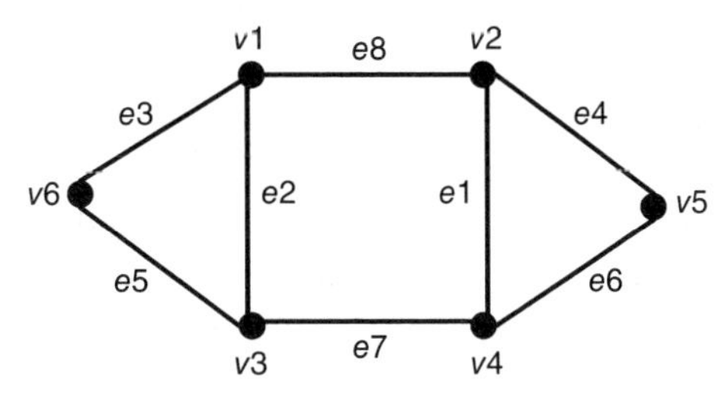

Figure 4.7 A simple network.

B. BFS

The basic idea of BFS is as follows. First, select any vertex in the graph as the root, and all the sides related to the root vertex are added, thus the other vertices related to these sides are taken as the first-level vertices of the tree. Second, sort the first-level vertices randomly and visit them one by one. When visiting a certain vertex, if no loop is formed, then the sides related to this vertex are added, and the other vertices related to these sides are taken as the second-level vertices of the tree. According to the same searching principle, the other levels of vertices may be found; thus the tree could be formed. Every vertex on the tree is a node of the electrical island.

4.3.1.2 Incidence Matrix Notation Method

Combining the physical meaning of the incidence matrix and the searching principle of BFS, the incidence matrix notation method is able to determine the connectivity of the power network topology according to the following steps.

Step 1: Suppose the first non-zero element in the incidence matrix **B** is b_{ij}, then the first element (vertex) in the electrical island $\boldsymbol{I}$ is v_i. Take v_i as the root and search for all the non-zero elements in the jth row in **B**. The sides corresponding to these non-zero elements are marked as tree branches, and the other terminal nodes of the tree branches are added to the electrical island, as the first-level vertices of the tree.

Step 2: Starting from the second element (vertex) in the electrical island $\boldsymbol{I}$, the sides and vertices related to the element are searched one by one. When vertex v_m is visited, search for the non-zero elements in the mth row in **B**. If the columns (sides) corresponding to the non-zero elements are not marked yet, and the other terminal nodes of these sides are not in the electrical island, then these columns (sides) are marked as tree branches and this vertex is added to the electrical island. If the columns (sides) corresponding to the non-zero elements are already marked, and the other terminal nodes of these sides are already in $\boldsymbol{I}$, then these columns (sides) are marked as link branches.

Step 3: When the number of nodes searched equals the total number of nodes in the network or when all the columns (sides) in the incidence matrix are marked, the notation analysis ends.

4.3.1.3 Static Topology Analysis Based on the Incidence Matrix Notation Method

By determining the connectivity of the graph, at the same time marking the tree branches and link branches in the network using the incidence matrix notation method, the network static electrical island may be analysed in the following steps.

Step 1: Form the power network incidence matrix **B** according to the system switch status.

Step 2: Take the vertex v_i corresponding to the row where the first non-zero element b_{im} in the first unmarked column m ($m \in [1,q]$) in **B** is the root, i.e. vertex v_i is the first element in the static electrical island $\boldsymbol{I}_g$ (the initial value of g is 1).

Step 3: Search for the other non-zero element b_{kj} in the jth column where the non-zero element b_{ij} in the ith row is. If the jth column is not marked yet and vertex v_k is not in the electrical island $\boldsymbol{I}_g$, then add v_k to the electrical island $\boldsymbol{I}_g$ and mark e_j as a tree branch (i.e. add e_j to the tree branch set $\boldsymbol{S}_g$). If the jth column is not marked yet and vertex v_k is in the electrical island $\boldsymbol{I}_g$, add e_j to the link branch set $\boldsymbol{A}_g$. Otherwise, go to Step 4.
Step 4: If all the vertices are in the electrical island, then continue to Step 6. Otherwise, go to Step 5.
Step 5: If all the elements in electrical island $\boldsymbol{I}_g$ are traversed, electrical island number $g = g + 1$ and go to Step 2. Otherwise, search the next element v_n of $\boldsymbol{I}_g$ and set $i = n$, then go back to Step 3.
Step 6: The power network static topology analysis ends.

Take the incidence matrix in Equation (4.5), for example. The power network static topology analysis is conducted using the incidence matrix notation method. The first non-zero element in **B** is b_{21}, thus according to Step 2, vertex v_2 is selected as the root. First, search for the non-zero elements in the second row in **B**, the corresponding columns being 1, 4, 8. The other non-zero elements in these columns are in rows 4, 5 and 1, respectively. Therefore, vertices v_4, v_5 and v_1 are added to the electrical island. Then, from the second node in the electrical island, i.e. v_4, continue searching using the same method until the end. In the above steps, the marking route is $e_1 \rightarrow e_4 \rightarrow e_8 \rightarrow e_6 \rightarrow e_7 \rightarrow \ldots$. All the nodes in the network are in the same electrical island. The marked tree branch set is $\boldsymbol{S}_1 = \{e_1, e_3, e_4, e_7, e_8\}$, and the link branch set is $\boldsymbol{A}_1 = \{e_2, e_5, e_6\}$, where the number i after the set name corresponds to the ith electrical island in the network.

4.3.2 Topology Update for a Power Network

Due to factors such as outage, overhaul and expansion, the initial power network topology may change, which mainly includes two situations: branch break-off and branch closure. For branch break-off, three cases are involved: (1) loop link branch break-off, (2) loop tree branch break-off and (3) when the break-off tree branch does not belong to any loop. For branch closure, there are also three cases involved: (1) branch terminal nodes in the same electrical island, (2) branch terminal nodes in different electrical islands and (3) one of the branch terminal nodes is a newly added node.

4.3.2.1 Loop Matrix

Suppose u and v are two vertices in graph **G**, and the route between u and v is l. If the route from u to v is different from the route from v to u, then l is called a track. If the vertices in l are also different, then l is called a path. If there exists a u–v path in **G**, then vertices u and v are connected. If the starting point and terminal point of a path are the same point, then this path is called a loop. In a loop, there is more than one path between any two points. Therefore, the break-off of a side in a loop will not affect the connectivity of the graph, which means that the electrical island distribution in the power network topology will remain the same.

In order to improve the topology update efficiency, consider the influence of the loop on the topology, and the loop matrix $\mathbf{L}_g$ is defined, where g is the number of the electrical island. The first column in $\mathbf{L}_g$ is the numbers of link branches in the basic loops. The other elements in the rows are the tree branches in the corresponding loop, and the vacancies are set to be 0. According to the link branches and tree branches marked in the static topology analysis, the loop matrix could be formed off-line. Take the system in Figure 4.9, for example. The link branches marked in the static topology analysis are e_2, e_5 and e_6, thus the loop matrix $\mathbf{L}_1$ formed off-line is shown in Equation (4.6), where the ith row represents the ith loop in the current topology.

$$\begin{matrix} \text{Loop} & & \text{Link branch} & \text{Tree branch} \\ \begin{matrix} 1 \\ 2 \\ 3 \end{matrix} & \mathbf{L}_1 = & \left[\begin{matrix} e_2 \\ e_5 \\ e_6 \end{matrix} \right. & \left. \begin{matrix} e_1 & e_7 & e_8 & 0 \\ e_1 & e_3 & e_7 & e_8 \\ e_1 & e_4 & 0 & 0 \end{matrix} \right] \end{matrix} \tag{4.6}$$

4.3.2.2 Topology Update in the Case of Branch Break-off

A. Loop Link Branch Break-Off

When a link branch breaks off, the electrical island distribution remains unchanged. Therefore, only the loop matrix needs to be updated, i.e. remove the break-off branch from the link branch set and the corresponding loop from the loop matrix $\mathbf{L}_i$.

Take the system in Figure 4.7, for example. If link branch e_2 breaks off, the electrical island remains unchanged. Thus, by removing e_2 from the marked link branch set and deleting loop 1 (the first row) in Equation (4.6), the updated loop matrix $\mathbf{L}_1$ may be formed, as shown in Equation (4.7).

$$\begin{matrix} & \text{Link branch} & \text{Tree branch} \\ \mathbf{L}_1 = & \left[\begin{matrix} e_5 \\ e_6 \end{matrix} \right. & \left. \begin{matrix} e_1 & e_3 & e_7 & e_8 \\ e_1 & e_4 & 0 & 0 \end{matrix} \right] \end{matrix} \tag{4.7}$$

B. Loop Tree Branch Break-Off

The basic idea of the broken circle method is as follows. Suppose there is at least one loop $\boldsymbol{C}$ in graph $\boldsymbol{G}$, and e is any side in $\boldsymbol{C}$, then $\boldsymbol{G} - e$ is still connected. Until the last side in the last loop is removed, the resulting graph $\boldsymbol{T}$ is a spanning tree of $\boldsymbol{G}$. Therefore, the exchange of any tree branch with the link branch in the same loop will not change the connectivity of the graph.

When a tree branch in the loop breaks off, the electrical island distribution remains unchanged. Therefore, only the loop matrix needs to be updated. According to the broken circle method, this can be achieved by changing the link branch of a certain loop where the break-off tree branch is to a tree branch, and modifying the elements in relevant loops. Two cases are considered in this situation.

Case 1: If the break-off tree branch belongs to only one loop, then the link branch in this loop is changed to a tree branch and this loop is deleted. Take the system in Figure 4.7, for example. If tree branch e_4 breaks off, then the loop matrix in Equation (4.6) could be updated off-line by deleting loop 3, where e_4 belongs and marking link branch e_6 in loop 3 as a tree branch. After the network topology change, the tree branch set and link branch set are, respectively $\boldsymbol{S}_1 = \{e_1, e_3, e_7, e_8, e_6\}$ and $\boldsymbol{A}_1 = \{e_2, e_5\}$.The updated loop matrix is shown in Equation (4.8).

$$\begin{array}{c} \text{Link branch} \quad \text{Tree branch} \\ \mathbf{L}_1 = \begin{bmatrix} e_2 & e_1 & e_7 & e_8 & 0 \\ e_5 & e_1 & e_3 & e_7 & e_8 \end{bmatrix} \end{array} \tag{4.8}$$

Case 2: If the break-off tree branch belongs to two or more loops, then select the loop with the smallest number of branches. Suppose loop i is selected; change the link branch in loop i to a tree branch and merge loop i with the other loops where the break-off tree branch belongs. On this basis, delete the common part of loop i and the other loops. Then remove loop i from the loop matrix. Take the system in Figure 4.7, for example. If tree branch e_1 breaks off, consider that e_1 belongs to loop 1, 2 and 3; loop 3, which has the fewest branches, is selected. Change link branch e_6 in loop 3 to a tree branch and merge loop 3 with loops 1 and 2 to form new loops. The updated loop matrix $\mathbf{L}_1$ is shown in Equation (4.9), which is in harmony with the power network topology.

$$\begin{array}{c} \text{Link branch} \quad \text{Tree branch} \\ \mathbf{L}_1 = \begin{bmatrix} e_2 & e_7 & e_8 & e_6 & e_4 & 0 \\ e_5 & e_3 & e_7 & e_8 & e_6 & e_4 \end{bmatrix} \end{array} \tag{4.9}$$

C. *When the Break-Off Tree Branch Does Not Belong to Any Loop*

If the break-off tree branch does not belong to any loop, then the connectivity of the power network will change, which will cause the electrical island to split. In this case, the radius search method is used here for topology update. Starting from the break-off node, search downward along the tree branches and stop when there is a new electrical island formed.

According to graph theory [3], if $\boldsymbol{G}$ is a weighted graph, and the weight for each side v_iv_j is $w_{ij} \geq 0$, the weight for each vertex v_i is $q(v_i) \geq 0$, $d(u, v)$ is the distance between vertex u and vertex v considering the side weight, then the radius of graph $\boldsymbol{G}$ is

$$rad\boldsymbol{G} = \min_{u \in V(G)} \max_{v \in V(G)} d(u,v) \tag{4.10}$$

For a particular connected graph, if the radius of the tree in the connected graph before the branch break-off is *rad*, then after the branch break-off, it requires a search depth of $n = rad - 1$ for relevant tree branches to reach the boundary nodes of the tree and form new electrical islands.

After the electrical island analysis is finished, the loop matrix can be updated in the following steps.

Step 1: According to the search route, separate the tree branches of the new electrical island from the original topology.
Step 2: Add the loops containing the tree branches of the new electrical island to the new loop matrix.
Step 3: Remove the link branches in the new loop matrix from the original topology.

Take the system in Figure 4.7, for example. After tree branch e_8 breaks off, the loop matrix is shown in Equation (4.11).

$$\begin{array}{c} \text{Link branch} \quad \text{Tree branch} \\ \mathbf{L}_1 = \begin{bmatrix} e_5 & e_3 & e_2 \\ e_6 & e_1 & e_4 \end{bmatrix} \end{array} \tag{4.11}$$

If tree branch e_7 also breaks off, since e_7 does not belong to any loop, the electrical island will split. According to the radius search method, starting from the break-off nodes v_1 and v_2, search for the one-level relevant tree branches, so that a new electrical island containing v_2, v_4 and v_5 is formed. After the topology update, the loop matrix can be modified off-line. Since tree branches e_1 and e_4 in the new electrical island both belong to loop 2, by separating loop 2 from $\mathbf{L}_1$, the new loop matrix $\mathbf{L}_2$ can be obtained, as shown in Equation (4.12).

$$\begin{array}{c} \text{Link branch} \quad \text{Tree branch} \\ \begin{cases} \mathbf{L}_1 = \begin{bmatrix} e_5 & e_3 & e_2 \end{bmatrix} \\ \mathbf{L}_2 = \begin{bmatrix} e_6 & e_1 & e_4 \end{bmatrix} \end{cases} \end{array} \tag{4.12}$$

4.3.2.3 Topology Update in the Case of Branch Closure

According to the location of the terminal nodes of the closed branch, three situations are considered in this case.

1) If the terminal nodes of the closed branch are in the same electrical island, the electrical island distribution will not change, and the closed branch is added as a link branch. The loop matrix may be modified off-line.
2) If the terminal nodes of the closed branch are in different electrical islands, then merge these two islands and the tree branches, link branches and loops in the islands. The closed branch is added as a tree branch.
3) If one terminal node of the closed branch is new, add the new node to the electrical island where the other terminal node belongs. The closed branch is added as a tree branch of the electrical island, while the link branches and loops do not change.

4.3.3 Scheme Verification

The New England 10-machine, 39-bus system (see Figure 4.8) is used for simulation verification here. The numbering of branches and corresponding nodes is shown in Table 4.11.

4.3.3.1 Static Topology Analysis

According to the branch-node relationship in Table 4.11, and the system switch status (all circuit breakers in the system are switched on), the incidence matrix is formed. Then, based on the incidence matrix notation method, power network static topology analysis is conducted. The result is that all the nodes in the system belong to the same electrical island $\boldsymbol{I}_1$, the marked tree branch set and link branch set are, respectively $\boldsymbol{S}_1 = \{1, 2, 3, 4, 5, 6, 7, 9, 10, 11, 12, 13, 14, 15, 16, 17, 18, 19, 20, 21, 23, 24, 25, 28, 30, 32, 33, 35, 36, 37, 38, 39, 40, 41, 42, 43, 45, 46\}$ and $\boldsymbol{A}_1 = \{8, 22, 26, 27, 29, 31, 34, 44\}$. The basic loop matrix $\mathbf{L}_1$ is then formed off-line, as shown in Equation (4.13).

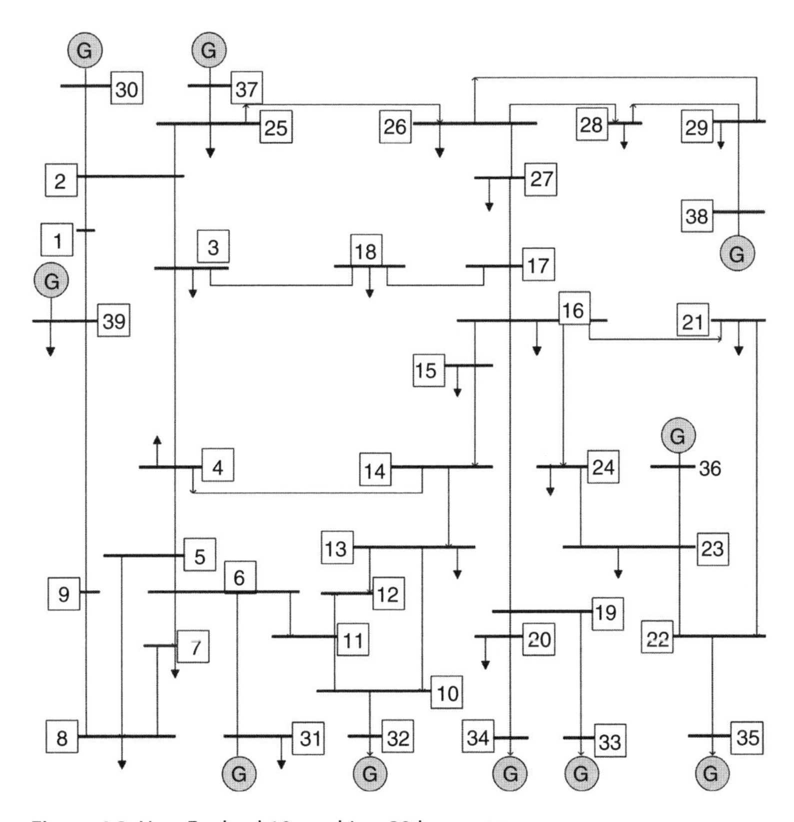

Figure 4.8 New England 10-machine, 39-bus system.

Table 4.11 Branches and corresponding nodes in the New England 10-machine, 39-bus system.

Branch number	Terminal nodes		Branch number	Terminal nodes		Branch number	Terminal nodes	
1	1	2	17	14	13	33	28	29
2	2	3	18	5	6	**34**	26	29
3	1	39	19	6	11	35	29	38
4	3	4	20	11	12	36	19	16
5	9	39	21	8	7	37	19	33
6	4	5	**22**	6	7	38	20	34
7	8	9	23	6	31	39	19	20
8	8	5	24	10	32	40	21	16
9	2	30	25	10	11	41	21	22
10	2	25	**26**	10	13	42	24	16
11	25	37	**27**	12	13	43	24	23
12	25	26	28	16	17	**44**	23	22
13	3	18	**29**	16	15	45	23	36
14	18	17	30	17	27	46	22	35
15	4	14	**31**	26	27	—	—	—
16	14	15	32	26	28	—	—	—

$$\mathbf{L}_1 = \begin{bmatrix} 8 & 1 & 2 & 3 & 4 & 5 & 6 & 7 & 0 & 0 \\ 22 & 1 & 2 & 3 & 4 & 5 & 6 & 7 & 18 & 21 \\ 26 & 6 & 15 & 17 & 18 & 19 & 25 & 0 & 0 & 0 \\ 27 & 6 & 15 & 17 & 18 & 19 & 20 & 0 & 0 & 0 \\ 29 & 4 & 13 & 14 & 15 & 16 & 28 & 0 & 0 & 0 \\ 31 & 2 & 12 & 13 & 14 & 30 & 0 & 0 & 0 & 0 \\ 34 & 32 & 33 & 0 & 0 & 0 & 0 & 0 & 0 & 0 \\ 44 & 40 & 41 & 42 & 43 & 0 & 0 & 0 & 0 & 0 \end{bmatrix} \quad (4.13)$$

(First column: Link branch; remaining columns: Tree branch)

4.3.3.2 Power Network Topology Update

A. Branch Break-Off

Case 1: Loop link branch break-off

When link branch 8 breaks off, the electrical island distribution remains unchanged. Remove branch 8 from the link branch set of the original topology, and delete loop 1, where branch 8 belongs in the loop matrix. The updated loop matrix is shown in Equation (4.14).

$$\mathbf{L}_1 = \overset{\text{Link branch} \qquad\qquad\qquad\qquad \text{Tree branch}}{\begin{bmatrix} 22 & 1 & 2 & 3 & 4 & 5 & 6 & 7 & 18 & 21 \\ 26 & 6 & 15 & 17 & 18 & 19 & 25 & 0 & 0 & 0 \\ 27 & 6 & 15 & 17 & 18 & 19 & 20 & 0 & 0 & 0 \\ 29 & 4 & 13 & 14 & 15 & 16 & 28 & 0 & 0 & 0 \\ 31 & 2 & 12 & 13 & 14 & 30 & 0 & 0 & 0 & 0 \\ 34 & 32 & 33 & 0 & 0 & 0 & 0 & 0 & 0 & 0 \\ 44 & 40 & 41 & 42 & 43 & 0 & 0 & 0 & 0 & 0 \end{bmatrix}} \tag{4.14}$$

Case 2: Loop tree branch break-off

After link branch 8 breaks off, if tree branch 6 also breaks off, the electrical island still remains unchanged. It can be seen in Equation (4.14) that the tree branch belongs to loops 1, 2 and 3. Select loop 2, which contains the fewest branches, and change the link branch 26 in loop 2 to a tree branch. Then, merge loop 2 with loops 1 and 3, and delete the common branches to form new loops. Finally, delete loop 2. The updated loop matrix is shown in Equation (4.15).

$$\mathbf{L}_1 = \overset{\text{Link branch} \qquad\qquad\qquad\qquad \text{Tree branch}}{\begin{bmatrix} 22 & 1 & 2 & 3 & 4 & 5 & 7 & 21 & 26 & 15 & 17 & 19 & 25 \\ 27 & 20 & 26 & 25 & 0 & 0 & 0 & 0 & 0 & 0 & 0 & 0 & 0 \\ 29 & 4 & 13 & 14 & 15 & 16 & 28 & 0 & 0 & 0 & 0 & 0 & 0 \\ 31 & 2 & 12 & 13 & 14 & 30 & 0 & 0 & 0 & 0 & 0 & 0 & 0 \\ 34 & 32 & 33 & 0 & 0 & 0 & 0 & 0 & 0 & 0 & 0 & 0 & 0 \\ 44 & 40 & 41 & 42 & 43 & 0 & 0 & 0 & 0 & 0 & 0 & 0 & 0 \end{bmatrix}} \tag{4.15}$$

Case 3: When the break-off tree branch does not belong to any loop

When tree branch 36 breaks off, start from the break-off nodes 16 and 19, and search downward for tree branches. According to the radius search method, only two tree branches need to be searched to form a new electrical island $\boldsymbol{I}_2 = \{19, 20, 33, 34\}$. The tree branch set of the new electrical island is $\boldsymbol{S}_2 = \{37, 38, 39\}$. As for the update of the loop matrix, since none of the branches in $\boldsymbol{S}_2$ belong to any loop, no loop or link branch needs to be separated from the original electrical island, i.e. $\boldsymbol{A}_2 = 0$. The loop matrices of the two electrical islands are shown in Equations (4.16) and (4.17).

$$\mathbf{L}_1 = \overset{\text{Link branch} \qquad\qquad\qquad\qquad \text{Tree branch}}{\begin{bmatrix} 22 & 1 & 2 & 3 & 4 & 5 & 6 & 7 & 18 & 21 \\ 26 & 6 & 15 & 17 & 18 & 19 & 25 & 0 & 0 & 0 \\ 27 & 6 & 15 & 17 & 18 & 19 & 20 & 0 & 0 & 0 \\ 29 & 4 & 13 & 14 & 15 & 16 & 28 & 0 & 0 & 0 \\ 31 & 2 & 12 & 13 & 14 & 30 & 0 & 0 & 0 & 0 \\ 34 & 32 & 33 & 0 & 0 & 0 & 0 & 0 & 0 & 0 \\ 44 & 40 & 41 & 42 & 43 & 0 & 0 & 0 & 0 & 0 \end{bmatrix}} \tag{4.16}$$

$$\mathbf{L}_2 = 0 \tag{4.17}$$

B. Branch Closure

If branch 6 in Equation (4.11) is closed, since the terminal nodes are in the same electrical island, the electrical island distribution remains unchanged. Because branch 6 is a link branch, a new loop is added to update the loop matrix, see Equation (4.18).

$$\mathbf{L}_1 = \begin{array}{c} \text{Link branch} \qquad\qquad\qquad\qquad \text{Tree branch} \\ \left[\begin{array}{c|cccccccccccc} 22 & 1 & 2 & 3 & 4 & 5 & 7 & 21 & 26 & 15 & 17 & 19 & 25 \\ 27 & 20 & 26 & 25 & 0 & 0 & 0 & 0 & 0 & 0 & 0 & 0 & 0 \\ 29 & 4 & 13 & 14 & 15 & 16 & 28 & 0 & 0 & 0 & 0 & 0 & 0 \\ 31 & 2 & 12 & 13 & 14 & 30 & 0 & 0 & 0 & 0 & 0 & 0 & 0 \\ 34 & 32 & 33 & 0 & 0 & 0 & 0 & 0 & 0 & 0 & 0 & 0 & 0 \\ 44 & 40 & 41 & 42 & 43 & 0 & 0 & 0 & 0 & 0 & 0 & 0 & 0 \\ 6 & 1 & 2 & 3 & 4 & 5 & 7 & 15 & 17 & 18 & 19 & 25 & 26 \end{array}\right] \end{array} \tag{4.18}$$

If branch 36 in Equations (4.16) and (4.17) is closed, since the terminal nodes 16 and 19 belong to different electrical islands, the two islands are merged. The topology update result is the same as the static topology analysis result in Equation (4.13).

4.4 False Topology Identification

A false topology identification method based on the road-loop equation is introduced in this section. First, use the road-loop equation established according to the network topology constraints and Kirchhoff's law to calculate the branch currents. On this basis, consider different cases of topology fault and undesirable data; the topology information and measured value of branch currents are checked according to the difference between the calculated value and measured value.

4.4.1 Road-Loop Equation

4.4.1.1 Road Matrix

In a power network, a road refers to a branch with direction that represents the flow of current. Thus, the road starts from the root of the system tree and goes to the top of the system tree along branches. The element in road matrix $\mathbf{T}$ ($l \times n$ matrix, where l is the number of branches and n is the number of nodes) is defined as

$$t_{ik} = \begin{cases} 1 & \text{if there is a road to branch } i \text{ via node } k \\ 0 & \text{if there is no road to branch } i \text{ via node } k \end{cases} \tag{4.19}$$

Take the power network shown in Figure 4.9, for example. Select branch L9 as the link branch and the other branches as tree branches, and take bus B2 as the root of the system tree; the roads are labelled with arrows. Thus, according to

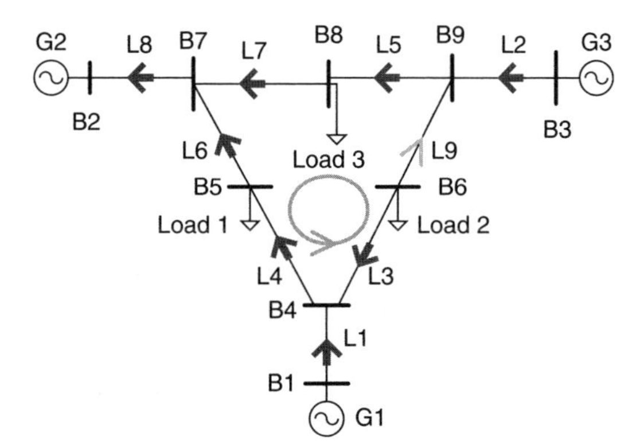

Figure 4.9 Road-loop labelled graph of IEEE 9-bus system.

Equation (4.19), the road matrix can be obtained, shown in Equation (4.20), where $\mathbf{T}_L$, which corresponds to the link branch, is a zero matrix.

$$\mathbf{T} = \begin{array}{c} \\ L1 \\ L2 \\ L3 \\ L4 \\ L5 \\ L6 \\ L7 \\ L8 \\ L9 \end{array} \begin{array}{c} \begin{array}{ccccccccc} B1 & B2 & B3 & B4 & B5 & B6 & B7 & B8 & B9 \end{array} \\ \begin{bmatrix} 1 & 0 & 0 & 0 & 0 & 0 & 0 & 0 & 0 \\ 0 & 0 & 1 & 0 & 0 & 0 & 0 & 0 & 0 \\ 0 & 0 & 0 & 0 & 0 & 1 & 0 & 0 & 0 \\ 1 & 0 & 0 & 1 & 0 & 1 & 0 & 0 & 0 \\ 0 & 0 & 1 & 0 & 0 & 0 & 0 & 0 & 1 \\ 1 & 0 & 0 & 1 & 1 & 1 & 0 & 0 & 0 \\ 0 & 0 & 1 & 0 & 0 & 0 & 0 & 1 & 1 \\ 1 & 0 & 1 & 1 & 1 & 1 & 1 & 1 & 1 \\ 0 & 0 & 0 & 0 & 0 & 0 & 0 & 0 & 0 \end{bmatrix} \end{array} \tag{4.20}$$

According to the current direction of tree branches in basic loops and the current direction of branches, the loop-branch correlation matrix $\mathbf{B}$ ($L \times l$ matrix, where L is the number of basic loops) can be obtained. Suppose the positive direction of a link branch is the positive direction of the loop, then the element in $\mathbf{B}$ is

$$b_{li} = \begin{cases} 1 & \text{Branch } i \text{ is in loop } l\text{, with the same direction to loop } l \\ -1 & \text{Branch } i \text{ is in loop } l\text{, with the reverse direction to loop } l \\ 0 & \text{Branch } i \text{ is not in loop } l \end{cases} \tag{4.21}$$

According to Equation (4.21), the loop-branch correlation matrix **B** of the network in Figure 4.9 is

$$\mathbf{B} = \begin{bmatrix} 0 & 0 & -1 & -1 & 1 & -1 & 1 & 0 & 1 \end{bmatrix} \tag{4.22}$$

4.4.1.2 Road-Loop Equation

The branch current consists of two parts: one is the contribution from the node injection current to the branch; the other is the contribution from the loop current to the branch. If the node injection current is $\dot{\boldsymbol{I}}_N$, the link branch current is $\dot{\boldsymbol{I}}_L$, then the branch current $\dot{\boldsymbol{I}}'_{\mathrm{b}}$ (the single prime "′" represents the calculated value) can be expressed with the road-loop equation as

$$\dot{\boldsymbol{I}}'_{\mathrm{b}} = \mathbf{T} \times \dot{\boldsymbol{I}}_N + \mathbf{B}^{\mathrm{T}} \times \dot{\boldsymbol{I}}_L \tag{4.23}$$

For a local network of a large power grid with a large number of nodes and complex wiring, the road-loop equation is still applicable. Suppose node i is the starting point of a road in the local network, and the terminal point of branch k and starting point of branch t are both node i. According to Equation (4.23), for starting node i, the equivalent node injection current $i_{Ni}{}'$ equals the sum of branch current i_{lk} and external current i_{Ni}, minus the loop current $\mathbf{B}_k \times \dot{\boldsymbol{I}}_L$ on branch k. When the equivalent injection current to the starting node of local network and the injection currents to the internal nodes are calculated, the node injection current vector $\dot{\boldsymbol{I}}_N$ of the local network can be formed, which is then applied to Equation (4.23) so that the branch current vector $\dot{\boldsymbol{I}}'_{\mathrm{b}}$ of the local network can be obtained.

$$\begin{aligned} i_{lt} &= \mathbf{T}_t \times \dot{\boldsymbol{I}}_N + \mathbf{B}_t \times \dot{\boldsymbol{I}}_L (\mathbf{T}_t \text{ and } \mathbf{B}_t \text{ are the } t\text{th rows in the respective matrices}) \\ &\qquad \text{The } i\text{th element} \\ &= \begin{bmatrix} \mathbf{T}_k & \ddot{\text{i}} & 0\ldots \end{bmatrix} \times \dot{\boldsymbol{I}}_N + \mathbf{B}_t \times \dot{\boldsymbol{I}}_L \\ &= \mathbf{T}_k \times \dot{\boldsymbol{I}}_N + \mathbf{B}_k \times \dot{\boldsymbol{I}}_L + i_{Ni} - \mathbf{B}_k \times \dot{\boldsymbol{I}}_L + \mathbf{B}_t \times \dot{\boldsymbol{I}}_L \\ &= \left(i_{lk} + i_{Ni} - \mathbf{B}_k \times \dot{\boldsymbol{I}}_L\right) + \mathbf{B}_t \times \dot{\boldsymbol{I}}_L \\ & i_{Ni} \text{ is the injection current to node } i, \mathbf{B}_k \text{ is the } k\text{th row in the matrix} \end{aligned} \tag{4.24}$$

4.4.2 Analysis of the Impacts of Topology Error and Undesirable Data on the Branch Current

After the branch currents are calculated according to the road-loop equation, the errors and undesirable data in the system topology can be identified using the following principles.

Principle 1 When there is no topology error or undesirable data, the calculation error of any branch current $|i_{LK} - i_{Lk}{}'| < \varepsilon 1$, where $\varepsilon 1$ is 0.3% of the absolute value of the maximum branch current, i.e. $\varepsilon 1 = 0.3\% \times |\max\{i_{Lk}\}|$.

Principle 2 The sum of the node injection currents should meet Kirchhoff's current law (KCL), i.e. $\left|\sum iBi\right| < \varepsilon 2, \varepsilon 2 = k \times n \times 0.2\% \times \max\{|i_{Bi}|\}$, where k is the reliability coefficient determined by the allowable error of the real measuring device (in this section $k = 1$), and n is the number of nodes. The relative error of PMU measurement is 0.2%, and $\max\{|i_n|\}$ is the maximum absolute value of the node injection current.

Principle 3 Identify whether a branch is switched off according to whether the absolute value of the branch current $|i_{Lk}| < \varepsilon 3$, where $\varepsilon 3$ is 0.08 times the rated branch current, i.e. $\varepsilon 3 = 0.08\dot{\boldsymbol{I}}_{LN}$.

According to the above principles, the road-loop equation could be used to identify topological error and analyse undesirable data. In the calculation process, the source current and load current are taken as node injection currents. Meanwhile, the branch-to-ground admittance is considered, and the currents on the equivalent π-branch-to-ground admittances are all converted to injection currents to the terminal bus of the branch.

4.4.2.1 Impact of Topology Error

Power network topology errors generally include three cases: the switch status is identified as 'on' but is actually off; the switch status is identified as 'off' but is actually on; false identification of double-bus wiring.

Suppose node i is the terminal node of branch k, the topology of which is in error, $i_{\Delta k}$ is the branch k-to-ground admittance current. Sort the nodes in **T**, the node currents in $\dot{\boldsymbol{I}}_N$ and the branches according to the distance from the root of the road. If node i is behind m nodes, then branch k is behind $m-1$ branches.

A. *Switch status identified as 'on' but is actually off*

When the switch status of a branch is identified as 'on' but is actually off, the calculated injection current to the terminal node of the branch $i_{\Delta k}$ will increase. In this case, the road-loop equation is

$$
\begin{aligned}
\dot{\boldsymbol{I}}_{\mathrm{b}}' &= \begin{bmatrix} \mathbf{T}_{(m-1)\times m} & \mathbf{T}_{m\times(n-m)} \\ \mathbf{T}_{(l+1-m)\times m} & \mathbf{T}_{(l+1-m)\times(n-m)} \end{bmatrix} \times \left[i_{B1}\ldots\ i_{Bk}+i_{\Delta k}\ldots\ i_{Bn} \right]^{\mathrm{T}} + \mathbf{B}^{\mathrm{T}} \times \dot{\boldsymbol{I}}_L \\
&= \begin{bmatrix} \mathbf{T}_{(m-1)\times m} & 0 \\ \mathbf{T}_{(l+1-m)\times m} & \mathbf{T}_{(l+1-m)\times(n-m)} \end{bmatrix} \times \dot{\boldsymbol{I}}_N + \mathbf{B}^{\mathrm{T}} \times \dot{\boldsymbol{I}}_L \\
&\quad + \begin{bmatrix} \mathbf{T}_{(m-1)\times m} & 0 \\ \mathbf{T}_{(l+1-m)\times m} & \mathbf{T}_{(l+1-m)\times(n-m)} \end{bmatrix} \times \begin{bmatrix} \underbrace{\overbrace{0\ldots0}^{m}\ i_{\Delta k}\ 0\ldots0}_{n} \end{bmatrix}^{\mathrm{T}} \\
&= \dot{\boldsymbol{I}}_{\mathrm{b}} + \begin{bmatrix} 0 & \mathbf{T}_{(l+1-m)\times(n-m)} \end{bmatrix} \begin{bmatrix} \underbrace{\overbrace{0\ldots0}^{m}\ i_{\Delta k}\ 0\ldots0}_{n} \end{bmatrix}^{\mathrm{T}}
\end{aligned}
\tag{4.25}
$$

According to Equation (4.25), when the switch status of a branch is identified as 'on' but is actually off, the current on the branch with topology error is smaller than $\varepsilon 3$, and none of the currents of branches in the road behind this branch satisfy Principle 1; the current difference has to do with the current of branch (with topology error) to ground admittance.

B. Switch status identified as 'off' but is actually on

When the switch status of a branch is identified as 'off' but is actually on, the calculated injection current to the terminal node of the branch $i_{\Delta k}$ will decrease.

$$
\begin{aligned}
\dot{I}_b' &= \begin{bmatrix} \mathbf{T}_{(m-1)\times m} & 0 \\ 0 & \mathbf{T}'_{(l+1-m)\times(n-m)} \end{bmatrix} \times \left[i_{B1} \ldots \ i_{Bk} - i_{\Delta k} \ldots \ i_{Bn} \right]^{\mathrm{T}} + \mathbf{B}^{\mathrm{T}} \times \dot{I}_L \\
&= \left(\mathbf{T} - \begin{bmatrix} 0 & 0 \\ \mathbf{T}_{(l+1-m)\times m} & \Delta\mathbf{T}_{(l+1-m)\times(n-m)} \end{bmatrix} \right) \times \dot{I}_N + \mathbf{B}^{\mathrm{T}} \times \dot{I}_L \\
&\quad + \begin{bmatrix} \mathbf{T}_{(m-1)\times m} & 0 \\ 0 & \mathbf{T}_{(l+1-m)\times(n-m)} \end{bmatrix} \times \Bigg[\underbrace{\overbrace{0 \ldots 0}^{m} \ -i_{\Delta k} \ 0 \ldots 0}_{n} \Bigg]^{\mathrm{T}} \\
&= \dot{I}_b - \left[\mathbf{T}_{(l+1-m)\times m} \ \ \Delta\mathbf{T}_{(l+1-m)\times(n-m)} \right] \dot{I}_N \\
&\quad + \left[0 \ \ \mathbf{T}_{(l+1-m)\times(n-m)} \right] \Bigg[\underbrace{\overbrace{0 \ldots 0}^{m} \ -i_{\Delta k} \ 0 \ldots 0}_{n} \Bigg]^{\mathrm{T}}
\end{aligned}
\tag{4.26}
$$

According to Equation (4.26), when the branch topology is in error, none of the branch currents satisfy Principle 1, and the current difference has to do with the current of the branch with topology error and its grounding current.

If a link branch that is switched on is identified as being switched off, according to Equation (4.26), matrix **B** will be affected, thus the branch current in the loop will not satisfy Principle 1, and the current difference has to do with the link branch current. In addition, the injection current to the terminal node of the link branch will be affected, and the currents of branches in the road behind this loop will not satisfy Principle 1, the current difference has to do with the current of link branch-to-ground admittance.

C. False identification of double-bus wiring

For plant-station outgoing lines with double-bus wiring, suppose the wiring of a bus is falsely identified. For example, suppose switching operation is available between node i and node u, the outgoing line is k, and there are s branches between two buses. The switch status shows that branch k is connected to node u, but actually branch k is connected to node i. According to Equation (4.27), both the calculated value and the measured value of the currents of branches between two buses will be different, and the current difference has to do with the current of branch k to ground admittance and the current of the loop where branch k is.

$$
\begin{aligned}
\dot{\boldsymbol{I}}_{\mathrm{b}}' &= \begin{bmatrix} \mathbf{T}_{m-\times m} & \mathbf{T}_{m\times(n-m)} \\ \mathbf{T}'_{s\times m} & \mathbf{T}'_{s\times(n-m)} \\ \mathbf{T}_{(l-s-m)\times m} & \mathbf{T}_{(l-s-m)\times(n-m)} \end{bmatrix} \times \left[i_{B1}\ldots\ i_{Bk}-i_{\Delta k}\ldots\ i_{Bu}+i_{\Delta k}\ldots\ i_{Bn}\right]^{\mathrm{T}} \\
&+ \left[\mathbf{B}_{L\times(m-1)}\ \ \mathbf{B}'_{L\times s}\ \ \mathbf{B}_{L\times(l+1-s-m)}\right]^{\mathrm{T}} \times \dot{\boldsymbol{I}}_L \\
&= \left(\mathbf{T} - \begin{bmatrix} 0 & 0 \\ \Delta\mathbf{T}_{s\times m} & \Delta\mathbf{T}_{s\times(n-m)} \\ 0 & 0 \end{bmatrix}\right) \times \dot{\boldsymbol{I}}_N + \begin{bmatrix} 0 & \cdots & 0 \\ \vdots & \overbrace{\mathbf{T}_{s\times(u-k)}}^{m+1,m+1} & \vdots \\ 0 & \cdots & 0 \end{bmatrix} \left[\underbrace{\overbrace{0\ldots0}^{m}\ -i_{\Delta k}\ 0\ldots0}_{n}\right]^{\mathrm{T}} \\
&+ \left(\mathbf{B}^{\mathrm{T}} - \left[\overbrace{0}^{m}\ \Delta\mathbf{B}_{L\times s} 0\right]^{\mathrm{T}}\right) \times \dot{\boldsymbol{I}}_L \\
&= \dot{\boldsymbol{I}}_{\mathrm{b}} + \begin{bmatrix} 0 & \cdots & 0 \\ \vdots & \mathbf{T}_{s\times(u-k)} & \vdots \\ 0 & \cdots & 0 \end{bmatrix} \left[\underbrace{\overbrace{0\ldots0}^{m}\ -i_{\Delta k}\ 0\ldots0}_{n}\right]^{\mathrm{T}} \\
&- \begin{bmatrix} 0 & 0 \\ \Delta\mathbf{T}_{s\times m} & \Delta\mathbf{T}_{s\times(n-m)} \\ 0 & 0 \end{bmatrix} \dot{\boldsymbol{I}}_N - [0\ \ \Delta\mathbf{B}_{L\times s}\ \ 0]^{\mathrm{T}}) \times \dot{\boldsymbol{I}}_L
\end{aligned}
\tag{4.27}
$$

4.4.2.2 Impact of Undesirable Data

To further improve the ability to identify topology faults, undesirable data need to be identified effectively, and corrected. Since the road-loop equation contains the link branch current and node currents, the node injection currents need to be checked first. According to Principle 2, if there are undesirable data in node injection currents, the sum of the node injection currents will be larger than $\varepsilon 2$. According to Equation (4.25), the calculated value and the measured value of branch currents connected to the node with undesirable data will be different, and the current difference is the error of the undesirable data. The node injection current could be corrected according to Principle 2.

If there are undesirable data in the link branch current, according to Equation (4.28), the calculated value and measured value of branch currents in the loop will be different, and the current difference is the error of undesirable data (where Δi represents the error of undesirable data, and $\mathbf{B}_k$ is the kth column of matrix $\mathbf{B}$).

$$
\begin{aligned}
\dot{\boldsymbol{I}}_{\mathrm{b}}{}' &= \mathbf{T} \times \dot{\boldsymbol{I}}_N + \mathbf{B}^{\mathrm{T}} \times \left[i_{L1}\ldots\ i_{Lk}+\Delta i\ldots\ i_{Lt}\right]^{\mathrm{T}} \\
&= \mathbf{T} \times \dot{\boldsymbol{I}}_N + \mathbf{B}^{\mathrm{T}} \times \dot{\boldsymbol{I}}_L + \mathbf{B}_k{}^{\mathrm{T}} \times \Delta i \\
&= \dot{\boldsymbol{I}}_{\mathrm{b}} + \mathbf{B}_k{}^{\mathrm{T}} \times \Delta i
\end{aligned}
\tag{4.28}
$$

If there are undesirable data in the tree branch current, the calculated value will not be affected, thus the measured value could be corrected according to the calculated value.

4.4.3 Topology Error Identification Method Based on the Road-Loop Equation

According to the analysis in Section 4.4.2, the consistency of switch status and electrical variables may be checked by first identifying the fault in topology and then modifying the undesirable data. Detailed steps are as follows:

Step 1 is to form **T**, **B**, $\dot{\boldsymbol{I}}_N$ and $\dot{\boldsymbol{I}}_L$ according to the power network, and calculate $\dot{\boldsymbol{I}}'_\text{b}$ using the road-loop equation, and then identify whether all branch currents satisfy Principle 1 and Principle 3.

Step 2 is that if there is current difference, identify whether there is a topology fault:

1) If a link branch is 'off' but the measured value of its current is bigger than $\varepsilon 3$, and the branch currents in this loop do not satisfy Principle 1, then the link branch is actually switched on. If a tree branch is 'off' but the calculated value of its current is bigger than $\varepsilon 3$, and none of the branch currents behind the terminal node of this tree branch in the road satisfy Principle 1, then the tree branch is actually switched on.
2) If a branch is 'on' but the calculated value of its current is smaller than $\varepsilon 3$, and none of the branch currents behind this branch in the road satisfy Principle 1, then the branch is actually switched off.
3) If a branch is 'on' and the calculated value of its current is smaller than $\varepsilon 3$, but when the current of this branch-to-ground admittance is removed, the sum of the node currents will increase, then this branch is identified as lightly loaded.
4) For branches where the switching operation is available, if a branch is 'on' and both the calculated value and the measured value of the branch current are bigger than $\varepsilon 3$, but the branch current between the two buses does not satisfy Principle 1, then the bus wiring of this branch is falsely identified.

According to the above methods, the topology of branches is checked. If a fault is found, then the topology is modified and **T**, **B**, $\dot{\boldsymbol{I}}_N$ and $\dot{\boldsymbol{I}}_L$ are updated. Then calculate $\dot{\boldsymbol{I}}'_\text{b}$ and check again if there is a topology fault.

Step 3 is that when there is no further topology fault (or the fault has been corrected), identify if there are any undesirable data:

1) Calculate the sum of system node injection currents; if it does not satisfy Principle 2, and none of the branch currents behind a certain source node or load node in the system road satisfy Principle 1, then there are undesirable data in the node injection current. This is corrected according to Principle 2 or the current difference.
2) If the branch current in a certain loop does not satisfy Principle 1, then there are undesirable data in the current of the link branch in this loop. The link branch current is corrected according to the current difference.
3) If a particular tree branch current does not satisfy Principle 1, the measured value may be corrected according to the calculated value of the tree branch current.

After the undesirable data have been corrected, continue to calculate $\dot{\boldsymbol{I}}'_\text{b}$, until all branch currents satisfy Principle 1.

Step 4 is that the topology fault identification and undesirable data correction are completed.

Table 4.12 Calculation results of road-loop equation in the IEEE three-machine, nine-bus system.

	Measured value (kA)		Calculated value (kA)		Difference (kA)	
Branch	Real part	Imaginary part	Real part	Imaginary part	Real part	Imaginary part
L1	0.7956	0.4556	0.7955	0.4556	0	0
L2	0.7939	0.4578	0.7939	0.4578	0	0
L3	−0.4363	−0.2164	−0.4355	−0.2179	7.6E−04	1.5E−03
L4	0.4145	0.2261	0.4148	0.2246	2.6E−04	1.4E−03
L5	0.4138	0.2269	0.4123	0.2284	1.5E−03	1.5E−03
L6	−0.4352	−0.2167	−0.4351	−0.2181	9.6E−05	1.4E−03
L7	−0.4359	−0.2164	−0.4337	−0.2148	2.3E−03	1.6E−03
L8	−0.8261	−0.4507	−0.8247	−0.4509	1.4E−03	1.3E−03
L9	−0.4363	−0.2164	−0.4363	−0.2164	0	0

4.4.4 Scheme Verification

4.4.4.1 IEEE Three-machine Nine-bus System

A model of the IEEE three-machine, nine-bus system was established using the PSCAD simulation software, as shown in Figure 4.8. Suppose switching operation is available with bus $B2$ and $B7$. The system source $S_N = 100\text{MVA}$, $U_N = 220\text{kV}$ thus $\dot{I}_{LN} = 0.223 + \text{j}0.138\text{kA}$, $\varepsilon 3 = 0.01785 + \text{j}0.0111\text{kA}$. Check Equation (4.23). The branch-to-ground admittance is $B/2 = 1.623 \times 10^{-4}\text{S}$. The node injection current is shown in Equation (4.29), loop current is $i_{L9} = -0.43628 - \text{j}0.216385\text{kA}$ and the calculation results are shown in Table 4.12. All branch currents in the system satisfy Principle 1, and there are no topology faults or undesirable data.

$$\dot{I}_N = \begin{bmatrix} 0.79557 \\ 0.8265 \\ 0.79387 \\ 0.05474 \\ -0.84991 \\ -0.87182 \\ 0.04408 \\ -0.84595 \\ 0.05471 \end{bmatrix} + \text{j} \begin{bmatrix} 0.45563 \\ 0.45128 \\ 0.45778 \\ -0.013094 \\ -0.44277 \\ -0.43429 \\ -0.01793 \\ -0.44316 \\ -0.013036 \end{bmatrix} \text{kA} \tag{4.29}$$

A. Switch status identified as 'off' but is actually on with undesirable data

In Figure 4.9, if all branches are switched on except for branch $L1$, then the system's $\mathbf{T}$ matrix and $\mathbf{B}$ matrix are shown in Equation (4.30), the node injection currents are shown in Equation (4.31), the link branch current $i_{L9} = -0.43628 - \text{j}0.216385\text{kA}$. The calculation results are shown in Table 4.13.

Table 4.13 Calculation results when branch $L1$ breaks off.

	Measured value (kA)		Calculated value (kA)		Difference (kA)	
Branch	Real part	Imaginary part	Real part	Imaginary part	Real part	Imaginary part
$L1$	0.7956	0.4556	0.7956	0.4556	3.0E−05	3.0E−05
$L2$	0.7939	0.4578	0.7939	0.4578	3.0E−05	2.0E−05
$L3$	−0.4363	−0.2164	−0.4355	−0.2179	7.9E−04	1.5E−03
$L4$	0.2073	0.1130	−0.3808	0.2179	0.5881	0.1049
$L5$	0.4138	0.2269	0.4123	0.2284	1.5E−03	1.6E−03
$L6$	−0.4352	−0.2167	−1.203	−0.6611	0.7678	0.4444
$L7$	−0.4359	−0.2164	−0.4372	−0.2148	1.3E−03	1.6E−03
$L8$	−0.8261	−0.4507	−1.6776	−0.8934	0.8515	0.4427
$L9$	−0.4363	−0.2164	−0.4363	−0.2164	0	0

$$\begin{cases} \mathbf{T} = \begin{matrix} \\ i_{L1} \\ i_{L2} \\ i_{L3} \\ i_{L4} \\ i_{L5} \\ i_{L6} \\ i_{L7} \\ i_{L8} \\ i_{L9} \end{matrix} \begin{matrix} \begin{matrix} B1 & B2 & B3 & B4 & B5 & B6 & B7 & B8 & B9 \end{matrix} \\ \begin{bmatrix} 1 & 0 & 0 & 0 & 0 & 0 & 0 & 0 & 0 \\ 0 & 0 & 1 & 0 & 0 & 0 & 0 & 0 & 0 \\ 0 & 0 & 0 & 0 & 0 & 1 & 0 & 0 & 0 \\ 0 & 0 & 0 & 1 & 0 & 1 & 0 & 0 & 0 \\ 0 & 0 & 1 & 0 & 0 & 0 & 0 & 0 & 1 \\ 0 & 0 & 0 & 1 & 1 & 1 & 0 & 0 & 0 \\ 0 & 0 & 1 & 0 & 0 & 0 & 0 & 1 & 1 \\ 0 & 0 & 1 & 1 & 1 & 1 & 1 & 1 & 1 \\ 0 & 0 & 0 & 0 & 0 & 0 & 0 & 0 & 0 \end{bmatrix} \end{matrix} \\ \mathbf{B} = \begin{bmatrix} 0 & 0 & -1 & -1 & 1 & -1 & 1 & 0 & 1 \end{bmatrix} \end{cases} \tag{4.30}$$

$$\dot{I}_N = \begin{bmatrix} 0.79557 \\ 0.8265 \\ 0.79387 \\ 0.02205 \\ -0.84991 \\ -0.87182 \\ 0.04408 \\ -0.84595 \\ 0.05471 \end{bmatrix} + \mathrm{j} \begin{bmatrix} 0.45563 \\ 0.45128 \\ 0.45778 \\ -0.00898 \\ -0.44277 \\ -0.43429 \\ -0.01793 \\ -0.44316 \\ -0.013036 \end{bmatrix} \text{kA} \tag{4.31}$$

Table 4.14 Calculation results after topology information is corrected.

Branch	Measured value (kA) Real part	Measured value (kA) Imaginary part	Calculated value (kA) Real part	Calculated value (kA) Imaginary part	Difference (kA) Real part	Difference (kA) Imaginary part
$L1$	0.7956	0.4556	0.7956	0.4556	0	0
$L2$	0.7939	0.4578	0.7939	0.4578	3.0E−05	2.0E−05
$L3$	−0.4363	−0.2164	−0.4355	−0.2179	7.9E−04	1.5E−03
$L4$	0.2073	0.1130	0.4148	0.2246	0.2075	0.1116
$L5$	0.4138	0.2268	0.4123	0.2284	1.5E−03	1.6E−03
$L6$	−0.4352	−0.2167	−0.4348	−0.2185	1.3E−04	1.4E−03
$L7$	−0.4359	−0.2164	−0.4376	−0.2144	1.2E−03	1.6E−03
$L8$	−0.8261	−0.4507	−0.8273	−0.4509	1.2E−03	1.6E−04
$L9$	−0.4363	−0.2164	−0.4363	−0.2164	0	0

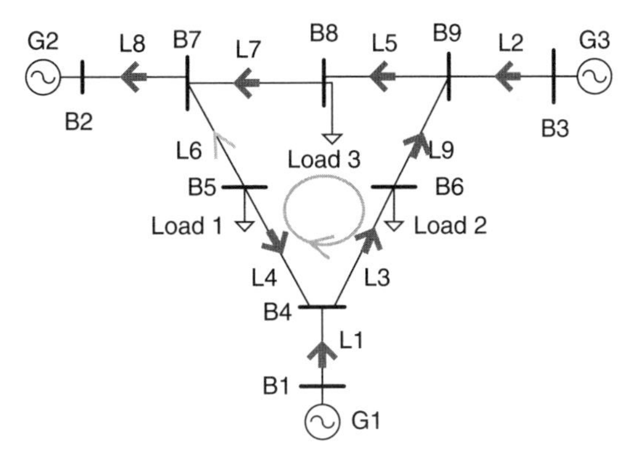

Figure 4.10 Road-loop labelled graph of system after $L6$ breaks off.

According to Equation (4.30) and Table 4.13, $\varepsilon 1 = (2.478 + j1.733) \times 10^{-3}\,\text{kA}$, $\varepsilon 2 = (1.255 + j0.659) \times 10^{-2}\,\text{kA}$, none of the branches behind $L1$ satisfy Principle 1, and the calculated value of current on $L1$ satisfies Principle 3, thus branch $L1$ is actually switched on. After the topology of the system is modified, the updated matrix **T**, matrix **B** and node injection currents are shown in Equations (4.20), (4.22) and (4.29) respectively, and the calculation results are shown in Table 4.14.

It can be seen from Table 4.14 that branch $L4$ does not satisfy Principle 1, and the switching operation is unavailable with the buses at two ends of the branch. Thus, the measured data of branch $L4$ are false, which could be corrected according to the calculated value.

B. Switch status identified as 'off' but is actually on with undesirable data

Suppose the system topology identifies that branch $L6$ is switched off; take the switched-off branch in the loop as the link branch, the direction of road is shown in Figure 4.10.

System matrices **T** and **B** are shown in Equation (4.32), the node injection currents are shown in Equation (4.33), and the calculation results are shown in Table 4.15.

Table 4.15 Calculation results after *L*6 breaks off.

	Measured value (kA)		Calculated value (kA)		Difference (kA)	
Branch	**Real part**	**Imaginary part**	**Real part**	**Imaginary part**	**Real part**	**Imaginary part**
$L1$	0.7956	0.4556	0.7956	0.4556	0	0
$L2$	0.7939	0.4578	0.7939	0.4578	3.0E−05	2.0E−05
$L3$	−0.4363	−0.2164	−0.4355	−0.2179	7.9E−04	1.5E−03
$L4$	0.2073	0.1130	0.4148	0.2246	0.2075	0.1116
$L5$	0.4138	0.2268	0.4123	0.2284	1.5E−03	1.6E−03
$L6$	−0.4352	−0.2167	−0.4348	−0.2185	1.3E−04	1.4E−03
$L7$	−0.4359	−0.2164	−0.4376	−0.2144	1.2E−03	1.6E−03
$L8$	−0.8261	−0.4507	−0.8273	−0.4509	1.2E−03	1.6E−04
$L9$	−0.4363	−0.2164	−0.4363	−0.2164	0	0

$$\begin{cases} \mathbf{T} = \begin{matrix} \\ i_{L1} \\ i_{L2} \\ i_{L3} \\ i_{L4} \\ i_{L5} \\ i_{L6} \\ i_{L7} \\ i_{L8} \\ i_{L9} \end{matrix} \begin{matrix} B1 & B2 & B3 & B4 & B5 & B6 & B7 & B8 & B9 \\ \end{matrix} \begin{bmatrix} 1 & 0 & 0 & 0 & 0 & 0 & 0 & 0 & 0 \\ 0 & 0 & 1 & 0 & 0 & 0 & 0 & 0 & 0 \\ 1 & 0 & 0 & 1 & 1 & 0 & 0 & 0 & 0 \\ 0 & 0 & 0 & 0 & 1 & 0 & 0 & 0 & 0 \\ 0 & 0 & 1 & 0 & 0 & 0 & 0 & 0 & 1 \\ 0 & 0 & 0 & 0 & 0 & 0 & 0 & 0 & 0 \\ 1 & 0 & 1 & 1 & 1 & 1 & 0 & 1 & 1 \\ 1 & 0 & 1 & 1 & 1 & 1 & 1 & 1 & 1 \\ 1 & 0 & 0 & 1 & 1 & 1 & 0 & 0 & 0 \end{bmatrix} \\ \mathbf{B} = 0 \end{cases} \tag{4.32}$$

$$\dot{I}_N = \begin{bmatrix} 0.79557 \\ 0.8265 \\ 0.79387 \\ 0.05473 \\ -0.87196 \\ 0.02205 \\ 0.02204 \\ -0.84595 \\ 0.05471 \end{bmatrix} + \mathrm{j} \begin{bmatrix} 0.45563 \\ 0.45128 \\ 0.45778 \\ -0.0131 \\ -0.43379 \\ -0.00898 \\ -0.00896 \\ -0.44316 \\ -0.013036 \end{bmatrix} \text{kA} \tag{4.33}$$

Table 4.16 Calculation results after the topology information of *L*6 is corrected.

	Measured value (kA)		Calculated value (kA)		Difference (kA)	
Branch	Real part	Imaginary part	Real part	Imaginary part	Real part	Imaginary part
*L*1	0.7956	0.4556	0.7956	0.4556	3.0E−05	3.0E−05
*L*2	0.7939	0.4578	0.7939	0.4578	3.0E−05	2.0E−05
*L*3	0.4141	0.2254	0.4136	0.2255	0.00048	9.0E−05
*L*4	−0.4367	−0.2170	−0.4367	−0.2171	2.7E−05	1.4E−04
*L*5	0.4138	0.2268	1.2842	0.6792	0.87036	0.4524
*L*6	−0.4352	−0.2167	−0.4352	−0.2167	3.1E−05	1.3E−05
*L*7	−0.4359	−0.2164	0.4383	0.236	0.8742	0.4524
*L*8	−0.8261	−0.4507	0.0471	0.0014	0.8732	0.4521
*L*9	−0.4363	−0.2164	0.4356	0.2344	−0.8719	0.4508

According to Equation (4.32) and Table 4.15, $\varepsilon 1 = (2.478 + j1.733) \times 10^{-3}$kA, $\varepsilon 2 = (1.2554 + j0.6592) \times 10^{-2}$kA, when branch *L*6 is switched off the calculated current is bigger than $\varepsilon 3$, and the loop current and none of the branch currents behind *L*6 in the road satisfy Principle 1, thus branch *L*6 is actually switched on. After the topology is modified, matrices **T** and **B** are as shown in Equations (4.20) and (4.22), the node injection currents are shown in Equation (4.34), and the calculation results are shown in Table 4.16.

$$\dot{I}_N = \begin{bmatrix} 0.79557 \\ 0.8265 \\ 0.79387 \\ 0.05474 \\ -0.84991 \\ 0.02205 \\ 0.04408 \\ -0.84595 \\ 0.05471 \end{bmatrix} + j \begin{bmatrix} 0.45563 \\ 0.45128 \\ 0.45778 \\ -0.013094 \\ -0.44277 \\ -0.008975 \\ -0.01793 \\ -0.44316 \\ -0.013036 \end{bmatrix} \text{kA} \tag{4.34}$$

According to Equation (4.34) and Table 4.16, all branch currents satisfy Principle 3, but the sum of the node injection currents does not satisfy Principle 2, and none of the branch currents behind load node *B*6 satisfy Principle 1, thus there are undesirable data in the injection current to node *B*6. After it is corrected, according to the sum of the node injection currents, the result is shown in Equation (4.35), and the calculation results of the road-loop equation are as shown in Table 4.17. It can be seen that all branch currents satisfy Principle 1, thus the system topology and measured values are correct.

Table 4.17 Calculation results after undesirable data of node *B6* are corrected.

	Measured value (kA)		Calculated value (kA)		Difference (kA)	
Branch	**Real part**	**Imaginary part**	**Real part**	**Imaginary part**	**Real part**	**Imaginary part**
L1	0.7956	0.4557	0.7956	0.4556	3.0E–05	3.0E–05
L2	0.7939	0.4578	0.7939	0.4578	3.0E–05	2.0E–05
L3	0.4141	0.2254	0.4136	0.2255	4.8E–04	9.0E–05
L4	–0.4367	–0.2170	–0.4367	–0.2171	2.7E–05	1.4E–04
L5	0.4138	0.2268	0.4106	0.2265	3.2E–03	3.5E–05
L6	–0.4352	–0.2167	–0.4352	–0.2167	3.1E–05	1.3E–05
L7	–0.4360	–0.2164	–0.4354	–0.2166	5.4E–04	2.0E–04
L8	–0.8261	–0.4507	–0.8265	–0.4513	4.3E–04	5.6 E–04
L9	–0.4363	–0.2164	–0.4380	–0.2168	1.7E–03	4.2E–04

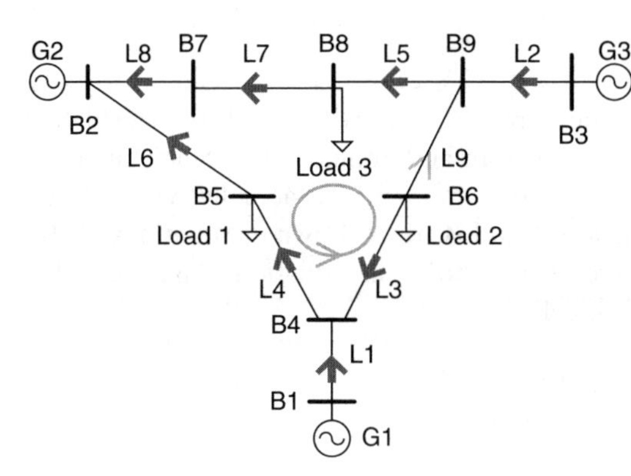

Figure 4.11 Road-loop labelled graph of system after *L6* is connected to *B2*.

$$\dot{I}_N = \begin{bmatrix} 0.79557 \\ 0.8265 \\ 0.79387 \\ 0.05474 \\ -0.84991 \\ -0.85155 \\ 0.04408 \\ -0.84595 \\ 0.05471 \end{bmatrix} + \mathrm{j} \begin{bmatrix} 0.45563 \\ 0.45128 \\ 0.45778 \\ -0.013094 \\ -0.44277 \\ -0.44368 \\ -0.01793 \\ -0.44316 \\ -0.013036 \end{bmatrix} \text{kA} \tag{4.35}$$

C. False identification of double-bus wiring

Suppose the system topology is as shown in Figure 4.11; the system matrices **T** and **B** are as shown in Equation (4.36), the node injection currents are as shown in Equation (4.37), and the calculation results are as shown in Table 4.18.

Table 4.18 Calculation results after $L6$ is connected to $B2$.

	Measured value (kA)		Calculated value (kA)		Difference (kA)	
Branch	Real part	Imaginary part	Real part	Imaginary part	Real part	Imaginary part
$L1$	0.7956	0.4556	0.7956	0.4556	3.0E − 05	3.0E − 05
$L2$	0.7939	0.4578	0.7939	0.4578	3.0E − 05	2.0E − 05
$L3$	−0.4363	−0.2164	−0.4355	−0.2179	7.9E − 04	1.5E − 03
$L4$	0.4145	0.2260	0.4148	0.2246	2.8E − 04	1.4E − 03
$L5$	0.4138	0.2268	0.4123	0.2284	0.001536	1.6E − 03
$L6$	−0.4352	−0.2167	−0.4351	−0.2181	0.00013	1.4E − 03
$L7$	−0.4360	−0.2164	−0.4372	−0.2148	1.3E − 03	1.6E − 03
$L8$	−0.8261	−0.4507	−0.4349	−0.2159	0.3912	0.2348
$L9$	−0.4363	−0.2164	−0.4363	−0.2164	0	0

According to Table 4.18, branch $L6$ satisfies Principle 3, but the branch currents behind $L6$ between bus $B2$ and $B7$ with which the switching operation is available do not satisfy Principle 1, the current difference is close to the difference between the loop current where $L6$ is and branch $L6$ to ground admittance current. Thus, the bus wiring of branch $L6$ is falsely identified. After the topology is modified, the updated matrix $\mathbf{T}$, matrix $\mathbf{B}$ and node injection currents are as shown in Equations (4.20), (4.22) and (4.29), respectively, and the calculation results are shown in Table 4.18. It can be seen that both the topology and measured values are correct.

$$\begin{cases} \mathbf{T} = \begin{matrix} \\ i_{L1} \\ i_{L2} \\ i_{L3} \\ i_{L4} \\ i_{L5} \\ i_{L6} \\ i_{L7} \\ i_{L8} \\ i_{L9} \end{matrix} \begin{matrix} \begin{matrix} B1 & B2 & B3 & B4 & B5 & B6 & B7 & B8 & B9 \end{matrix} \\ \begin{bmatrix} 1 & 0 & 0 & 0 & 0 & 0 & 0 & 0 & 0 \\ 0 & 0 & 1 & 0 & 0 & 0 & 0 & 0 & 0 \\ 0 & 0 & 0 & 0 & 0 & 1 & 0 & 0 & 0 \\ 1 & 0 & 0 & 1 & 0 & 1 & 0 & 0 & 0 \\ 0 & 0 & 1 & 0 & 0 & 0 & 0 & 0 & 1 \\ 1 & 0 & 0 & 1 & 1 & 1 & 0 & 0 & 0 \\ 0 & 0 & 1 & 0 & 0 & 0 & 0 & 1 & 1 \\ 1 & 0 & 1 & 1 & 1 & 1 & 0 & 1 & 1 \\ 0 & 0 & 0 & 0 & 0 & 0 & 0 & 0 & 0 \end{bmatrix} \end{matrix} \\ \mathbf{B} = \begin{bmatrix} 0 & 0 & -1 & -1 & 1 & -1 & 1 & 1 & 1 \end{bmatrix} \end{cases} \tag{4.36}$$

Table 4.19 Calculation results when all branches are switched on.

	Measured value (kA)		Calculated value (kA)		Difference (kA)	
Branch	Real part	Imaginary part	Real part	Imaginary part	Real part	Imaginary part
$L1$	0.9311	0.43963	0.9311	0.4396	1.2E−05	3.1E−05
$L2$	0.7626	0.4414	0.7626	0.4414	4.0E−05	1.1E−05
$L3$	−0.3262	−0.2091	−0.3243	−0.2083	0.00185	7.8E−04
$L4$	0.6571	0.2181	0.6586	0.2189	1.5E−03	8.3E−04
$L5$	0.3065	0.2170	0.3042	0.2164	2.3E−03	6.4E−04
$L6$	0	0	0.0013	9.2E−04	1.3E−03	9.2E−04
$L7$	−0.5903	−0.3238	−0.5928	−0.3244	2.5E−03	6.3E−04
$L8$	−0.5644	−0.3317	−0.5444	−0.3389	0.01998	0.00718
$L9$	−0.5134	−0.2125	−0.5134	−0.2125	2.0E−05	0

$$\dot{I}_N = \begin{bmatrix} 0.79557 \\ 0.84854 \\ 0.79387 \\ 0.05474 \\ -0.84991 \\ -0.87182 \\ 0.02204 \\ -0.84595 \\ 0.05471 \end{bmatrix} + j \begin{bmatrix} 0.45563 \\ 0.44232 \\ 0.45778 \\ -0.013094 \\ -0.44277 \\ -0.43429 \\ -0.008962 \\ -0.44316 \\ -0.013036 \end{bmatrix} \text{kA} \tag{4.37}$$

D. Switch status identified as 'on' but is actually off

In Figure 4.9, suppose all branches are switched on, matrix **T** and matrix **B** are as shown in Equations (4.20) and (4.22), and the node injection currents are as shown in Equation (4.38). The threshold values $\varepsilon 1 = (2.793 + j1.324) \times 10^{-3}$kA, $\varepsilon 2 = (1.676 + j0.9734) \times 10^{-2}$kA, and the calculation results are as shown in Table 4.19. It can be seen that both the measured value and the calculated value of branch $L6$ are smaller than $\varepsilon 3$, and the branch currents behind the terminal node of $L6$ do not satisfy Principle 1, thus branch $L6$ is actually switched off.

After the topology information of $L6$ is modified, the system road is shown in Figure 4.10, matrix **T** and matrix **B** are as shown in Equation (4.32), the node injection currents are as shown in Equation (4.39), and the calculation results are as shown in Table 4.20. It can be seen that, both the topology and measured values are correct.

Table 4.20 Calculation results after $L6$ breaks off.

	Measured value (kA)		Calculated value (kA)		Difference (kA)	
Branch	Real part	Imaginary part	Real part	Imaginary part	Real part	Imaginary part
$L1$	0.9311	0.43963	0.9311	0.43963	0	0
$L2$	0.7626	0.4414	0.7626	0.4414	0	0
$L3$	−0.3262	−0.2091	0.3051	0.2178	3.8E−04	9.3E−05
$L4$	0.6571	0.2181	−0.6749	−0.2097	5E−06	5E−06
$L5$	0.3065	0.2170	0.3055	0.2173	9.6E−04	2.5E−04
$L6$	0	0	0	0	0	0
$L7$	−0.5903	−0.3238	−0.5915	−0.3235	1.1E−03	3.2E−04
$L8$	−0.5644	−0.3317	−0.5657	−0.3313	1.3E−03	3.8E−04
$L9$	−0.5134	−0.2125	−0.5121	−0.2116	1.3E−03	9.2E−04

$$\dot{I}_N = \begin{bmatrix} 0.93109 \\ 0.56411 \\ 0.76264 \\ 0.05181 \\ -0.65728 \\ -0.83773 \\ 0.04707 \\ -0.8970 \\ 0.05501 \end{bmatrix} + j \begin{bmatrix} 0.43963 \\ 0.33174 \\ 0.44141 \\ -0.01239 \\ -0.2180 \\ -0.42081 \\ -0.01549 \\ -0.54075 \\ -0.01254 \end{bmatrix} \text{kA} \tag{4.38}$$

$$\dot{I}_N = \begin{bmatrix} 0.93109 \\ 0.56411 \\ 0.76264 \\ 0.05181 \\ -0.65728 \\ -0.83773 \\ 0.04707 \\ -0.8970 \\ 0.05501 \end{bmatrix} + j \begin{bmatrix} 0.43963 \\ 0.33174 \\ 0.44141 \\ -0.01239 \\ -0.2180 \\ -0.42081 \\ -0.01549 \\ -0.54075 \\ -0.01254 \end{bmatrix} \text{kA} \tag{4.39}$$

4.4.4.2 2 IEEE 10-machine 39-Bus System

In this section, the correctness and effectiveness of the proposed method is further verified based on the IEEE 10-machine, 39-bus system. The structure and road-loop labelled graph of the system is shown in Figure 4.12, where the branch labelled

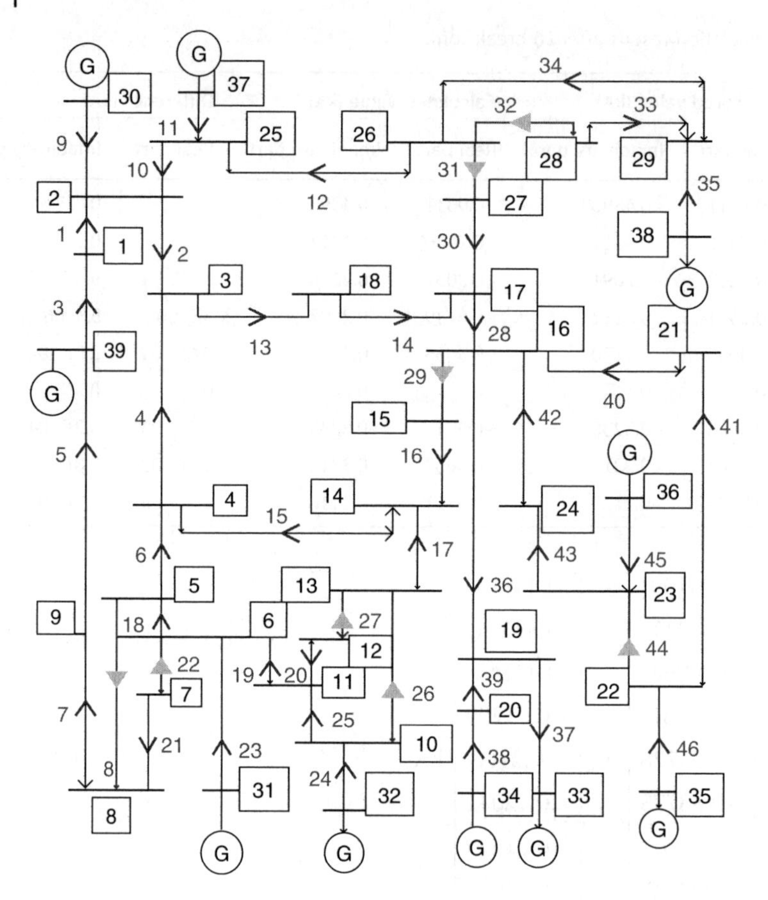

Figure 4.12 Road-loop labelled graph of IEEE 39-bus system.

with Δ and arrows is the link branch, and the other branches are tree branches. The minimum rated value of the branch current is 0.2693 + j0.1617 kA, thus $\varepsilon 3 = 0.021544 + j0.012933$ kA.

For a power network with a large number of nodes and complex wiring, topology fault identification could be conducted in local areas. In this section, the two local networks marked in Figure 4.12 are taken as an example.

A. Switch status identified as 'off' but is actually on

Figure 4.13 shows a local network in the IEEE 39-bus system, where branch 29 is switched off. Matrix **T** and matrix **B** of the local network are shown in Equation (4.40), the node injection currents can be calculated according to Equation (4.24), and are shown in Equation (4.41), and the calculation results are shown in Table 4.21.

It can be seen from Figure 4.13 that node 15 is the starting point of the road and is not connected to any other node in the local network, thus it should be verified alone according to Kirchhoff's current law (KCL). For this local network, $\varepsilon 1 = (7.988 + j6.904) \times 10^{-3}$ kA; for node 15, $\varepsilon 2 = (4.01 + j3.45.) \times 10^{-3}$ kA; for the other nodes, $\varepsilon 2 = (0.64 + j1.59) \times 10^{-2}$ kA. According to Table 4.21, the measured value of link branch 29 satisfies Principle 3, which is inconsistent with the

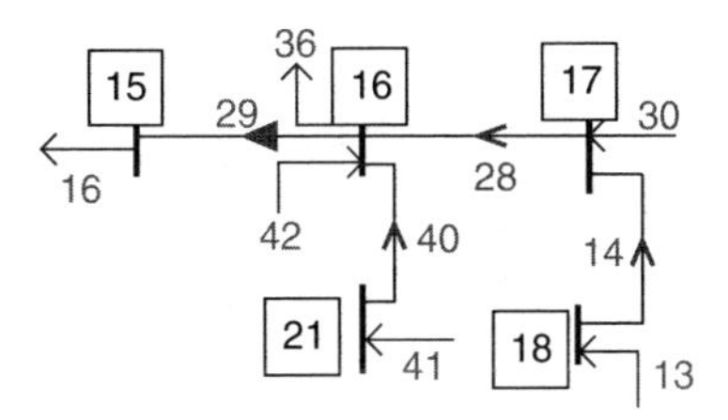

Figure 4.13 Road-loop labelled graph of local network in the IEEE 39-bus system.

Table 4.21 Calculation results of the network in Figure 4.13.

	Measured value (kA)		Calculated value (kA)		Difference (kA)	
Branch	**Real part**	**Imaginary part**	**Real part**	**Imaginary part**	**Real part**	**Imaginary part**
14	−0.7907	−0.3045	−2.3882	−1.4724	1.5975	1.16789
28	0.1698	−0.6476	−1.4278	−1.8155	1.5975	1.16784
29	1.5975	1.1679	0	0	1.5975	1.16789
40	0.8421	1.3808	0.8421	1.3808	9.5E−07	1.2E−06

Table 4.22 Calculation results after undesirable data of the current on branch 29 have been corrected.

	Measured value (kA)		Calculated value (kA)		Difference (kA)	
Branch	**Real part**	**Imaginary part**	**Real part**	**Imaginary part**	**Real part**	**Imaginary part**
14	−0.7907	−0.3045	−2.3882	−1.4724	1.5975	1.16789
28	0.1698	−0.6476	−1.4278	−1.8155	1.5975	1.16784
29	1.5975	1.1679	0	0	1.5975	1.16789
40	0.8421	1.3808	0.8421	1.3808	9.5E−07	1.2E−06

topology information. In addition, none of the branch currents in this loop satisfy Principle 1, thus the topology of branch 29 is falsely identified. In Equation (4.40), rectify the first row of matrix **B** as $[1 \;\; 1 \;\; 1 \;\; 0]$. After the topology is corrected, the calculation results are shown in Table 4.22. It can be seen that the branch currents satisfy Principle 1 and Principle 3, and the node injection currents satisfy Principle 2. Therefore, there are no more topology faults or undesirable data in this local network.

$$\left\{\begin{aligned} \mathbf{T} &= \begin{array}{c} \\ 14 \\ 28 \\ 29 \\ 40 \end{array}\!\!\begin{array}{c} \boxed{15}-\boxed{16} \;\; \boxed{17}-\boxed{18} \;\; \boxed{21} \\ \begin{bmatrix} 0 & 0 & 0 & 1 & 0 \\ 0 & 0 & 1 & 1 & 0 \\ 0 & 0 & 0 & 0 & 0 \\ 0 & 0 & 0 & 0 & 1 \end{bmatrix} \end{array} \\ \mathbf{B} &= \begin{array}{c} \\ 29 \\ 31 \\ 44 \end{array}\!\!\begin{array}{c} 14 \;\; 28 \;\; 29 \;\; 40 \\ \begin{bmatrix} 0 & 0 & 0 & 0 \\ -1 & 0 & 0 & 0 \\ 0 & 0 & 0 & -1 \end{bmatrix} \end{array} \end{aligned}\right. \tag{4.40}$$

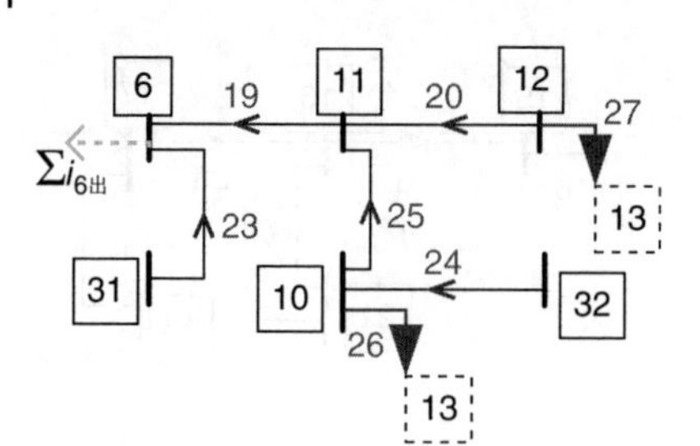

Figure 4.14 Road-loop labelled graph of local network with node 6 as the terminal point.

Table 4.23 Calculation results of the network in Figure 4.14.

	Measured value (kA)		Calculated value (kA)		Difference (kA)	
Branch	**Real part**	**Imaginary part**	**Real part**	**Imaginary part**	**Real part**	**Imaginary part**
19	0.2900	1.3187	1.2881	2.5379	0.9982	1.2192
20	−0.3671	−0.2724	−0.0364	−0.0142	0.3307	0.2583
23	2.1507	3.2173	2.1507	3.2173	4.8E−07	3.8E−07
24	1.9536	3.5179	1.9536	3.5180	0	0
25	0.6187	1.5961	1.2861	2.5570	0.6675	0.9609
26	0.6675	0.9609	0.6675	0.9609	0	0

$$\dot{I}_N = \begin{matrix} \boxed{15} \\ \boxed{16} \\ \boxed{17} \\ \boxed{18} \\ \boxed{21} \end{matrix} \begin{bmatrix} -2.00428 \\ 0.62621 \\ -0.41452 \\ 0.58427 \\ 0.80161 \end{bmatrix} + \mathrm{j} \begin{bmatrix} -1.72399 \\ -0.16704 \\ -1.67187 \\ 1.02429 \\ 1.98273 \end{bmatrix} \mathrm{kA} \tag{4.41}$$

B. Undesirable data at multiple locations

The other local network with node 6 as the terminal node is shown in Figure 4.14, matrix **T** and **B** of which are shown in Equation (4.42). According to Equation (4.24), the injection current to a node can be equated with the sum of branch currents connected with the node (if the node is in a loop, then the effect of the loop current should be eliminated). The equivalent node injection currents of this local network are shown in Equation (4.43). Link branch currents $i_{26} = 0.6675 + \mathrm{j}0.9609\,\mathrm{kA}$, $i_{27} = -0.2943 - \mathrm{j}0.2441\,\mathrm{kA}$. The calculation results are shown in Table 4.23.

According to Equation (4.43) and Table 4.21, the threshold values $\varepsilon 1 = (6.452 + \mathrm{j}9.652) \times 10^{-3}\,\mathrm{kA}$, $\varepsilon 2 = (4.3 + \mathrm{j}6.4) \times 10^{-2}\,\mathrm{kA}$. The total output current of node 6 $\sum i_{6\mathrm{out}} = 2.5127 + \mathrm{j}4.5255\,\mathrm{kA}$. All branch currents satisfy Principle 3, thus there is no topology fault. According to KCL, the absolute value of the difference between $\sum i_{6\mathrm{out}} + i_{26} + i_{27}$ and the sum of the node injection currents is bigger than $\varepsilon 2$. The currents of branch 19, 20 and 25 do not satisfy Principle 1. Since node 10 is neither load node nor source node, the injection current to node 12 will be different. Considering that the link branch current may also be

Table 4.24 Calculation results before undesirable data of the injection current to node 12 are corrected.

	Measured value (kA)		Calculated value (kA)		Difference (kA)	
Branch	Real part	Imaginary part	Real part	Imaginary part	Real part	Imaginary part
19	0.2900	1.3187	0.2900	1.3187	0	0
20	−0.3671	−0.2724	−0.3671	−0.2725	3.7E−06	3.8E−05
23	2.1507	3.2173	2.15067	3.2173	4.8E−07	3.8E−07
24	1.9536	3.5179	1.9536	3.5179	0	0
25	0.6187	1.5961	0.6187	1.5961	4.3E−06	3.2E−06
26	1.3349	1.9218	1.3349	1.9218	0	0
19	0.2900	1.3187	0.2900	1.3187	0	0

incorrect, the node injection current cannot be corrected directly according to KCL. Comparing the factors that result in line current difference, the difference of injection current to node 12 equals the current difference of branch 19 minus the current difference of branch 25. Use this difference to modify the node injection current, thus $\dot{i}_{12} = -0.6614 - j0.5166\text{kA}$. The calculation results are shown in Table 4.24.

$$\begin{cases} \mathbf{T} = \begin{array}{c} \\ 19 \\ 20 \\ 23 \\ 24 \\ 25 \\ 26 \\ 27 \end{array} \begin{array}{c} \boxed{6}-\boxed{10}\ \ \boxed{11}-\boxed{12}\ \ \boxed{21}-\boxed{32} \\ \begin{bmatrix} 0 & 1 & 1 & 1 & 0 & 1 \\ 0 & 0 & 0 & 1 & 0 & 0 \\ 0 & 0 & 0 & 0 & 1 & 0 \\ 0 & 0 & 0 & 0 & 0 & 1 \\ 0 & 1 & 0 & 0 & 0 & 1 \\ 0 & 0 & 0 & 0 & 0 & 0 \\ 0 & 0 & 0 & 0 & 0 & 0 \end{bmatrix} \end{array} \\ \mathbf{B} = \begin{array}{c} \\ 26 \\ 27 \end{array} \begin{array}{c} 19\ \ 20\ \ 23\ \ 24\ \ 25\ \ 26\ \ 27 \\ \begin{bmatrix} -1 & 0 & 0 & 0 & -1 & 1 & 0 \\ -1 & -1 & 0 & 0 & 0 & 0 & 1 \end{bmatrix} \end{array} \end{cases} \tag{4.42}$$

$$\dot{I}_N = \begin{array}{c} \boxed{6} \\ \boxed{10} \\ \boxed{11} \\ \boxed{12} \\ \boxed{31} \\ \boxed{32} \end{array} \begin{bmatrix} 0.070275 \\ 0 \\ 0.038359 \\ -0.33069 \\ 2.15067 \\ 1.95360 \end{bmatrix} + j \begin{bmatrix} -0.010483 \\ 0 \\ -0.00495 \\ -0.25828 \\ 3.21729 \\ 3.51794 \end{bmatrix} \text{kA} \tag{4.43}$$

It can be seen from Table 4.24 that none of the tree branches connected to link branch 26 satisfy Principle 1, and the node injection currents of the local network do not satisfy Principle 2, and the current differences of the two are the same.

Table 4.25 Calculation results after undesirable data of the injection current to node 12 have been corrected.

	Measured value (kA)		Calculated value (kA)		Difference (kA)	
Branch	Real part	Imaginary part	Real part	Imaginary part	Real part	Imaginary part
19	0.2900	1.3187	1.2881	2.5379	0.9982	1.2192
20	−0.3671	−0.2724	−0.0364	−0.0142	0.3307	0.2583
23	2.1507	3.2173	2.1507	3.2173	4.8E−07	3.8E−07
24	1.9536	3.5179	1.9536	3.5180	0	0
25	0.6187	1.5961	1.2861	2.5570	0.6675	0.9609
26	0.6675	0.9609	0.6675	0.9609	0	0
27	−0.2943	−0.2441	−0.2943	−0.2441	3.7E−06	1.8E−06

Table 4.26 Calculation results after branch 24 breaks off.

	Measured value (kA)		Calculated value (kA)		Difference (kA)	
Branch	Real part	Imaginary part	Real part	Imaginary part	Real part	Imaginary part
19	0.2900	1.3187	−1.6636	−2.2093	0.6674	0.9609
20	−0.1835	−0.1362	−0.3671	−0.2725	0.1835	0.1363
23	2.1507	3.2173	2.1507	3.2173	4.8E−07	3.8E−07
24	1.9536	3.5179	1.9536	3.5179	0	0
25	0.6187	1.5961	−1.3349	−1.9218	0.6674	0.9609
26	1.3349	1.9218	1.3349	1.9218	0	0
27	−0.2943	−0.2441	−0.2943	−0.2441	3.7E−06	1.8E−06

Thus, the current of branch 26 is incorrect, which is corrected with the current difference of branch 19, i.e. $i_{26} = 1.33495 + j1.92183\,\text{kA}$. Then, using the corrected value for calculation, the results are shown in Table 4.25. It can be seen that all the electrical variables satisfy Principle 1 to Principle 3, thus there are no more topology faults or undesirable data.

C. Switch status identified as 'off' but is actually on with undesirable data

In Figure 4.14, suppose the switch status shows that branch 24 is switched off, then matrices **T** and **B** of the local network are as shown in Equation (4.44), the node injection currents are as shown in Equation (4.43), $i_{26} = 1.3349 + j1.9218\,\text{kA}$, $i_{27} = -0.2943 - j0.2441\,\text{kA}$. According to the road matrix and loop matrix, the branch currents can be calculated, and the results are as shown in Table 4.26.

It can be seen from Table 4.26 that branch 24 satisfies Principle 3, and none of the branch currents behind the terminal node of branch 24 satisfy Principle 1, thus the topology information of branch 24 is incorrect. After the topology is corrected,

Table 4.27 Calculation results after topology information of branch 24 is corrected.

	Measured value (kA)		Calculated value (kA)		Difference (kA)	
Branch	**Real part**	**Imaginary part**	**Real part**	**Imaginary part**	**Real part**	**Imaginary part**
19	0.2900	1.3187	0.2900	1.3187	5.4E−06	3.2E−06
20	−0.1835	−0.1362	−0.3671	−0.2725	0.1835	0.1362
23	2.1507	3.2173	2.1507	3.2173	4.8E−07	3.8E−07
24	1.9536	3.5179	1.9536	3.5179	0	0
25	0.6187	1.5961	0.6187	1.5961	1.2E−06	1E−08
26	1.3349	1.9218	1.3349	1.9218	0	0
27	−0.2943	−0.2441	−0.2943	−0.2441	3.7E−06	1.8E−06

matrices **T** and **B** are as shown in Equation (4.42), and the updated calculated values are as shown in Table 4.27.

$$
\begin{cases}
\mathbf{T} = \begin{array}{c} 19\\ 20\\ 23\\ 24\\ 25\\ 26\\ 27 \end{array}
\overset{\begin{array}{cccccc} \boxed{6} & \boxed{10} & \boxed{11} & \boxed{12} & \boxed{31} & \boxed{32} \end{array}}{\begin{bmatrix}
0 & 1 & 1 & 1 & 0 & 0\\
0 & 0 & 0 & 1 & 0 & 0\\
0 & 0 & 0 & 0 & 1 & 0\\
0 & 0 & 0 & 0 & 0 & 1\\
0 & 1 & 0 & 0 & 0 & 0\\
0 & 0 & 0 & 0 & 0 & 0\\
0 & 0 & 0 & 0 & 0 & 0
\end{bmatrix}} \\
\mathbf{B} = \begin{array}{c} 26\\ 27 \end{array}
\overset{\begin{array}{ccccccc} 19 & 20 & 23 & 24 & 25 & 26 & 27 \end{array}}{\begin{bmatrix}
-1 & 0 & 0 & 0 & -1 & 1 & 0\\
-1 & -1 & 0 & 0 & 0 & 0 & 1
\end{bmatrix}}
\end{cases} \tag{4.44}
$$

It can be seen that the current of branch 20 does not satisfy Principle 1, thus there are undesirable data which could be corrected according to the calculated value.

4.5 Summary

In this chapter, a substation topology update scheme and inter-station topology update scheme are introduced. When the switch status changes, the part of the topology with changes is modified based on the radius search method, thus slow updating due to topology reformation can be avoided. The proposed schemes have obvious advantages concerning the amount of computation, and can guarantee fast tracking of the switch status. Compared with existing topology analysis methods, the proposed methods can improve the speed and accuracy of power network topology analysis. Also, a new method for topology fault identification is introduced. Verification results show that the proposed method is not affected by topology faults or undesirable data with high accuracy and reliability of identification.

References

1 Yehsakul, P. D. and Dabbaghchi, I. (1995) A topology-based algorithm for tracking network connectivity, *IEEE Trans. Power Syst*, **10** (1), 339–345.
2 Korres, G. N. and Katsikas, P. J. (2002) Identification of circuit breaker statuses in WLS state estimator, *IEEE Trans. Power Syst*, **17** (3), 818–825.
3 Singh, D., Pandey, J. P. and Chauhan, D. S. (2005) Topology identification, bad data processing, and state estimation using fuzzy pattern matching, *IEEE Trans. Power Del.*, **20** (3), 1570–1579.
4 Krstulovic, J., Miranda, V., Simões Costa, A. J. A. and Pereira, J. (2013) Towards an auto-associative topology state estimator, *IEEE Trans. Power Syst*, **28** (3), 3311–3318.
5 Goderya, F., Metwally, A. A. and Mansour, O. (1980) Fast detection and identification of islands in power networks, *IEEE Transactions on Power Apparatus and Systems*, **99** (1), 217–221.
6 Munkres, J. R. (1975) *Topology: A First Course*, Prentice-Hall, Inc, New Jersey.
7 Armstrong, M. A. (1983) *Basic Topology*, Springer, New York.
8 Diestel, R. (2000) *Graph Theory*, Springer-Verlag, NewYork.
9 Armstrong, M. A. (1979) *Basic Topology*, McGraw-Hill, New York.

5

Substation Area Protection

5.1 Introduction

Due to constraints such as microcomputer protection hardware, CPU processing ability and specification, traditional relay protection is based on independent and decentralized configuration. The protection devices can only obtain information such as the voltage, current, circuit breaker and switch about a single space, which makes it difficult for protection to strike a balance between selectivity, speed and sensitivity [1,2]. At the same time, in order to meet the requirements of protection, coordination between multiple protective devices is essential, where problems such as difficult value setting, long operational time of backup protection, complicated criteria and difficult maintenance are common. With the continuous development of computer and communication technology, substation backup protection is no longer limited to single spaces or local information, but could acquire information of the whole station to realize substation area protection. Research on substation area protection [3,4] is of great significance to ensuring the safe operation of a power grid.

In this chapter, first a trans-space substation area differential protection scheme based on electrical variable information is introduced. Through coordination between multi-level trans-space differential regions, the substation area protection is fault tolerant and could isolate the fault quickly and selectively, thus improving the performance of substation component protection. To reduce the dependence of a substation area protection scheme on information synchronization and communication level, a substation area protection scheme based on logical variable information is introduced from the perspective of fully fusing the fault characteristic information contained in the protection operation signal. The proposed scheme could deal with information loss and error, with good adaptability and fault tolerance.

5.2 Substation Area Protection Based on Electrical Information

The decision centre in centralized substation area protection could acquire more information and locate the fault from the whole station level [5]. In this section, a

Hierarchical Protection for Smart Grids, First Edition. Jing Ma and Zengping Wang.

trans-space substation area differential protection scheme in the centralized substation protection framework is introduced, which is completely independent of existing protection. If a fault occurs within the substation area and the primary protection does not operate, the proposed scheme could isolate the fault quickly and selectively, effectively improving the backup protection performance of high-voltage, medium-voltage and low-voltage components. Problems in traditional stage backup protection such as difficult coordination and long operational time delay are solved. In addition, it is simple, reliable and easy to implement.

5.2.1 Substation Area Regionalization

Take a typical substation and the adjacent connection lines, for example. As shown in Figure 5.1, the voltage levels of three sides of the transformer are 220 kV, 110 kV and 35 kV. Substation buses are Bus03, Bus04, Bus07, Bus08, Bus11 and Bus12. The substation is connected to the adjacent substation via line L1–L8. Load1–Load3 are equivalent loads. G1–G8 are equivalent power sources. CB01–CB33 are circuit breakers. According to the scope of effect, three types of finite overlapping regions are defined, i.e. the bus–line fusion region, bus–transformer fusion region and multi-component coordinated fusion region [6,7]. Taking Figure 5.1, for example, the principle of division of these regions is illustrated.

1) To ensure that any circuit breaker in the substation is included in at least one region, the three types of finite overlapping regions all contain two or more components. The bus–line fusion region contains a substation outgoing line and the connected substation bus, e.g. the bus–line fusion region of line L1 contains L1, Bus03 and boundary circuit breakers CB01, CB05, CB06, i.e. the Region C in Figure 5.1.

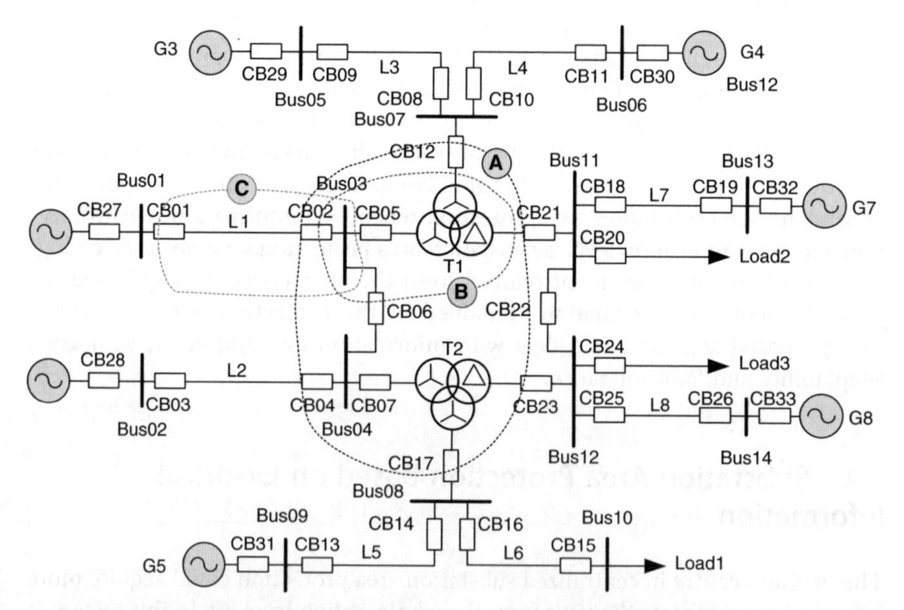

Figure 5.1 Wiring diagram of a typical substation and adjacent power grid.

2) The bus–transformer fusion region contains a substation transformer and the connected substation bus, e.g. the bus–transformer fusion region of the high-voltage side of transformer T1 contains T1, Bus02 and boundary circuit breakers CB02, CB21, CB12 (CB06 is switched off and is not counted as a boundary circuit breaker), i.e. the Region B in Figure 5.1. In this case, the sectional circuit breaker is switched off, or there is no sectional circuit breaker on this voltage side of the transformer.
3) When the sectional circuit breaker is switched on, according to the protection range of the bus–transformer fusion region, sectional circuit breaker CB06 is the boundary circuit breaker (not in the region). Meanwhile, CB06 is also the boundary circuit breaker of the bus–line fusion region (not in the region). Since the boundary circuit breaker is not included in two regions, which does not meet the division principle, the multi-component coordinated fusion region is designed, which covers two bus–transformer fusion regions with the same voltage level, and contains two substation transformers and the corresponding buses with the same voltage level. For example, the high-voltage side substation multi-component coordinated fusion region contains T1, T2, Bus03, Bus04 and boundary circuit breakers CB02, CB04, CB17, CB23, CB21 and CB12, i.e. the Region A in Figure 5.1. When the sectional circuit breaker in the multi-component coordinated fusion region is tripped, two bus–transformer fusion regions will result.

According to the above division principle, in the substation shown in Figure 5.1, there are two multi-component coordinated fusion regions, the names, protection objects and boundary circuit breaker numbers of which are shown in Table 5.1. There are six bus–transformer fusion regions, the names, protection objects and boundary circuit breaker numbers of which are shown in Table 5.2. The number of bus–line fusion regions is the same as the number of lines, and bus–line fusion regions are named after the corresponding protected line as C1, C2...C8. The flowchart of the substation area backup protection scheme is shown in Figure 5.2. The protection scheme is composed of three parts: the bus–line fusion region, the bus–transformer fusion region and the multi-component coordinated fusion region. Whether the current differential corresponding to the boundary circuit breaker of a particular fusion region satisfies the startup conditions will determine whether the fusion region starts or not. There are two time delays in the scheme, $t1$ and $t2$, and $t1 = 0.5$ s is the operational time of backup protection, while $t2$ is bigger than $t1$ by Δt, $t2 = 1.0$ s. It can be seen from Figure 5.1 that the protection range of the multi-component coordinated fusion region includes

Table 5.1 Protection object and boundary circuit breaker numbers for the substation area multi-component coordinated fusion region.

Number	Protection object	Boundary circuit breaker number
A1	T1, T2, Bus03, Bus04	02, 04, 12, 17, 21, 23
A2	T1, T2, Bus11, Bus12	05, 07, 12, 17, 18, 20, 24, 25

Table 5.2 Protection object and boundary circuit breaker numbers for the bus–transformer fusion region.

Number	Protection object	Boundary circuit breaker number
B1	T1, Bus03	02, 12, 21
B2	T1, Bus07	05, 08, 10, 21
B3	T1, Bus11	05, 12, 18, 20
B4	T2, Bus04	04, 17, 23
B5	T2, Bus08	07, 14, 16, 23
B6	T2, Bus12	07, 17, 24, 25

the protection range of the bus–transformer fusion region, and the protection ranges of the bus–transformer fusion region and the bus–line fusion region overlap with each other. Thus, fault components in the substation area may be cleared according to the inclusion relationship and overlapping relationship between these fusion regions.

5.2.1.1 Protection Operation Analysis and Scheme for the Bus–Line Fusion Region

The bus–line fusion region C1 is shown in Figure 5.3. If C1 starts, then the fault may be at Bus03 or on L1. If the fault is at Bus03, then circuit breakers CB02, CB05 and CB06 are tripped. If the fault is on L1, then circuit breakers CB01 and CB02 are tripped. Whether the fault is at Bus03 or on L1, circuit breaker CB02 needs to be tripped. If CB02 is tripped, then line L1 cannot provide power supply to the substation, nor can the substation transmit power to other stations via L1. Even if the fault is at the bus, tripping CB01 will not cause any effect. Therefore, when the bus–line fusion region operates and trips the circuit breakers on two line ends of the line in the region after time delay $t1$, if the fault is on the line, then the fault is cleared; if the fault is at the bus, then one of the bus coupler circuit breakers is tripped.

5.2.1.2 Protection Operation Analysis and Scheme for the Bus–Transformer Fusion Region

Bus–transformer fusion region B1 is shown in Figure 5.4, when the sectional circuit breaker is switched off. If B1 starts, then the fault may be at Bus03 or T1. If the fault is at Bus03, then circuit breakers CB02 and CB05 are tripped. If the fault is at T1, then circuit breakers CB05, CB12 and CB21 are tripped. Whether the fault is at Bus03 or T1, circuit breaker CB05 needs to be tripped. If the fault is at the transformer and CB02 is tripped, or if the fault is at the bus and CB12 and CB21 are tripped, then the fault clearance range is enlarged. Therefore, the first bus–transformer fusion region operates and trips the bus–transformer coupler circuit breaker in the region after time delay $t1$, and then through cooperation between the fusion regions, the fault can be cleared effectively. This will be discussed in detail in Section 5.2.2.

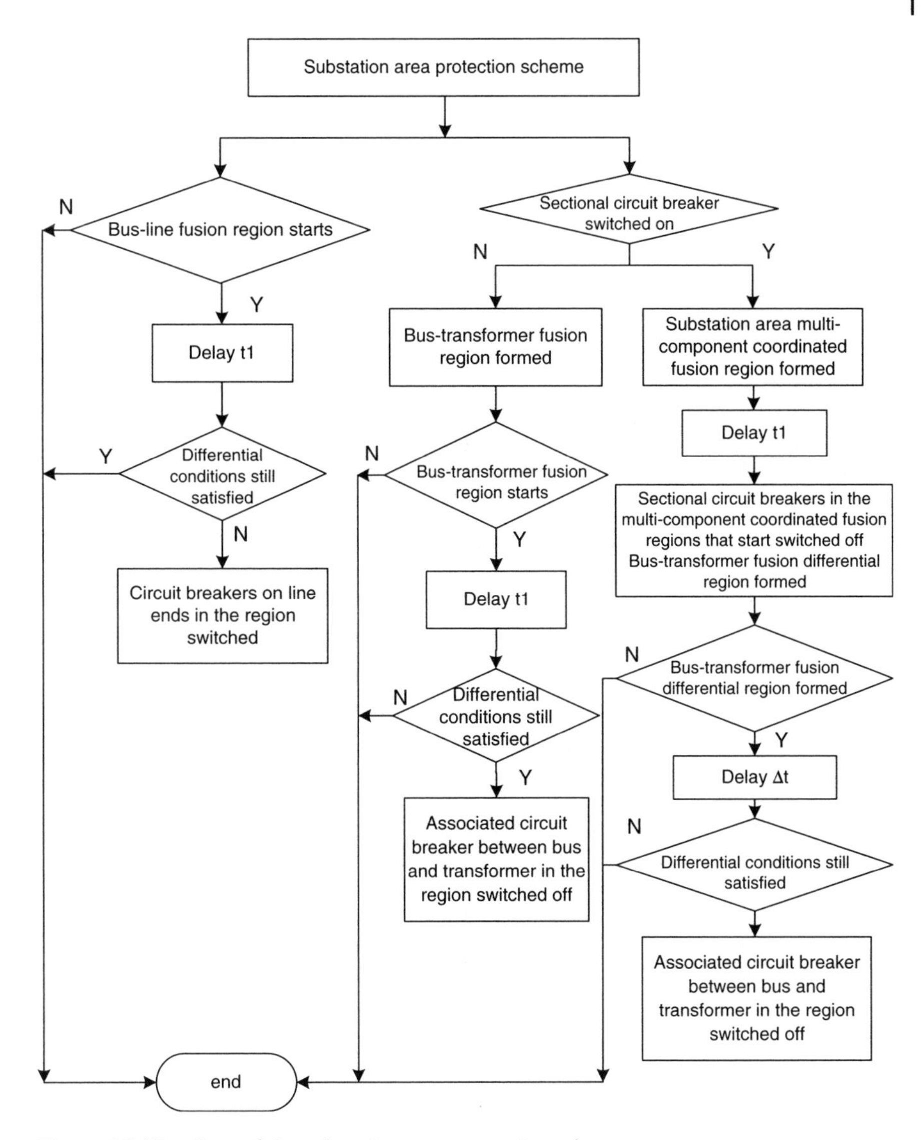

Figure 5.2 Flowchart of the substation area protection scheme.

Figure 5.3 Operational scheme of bus–line fusion region C1.

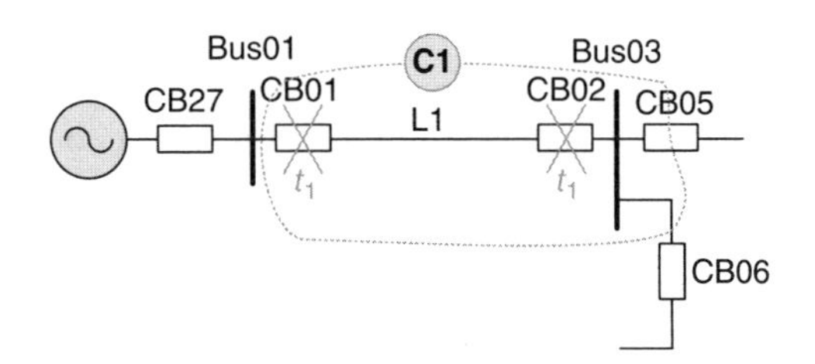

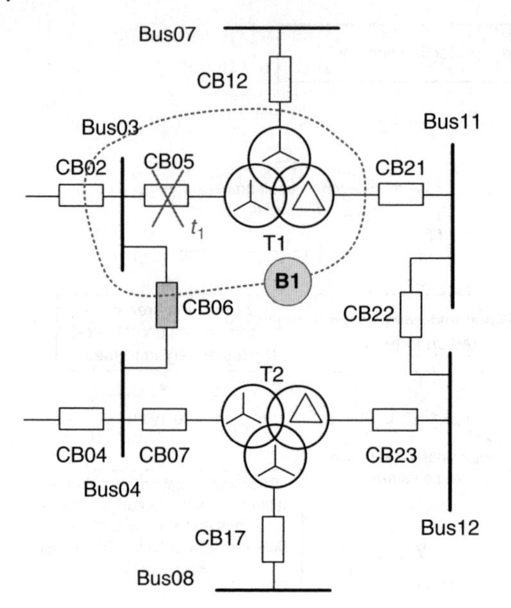

Figure 5.4 Operational scheme of bus–transformer fusion region B1.

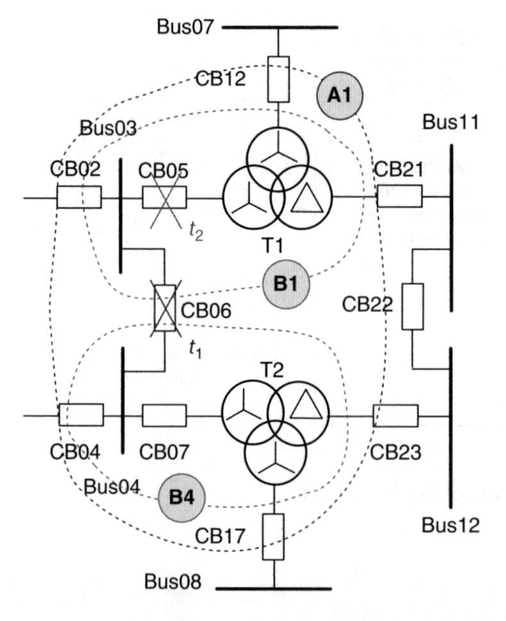

Figure 5.5 Operational scheme of substation area multi-component coordinated fusion region A1.

5.2.1.3 Protection Operation Analysis and Scheme for the Substation Area Multi-Component Coordinated Fusion Region

The substation area multi-component coordinated fusion region A1 is shown in Figure 5.5, when the sectional circuit breaker is switched on. If A1 starts, then the fault may be at T1, T2, Bus02 or Bus04. For any bus fault, CB06 needs to be tripped; and for a transformer fault, the circuit breakers on three sides of the transformer need to be tripped. If CB06 is tripped, then the connection between the power source and the load on three sides of the fault transformer and the non-fault transformers is cut off, which will cause certain effects, but the heavy overload of the non-fault transformer can be avoided. In addition, substation area protection has the function of a 'local backup automatic switch'. In this case, compared with the effects of tripping CB06, the protection of a fault transformer refusing to operate will cause more serious problems. Therefore, the multi-component coordinated fusion region operates and trips the sectional circuit breaker in the region after time delay $t1$.

In Figure 5.5, after A1 operates and trips CB06, two bus–transformer fusion regions B1 and B4 are formed. Then, according to the protection scheme of the bus–transformer fusion region, we determine whether each bus–transformer fusion region satisfies the startup conditions. If yes, then the bus–transformer coupler circuit breaker in the region is tripped after another time delay Δt, i.e. after time delay $t2$ ($t2 = t1 + \Delta t$).

5.2.2 Typical Fault Cases

5.2.2.1 Transformer Fault

A. Sectional Circuit Breakers All Switched Off

Suppose a fault occurs at T1 and the primary protection of the transformer refuses to operate, thus the transformer fault is not cleared, as shown in Figure 5.6. If high-voltage sectional circuit breaker CB06 and low-voltage sectional circuit breaker CB22 are both switched off, then bus–transformer

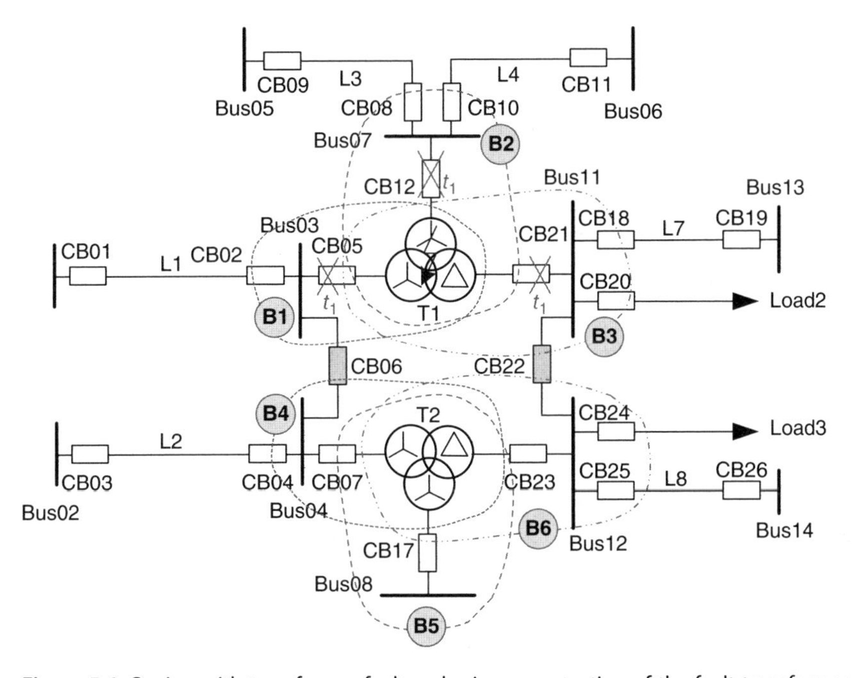

Figure 5.6 Coping with transformer fault and primary protection of the fault transformer refusing to operate (sectional circuit breakers all switched off).

fusion regions B1–B6 can be formed. Three of the fusion regions, B1, B2 and B3, will start and operate after time delay $t1$ to trip the bus–transformer coupler circuit breaker CB05 in B1, the bus–transformer coupler circuit breaker CB12 in B2, and the bus–transformer coupler circuit breaker CB21 in B3. Thus, the transformer fault may be cleared after time delay $t1$ without enlarging the fault clearance range.

B. Sectional Circuit Breakers Not All Switched Off

If two sectional circuit breakers are not both switched off, suppose high-voltage sectional circuit breaker CB06 is switched on and low-voltage sectional circuit breaker CB22 is switched off, as shown in Figure 5.7. In this case, multi-component coordinated fusion region A1 and bus–transformer fusion regions B2–B6 are formed. Among them, A1, B2 and B3 will start and operate after time delay $t1$ to trip the sectional circuit breaker CB06 in A1, the bus–transformer coupler circuit breaker CB12 in B2 and the bus–transformer coupler circuit breaker CB21 in B3. Then A1 is divided into two bus–transformer fusion regions B1 and B4. After another time delay Δt, B1 will operate to trip the bus–transformer coupler circuit breaker CB05 in the region. Thus, the transformer fault may be cleared after time delay $t2$. If two sectional circuit breakers are switched on, then on the low-voltage side, multi-component coordinated fusion region A2 will be formed first, which will operate in a similar way to A1, and finally the transformer fault will also be cleared after time delay $t2$.

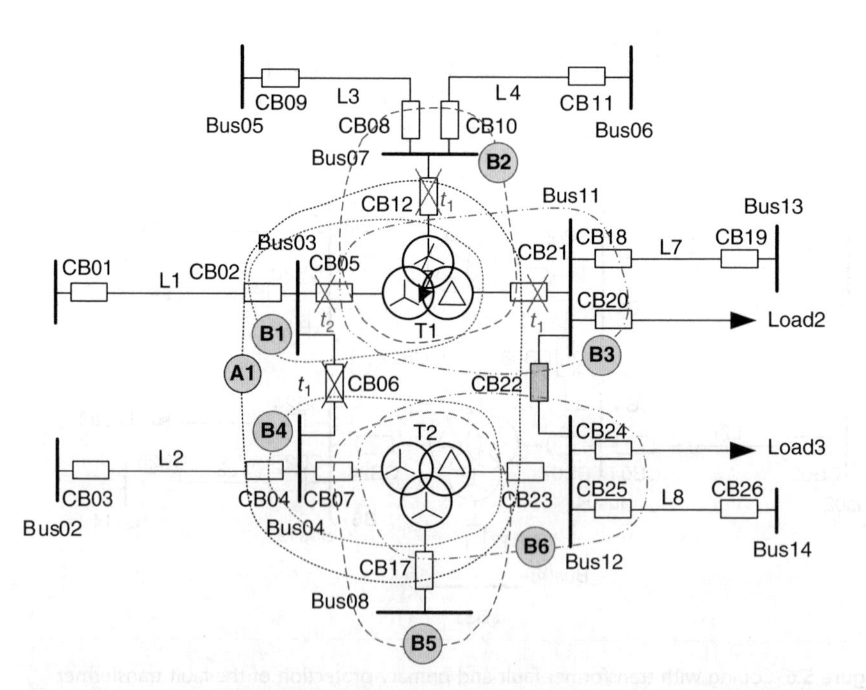

Figure 5.7 Coping with transformer fault and primary protection of the fault transformer refusing to operate (sectional circuit breakers not all switched off).

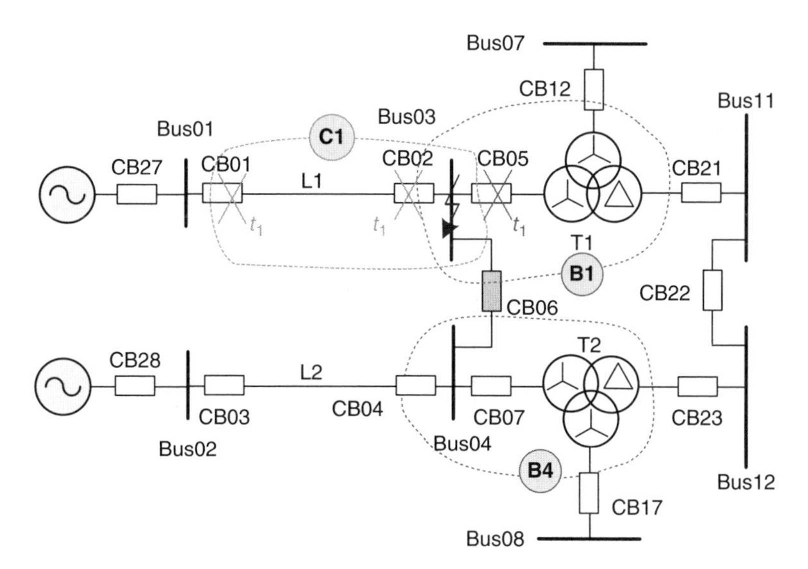

Figure 5.8 Coping with bus fault and primary protection of the fault bus refusing to operate (sectional circuit breaker switched off).

5.2.2.2 Bus Fault

A. Sectional Circuit Breaker Switched Off

Suppose a fault occurs at Bus03 and the corresponding bus primary protection refuses to operate, thus the bus fault is not cleared, as shown in Figure 5.8. If the sectional circuit breaker CB06 connected to the faulty bus is switched off, then bus–transformer fusion regions B1 and B4 are formed. After time delay $t1$, B1 will operate to trip CB05, and bus–line fusion region C1 will trip CB01 and CB02 on two line ends of the line in the region. Thus, finally the fault at Bus03 will be cleared after time delay $t1$. The tripping of CB01 does not enlarge the fault clearance range.

B. Sectional Circuit Breaker Switched On

If the sectional circuit breaker CB06 connected to the faulty bus is switched on, as shown in Figure 5.9, then multi-component coordinated fusion region A1 is formed. After time delay $t1$, A1 will operate to trip sectional circuit breaker CB06, and bus–line fusion region C1 will trip CB01 and CB02 on two line ends of the line in the region. Then A1 is divided into two bus–transformer fusion regions B1 and B4. After another time delay Δt, B1 will operate to trip the bus–transformer coupler circuit breaker CB05 in the region. Thus, the bus fault will be cleared after time delay $t2$. The tripping of CB01 does not enlarge the fault clearance range.

5.2.2.3 Line Fault

As shown in Figure 5.10, if a fault occurs on line L1 and the primary protection refuses to operate, in this case only bus–line fusion region C1 will start and, after time delay $t1$, will operate to trip CB01 and CB02 on two line ends of L1 in the region. Finally, the line fault will be cleared after time delay $t1$ without enlarging the fault clearance range.

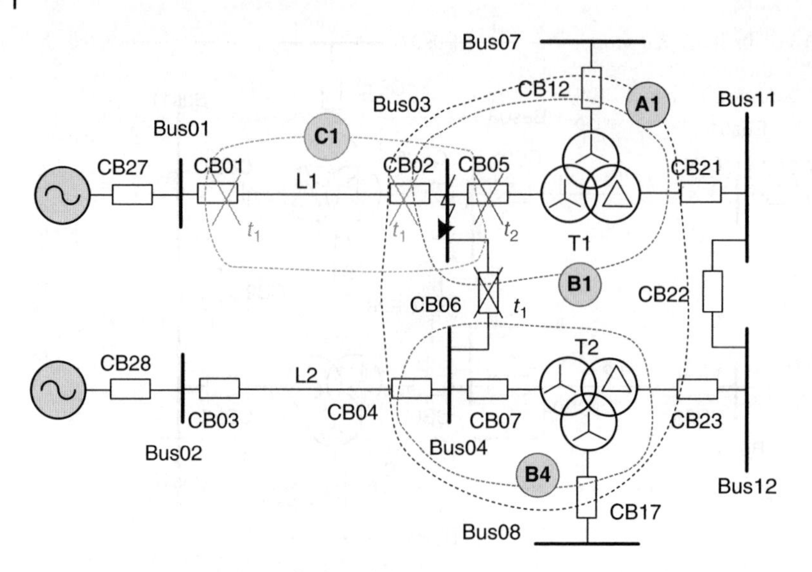

Figure 5.9 Coping with bus fault and the primary protection of the faulty bus refusing to operate (sectional circuit breaker switched on).

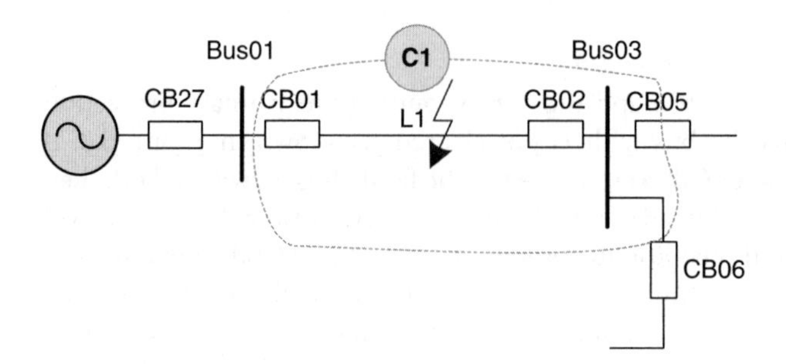

Figure 5.10 Coping with line fault and the primary protection of the faulty line refusing to operate.

If there are multiple lines connected to the bus and a fault occurs on one of them, then only the bus–line fusion region where the faulty line is will start and operate to trip the circuit breakers on two line ends of the line in the region and clear the fault.

5.2.3 Scheme Performance Analysis

The proposed substation area protection scheme is completely independent of existing protection schemes in the acquisition of information, data processing and tripping. It is not dependent on the information on primary protection, and the time delay of tripping is longer than that of primary protection. Therefore, the proposed scheme is based on when primary protection fails to clear the fault, as an effective complement to primary protection refusing to operate, at the same time improving the performance of backup protection.

1) In the configuration of traditional backup protection, the backup protection of medium-voltage and low-voltage components is usually stage over-current protection, and the backup protection of high-voltage components is usually distance protection [8]. In this section, considering that substation area backup protection could directly acquire electrical variables at multiple locations, the protection scheme based on the current differential principle is used as backup protection, which simplifies the setting and cooperation of backup protection and ensures the selectivity of protection.
2) Traditional backup protection applies stage cooperation [9], the operational time of which is usually long. Take the power grid in Figure 5.1, for example; if traditional backup protection is applied, the shortest operational time of local backup protection for line L7 is 0.5 s, and the shortest operational time of remote backup protection is 1.0 s. The shortest operational time of local backup protection for the transformer is 1.0 s. In a real power system, since there are multiple levels in the outgoing transmission line, the operational time of backup protection could be 1.5 s or longer – even 3 s. For the substation area protection scheme introduced in this section, the operational time of the line fault is 0.5 s, and the longest operational time of the transformer fault and bus fault is 1.0 s. Thus, the speed of protection is guaranteed.

5.3 Substation Area Protection Based on Operating Signals

In this section, a substation area backup protection scheme based on the current protection information fusion factor is introduced. The scheme is not affected by the system operating mode, and could identify the fault components quickly and accurately in the case of asynchronous information, without the interaction of analogue information. Even in the case of information loss and error, it is easily adaptable and fault-tolerant.

5.3.1 Setting Principle of Adaptive Current Protection

The equivalent structure of a substation area and the names of protection components are shown in Figure 5.11. T_g, T_z and T_d are the protection of the high-voltage side, medium-voltage side and low-voltage side of transformer. g_m, z_m and d_m ($m = 2a$, $a = 1, 2, 3\ldots$) are the protection on the high-voltage,

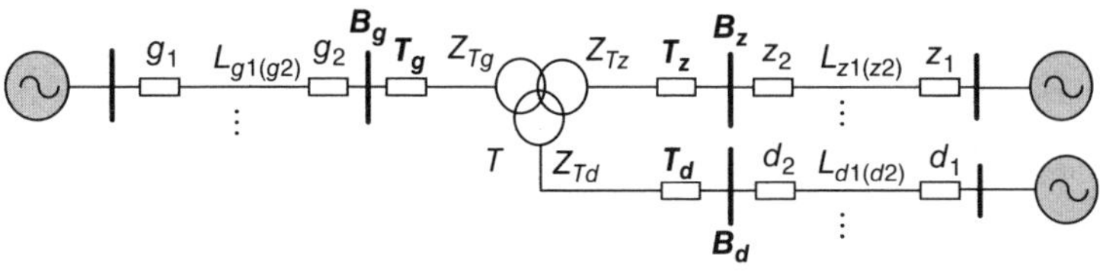

Figure 5.11 Structure of the substation area.

medium-voltage and low-voltage outgoing lines of the substation. g_n, z_n and d_n ($n = 2a-1, a = 1, 2, 3...$) are the protection on the opposite side of the high-voltage, medium-voltage and low-voltage outgoing lines of the substation. The set of protection names is $\boldsymbol{bh} = \{T_g, T_z, T_d, g_m, z_m, d_m, g_n, z_n, d_n\}$, and the set of component names is $\boldsymbol{yj} = \{T, L_{gn}, L_{zn}, L_{dn}\}$.

Adaptive protection could change the protection performance, characteristic and setting value according to a change in system operating mode and fault state, thus effectively improving the performance of relay protection and contributing to the safe and stable operation of the power system. Therefore, it is of great theoretical and practical value to conduct research on adaptive protection. In this section, based on the principle of adaptive current protection, the protection setting value of each component is calculated, and $Z_s.bh$ is defined as the system impedance on the back side of protection bh in the current system operating mode.

5.3.1.1 Transformer Setting Method

A. Transformer Zone-I

If transformer zone-I operates, it means that the fault has occurred at the transformer. Thus, the setting value of transformer zone-I $\dot{\boldsymbol{I}}^{\mathrm{I}}_{\mathrm{set.T}_i}$ should be smaller than the short circuit current at the protection device in the case of transformer fault, i.e.

$$\dot{I}^{\mathrm{I}}_{\mathrm{set.T}_i} = K_{\mathrm{rel}} \max\left\{ \frac{K_{\mathrm{f}}\dot{E}}{K_{\mathrm{bT}_j-\mathrm{T}_i}\left(Z_{m.\mathrm{T}_j} + Z_{\mathrm{T}_j}\right)}, \frac{K_{\mathrm{f}}\dot{E}}{K_{\mathrm{bT}_k-\mathrm{T}_i}\left(Z_{m.\mathrm{T}_k} + Z_{\mathrm{T}_k}\right)} \right\} \tag{5.1}$$

where K_{rel} is the reliability coefficient. K_{f} is the fault type coefficient. $\dot{E}$ ($\dot{E} = \dot{U}_{\mathrm{m}} + \dot{I}_{\mathrm{m}}\dot{Z}_S$) is the equivalent phase emf of the system. The corresponding relationships of numbers i, j and k are shown in Table 5.3. Z_{T_i}, Z_{T_j} and Z_{T_k} are the impedances of the corresponding sides of the transformer. $Z_{m.\mathrm{T}_j}$ is the equivalent impedance of the i-side and k-side systems to Z_{T_j}, which equals the parallel connection of $Z_{S.\mathrm{T}_i} + Z_{\mathrm{T}_i}$ and $Z_{S.\mathrm{T}_k} + Z_{\mathrm{T}_k}$. $Z_{m.\mathrm{T}_k}$ is the equivalent impedance of the i-side and j-side systems to Z_{T_k}, which equals the parallel connection of $Z_{S.\mathrm{T}_i} + Z_{\mathrm{T}_i}$ and $Z_{S.\mathrm{T}_j} + Z_{\mathrm{T}_j}$. $K_{\mathrm{bT}_j-\mathrm{T}_i}$ and $K_{\mathrm{bT}_k-\mathrm{T}_i}$ are branch coefficients.

Table 5.3 Corresponding relationships between numbers *i*, *j* and *k*.

Number	*i*	*j*	*k*
Value 1	g	z	d
Value 2	z	d	g
Value 3	d	g	z

B. Transformer Zone-II

Transformer zone-II should ensure that the protection operates reliably when a fault occurs at the transformer; the setting value $\dot{I}^{\mathrm{II}}_{\mathrm{set.T}_i}$ is

$$\dot{I}^{\mathrm{II}}_{\mathrm{set.T}_i} = \frac{\min\left\{\dfrac{K_f\dot{E}}{K_{\mathrm{bT}_j-\mathrm{T}_i}\left(Z_{m.\mathrm{T}_j}+Z_{\mathrm{T}_j}\right)}, \dfrac{K_f\dot{E}}{K_{\mathrm{bT}_k-\mathrm{T}_i}\left(Z_{m.\mathrm{T}_k}+Z_{\mathrm{T}_k}\right)}\right\}}{K_{\mathrm{sen}}} \tag{5.2}$$

where i = g, z, d. K_{sen} is the sensitivity degree.

C. Transformer Zone-III

Transformer zone-III should reflect faults at the transformer and on the adjacent line, so the setting value $\dot{I}^{\mathrm{III}}_{\mathrm{set.T}_i}$ should be smaller than the short circuit current at the protection device when a fault occurs at the end of the adjacent line, i.e.

$$\dot{I}^{\mathrm{III}}_{\mathrm{set.T}_i} = \frac{\min\left\{\dfrac{K_f\dot{E}}{K_{\mathrm{b}jm-\mathrm{T}_i}\left(Z_{S.jm}+Z_{L.jm}\right)}, \dfrac{K_f\dot{E}}{K_{\mathrm{b}km-\mathrm{T}_i}\left(Z_{S.km}+Z_{L.km}\right)}\right\}}{K_{\mathrm{sen}}} \tag{5.3}$$

where $Z_{L.i_m}$, $Z_{L.j_m}$ and $Z_{L.k_m}$ are, respectively, the impedance of the line where protection i_m, j_m and k_m are. $Z_{S.j_m}$ and $Z_{S.k_m}$ are the system impedance on the back side of protection j_m and k_m in the current system operating mode. $K_{\mathrm{b}j_m-\mathrm{T}_i}$ and $K_{\mathrm{b}k_m-\mathrm{T}_i}$ are branch coefficients. Protection j_m and k_m are the lower-stage protection devices in the forward direction of protection T_i.

5.3.1.2 Line Setting Method

A. Line Zone-I

The setting value of line protection zone-I $\dot{I}^{\mathrm{I}}_{\mathrm{set}.i_n}$ should be smaller than the short circuit current at the protection device when a fault occurs at the end of the line, i.e.

$$\dot{I}^{\mathrm{I}}_{\mathrm{set}.i_n} = \frac{K_{\mathrm{rel}}K_f\dot{E}}{Z_{S.i_n}+Z_{L.i_n}} \tag{5.4}$$

where i = g, z, d. $Z_{S.i_n}$ is the system impedance on the back side of protection i_n in the current system operating mode. $Z_{L.i_n}$ is the impedance of the line where protection i_n is. The setting of protection i_m zone-I is the same as that of protection i_n.

B. Line Zone-II

Line protection zone-II should ensure sufficient sensitivity of protection when a fault occurs at the end of the line, and the setting value $\dot{I}^{\mathrm{II}}_{\mathrm{set}.i_n}$ is

$$\dot{I}^{\mathrm{II}}_{\mathrm{set}.i_n} = \frac{K_f\dot{E}}{K_{\mathrm{sen}}\left(Z_{S.i_n}+Z_{L.i_n}\right)} \tag{5.5}$$

C. Line Zone-III

Line protection zone-III should ensure sufficient sensitivity of protection when a fault occurs on the adjacent line, and the setting value $\dot{I}^{\mathrm{III}}_{\mathrm{set.T}_n}$ is

$$\dot{I}^{\text{III}}_{\text{set.T}_i} = \frac{\min\left\{\dfrac{K_{\text{f}}\dot{E}}{K_{\text{b}jm-\text{T}_i}\left(Z_{S.jm}+Z_{L.jm}\right)}, \dfrac{K_{\text{f}}\dot{E}}{K_{\text{b}km-\text{T}_i}\left(Z_{S.km}+Z_{L.km}\right)}\right\}}{K_{\text{sen}}} \tag{5.6}$$

where protection T_{i} is the lower-stage protection device in the forward direction of protection i_n. $K_{\text{bT}_i-i_n}$ and $K_{\text{bT}_k-\text{T}_i}$ are branch coefficients.

The adaptive current protection set according to the above principles could reflect the relationship between the operation of protection and the fault location in the forward direction of the protection. If zone-I operates, then the fault can only be at the component where the protection is. If zone-II operates, then the fault can be at the component where the protection is or part of the lower-stage component. If zone-III operates, then the fault can be at the component where the protection is, the lower-stage component or part of the next lower-stage component.

5.3.2 Supporting Degree Calculation Method

To reflect the supporting degree of protection operation to fault component identification, define the current protection supporting degree $C_{bh\text{-}yj}$, which refers to the probability that a fault occurs at component yj when protection bh operates. When a fault occurs at the boundary of the protection range, the current at the relaying point equals the protection setting value $I_{\text{set}\cdot bh}$, i.e.

$$I_{\text{set}.bh} = \frac{K_{\text{f}}E}{K_{\text{b}}\left(Z_{S.yj}+Z_{\alpha bh-yj}\right)} \tag{5.7}$$

where $Z_{\alpha bh-yj}$ represents the effective protection range that protection bh has on component yj. K_{b} is the branch coefficient. Therefore, the effective protection range $Z_{\alpha bh-yj}$ could be quantified according to the protection setting value.

$$Z_{\alpha bh-yj} = \frac{K_{\text{f}}E}{K_{\text{b}}I_{\text{set}.bh}} - Z_{S.yj} \tag{5.8}$$

Based on Equation (5.8), the effective protection range that each protection zone has on the component can be calculated, so the supporting degree of each protection zone to the component can be calculated.

5.3.2.1 Calculation of Transformer Protection Supporting Degree

A. Transformer Protection Zone-I Supporting Degree

When $\dot{I}^{\text{I}}_{\text{set.T}_i} > \frac{K_{\text{f}}\dot{E}}{Z_{S.\text{T}_i}+Z_{\text{T}_i}}$, the effective protection range that zone-I of transformer protection T_i has on transformer T is part of Z_{T_i}, i.e. $Z^{\text{I}}_{\alpha\text{T}_i-\text{T}}(i)$.

$$Z^{\text{I}}_{\alpha\text{T}_i-\text{T}}(i) = \frac{K_{\text{f}}\dot{E}}{\dot{I}^{\text{I}}_{\text{set.T}_i}} - Z_{S.\text{T}_i} \tag{5.9}$$

When $\dot{I}^{\rm I}_{{\rm set.T}_i} < \frac{K_{\rm f}\dot{E}}{Z_{S.{\rm T}_i} + Z_{{\rm T}_i}}$, the effective protection range that zone-I of transformer protection T_i has on transformer T is part of $Z_{{\rm T}_i}$, $Z_{{\rm T}_j}$ and $Z_{{\rm T}_k}$, i.e. $Z^{\rm I}_{\alpha{\rm T}_i-{\rm T}}(i)$, $Z^{\rm I}_{\alpha{\rm T}_i-{\rm T}}(j)$ and $Z^{\rm I}_{\alpha{\rm T}_i-{\rm T}}(k)$.

$$\begin{cases} Z^{\rm I}_{\alpha{\rm T}_i-{\rm T}}(i) = Z_{{\rm T}_i} \\ Z^{\rm I}_{\alpha{\rm T}_i-{\rm T}}(j) = \dfrac{K_{\rm f}\dot{E}}{K_{{\rm bT}_j-{\rm T}_i}\dot{I}^{\rm I}_{set.{\rm T}_i}} - Z_{m.{\rm T}_j} \\ Z^{\rm I}_{\alpha{\rm T}_i-{\rm T}}(k) = \dfrac{K_{\rm f}\dot{E}}{K_{{\rm bT}_k-{\rm T}_i}\dot{I}^{\rm I}_{set.{\rm T}_i}} - Z_{m.{\rm T}_k} \end{cases} \tag{5.10}$$

The effective protection range that protection bh has on the transformer $Z_{\alpha bh-{\rm T}}$ is the sum of the effective protection ranges on the equivalent impedances on three sides of the transformer.

$$Z_{\alpha bh-{\rm T}} = \sum_{i=1}^{3} Z_{\alpha bh-{\rm T}}(i) \tag{5.11}$$

According to Equation (5.11), the effective protection range that zone-I of transformer protection T_i has on transformer T can be calculated. When zone-I of transformer protection T_i operates, it only has a supporting degree to the transformer fault, and has no supporting degree to faults on other components. Therefore, the supporting degree $C^{\rm I}_{{\rm T}_i-yj}$ of zone-I of protection T_i to component yj is

$$C^{\rm I}_{{\rm T}_i-yj} = \begin{cases} 1 & yj \text{ is T} \\ 0 & yj \text{ is a component other than T} \end{cases} \tag{5.12}$$

B. Transformer Protection Zone-II Supporting Degree

The effective protection range that zone-II of transformer protection T_i has on transformer T covers the whole of T, i.e. the effective protection ranges $Z^{\rm II}_{\alpha{\rm T}_i-{\rm T}}(i)$, $Z^{\rm II}_{\alpha{\rm T}_i-{\rm T}}(j)$ and $Z^{\rm II}_{\alpha{\rm T}_i-{\rm T}}(k)$ are

$$\begin{cases} Z^{\rm II}_{\alpha{\rm T}_i-{\rm T}}(i) = Z_{{\rm T}_i} \\ Z^{\rm II}_{\alpha{\rm T}_i-{\rm T}}(j) = Z_{{\rm T}_j} \\ Z^{\rm II}_{\alpha{\rm T}_i-{\rm T}}(k) = Z_{{\rm T}_k} \end{cases} \tag{5.13}$$

The effective protection ranges that zone-II of transformer protection T_i has on the adjacent lines L_{jm} and L_{km} are respectively $Z^{\rm II}_{\alpha{\rm T}_i-L_{jm}}$ and $Z^{\rm II}_{\alpha{\rm T}_i-L_{km}}$:

$$\begin{cases} Z^{\rm II}_{\alpha{\rm T}_i-L_{jm}} = \dfrac{K_{\rm f}\dot{E}}{K_{{\rm b}_{jm}-{\rm T}_i}\dot{I}^{\rm II}_{{\rm set.T}_i}} - Z_{S.L_{jm}} \\ Z^{\rm II}_{\alpha{\rm T}_i-L_{km}} = \dfrac{K_{\rm f}\dot{E}}{K_{{\rm b}_{km}-{\rm T}_i}\dot{I}^{\rm II}_{{\rm set.T}_i}} - Z_{S.L_{km}} \end{cases} \tag{5.14}$$

As for the supporting degree of zone-II of transformer protection T_i to a component, if component yj is within the protection range of zone-II of T_i, the supporting degree equals the ratio of the effective protection range that zone-II has on component yj to the sum of effective protection ranges that zone-II has on all the components within the protection range of zone-II. If component yj is not in the protection range of zone-II of T_i, the supporting degree of zone-II to yj is 0. Therefore, the supporting degree $C^{\mathrm{II}}_{\mathrm{T}_i-yj}$ of zone-II of protection T_i to component yj is

$$C^{\mathrm{II}}_{\mathrm{T}_i-yj}=\begin{cases}\dfrac{Z^{\mathrm{II}}_{\alpha\mathrm{T}_i-yj}}{\sum Z^{\mathrm{II}}_{\alpha\mathrm{T}_i}} & yj \text{ is within the protection range of zone-II of } \mathrm{T}_i\\ 0 & yj \text{ is not in the protection range of zone-II of } \mathrm{T}_i\end{cases} \tag{5.15}$$

C. *Transformer Protection Zone-III Supporting Degree*

The effective protection range that zone-III of transformer protection T_i has on transformer T covers the whole of T, i.e. the effective protection ranges $Z^{\mathrm{III}}_{\alpha\mathrm{T}_i-\mathrm{T}}(i)$, $Z^{\mathrm{III}}_{\alpha\mathrm{T}_i-\mathrm{T}}(j)$ and $Z^{\mathrm{III}}_{\alpha\mathrm{T}_i-\mathrm{T}}(k)$ are

$$\begin{cases}Z^{\mathrm{III}}_{\alpha\mathrm{T}_i-\mathrm{T}}(i)=Z_{\mathrm{T}i}\\ Z^{\mathrm{III}}_{\alpha\mathrm{T}_i-\mathrm{T}}(j)=Z_{\mathrm{T}_j}\\ Z^{\mathrm{III}}_{\alpha\mathrm{T}_i-\mathrm{T}}(k)=Z_{\mathrm{T}_k}\end{cases} \tag{5.16}$$

The effective protection ranges that zone-III of transformer protection T_i has on the adjacent lines L_{jm} and L_{km} cover the whole component, i.e. the effective protection range $Z^{\mathrm{III}}_{\alpha\mathrm{T}_i-L_{jm}}$ and $Z^{\mathrm{III}}_{\alpha\mathrm{T}_i-L_{km}}$ are

$$\begin{cases}Z^{\mathrm{III}}_{\alpha\mathrm{T}_i-L_{jm}}=Z_{L_{jm}}\\ Z^{\mathrm{III}}_{\alpha\mathrm{T}_i-L_{km}}=Z_{L_{km}}\end{cases} \tag{5.17}$$

As for the supporting degree of zone-III of transformer protection T_i to a component, if component yj is within the protection range of zone-III of T_i, the supporting degree equals the ratio of the effective protection range that zone-III has on component yj to the sum of effective protection ranges that zone-III has on all the components; if component yj is not in the protection range of zone-III of T_i, the supporting degree of zone-III to yj is 0. Therefore, the supporting degree $C^{\mathrm{II}}_{\mathrm{T}_i-yj}$ of zone-III of protection T_i to component yj is

$$C^{\mathrm{III}}_{\mathrm{T}_i-yj}=\begin{cases}\dfrac{Z^{\mathrm{III}}_{\alpha\mathrm{T}_i-yj}}{\sum Z^{\mathrm{III}}_{\alpha\mathrm{T}_i}} & yj \text{ is within the protection range of zone-III of } \mathrm{T}_i\\ 0 & yj \text{ is not in the protection range of zone-III of } \mathrm{T}_i\end{cases} \tag{5.18}$$

5.3.2.2 Calculation of Line Protection Supporting Degree

A. *Line Protection Zone-I Supporting Degree*

Taking line protection i_n, for example, the effective protection range that zone-I of i_n has on line L_{in} is part of the line impedance in the forward direction of protection, i.e. $Z^{\mathrm{I}}_{\alpha i_n - L_{in}}$ is

$$Z^{\mathrm{I}}_{\alpha i_n - L_{in}} = \frac{K_{\mathrm{f}}\dot{E}}{\dot{I}^{\mathrm{I}}_{\mathrm{set}.i_n}} - Z_{S.i_n} \tag{5.19}$$

The supporting degree $C^{\mathrm{I}}_{i_n - yj}$ of zone-I of protection i_n to component yj is

$$C^{\mathrm{I}}_{i_n - yj} = \begin{cases} 1 & yj \text{ is } L_{in} \\ 0 & yj \text{ is a component other than } L_{in} \end{cases} \tag{5.20}$$

B. *Line Protection Zone-II Supporting Degree*

The effective protection range that zone-II of i_n has on line L_{in} covers the whole line length of L_{in}, i.e. the effective protection range $Z^{\mathrm{II}}_{\alpha i_n - L_{in}}$ is

$$Z^{\mathrm{II}}_{\alpha i_n - L_{in}} = Z_{L_{in}} \tag{5.21}$$

The effective protection ranges that zone-II of line protection i_n has on the adjacent transformer $Z^{\mathrm{II}}_{\alpha i_n - \mathrm{T}}(i)$, $Z^{\mathrm{II}}_{\alpha i_n - \mathrm{T}}(j)$ and $Z^{\mathrm{II}}_{\alpha i_n - \mathrm{T}}(k)$ are calculated as follows:

When $\dot{I}^{\mathrm{II}}_{\mathrm{set}.i_n} > \frac{K_{\mathrm{f}}\dot{E}}{K_{\mathrm{bT}_i - i_n}\left(Z_{S.\mathrm{T}_i} + Z_{\mathrm{T}_i}\right)}$, the effective protection range that zone-II of protection i_n has on the adjacent transformer T is part of Z_{T_i}

$$Z^{\mathrm{II}}_{\alpha i_n - \mathrm{T}}(i) = \frac{K_{\mathrm{f}}\dot{E}}{K_{\mathrm{bT}_i - i_n}\dot{I}^{\mathrm{II}}_{\mathrm{set}.i_n}} - Z_{S.\mathrm{T}_i} \tag{5.22}$$

When $\dot{I}^{\mathrm{II}}_{\mathrm{set}.i_n} < \frac{K_{\mathrm{f}}\dot{E}}{K_{\mathrm{bT}_i - i_n}\left(Z_{S.\mathrm{T}_i} + Z_{\mathrm{T}_i}\right)}$, the effective protection range that zone-II of protection i_n has on the adjacent transformer T is part of Z_{T_i}, Z_{T_j} and Z_{T_k}

$$\begin{cases} Z^{\mathrm{II}}_{\alpha i_n - \mathrm{T}}(i) = Z_{\mathrm{T}_i} \\ Z^{\mathrm{II}}_{\alpha i_n - \mathrm{T}}(j) = \dfrac{K_{\mathrm{f}}\dot{E}}{K_{\mathrm{bT}_j - \mathrm{T}_i}K_{\mathrm{bT}_i - i_n}\dot{I}^{\mathrm{II}}_{\mathrm{set}.i_n}} - Z_{m.\mathrm{T}_j} \\ Z^{\mathrm{II}}_{\alpha i_n - \mathrm{T}}(k) = \dfrac{K_{\mathrm{f}}\dot{E}}{K_{\mathrm{bT}_k - T_i}K_{\mathrm{bT}_i - i_n}\dot{I}^{\mathrm{II}}_{\mathrm{set}.i_n}} - Z_{m.\mathrm{T}_k} \end{cases} \tag{5.23}$$

The effective protection range that zone-II of protection i_n has on the adjacent line L_{im} is $Z^{\mathrm{II}}_{\alpha i_n - L_{im}}(i)$, i.e.

$$Z^{\mathrm{II}}_{\alpha i_n - L_{im}}(i) = \frac{K_{\mathrm{f}}\dot{E}}{K_{\mathrm{b}i_m - i_n}\dot{I}^{\mathrm{II}}_{\mathrm{set}.i_n}} - Z_{S.i_m} \tag{5.24}$$

Similar to the calculation of the supporting degree of transformer protection zone-II to the component, the supporting degree $C^{\text{II}}_{i_n-yj}$ of zone-II of line protection i_n to component yj is

$$C^{\text{II}}_{i_n-yj}=\begin{cases}\dfrac{Z^{\text{II}}_{\alpha i_n-yj}}{\sum Z^{\text{II}}_{\alpha i_n}} & yj \text{ is within the protection range of zone-II of } i_n\\ 0 & yj \text{ is not in the protection range of zone-II of } i_n\end{cases} \tag{5.25}$$

C. Line Protection Zone-III Supporting Degree

The effective protection ranges that zone-III of protection i_n has on L_{in} and the adjacent transformer T, i.e. $Z^{\text{III}}_{\alpha i_n-L_{in}}$, $Z^{\text{III}}_{\alpha i_n-\text{T}}(i)$, $Z^{\text{III}}_{\alpha i_n-\text{T}}(j)$ and $Z^{\text{III}}_{\alpha i_n-\text{T}}(k)$ cover the whole length of L_{in} and T:

$$\begin{cases}Z^{\text{III}}_{\alpha i_n-L_{in}}=Z_{L_{in}}\\ Z^{\text{III}}_{\alpha i_n-\text{T}}(i)=Z_{\text{T}_i}\\ Z^{\text{III}}_{\alpha i_n-\text{T}}(j)=Z_{\text{T}_j}\\ Z^{\text{III}}_{\alpha i_n-\text{T}}(k)=Z_{\text{T}_k}\end{cases} \tag{5.26}$$

The effective protection ranges that zone-III of protection i_n has on the adjacent lines L_{jm} and L_{km} are $Z^{\text{III}}_{\alpha i_n-L_{jm}}$ and $Z^{\text{III}}_{\alpha i_n-L_{km}}$, respectively, i.e.

$$\begin{cases}Z^{\text{III}}_{\alpha i_n-L_{jm}}=\dfrac{K_{\text{f}}\dot{E}}{K_{\text{b}jm-i_n}\dot{I}^{\text{III}}_{\text{set}.i_n}}-Z_{S.jm}\\ Z^{\text{III}}_{\alpha i_n-L_{km}}=\dfrac{K_{\text{f}}\dot{E}}{K_{\text{b}jm-i_n}\dot{I}^{\text{III}}_{\text{set}.i_n}}-Z_{S.km}\end{cases} \tag{5.27}$$

The supporting degree $C^{\text{III}}_{i_n-yj}$ of zone-III of line protection i_n to component yj is

$$C^{\text{III}}_{i_n-yj}=\begin{cases}\dfrac{Z^{\text{III}}_{\alpha i_n-yj}}{\sum Z^{\text{III}}_{\alpha i_n}} & yj \text{ is within the protection range of zone-III of } i_n\\ 0 & yj \text{ is not in the protection range of zone-III of } i_n\end{cases} \tag{5.28}$$

5.3.3 Substation Area Current Protection Algorithm

5.3.3.1 Construct the Information Fusion Degree Function of Substation Area Current Protection

When the supporting degree of protection to component fault identification is fused with the protection operation information as the weight, the discrimination between faulty component and non-faulty component can be even more evident. The information fusion degree function of substation area current protection is constructed with the supporting degree and the actual operational status of protection:

$$\begin{cases} E_R(\mathrm{T}) = \sum_{1}^{N_{\mathrm{I}}} C_{bh-\mathrm{T}}^{\mathrm{I}} R_{bh}^{\mathrm{I}} + \sum_{1}^{N_{\mathrm{II}}} C_{bh-\mathrm{T}}^{\mathrm{II}} R_{bh}^{\mathrm{II}} + \sum_{1}^{N_{\mathrm{III}}} C_{bh-\mathrm{T}}^{\mathrm{III}} R_{bh}^{\mathrm{III}} \\ E_R(L_{in}) = \sum_{1}^{N_{\mathrm{I}}} C_{bh-L_{im}}^{\mathrm{I}} R_{bh}^{\mathrm{I}} + \sum_{1}^{N_{\mathrm{II}}} C_{bh-L_{im}}^{\mathrm{II}} R_{bh}^{\mathrm{II}} + \sum_{1}^{N_{\mathrm{III}}} C_{bh-L_{im}}^{\mathrm{III}} R_{bh}^{\mathrm{III}} \\ \qquad + \sum_{1}^{N_{\mathrm{I}}} C_{bh-L_{in}}^{\mathrm{I}} R_{bh}^{\mathrm{I}} + \sum_{1}^{N_{\mathrm{II}}} C_{bh-L_{in}}^{\mathrm{II}} R_{bh}^{\mathrm{II}} + \sum_{1}^{N_{\mathrm{III}}} C_{bh-L_{in}}^{\mathrm{III}} R_{bh}^{\mathrm{III}} \end{cases} \tag{5.29}$$

where $E_R(\mathrm{T})$ and $E_R(L_{in})$ are the information fusion degree functions of the transformer and line, respectively. R_{bh}^{I}, R_{bh}^{II} and R_{bh}^{III} are the actual operational statuses of the protection zones. N_{I}, N_{II} and N_{III} are the total number of protection zones-I, zones-II and zones-III.

5.3.3.2 Construct the Information Fusion Degree Expectation Function of Substation Area Current Protection

The information fusion degree expectation function of substation area current protection is the fusion of the expected operational statuses of different protection devices supposing that the fault occurs on a particular component. Since the expected operational status of a particular protection has to do with the fault location, the component can be divided into some protection sections D_x according to the effective protection range that the protection device has on the component. The expected operational status of a particular protection concerning different protection sections is different. In each protection section, with the supporting degree as the fault weight, we calculate the fusion of the expected operational status of protection $H(D_x)$. Then, with the fault probability of each protection section $\mathrm{p}D_x$ as the weight of $H(D_x)$, the final fusion result is the information fusion degree expectation function of the component.

Take the division of the transformer by protection zone-II, for example. According to the protection ranges that zone-II of protection g_1, z_1 and d_1 have on the transformer (as shown in the fold gridline diagram in Figure 5.12), the equivalent transformer is divided into four protection sections $D_{\mathrm{T1}}^{\mathrm{II}}$, $D_{\mathrm{T2}}^{\mathrm{II}}$, $D_{\mathrm{T3}}^{\mathrm{II}}$ and $D_{\mathrm{T4}}^{\mathrm{II}}$, the expected operational status of zone-II of each protection shown in Table 5.4.

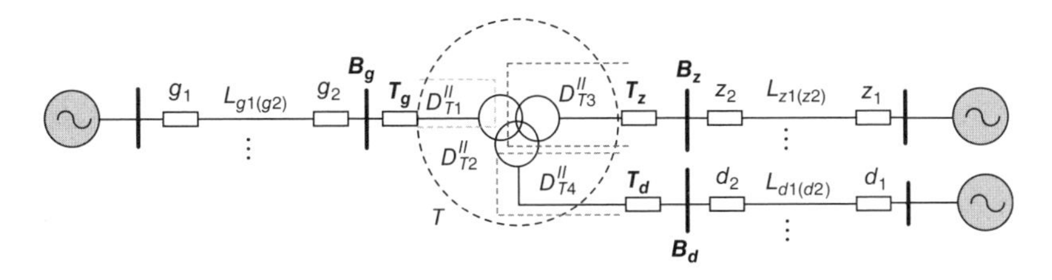

Figure 5.12 Example of division of the transformer by protection zone-II.

Table 5.4 Expected operational status of zone-II when a fault occurs in different protection sections.

Protection component	g_1	g_2	T_g	T_z	T_d	z_1	z_2	d_1	d_2
Fault occurs in D_{T1}^{II}	1	0	1	1	1	0	0	0	0
Fault occurs in D_{T2}^{II}	0	0	1	1	1	0	0	0	0
Fault occurs in D_{T3}^{II}	0	0	1	1	1	1	0	0	0
Fault occurs in D_{T4}^{II}	0	0	1	1	1	0	0	1	0

The fault probability of each protection section can be calculated according to the effective protection range that zone-II of each protection has on the transformer:

$$\begin{cases} p_{D_{T1}}^{II} = Z_{\alpha g1-T}^{II} / \left(Z_{Tg} + Z_{Tz} + Z_{Td}\right) \\ p_{D_{T3}}^{II} = Z_{\alpha z1-T}^{II} / \left(Z_{Tg} + Z_{Tz} + Z_{Td}\right) \\ p_{D_{T4}}^{II} = Z_{\alpha d1-T}^{II} / \left(Z_{Tg} + Z_{Tz} + Z_{Td}\right) \\ p_{D_{T2}}^{II} = \left(1 - Z_{\alpha g1-T}^{II} - Z_{\alpha z1-T}^{II} - Z_{\alpha d1-T}^{II}\right) / \left(Z_{Tg} + Z_{Tz} + Z_{Td}\right) \end{cases} \tag{5.30}$$

Thus, when a fault occurs at the transformer, the value of zone-II information fusion degree expectation function $E_H^{II}(\mathrm{T})$ can be calculated as

$$E_H^{II}(\mathrm{T}) = \sum_{1}^{ND_{\mathrm{T}}^{II}} \left(p_{D_{Tx}}^{II} \sum_{1}^{N_{II}} C_{bh-T}^{II} R_{bh-D_{Tx}}^{*II} \right) \tag{5.31}$$

where ND_{T}^{II} is the number of protection sections the transformer is divided into by protection zone-II. $R_{bh-D_{Tx}}^{*II}$ is the expected operational status of zone-II of each protection when a fault occurs in protection section D_{TX}^{II}.

Similarly, the transformer can be divided into some zone-I protection sections according to the effective protection range that zone-I of each protection has on the transformer. As for protection zone-III, since the effective protection range that zone-III of each protection has on the transformer covers the whole transformer, there is no necessity for partitioning. When a fault occurs at the transformer, the values of the zone-I and zone-III information fusion degree expectation functions $E_H^{I}(\mathrm{T})$ and $E_H^{III}(\mathrm{T})$ can be calculated as

$$\begin{cases} \mathrm{E}_H^{I}(\mathrm{T}) = \sum_{1}^{ND_{\mathrm{T}}^{I}} \left(p_{D_{Tx}}^{I} \sum_{1}^{N_{I}} C_{bh-T}^{I} R_{bh-D_{Tx}}^{*I} \right) \\ E_H^{III}(\mathrm{T}) = \sum_{1}^{N_{III}} C_{bh-T}^{III} R_{bh}^{*III} \end{cases} \tag{5.32}$$

where ND_{T}^{I} is the number of protection sections the transformer is divided into by protection zone-I. $R_{bh-D_{Tx}}^{*I}$ is the expected operational status of zone-I of each protection when a fault occurs in protection section $D^{\mathrm{I}}_{\mathrm{TX}}$. $R^{*\mathrm{III}}_{bh}$ is the expected operational status of zone-III of each protection when a fault occurs at the transformer.

Similarly, a line can also be divided into some zone-I, zone-II and zone-III protection sections $D^{\mathrm{I}}_{Lin.X}$, $D^{\mathrm{II}}_{Lin.X}$ and $D^{\mathrm{III}}_{Lin.X}$, according to the effective protection range that zone-I, zone-II and zone-III of each protection has on the line. The numbers of $D^{\mathrm{I}}_{Lin.X}$, $D^{\mathrm{II}}_{Lin.X}$ and $D^{\mathrm{II}}_{Lin.X}$ are ND^{I}_{Lin}, ND^{II}_{Lin} and ND^{III}_{Lin}, respectively, and the fault probabilities of $D^{\mathrm{I}}_{Lin.X}$, $D^{\mathrm{II}}_{Lin.X}$ and $D^{\mathrm{III}}_{Lin.X}$ are denoted $p^{\mathrm{I}}_{D_{Lin.x}}$, $p^{\mathrm{II}}_{D_{Lin.x}}$ and $p^{\mathrm{III}}_{D_{Lin.x}}$. Thus, when a fault occurs on the line, the value of zone-I, zone-II and zone-III information fusion degree expectation functions $E^{\mathrm{I}}_{H}(L_{in})$, $E^{\mathrm{II}}_{H}(L_{in})$ and $E^{\mathrm{III}}_{H}(L_{in})$ can be calculated as

$$\begin{cases} E^{\mathrm{I}}_{H}(L_{in}) = \sum\limits_{1}^{ND^{\mathrm{I}}_{Lin}} \left(p^{\mathrm{I}}_{D_{Lin.x}} \sum\limits_{1}^{N_{\mathrm{I}}} \left(C^{\mathrm{I}}_{bh-L_{in}} + C^{\mathrm{I}}_{bh-L_{im}} \right) R^{*\mathrm{I}}_{bh-D_{Lin.x}} \right) \\ E^{\mathrm{II}}_{H}(L_{in}) = \sum\limits_{1}^{ND^{\mathrm{II}}_{Lin}} \left(p^{\mathrm{II}}_{D_{Lin.x}} \sum\limits_{1}^{N_{\mathrm{II}}} \left(C^{\mathrm{II}}_{bh-Li_{n}} + C^{\mathrm{II}}_{bh-L_{im}} \right) R^{*\mathrm{II}}_{bh-D_{Lin.x}} \right) \\ E^{\mathrm{III}}_{H}(L_{in}) = \sum\limits_{1}^{ND^{\mathrm{III}}_{Lin}} \left(p^{\mathrm{III}}_{D_{Lin.x}} \sum\limits_{1}^{N_{\mathrm{I}}} \left(C^{\mathrm{III}}_{bh-L_{in}} + C^{\mathrm{III}}_{bh-L_{im}} \right) R^{*\mathrm{III}}_{bh-D_{Lin.x}} \right) \end{cases} \tag{5.33}$$

The value of the information fusion degree expectation function of a component is the sum of the values of zone-I, zone-II and zone-III information fusion degree expectation functions when a fault occurs at the component.

$$\begin{cases} E_{H}(\mathrm{T}) = E^{\mathrm{I}}_{H}(\mathrm{T}) + E^{\mathrm{II}}_{H}(\mathrm{T}) + E^{\mathrm{III}}_{H}(\mathrm{T}) \\ E_{H}(L_{in}) = E^{\mathrm{I}}_{H}(L_{in}) + E^{\mathrm{II}}_{H}(L_{in}) + E^{\mathrm{III}}_{H}(L_{in}) \end{cases} \tag{5.34}$$

where $E_H(\mathrm{T})$ and $E_H(L_{in})$ are the information fusion degree expectation functions of the transformer and the line, respectively.

5.3.3.3 Fault Identification Based on the Information Fusion Factor

Define the ratio of the substation area current protection information fusion degree function of a component to the information fusion degree expectation function of the component as the information fusion factor of the component:

$$\begin{cases} F(\mathrm{T}) = \dfrac{E_R(\mathrm{T})}{E_H(\mathrm{T})} \\ F(L_{in}) = \dfrac{E_R(L_{in})}{E_H(L_{in})} \end{cases} \tag{5.35}$$

where $F(\mathrm{T})$ and $F(L_{in})$ are the information fusion factors of the transformer and the line, respectively. The bigger the information fusion factor of a component, the closer the value of the information fusion degree function is to the value of the information fusion degree expectation function, and the higher the probability that the fault has occurred at the component. Therefore, the information fusion factor of the fault component should be the biggest of the information fusion factors of all the components. The component corresponding to the biggest information fusion factor is identified as the faulty component, i.e. the protection criterion is

$$F(yj_{\text{fault}}) = \max(F(\text{T}), F(L_{in})) \tag{5.36}$$

where yj_{fault} represents the faulty component.

5.3.4 Scheme Verification

The data of a real substation were used to verify the proposed scheme; the system structure is shown in Figure 5.13. Parameters of the transformer are Y/Y/Δ-12-11, SN = 400 MVA, U1N/U2N/U3N = 220 kV/110 kV/35 kV. High-voltage/medium-voltage short circuit impedance is 13.27%; high-voltage/low-voltage short circuit impedance is 23.14%; medium-voltage/low-voltage short circuit impedance is 10.21%. The high-voltage side line model is LGJ-400/50; the medium-voltage side line model is LGJ-300/40; the low-voltage side line model is LGJ-185/30. Distance protection is installed on the high-voltage side line. In Figure 5.13, protection T_{g}, T_{z} and T_{d} are on the high-voltage, medium-voltage and low-voltage side of the transformer, respectively.

5.3.4.1 Examples of Information Fusion Factor Calculation

If CB is switched on and a fault occurs at f_1 on the high-voltage side of the transformer, the substation area protection scheme based on the current protection information fusion factor is as illustrated. First, according to the current setting value, calculate the effective protection range that protection has on each component, as well as the supporting degree of each protection zone to the component. The supporting degree of each protection zone to the identification of the transformer fault is shown in Table 5.5.

Second, according to the effective protection range that protection has on each component, calculate the fault probabilities of the protection sections of each

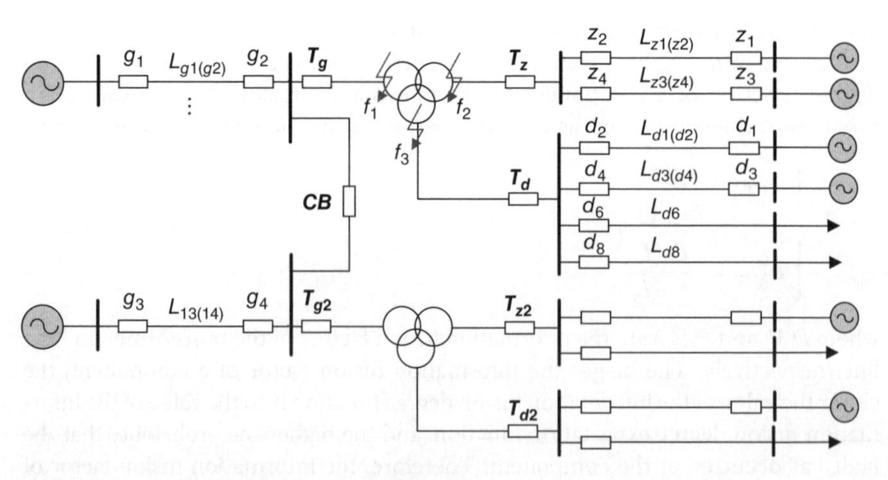

Figure 5.13 System used for scheme verification.

Table 5.5 Supporting degree of protection to transformer when CB is switched on and a fault occurs at f_1.

Protection name	g_1	g_2	g_3	g_4	z_1	z_2
Zone-I supporting degree	0	0	0	0	0	0
Zone-II supporting degree	0.0686	0	0.0572	0	0.1474	0
Zone-III supporting degree	0.1824	0	0.1849	0	0.2090	0
Protection name	z_3	z_4	d_1	d_2	d_3	d_4
Zone-I supporting degree	0	0	0	0	0	0
Zone-II supporting degree	0.1633	0	0.1308	0	0.1279	0
Zone-III supporting degree	0.2235	0	0.1333	0	0.1326	0
Protection name	d_6	d_8	T_g	T_z	T_d	
Zone-I supporting degree	0	0	1	1	1	
Zone-II supporting degree	0	0	0.3549	0.5145	0.4144	
Zone-III supporting degree	0	0	0.0684	0.0760	0.1254	

Table 5.6 Information fusion degree expectation function value of each component when CB is switched on.

Component name	L_{g1}	L_{g3}	L_{z1}	L_{z3}	L_{d1}
Information fusion degree expectation function value	4.6610	4.5070	4.8574	4.7961	4.3629
Component name	L_{d3}	L_{d6}	L_{d8}	T	
Information fusion degree expectation function value	4.3927	3.3501	3.3441	4.2800	

component and the expected operational status of protection. Then, with the supporting degree as the weight, calculate the value of the information fusion degree expectation function of each component, the results being as shown in Table 5.6.

With the supporting degree of protection to each component as the weight, according to the actual operational status of each protection (as shown in Table 5.7), calculate the value of the information fusion degree function of each component, the results being as shown in Table 5.8. Then, according to the ratio of the value of the information fusion degree function to the value of the information fusion degree expectation function, the information fusion factor of each component can be obtained, the results being as shown in Table 5.9.

Table 5.7 Actual operational status of each protection when CB is switched on and a fault occurs at f_1.

Protection name	g_1	g_2	g_3	g_4	z_1	z_2
Zone-I operational status	0	0	0	0	0	0
Zone-II operational status	1	0	1	0	0	0
Zone-III operational status	1	0	1	0	1	0
Protection name	z_3	z_4	d_1	d_2	d_3	d_4
Zone-I operational status	0	0	0	0	0	0
Zone-II operational status	0	0	0	0	0	0
Zone-III operational status	1	0	1	0	1	0
Protection name	d_6	d_8	T_g	T_z	T_d	
Zone-I operational status	0	0	1	0	0	
Zone-II operational status	0	0	1	1	1	
Zone-III operational status	0	0	1	1	1	

Table 5.8 Calculated value of the information fusion degree function of each component when CB is switched on and a fault occurs at f_1.

Component name	L_{g1}	L_{g3}	L_{z1}	L_{z3}	L_{d1}
Information fusion degree function value	1.9985	1.8971	1.8232	1.7755	1.1622
Component name	L_{d3}	L_{d6}	L_{d8}	T	
Information fusion degree function value	1.1588	0.7905	0.7800	3.7451	

Table 5.9 Calculated value of the information fusion factor of each component when CB is switched on and a fault occurs at f_1.

Component name	L_{g1}	L_{g3}	L_{z1}	L_{z3}	L_{d1}
Information fusion factor	0.4288	0.4209	0.3753	0.3702	0.2664
Component name	L_{d3}	L_{d6}	L_{d8}	T	
Information fusion factor	0.2638	0.2360	0.2333	0.8750	

The information fusion factor of the transformer is 0.8750, which is the biggest value among the information fusion factors of all the components. Thus, protection will identify the transformer as the faulty component and trip the corresponding circuit breakers.

5.3.4.2 Coping with Information Loss and Error

If CB is switched on and a fault occurs at f_1, f_2 and f_3 of the transformer and the outlet of line L_{g3}, L_{z3} and L_{d6} respectively, and seven random digits of information are missing or in error (this means a distortion rate of 14% for 51-digit protection operational information), the statistical mean value of 100 tests of the information fusion factor is shown in Figures 5.14–5.19. It can be seen that, when part of the information is missing or in error, the component corresponding to the maximum information fusion factor is still consistent with the preset fault component. Therefore, the proposed scheme could identify the faulty component correctly, with good tolerance of information fault.

5.3.4.3 Coping with Topology Change

If CB is switched off after a fault occurs, then the protection setting value will change, and so will the value of the expectation function of each component. The calculated values of the information fusion degree function, the information fusion degree expectation function and the information fusion factor of each

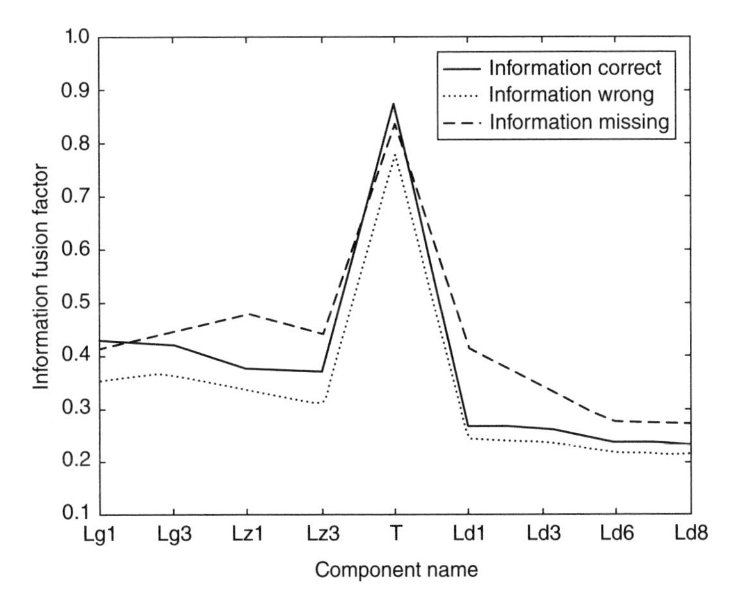

Figure 5.14 Information fusion factor considering information loss and error when CB is switched on and a fault occurs at f_1.

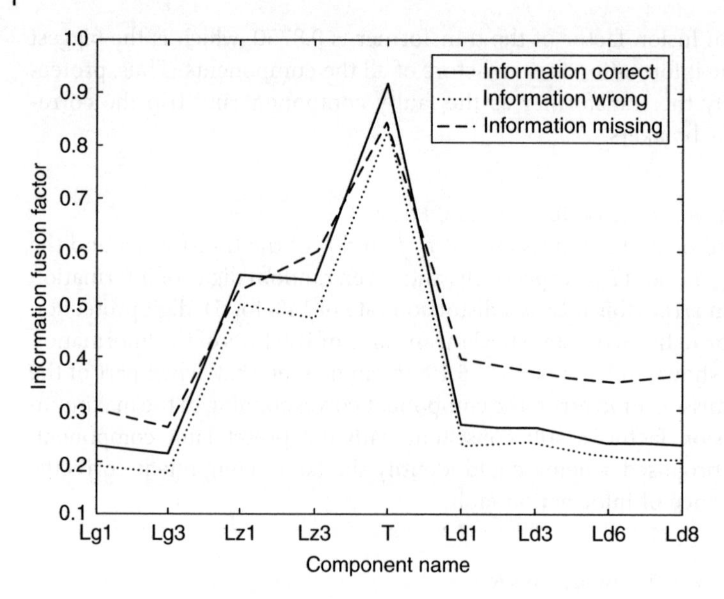

Figure 5.15 Information fusion factor considering information loss and error when CB is switched on and a fault occurs at f_2.

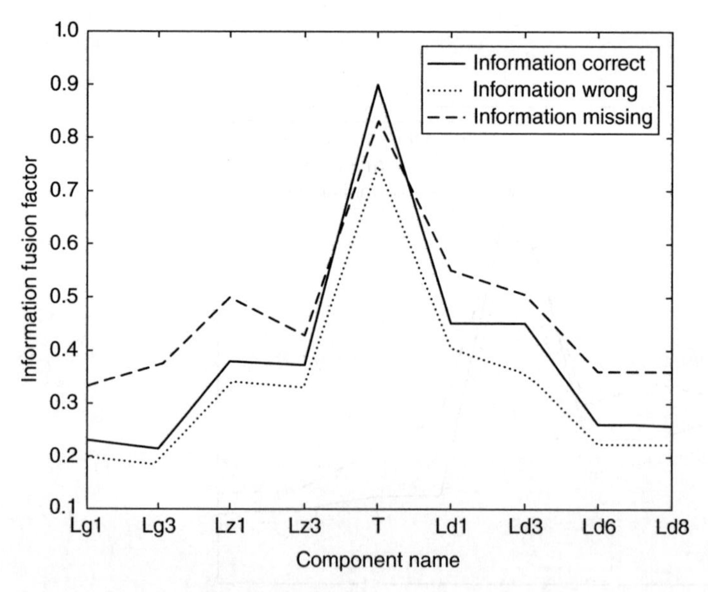

Figure 5.16 Information fusion factor considering information loss and error when CB is switched on and a fault occurs at f_3.

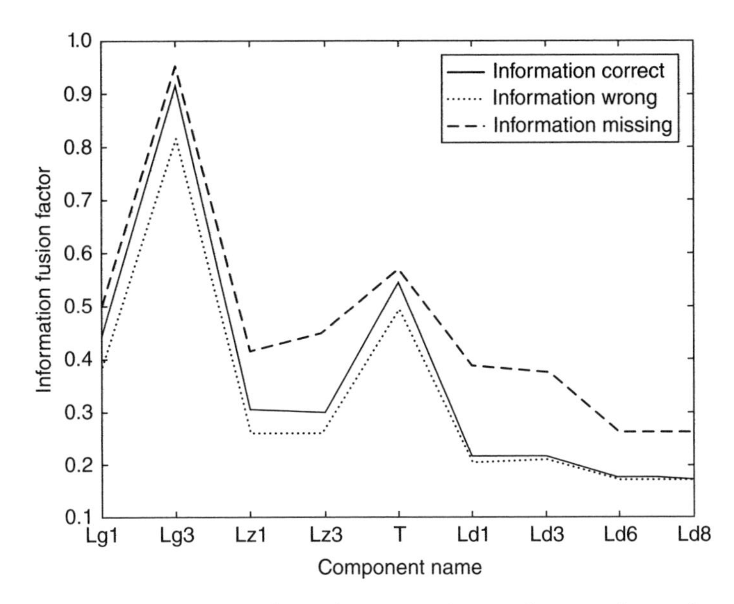

Figure 5.17 Information fusion factor considering information loss and error when CB is switched on and a fault occurs at the exit of high-voltage side line L_{g3}.

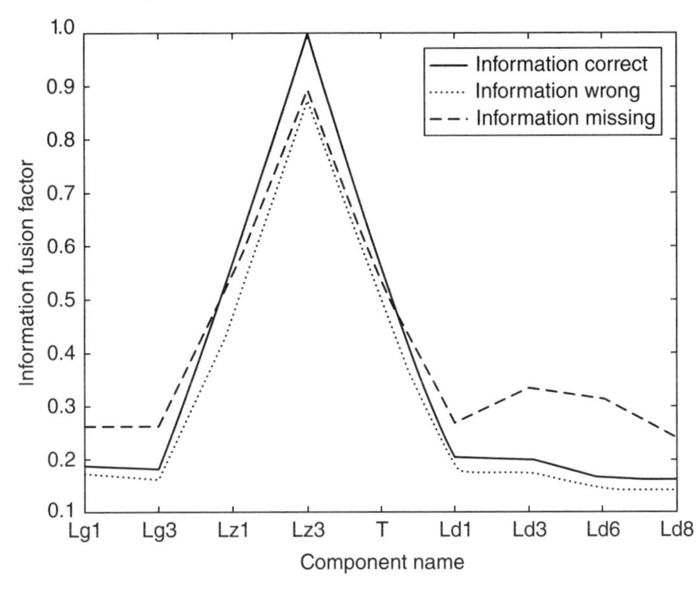

Figure 5.18 Information fusion factor considering information loss and error when CB is switched on and a fault occurs at the exit of medium-voltage side line L_{z3}.

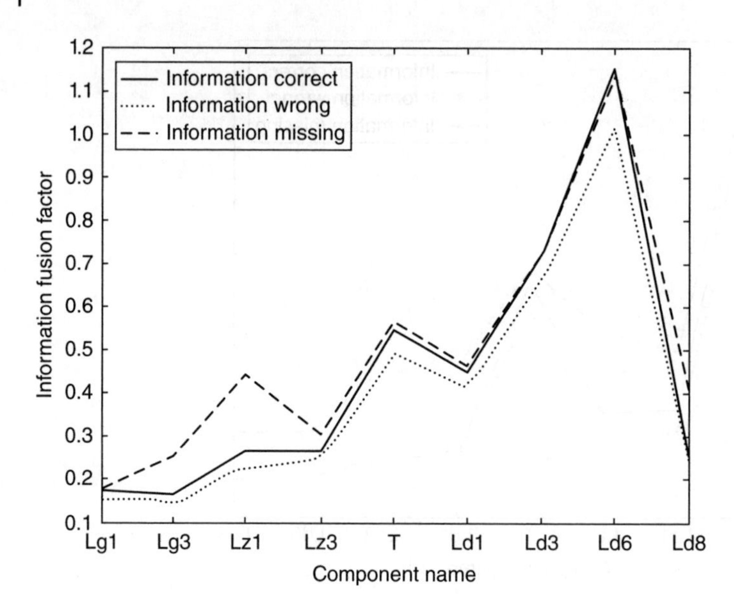

Figure 5.19 Information fusion factor considering information loss and error when CB is switched on and a fault occurs at the exit of low-voltage side line L_{d6}.

Table 5.10 Calculated values of the information fusion factor of each component when CB is switched off and a fault occurs at f_1.

Component name	L_{g1}	L_{z1}	L_{z3}	L_{d1}
Information fusion degree function	1.9946	1.8201	1.7362	1.2375
Information fusion degree expectation function	4.483	4.7905	4.7011	4.386
Information fusion factor	0.4449	0.3799	0.3693	0.2822
Component name	L_{d3}	L_{d6}	L_{d8}	T
Information fusion degree function	1.1924	0.8862	0.8744	3.3846
Information fusion degree expectation function	4.394	3.4161	2.5323	3.6065
Information fusion factor	0.2714	0.2594	0.3453	0.9385

component are shown in Table 5.10. It can be seen that, when a fault occurs at f_1, and CB is switched off, the information fusion factor of the transformer is 0.9385, which is the biggest value. Thus, protection will identify the transformer as the faulty component and trip the corresponding circuit breakers.

When CB is switched off and a fault occurs at f_1 of the transformer and the outlet of high-voltage line L_{g1}, suppose seven random digits of information are missing or in error; the statistical mean value of 100 tests of the information fusion factor is shown in Figures 5.20 and 5.21. It can be seen that the component

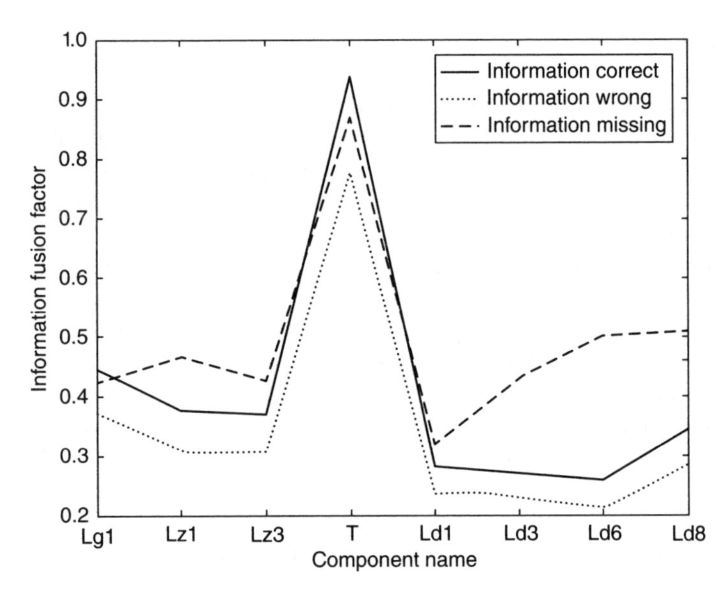

Figure 5.20 Information fusion factor considering information loss and error when CB is switched off and a fault occurs at f_1.

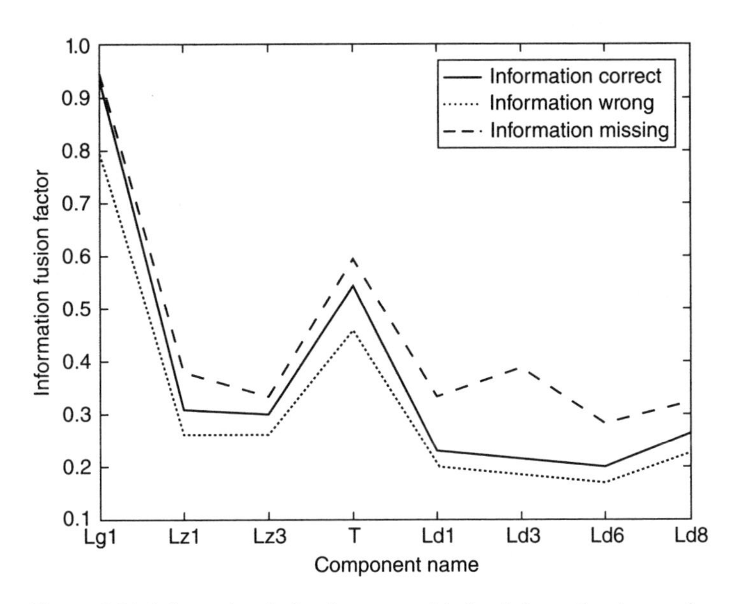

Figure 5.21 Information fusion factor considering information loss and error when CB is switched off and a fault occurs at the exit of high-voltage side line L_{g1}.

corresponding to the maximum information fusion factor is the preset fault component. Therefore, when the network topology changes, the proposed scheme can still identify the faulty component correctly, with good tolerance of information fault.

5.4 Summary

In this chapter, two substation area protection schemes have been introduced, which are based on the upload of protection operational logical signals and global electrical variables respectively, and which could simplify the cooperation between different protection devices in traditional current protection and reduce the operational time of backup protection.

In the substation area protection scheme based on logical variables, the operational information of current protection is fused, and the effective protection range of current protection on the component is quantified, thus the scheme is not affected by the system operating mode or the fault type, etc., and could reflect the relationship between the operation of protection and component fault more specifically. In addition, the scheme is effective even when the protection operational information is missing or in error, thus it is fault-tolerant.

In the substation area protection scheme based on electrical variables, three types of finite overlapping regions are defined according to the scope of effect and the protection range of each region overlapping with each other, thus the fault may be cleared selectively within the minimum range. Compared with traditional backup protection, the scheme significantly improves the backup protection performance of medium-voltage and low-voltage components, and effectively improves the backup protection performance of high-voltage components. In addition, the scheme is reliable and easy to implement.

References

1 Kezunovic, M., Ren, J. and Lotfifard, S. (2016) Basics of Protective Relaying and Design Principles, *Design, Modeling and Evaluation of Protective Relays for Power Systems.*

2 Eremia, M. and Shahidehpour, M. (2013) Power System Stability and Protection, in *Handbook of Electrical Power System Dynamics: Modeling, Stability, and Control,* John Wiley & Sons, Ltd, Chichester, pp. 451–452.

3 Zare, J., Aminifar, F. and Majid, S. P. (2015) Synchrophasor-Based Wide-Area Backup Protection Scheme with Data Requirement Analysis, *IEEE Trans on Power Delivery,* **30** (3), 1410–1419.

4 Neyestanaki, M. K. and Ranjbar, A. M. (2015) An adaptive PMU-based wide area backup protection scheme for power transmission lines, *IEEE Trans on Smart Grids,* **6** (3), 1550–1559.

5 Eissa, M. M., Masoud, M. E. and Elanwar, M. M. M. (2010) A novel back up wide area protection technique for power transmission grids using phasor measurement unit, *IEEE Trans on Power Delivery,* **25** (1), 270–278.

6 IEC 61850-5 Communication networks and systems in substations – Part 5: Communication requirements for functions and device models.

7 Brand, K. P., Brunner, C. and De Mesmaeker, I. (2005) How to use IEC 61850 in protection and automation, *Electra,* **222**, 11–21.

8 Zhu, S. S. (2005) *Principles and technology of high voltage power network protective relaying,* China Electric Power Press, Beijing.

9 Phadke, A. G. and Thorp, J. S. (2009) *Computer Relaying for Power Systems,* Second edition, John Wiley & Sons, Inc., New York.

6

Wide Area Protection

6.1 Introduction

With the integration of large-scale renewable power generation and the development of AC/DC hybrid EHV/UHV interconnected power grids, the scale of the power grid in China is ever expanding, the structure becoming ever more complex and the operating mode ever more changeable. In this context, traditional relay protection that utilizes only local information has problems such as difficult setting and cooperation, long operational time delay, lack of effective overload identification, fault during oscillation, difficulty in tracking the variation in system operating mode, etc. which greatly endanger the safe and stable operation of a power system. Therefore, it is imperative to conduct research on wide area protection which makes full use of multi-source wide area information at multiple locations in the power grid.

In this chapter, first wide area fault identification schemes based on electrical variables are introduced. By defining the fault correlation area, wide area electrical variable information is used to identify the fault directly, without considering the cooperation between different protection devices, thus the security and reliability of protection are improved. In order to lower the requirement of wide area protection schemes on information synchronization and the communication amount, wide area protection schemes based on logical variables are introduced which could allow reliable and fast identification of the faulty component in the case of asynchronous information by extracting the fault characteristic information contained in logical variables. After the fault has been located, it is important to determine the tripping strategy. A wide area protection tripping strategy based on directional weight is also introduced in this chapter, and this covers the functions of local backup protection, remote backup protection and breaker failure protection, simplifies the cooperation between protection devices, shortens the protection tripping time and could limit fault isolation to the minimum range with a certain degree of fault tolerance.

6.2 Wide Area Protection Using Electrical Information

Considering that the measurement information of wide area protection based on electrical quantities is more comprehensive, this section introduces a wide area fault identification scheme based on electrical quantities. The fault components of

Hierarchical Protection for Smart Grids, First Edition. Jing Ma and Zengping Wang.

the voltage and current are introduced in the scheme. On the basis of the quickly formed protection correlation region (PCR), by means of the fault source information [1] and the fault network information [2,3], accurate single and multiple fault identification is realized under the condition of sparse PMU placement.

6.2.1 Wide Area Protection Using Fault Power Source Information

6.2.1.1 Protection Correlation Region

A. PMU Placement Rules under Fault Conditions

It is no longer sufficient to know the bus voltage and the branch current determining the voltage at the other end of the branch for a faulty branch. Knowledge of the voltage at both ends of a branch is also insufficient to determine the branch current if the fault is on that branch. With a fault on the system, the following two PMU placement rules are no longer valid, as the fault location is unknown and the fault may occur on any branch:

- If the bus voltage and branch current at one end of a branch are known, the bus voltage at the other end of the branch can be calculated.
- If the bus voltages at both the ends of a branch are known, the branch current can be calculated.

Alternatively, the following rules are valid even in the presence of a fault [4,5]:

- If there is a PMU at a bus, the bus voltage and currents of all the branches connected to the bus can be measured.
- If a zero injection bus has n branches connected to it and $n - 1$ of the currents are known, then the remaining current can be calculated.
- If n non-zero injection buses with PMUs are adjacent to a zero injection network (no circuit loop in the network), voltages and currents in the network can be calculated. If there is a loop in the network, a PMU is added on the loop to make the system observable.

B. Formation of Protection Correlation Region

On the basis of network topology and PMU placement, the protection correlation region (PCR) is formed as follows:

Step 1: Generate the bus-bus incidence matrix **A** of the whole network in the order of buses with PMUs in the former lines and columns, and buses without PMUs in the latter.

$$\mathbf{A} = \begin{bmatrix} \mathbf{A}_{11} & \mathbf{A}_{12} \\ \mathbf{A}_{12}^{\mathrm{T}} & \mathbf{A}_{22} \end{bmatrix} \tag{6.1}$$

where $\mathbf{A}_{11}$ is the incidence matrix of buses with PMUs. $\mathbf{A}_{22}$ is the incidence matrix of buses without PMUs. $\mathbf{A}_{12}$ is the incidence matrix between the buses with and without PMUs.

Step 2: Any two connected buses where both have PMUs (hence in $\mathbf{A}_{11}$) are formed into a protection correlation region, defined as a specialized PCR.

Step 3: Connected buses in $\mathbf{A}_{22}$ along with the buses to which they are connected in $\mathbf{A}_{11}$ form a generalized PCR.

The calculation of PCRs in $\mathbf{A}_{22}$ is straightforward. The incidence matrix is formed from the branch data with $a_{ii} = 1$ and $a_{ij} = a_{ji} = 1$ if there is a branch connecting i and j. Specialized PCRs are formed from all off-diagonal entries in $\mathbf{A}_{11}$ which are not zero.

To find the generalized PCRs, the matrix $\mathbf{A}_{22}$ is repeatedly multiplied by itself.

$$\mathbf{M}_K = \text{sign}\left(\mathbf{A}_{22}^k\right) \tag{6.2}$$

until

$$\mathbf{M}_{K+1} = \mathbf{M}_K \tag{6.3}$$

The ijth entry of the Kth power of an incidence matrix is known to give the number of different paths of length K beginning at i and ending at j [X, Y]. Multiply $\mathbf{M}_K$ by a column of ones, and then give a vector of integers indicating the connectivity of the non-PMU buses:

$$\boldsymbol{n} = \mathbf{M}_K 1 \tag{6.4}$$

For the example in Section 6.2.1.3, the resulting column has four 1s, two 2s and six 6s. The interpretation is that four of the non-PMU buses are not connected to any other non-PMU buses, that two of the non-PMU buses are connected to each other and that the remaining six buses are connected to each other. The matrix $\mathbf{A}_{12}$ then gives the PMU buses that are connected to these groups of non-PMU buses and defines the generalized PCRs. For example, if the six interconnected non-PMU buses are (5,6,10,11,13,14), then all the PMU buses connected to these six buses (4,7,8,12,15,31,32), combined with the six non-PMU buses, form a large generalized PCR. More details are in the example in Section 6.2.1.3.

6.2.1.2 Introduction to the Fault Branch Location Algorithm

A. Basic Theory

When a fault occurs in the power system, the superposition theorem [6,7] allows us to consider the currents and voltages as containing a prefault component, a fault transient component and a fault steady-state component.

Assuming generators and loads to be current injection sources, the bus voltage vector $\boldsymbol{U}$ and branch current vector $\boldsymbol{I}$ in the prefault network are given by

$$\begin{cases} \boldsymbol{U} = \mathbf{Y}^{-1}\boldsymbol{J} \\ \boldsymbol{I} = \mathbf{Y}_\mathbf{B}\mathbf{A}_\mathbf{A}^\mathrm{T}\boldsymbol{U} \end{cases} \tag{6.5}$$

where $\mathbf{Y}$ is the bus admittance matrix, $\mathbf{Y}_\mathbf{B}$ is the branch admittance matrix, $\mathbf{A}_\mathbf{A}$ is the bus-branch incidence matrix and $\boldsymbol{J}$ is the bus injection current vector.

The variation in the injection currents after the fault inception results in the transient process. Therefore, the bus voltage vector $\boldsymbol{U}_\mathrm{T}$ and branch current vector $\boldsymbol{I}_\mathrm{T}$ in the fault transient state network are given by

$$\begin{cases} \boldsymbol{U}_\mathrm{T} = \mathbf{Y}^{-1}\boldsymbol{J}_\mathrm{T} \\ \boldsymbol{I}_\mathrm{T} = \mathbf{Y}_\mathbf{B}\mathbf{A}_\mathbf{A}^\mathrm{T}\boldsymbol{U}_\mathrm{T} \end{cases} \tag{6.6}$$

where $\boldsymbol{J}_{\mathrm{T}}$ is the post-fault variation of injection current vector.

In the fault steady-state network, the bus voltage vector $\boldsymbol{U}_{\mathrm{S}}$ and branch current vector $\boldsymbol{I}_{\mathrm{S}}$ are given by

$$\begin{cases} \boldsymbol{U}_{\mathrm{S}} = \boldsymbol{U}_{\mathrm{M}} - \boldsymbol{U} - \boldsymbol{U}_{\mathrm{T}} \\ \boldsymbol{I}_{\mathrm{S}} = \boldsymbol{I}_{\mathrm{M}} - \boldsymbol{I} - \boldsymbol{I}_{\mathrm{T}} \end{cases} \tag{6.7}$$

where $\boldsymbol{U}_{\mathrm{M}}$ is the faulty bus voltage vector, and $\boldsymbol{I}_{\mathrm{M}}$ is the faulty branch current vector.

During normal conditions, i_{sd} the fault steady-state component of differential current injecting into each PCR is calculated to monitor the system status. If a fault occurs somewhere in the system, the primary protection should determine the fault within typically 1–1.5 cycles. After this period plus breaker time, if the fault still persists, indicated by $|i_{\mathrm{sd}}|$ > threshold in some PCR, the decision-making unit will take over and perform its backup function. Then a fault branch location mechanism based on the fault correlation factor (FCF) is activated in this PCR until the fault branch is inferred.

B. Fault Correlation Factor

A fault steady-state network of a generalized PCR containing m interconnected non-PMU buses along with n PMU buses is shown in Figure 6.1.

The n PMU buses are connected to the m non-PMU buses via h branches. z_i and y_i are the branch impedance and the ground admittance of the ith branch, respectively, where $i = 1 \ldots h$. $u_1 \ldots u_k \ldots u_n$ are voltages of the n PMU buses. $u_{n+1} \ldots u_{n+g} \ldots u_{n+m}$ are voltages of the m non-PMU buses. $i_1 \ldots\ i_w \ldots i_h$ are the branch currents flowing from the n PMU buses. $i_{n+1} \ldots i_{n+w} \ldots\ i_{n+h}$ are the branch

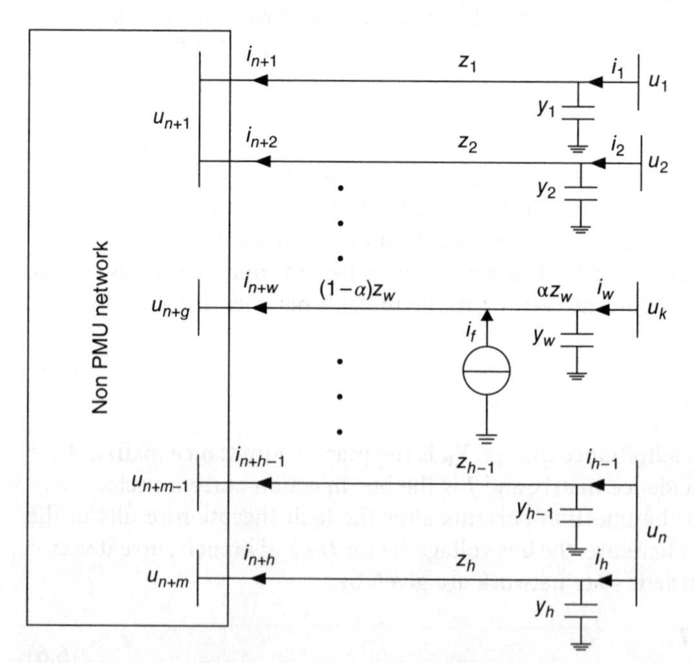

Figure 6.1 Fault steady-state network of a generalized PCR.

currents flowing into the m non-PMU buses. Assume that a fault occurs on the branch w, and the distance between the fault point and the bus k accounts for α per cent of the total branch length. The current injection from the fault point to the network is i_f.

To determine the faulty branch, the m non-PMU buses and the n PMU buses need to be expanded to h buses. As shown in Figure 6.2, the network has been changed by connecting the new buses with the original buses via virtual zero-impedance branches.

Consider $i_{n+1}\ldots i_{n+w}\ldots\ i_{n+h}$ as injection currents of the non-PMU network (blocks in Figures 6.1 and 6.2). The bus voltage vector and branch current vector can be expressed as

$$\begin{bmatrix} \boldsymbol{U}_{\mathrm{A}} \\ \boldsymbol{U}_{\mathrm{B}} \end{bmatrix} = \begin{bmatrix} \mathbf{Z}_{\mathrm{A}} & \mathbf{Z}_{\mathrm{B}} \\ \mathbf{Z}_{\mathrm{C}} & \mathbf{Z}_{\mathrm{D}} \end{bmatrix} \begin{bmatrix} \boldsymbol{J}_{\mathrm{A}} \\ \boldsymbol{J}_{\mathrm{B}} \end{bmatrix} \tag{6.8}$$

where $\mathbf{Z}_{\mathrm{A}}$, $\mathbf{Z}_{\mathrm{B}}$, $\mathbf{Z}_{\mathrm{C}}$ and $\mathbf{Z}_{\mathrm{D}}$ constitute the partitioned bus impedance matrix. $\boldsymbol{J}_{\mathrm{A}} = \left[i_{n+1}\cdots i_{n+h}\right]^{\mathrm{T}}$, $\boldsymbol{J}_{\mathrm{B}} = \left[i_{n+h+1}\cdots i_{n+h+p-m}\right]^{\mathrm{T}}$, $\boldsymbol{U}_{\mathrm{A}} = \left[u_{n+1}\cdots u_{n+h}\right]^{\mathrm{T}}$, $\boldsymbol{U}_{\mathrm{B}} = \left[u_{n+h+1}\cdots u_{n+h+p-m}\right]^{\mathrm{T}}$, $\mathbf{Z}_{\mathrm{A}} = \begin{bmatrix} z_{n+1,n+1} & \cdots & z_{n+1,n+w} & \cdots & z_{n+1,n+h} \\ \vdots & & \vdots & & \vdots \\ z_{n+w,n+1} & \cdots & z_{n+w,n+w} & \cdots & z_{n+w,n+h} \\ \vdots & & \vdots & & \vdots \\ z_{n+h,n+1} & \cdots & z_{n+h,n+w} & \cdots & z_{n+h,n+h} \end{bmatrix}^{\mathrm{T}}$.

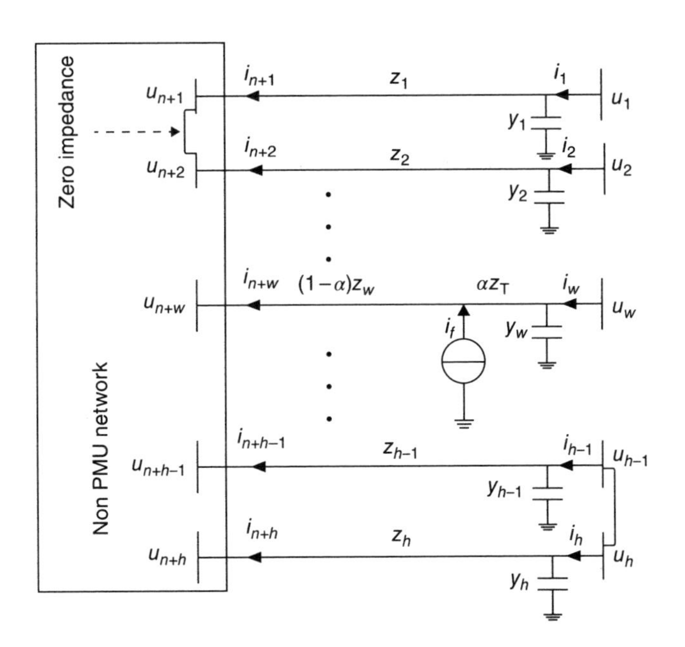

Figure 6.2 Fault steady-state expanded network of the generalized PCR.

Injection currents corresponding to non-faulty branches are

$$i_{n+i} = i_i - y_i u_i \tag{6.9}$$

where $i = 1...h$, and $i \neq w$.

The injection current corresponding to the faulty branch is

$$i_{n+w} = i_w - y_w u_w + i_f \tag{6.10}$$

Substitute Equations (6.9) and (6.10) into $\boldsymbol{U}_A = \mathbf{Z}_A \boldsymbol{J}_A + \mathbf{Z}_B \boldsymbol{J}_B$, and the voltage vector of buses $n+1...n+h$ can be expressed as

$$\begin{bmatrix} u_{n+1} \\ \vdots \\ u_{n+w} \\ \vdots \\ u_{n+h} \end{bmatrix} = i_f \begin{bmatrix} z_{n+1,n+w} \\ \vdots \\ z_{n+w,n+w} \\ \vdots \\ z_{n+h,n+w} \end{bmatrix} + \mathbf{Z}_A \begin{bmatrix} i_1 - y_1 u_1 \\ \vdots \\ i_w - y_w u_w \\ \vdots \\ i_h - y_h u_h \end{bmatrix} \tag{6.11}$$

Assume the faulty branch to be non-existent and the injection currents of all branches to be calculated as in Equation (6.9). Then the virtual voltage vector of buses $n+1...n+h$ is obtained as

$$\begin{bmatrix} u'_{n+1} \\ \vdots \\ u'_{n+w} \\ \vdots \\ u'_{n+h} \end{bmatrix} = \mathbf{Z}_A \begin{bmatrix} i_1 - y_1 u_1 \\ \vdots \\ i_w - y_w u_w \\ \vdots \\ i_h - y_h u_h \end{bmatrix} \tag{6.12}$$

Also, the virtual branch voltage vector $\boldsymbol{U}_X$ is calculated as

$$\boldsymbol{U}_X = \begin{bmatrix} u_1 \\ \vdots \\ u_w \\ \vdots \\ u_h \end{bmatrix} - \begin{bmatrix} u'_{n+1} \\ \vdots \\ u'_{n+w} \\ \vdots \\ u'_{n+h} \end{bmatrix} = \mathbf{Z}_G \begin{bmatrix} i_1 - y_1 u_1 \\ \vdots \\ i_w - y_w u_w + (1-\alpha) i_f \\ \vdots \\ i_h - y_h u_h \end{bmatrix} + i_f \begin{bmatrix} z_{n+1,n+w} \\ \vdots \\ z_{n+w,n+w} \\ \vdots \\ z_{n+h,n+w} \end{bmatrix} \tag{6.13}$$

where $\mathbf{Z}_G = \text{diag}\{z_1 ... z_w ... z_h\}$.

Using Equation (6.9), the normal branch voltage vector $\boldsymbol{U}_Z$ is given by

$$\boldsymbol{U}_Z = \mathbf{Z}_G [i_1 - y_1 u_1 \cdots i_w - y_w u_w \cdots i_h - y_h u_h]^T \tag{6.14}$$

The difference between the virtual branch voltage vector $\boldsymbol{U}_{\mathrm{X}}$ and the normal branch voltage vector $\boldsymbol{U}_{\mathrm{Z}}$ is

$$\Delta\boldsymbol{U}_{\mathrm{X}} = \boldsymbol{U}_{\mathrm{X}} - \boldsymbol{U}_{\mathrm{Z}} = i_{\mathrm{f}} \begin{bmatrix} z_{n+1,n+w} \\ \vdots \\ z_{n+w,n+w} \\ \vdots \\ z_{n+h,n+w} \end{bmatrix} + i_{\mathrm{f}} \begin{bmatrix} 0 \\ \vdots \\ (1-\alpha)z_{W} \\ \vdots \\ 0 \end{bmatrix} \tag{6.15}$$

Vector $\boldsymbol{E}$ is defined as

$$\boldsymbol{E} = \mathbf{Z}_{\mathbf{Ai}}^{-1}\Delta\boldsymbol{U}_{\mathrm{X}} = i_{\mathrm{f}} \begin{bmatrix} z_{n+1,n+w}/z_{n+1,n+i} \\ \vdots \\ z_{n+w,n+w}/z_{n+w,n+i} \\ \vdots \\ z_{n+h,n+w}/z_{n+h,n+i} \end{bmatrix} + i_{\mathrm{f}} \begin{bmatrix} 0 \\ \vdots \\ \dfrac{(1-\alpha)z_{W}}{z_{n+w,n+i}} \\ \vdots \\ 0 \end{bmatrix} \tag{6.16}$$

where $\mathbf{Z}_{\mathbf{Ai}}$ is a diagonal matrix and is formed by the ith column of $\mathbf{Z}_{\mathbf{A}}$. $\mathbf{Z}_{\mathbf{Ai}} = \mathrm{diag}\{z_{n+1,n+i} \ldots z_{n+w,n+i} \ldots z_{n+h,n+i}\}$.

Let $\delta = \frac{z_i}{h-1}\sum_{j\neq 1}\boldsymbol{E}(j)$, $j = 1 \ldots h$. Vector $\boldsymbol{S}$ is defined as

$$\boldsymbol{S} = z_{n+1,n+i}\boldsymbol{E}/(z_i\delta) \tag{6.17}$$

If $i \neq w$,

$$\boldsymbol{S} = \frac{1}{c_i z_i^2} \begin{bmatrix} z_{n+i,n+i}z_{n+1,n+w}/z_{n+1,n+i} \\ \vdots \\ z_{n+i,n+i}z_{n+w,n+w}/z_{n+w,n+i} \\ \vdots \\ z_{n+i,n+i}z_{n+h,n+w}/z_{n+h,n+i} \end{bmatrix} + \begin{bmatrix} 0 \\ \vdots \\ \dfrac{(1-\alpha)z_w z_{n+i,n+i}}{c_i z_{n+w,n+i} z_i^2} \\ \vdots \\ 0 \end{bmatrix} \tag{6.18}$$

where $c_i = \dfrac{1}{h-1}\left[\sum_{j\neq 1}\dfrac{z_{n+j,n+w}}{z_{n+j,n+\mathrm{i}}} + (1-\alpha)\dfrac{z_w}{z_{n+w,n+\mathrm{i}}}\right]$.

If $i = w$,

$$\boldsymbol{S} = \begin{bmatrix} z_{n+w,n+w}/z_w^2 \\ \vdots \\ z_{n+w,n+w}/z_w^2 \\ \vdots \\ z_{n+w,n+w}/z_w^2 \end{bmatrix} + i_{\mathrm{f}} \begin{bmatrix} 0 \\ \vdots \\ (1-\alpha)/z_W \\ \vdots \\ 0 \end{bmatrix} \tag{6.19}$$

From Equations (6.18) and (6.19), we can find that if branch i is not the faulty branch, all entries in the vector $\boldsymbol{S}$ are different from each other. Or else, all entries in the vector $\boldsymbol{S}$ are identical except the ith entry. The maximum absolute difference between entries in the vector $\boldsymbol{S}$ (except the ith entry) is given by

$$D = \max\{\boldsymbol{S}(j_1) - \boldsymbol{S}(j_2)\}, j_1, j_2 \neq i \quad (6.20)$$

D is employed as the FCF to identify the faulty branch. If only D of some branch is less than a threshold, the decision-making unit determines that a fault has occurred on that branch and trips the fault. If D of each branch is more than a threshold, the decision-making unit determines that there is a fault in the non-PMU network. In this case, voltages of buses $n+1...n+h$ and corresponding branch currents are calculated by Equations (6.9) and (6.14). Then the above procedure is executed repeatedly until the faulty branch is located.

6.2.1.3 Test Results and Analysis

The 10-generator, 39-bus New England test system was used to demonstrate the effectiveness of the proposed PCR formation method and wide area backup protection algorithm. The system structure is shown in Figure 6.3 and there is no circuit loop in any zero-injection network.

By adopting the three rules in Section 6.2.1.1, PMUs only need to be placed at the non-zero injection buses to make the system observable. The result of PMU placement is shown in Table 6.1. On the basis of the power network topology and the PMU placement result, bus–bus incidence matrix $\mathbf{A}$ is obtained. $\mathbf{A}_{11}$, $\mathbf{A}_{12}$ and $\mathbf{A}_{22}$ are given in Equations (6.21)–(6.23).

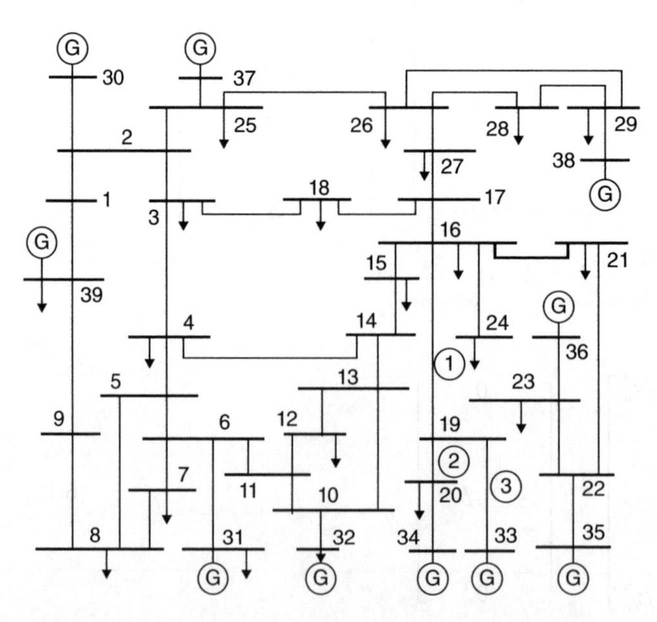

Figure 6.3 10-generator, 39-bus New England test system.

Table 6.1 PMU placement result.

Buses with PMU	Buses without PMU
3,4,7,8,12,15,16,18,20,21,23,24,25,26,27,28,29,30,31,32,33, 34,35,36,37,38,39	1,2,5,6,9,10,11,13,14,17, 19,22

$$
\mathbf{A}_{11}=\begin{bmatrix}
1&1&0&0&0&0&0&1&0&0&0&0&0&0&0&0&0&0&0&0&0&0&0&0&0&0&0\\
1&1&0&0&0&0&0&0&0&0&0&0&0&0&0&0&0&0&0&0&0&0&0&0&0&0&0\\
0&0&1&1&0&0&0&0&0&0&0&0&0&0&0&0&0&0&0&0&0&0&0&0&0&0&0\\
0&0&1&1&0&0&0&0&0&0&0&0&0&0&0&0&0&0&0&0&0&0&0&0&0&0&0\\
0&0&0&0&1&0&0&0&0&0&0&0&0&0&0&0&0&0&0&0&0&0&0&0&0&0&0\\
0&0&0&0&0&1&1&0&0&0&0&0&0&0&0&0&0&0&0&0&0&0&0&0&0&0&0\\
0&0&0&0&0&1&1&0&0&1&0&1&0&0&0&0&0&0&0&0&0&0&0&0&0&0&0\\
1&0&0&0&0&0&0&1&0&0&0&0&0&0&0&0&0&0&0&0&0&0&0&0&0&0&0\\
0&0&0&0&0&0&0&0&1&0&0&0&0&0&0&0&0&0&0&0&0&1&0&0&0&0&0\\
0&0&0&0&0&0&1&0&0&1&0&0&0&0&0&0&0&0&0&0&0&0&0&0&0&0&0\\
0&0&0&0&0&0&0&0&0&0&1&1&0&0&0&0&0&0&0&0&0&0&0&1&0&0&0\\
0&0&0&0&0&0&1&0&0&0&1&1&0&0&0&0&0&0&0&0&0&0&0&0&0&0&0\\
0&0&0&0&0&0&0&0&0&0&0&0&1&1&0&0&0&0&0&0&0&0&0&0&1&0&0\\
0&0&0&0&0&0&0&0&0&0&0&0&1&1&1&1&1&0&0&0&0&0&0&0&0&0&0\\
0&0&0&0&0&0&0&0&0&0&0&0&0&1&1&0&0&0&0&0&0&0&0&0&0&0&0\\
0&0&0&0&0&0&0&0&0&0&0&0&0&1&0&1&1&0&0&0&0&0&0&0&0&0&0\\
0&0&0&0&0&0&0&0&0&0&0&0&0&1&0&1&1&0&0&0&0&0&0&0&0&1&0\\
0&0&0&0&0&0&0&0&0&0&0&0&0&0&0&0&0&1&0&0&0&0&0&0&0&0&0\\
0&0&0&0&0&0&0&0&0&0&0&0&0&0&0&0&0&0&1&0&0&0&0&0&0&0&0\\
0&0&0&0&0&0&0&0&0&0&0&0&0&0&0&0&0&0&0&1&0&0&0&0&0&0&0\\
0&0&0&0&0&0&0&0&0&0&0&0&0&0&0&0&0&0&0&0&1&0&0&0&0&0&0\\
0&0&0&0&0&0&0&0&1&0&0&0&0&0&0&0&0&0&0&0&0&1&0&0&0&0&0\\
0&0&0&0&0&0&0&0&0&0&0&0&0&0&0&0&0&0&0&0&0&0&1&0&0&0&0\\
0&0&0&0&0&0&0&0&0&0&1&0&0&0&0&0&0&0&0&0&0&0&0&1&0&0&0\\
0&0&0&0&0&0&0&0&0&0&0&0&1&0&0&0&0&0&0&0&0&0&0&0&1&0&0\\
0&0&0&0&0&0&0&0&0&0&0&0&0&0&0&0&1&0&0&0&0&0&0&0&0&1&0\\
0&0&0&0&0&0&0&0&0&0&0&0&0&0&0&0&0&0&0&0&0&0&0&0&0&0&1
\end{bmatrix} \tag{6.21}
$$

$$\mathbf{A}_{12} = \begin{bmatrix} 0&0&0&0&0&0&0&0&0&0&0&0&0&0&0&0&0&0&0&0&0&0&0&0&0&0&1 \\ 1&0&0&0&0&0&0&0&0&0&0&0&1&0&0&0&0&1&0&0&0&0&0&0&0&0&0 \\ 0&1&0&1&0&0&0&0&0&0&0&0&0&0&0&0&0&0&0&0&0&0&0&0&0&0&0 \\ 0&0&1&0&0&0&0&0&0&0&0&0&0&0&0&0&0&0&1&0&0&0&0&0&0&0&0 \\ 0&0&0&1&0&0&0&0&0&0&0&0&0&0&0&0&0&0&0&0&0&0&0&0&0&0&1 \\ 0&0&0&0&0&0&0&0&0&0&0&0&0&0&0&0&0&0&0&1&0&0&0&0&0&0&0 \\ 0&0&0&0&1&0&0&0&0&0&0&0&0&0&0&0&0&0&0&0&0&0&0&0&0&0&0 \\ 0&0&0&0&1&0&0&0&0&0&0&0&0&0&0&0&0&0&0&0&0&0&0&0&0&0&0 \\ 0&1&0&0&0&1&0&0&0&0&0&0&0&0&0&0&0&0&0&0&0&0&0&0&0&0&0 \\ 0&0&0&0&0&0&1&1&0&0&0&0&0&0&1&0&0&0&0&0&0&0&0&0&0&0&0 \\ 0&0&0&0&0&0&1&0&1&0&0&0&0&0&0&0&0&0&0&0&1&0&0&0&0&0&0 \\ 0&0&0&0&0&0&0&0&0&1&1&0&0&0&0&0&0&0&0&0&0&0&1&0&0&0&0 \end{bmatrix}^{T} \tag{6.22}$$

$$\mathbf{A}_{22} = \begin{bmatrix} 1&1&0&0&0&0&0&0&0&0&0&0 \\ 1&1&0&0&0&0&0&0&0&0&0&0 \\ 0&0&1&1&0&0&0&0&0&0&0&0 \\ 0&0&1&1&0&0&1&0&0&0&0&0 \\ 0&0&0&0&1&0&0&0&0&0&0&0 \\ 0&0&0&0&0&1&1&1&0&0&0&0 \\ 0&0&0&1&0&1&1&0&0&0&0&0 \\ 0&0&0&0&0&1&0&1&1&0&0&0 \\ 0&0&0&0&0&0&0&1&1&0&0&0 \\ 0&0&0&0&0&0&0&0&0&1&0&0 \\ 0&0&0&0&0&0&0&0&0&0&1&0 \\ 0&0&0&0&0&0&0&0&0&0&0&1 \end{bmatrix} \tag{6.23}$$

Any two connected buses, where both have PMUs (hence in $\mathbf{A}_{11}$), are formed into a specialized PCR, as shown in Table 6.2.

To find the generalized PCRs, the matrix $\mathbf{A}_{22}$ is repeatedly multiplied by itself. When $K = 6$, $\mathbf{M}_{K+1} = \mathbf{M}_K$. Multiply $\mathbf{M}_6$ by a column of ones, and then the vector of integers indicating the connectivity of the non-PMU buses is

$$\boldsymbol{n} = [2\ 2\ 6\ 6\ 1\ 6\ 6\ 6\ 6\ 1\ 1\ 1]^{T} \tag{6.24}$$

From Equation (6.24), it can be found that non-PMU buses 9, 17, 19 and 22 are not connected to any other non-PMU buses, non-PMU buses 1 and 2 are connected to each other and non-PMU buses 5, 6, 10, 11, 13 and 14 are connected to each other. The matrix $\mathbf{A}_{12}$ then gives PMU buses that are connected to these groups of non-PMU buses and defines generalized PCRs as shown in Table 6.3.

Table 6.2 Topology analysis of the specialized PCRs.

PCR	Bus	PCR	Bus	PCR	Bus	PCR	Bus
1	3, 4	5	16, 21	9	23, 36	13	26, 28
2	3, 18	6	16, 24	10	25, 26	14	26, 29
3	7, 8	7	20, 34	11	25, 37	15	28, 29
4	15, 16	8	23, 34	12	26, 27	16	29, 38

Table 6.3 Topology analysis of the generalized PCRs.

PCR	PMU buses	Non-PMU buses	PCR	PMU buses	Non-PMU buses
17	3, 25, 30, 39	1, 2	20	16, 18, 27	17
18	4, 7, 8, 12, 15, 31, 32	5, 6, 10, 11, 13, 14	21	16, 20, 33	19
19	8, 39	9	22	21, 23, 35	22

A. *Fault Correlation Region Determination*

There are three branches in PCR 21: ① (16–19), ② (20–19) and ③ (33–19). A single-phase earth fault and a three-phase fault were applied separately at branch ① (16–19) at the fifth cycle (0.1 s). The fault steady-state components of the differential currents of all PCRs in both cases were calculated and are shown in Figures 6.4 and 6.5.

In each case, the calculated steady-state component of the differential current of PCR 21 is very noticeable, whereas steady-state components of different currents in other PCRs are negligible and have little variation resulting from the measurement and calculation errors. These results are in accordance with the practical

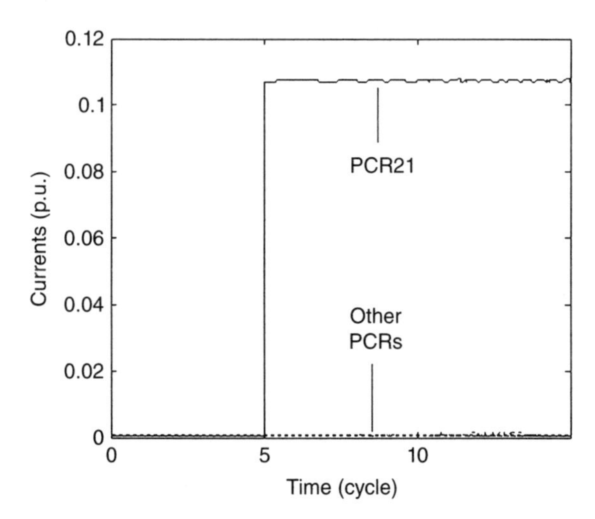

Figure 6.4 Fault steady-state differential currents in different PCRs when a single-phase fault occurs.

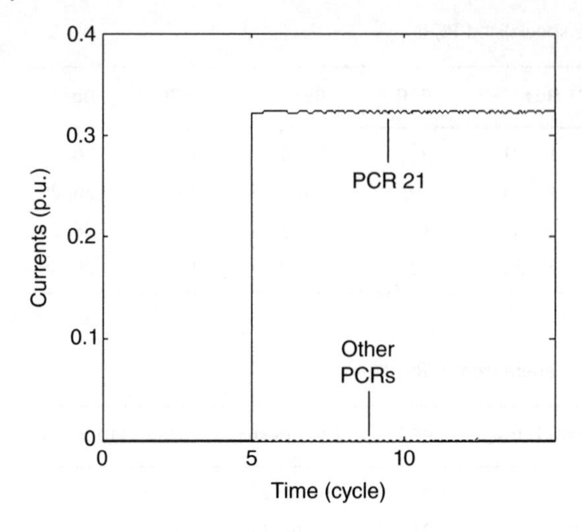

Figure 6.5 Fault steady-state differential currents in different PCRs when a three-phase fault occurs.

Table 6.4 The calculated FCFs of branches ①, ② and ③.

Fault type	①	②	③	Fault type	①	②	③
Single-phase grounding	0.005	47.561	44.098	Two-phase earth fault	0.006	47.538	44.077
Two-phase short circuit	0.007	47.551	44.069	Three-phase short circuit	0.004	47.535	44.072

state of the power system and prove that the method is effective for identifying the fault correlation region no matter what type of fault.

B. Fault Branch Location

Various types of faults with values of fault resistances 0 Ω and 300 Ω were adopted to test the effectiveness and reliability of the FCFs. The calculated FCFs of branches ①, ② and ③ are shown in Table 6.4. The values of FCFs of the faulted branch ① are no more than 0.238, whereas the values of FCFs of non-fault branches ② and ③ are no less than 43.997. These results prove that this method is effective and sensitive for locating the fault at branch ①.

6.2.2 Wide Area Protection Using Fault Network Information

6.2.2.1 Wide Area Protection Based on a Lumped Parameter Model

A. Introduction to the Protection Scheme

The lumped parameter model of transmission line including the influence of distributed capacitance is adopted [8]. When a fault occurs in the power system, the superposition theorem allows us to consider the currents and voltages as containing a prefault component and a superimposed component. The fault branch in the superimposed network is shown in Figure 6.6. z is the branch impedance. y_1 and y_2 are the equivalent ground admittances on each bus. The distance between the fault point and bus $K1$ accounts for α per cent of the total branch length. The current injection at

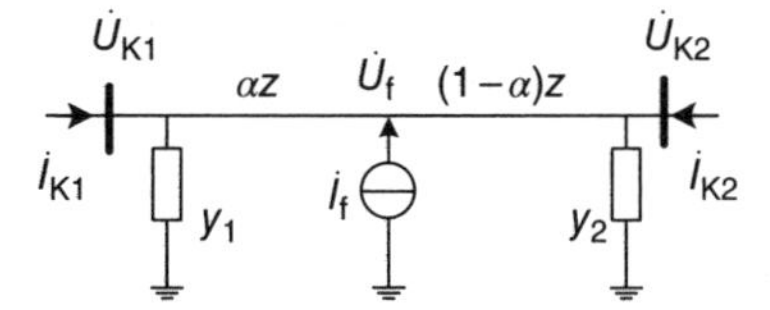

Figure 6.6 Fault branch.

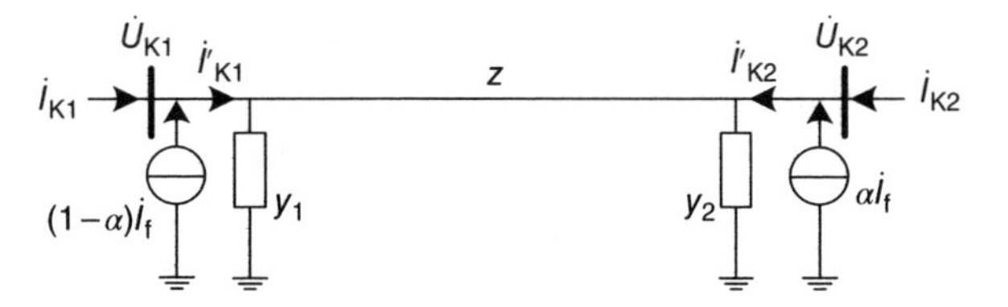

Figure 6.7 Fault model.

fault point is $\dot{I}_f$, and the fault point voltage is assumed to be $\dot{U}_3$. $\dot{I}_{K1}$ and $\dot{I}_{K2}$ are currents injected from other parts of the network. The voltage equation of fault branch is

$$\begin{cases} \left(y_1 + \frac{1}{\alpha z}\right)\dot{U}_{K1} - \frac{1}{\alpha z}\dot{U}_3 = \dot{I}_{K1} \\ \left(y_2 + \frac{1}{(1-\alpha)z}\right)\dot{U}_{K2} - \frac{1}{(1-\alpha)z}\dot{U}_3 = \dot{I}_{K2} \\ -\frac{1}{\alpha z}\dot{U}_{K1} - \frac{1}{(1-\alpha)z}\dot{U}_{K2} + \left(\frac{1}{\alpha z} + \frac{1}{(1-\alpha)z}\right)\dot{U}_3 = \dot{I}_f \end{cases} \tag{6.25}$$

Eliminating $\dot{U}_3$, we get the equivalent matrix equation,

$$\begin{bmatrix} y_1 + 1/z & -1/z \\ -1/z & y_2 + 1/z \end{bmatrix} \begin{bmatrix} \dot{U}_{K1} \\ \dot{U}_{K2} \end{bmatrix} = \begin{bmatrix} \dot{I}_{K1} + (1-\alpha)\dot{I}_f \\ \dot{I}_{K1} + \alpha\dot{I}_f \end{bmatrix} \tag{6.26}$$

Equation (6.26) is the equivalent voltage equation of the fault branch, in which the bus admittance matrix is equal to the one without fault, but bus injections change into superpositions of the original currents and additional currents $(1-\alpha)\dot{I}_f$ and $\alpha\dot{I}_f$. The fault model derived from Equation (6.26) is shown in Figure 6.7. It transforms fault point injection into bus injection, but does not change the branch structure or parameters, so it does not change the bus admittance matrix of the whole network. It also translates the fault distance into partition coefficients of bus injection.

1. Fault Location Algorithm A fault occurs on line b_0, and the fault distance accounts for α per cent of the total line length. According to the fault model, the bus admittance matrix of the network remains unchanged pre/post-fault. The voltage equation of the superimposed network is

$$\boldsymbol{U} = \mathbf{Y}^{-1}\boldsymbol{J} \tag{6.27}$$

where $\mathbf{Y}$ is the bus admittance matrix of the whole system, $\boldsymbol{J}$ is the bus injection current vector and $\boldsymbol{U}$ is the bus voltage vector.

Considering that current variation is more distinct after a fault, the fault component of the branch current is adopted to locate the fault. The branch current vector [9,10] is

$$I = \left(\mathbf{Y_B A}^{\mathrm{T}} + \mathbf{Y_{B1} A_1^{\mathrm{T}}}\right)\mathbf{Y}^{-1} J \tag{6.28}$$

where $\mathbf{Y_B}$ is the branch admittance matrix, $\mathbf{A}$ is the bus-branch incident matrix, $\mathbf{Y_{B1}}$ is the ground admittance matrix, $\mathbf{A_1}$ is the bus-grounding branch incident matrix, and T represents matrix transposition.

Considering the fault branch, the actual branch current vector is

$$I = \left(\mathbf{Y_B A}^{\mathrm{T}} + \mathbf{Y_{B1} A_1^{\mathrm{T}}}\right)\mathbf{Y}^{-1} J - (1-\alpha)\dot{I}_{\mathrm{f}} e_{\mathrm{K1}} - \alpha \dot{I}_{\mathrm{f}} e_{\mathrm{K2}} \tag{6.29}$$

where e_{K1} is a unit vector in which only the K_1th element is 1, and e_{K2} is a unit vector in which only the K_2th element is 1.

Define the network correlation coefficient matrix (NCCM) as

$$\mathbf{C} = \left(\mathbf{Y_B A}^{\mathrm{T}} + \mathbf{Y_{B1} A_1^{\mathrm{T}}}\right)\mathbf{Y}^{-1} \tag{6.30}$$

The matrix reflects the proportional coefficients between branch currents and bus injection currents. It is determined by the topology and parameters of the network, and thus can be calculated before the fault occurs in the network. The ith column of the NCCM is the branch current vector when only bus i has a unit injection, and injections on other buses are all zero.

Only bus $K1$ and $K2$ have injection currents, thus

$$J = \left[0 \cdots (1-\alpha)\dot{I}_{\mathrm{f}} \cdots \alpha \dot{I}_{\mathrm{f}} \cdots 0\right] \tag{6.31}$$

Defining $\mathbf{C_{K1}}$ and $\mathbf{C_{K2}}$ as the K_1th and K_2th columns of matrix $\mathbf{C}$, respectively gives

$$I = \left[\mathbf{C_{K1}} - e_{\mathrm{K1}}, \mathbf{C_{K2}} - e_{\mathrm{K2}}\right] \left[(1-\alpha) \;\; \alpha\right]^{\mathrm{T}} \dot{I}_{\mathrm{f}} \tag{6.32}$$

To branch b, let $d \in [0,1]$, then function $\mathbf{p}(b, d)$ is calculated as

$$\mathbf{p}(b,d) = I \cdot / \left[(1-d)(\mathbf{C_{K1}} - e_{\mathrm{K1}}) + d(\mathbf{C_{K2}} - e_{\mathrm{K2}})\right] \tag{6.33}$$

where $\cdot/$is the corresponding element division of the two vectors.

If the calculated branch is just the fault branch, i.e. $b = b_0$, and $d = \alpha$, all elements in $\mathbf{p}(b_0, \alpha)$ are equal to $\dot{I}_{\mathrm{f}}$; otherwise a great difference exists between its elements. To utilize the closing degree of the elements in $\mathbf{p}(b, d)$, the location function is defined as the first order central moment of $\mathbf{p}(b, d)$.

$$g(b,d) = \sum_{i=1}^{m} \left| \mathbf{p}[i] - \frac{1}{m}\sum_{j=1}^{m} \mathbf{p}[j] \right| \tag{6.34}$$

where m is the element number of $\mathbf{p}(b, d)$, $\mathbf{p}[i]$ is the ith element of $\mathbf{p}(b, d)$ and b is the calculated branch.

According to previous analysis, the location function is a binary function of the calculated branch and distance. If the calculated branch is the faulty one, with the calculated distance, then $g = 0$ is tenable. Calculation error makes $g \approx 0$. Otherwise the value of g is large. Once a fault is detected, making variable d vary in closed interval [0,1], we search all the branches one by one. The point (b_0,d_0) where location function g takes its minimum value is determined as the fault position. Thus, b_0 is the faulty branch and d_0 is the fault distance. The fault model makes the effect of the fault location equivalent to bus injection variation, and makes the network topology and bus admittance matrix invariant, thus matrix modification and inversion are dispensed with in each search step, and the calculation burden is greatly reduced. Combining with variable step search, which applies a large step size when g is big and a small step size when g is small, fault location time is significantly reduced, while location accuracy is guaranteed.

2. PMU Placement If elements at the same position in $\mathbf{C}_{K1} - \boldsymbol{e}_{K1}$ and $\mathbf{C}_{K2} - \boldsymbol{e}_{K2}$ are both equal to 0, the corresponding branch current caused by injection currents on buses $K1$ and $K2$ is identically vanishing. Partition coefficients caused by fault distance fail and the fault model is not applicable. Meanwhile, invalid element $\frac{0}{0}$ appears in vector $\boldsymbol{p}$ due to Equation (6.33). $\frac{0}{0}$ can be any random value in practical calculation, and invalidates the location function. According to the analysis above, PMUs should measure the branch currents that are permanently nonzero whichever bus the injection is on, i.e. the branch whose corresponding row in the NCCM does not contain 0. Physically, these branches are connected to ground through a small impedance in the superimposed network. In a practical superimposed network, branches are connected to ground via transient reactance of generators and load impedance, which is very large. PMUs on generator terminals – which measure the currents of branches connected to generators directly – can meet the requirements. Such PMU configuration is also suitable for the present installation situation.

B. Scheme Verification

The IEEE 9-bus test system is used to demonstrate the effectiveness of the proposed fault location algorithm. The system structure is shown in Figure 6.8.

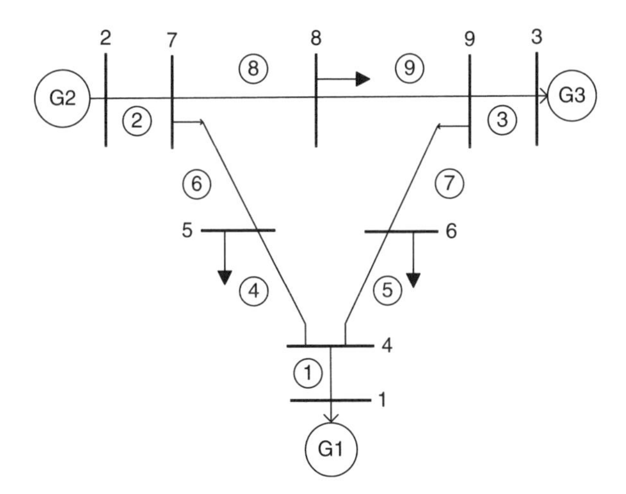

Figure 6.8 IEEE 9-bus test system.

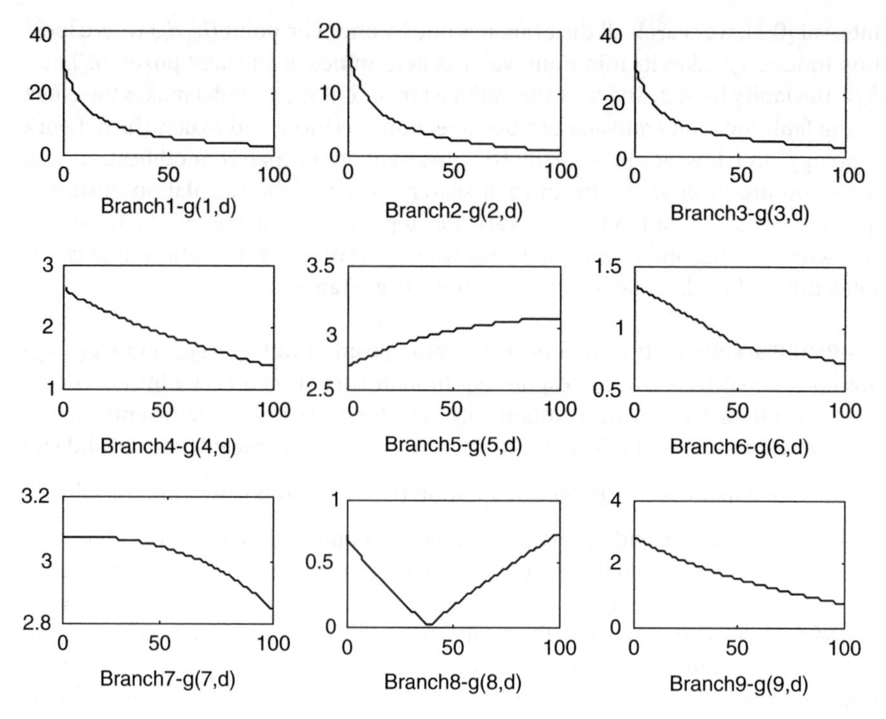

Figure 6.9 Location function value on different branches.

The system voltage is 220 kV. The length of each transmission line is 100 km. The positive and negative sequence parameters of each line are $R_1 = 0.035\ \Omega/\text{km}$, $X_1 = 0.5077\ \Omega/\text{km}$, $G_1 = 0$, $B_1 = 3.271\text{E} - 6\ \text{S/km}$; and zero-sequence parameters are $R_0 = 0.03631\ \Omega/\text{km}$, $X_0 = 0.1326\ \Omega/\text{km}$, $G_0 = 0$, $B_0 = 2.322\text{E} - 6\ \text{S/km}$. Generator parameters are G1: $Z_1 = 0.155 + \text{j}5.95\ \Omega$, $Z_0 = 1.786 + \text{j}7.58\ \Omega$;G2: $Z_1 = 0.238 + \text{j}6.19\ \Omega$, $Z_0 = 0.833 + \text{j}5.12\ \Omega$; G3: $Z_1 = 0.420 + \text{j}5.95\ \Omega$, $Z_0 = 1.785 + \text{j}7.54\ \Omega$. Load power is 30 + j20 MVA on each load bus.

A single-phase fault is applied on line ⑧, and the fault position is 40 km from bus 7. The calculation results of the location function are shown in Figure 6.9.

In Figure 6.9, it is clear that the value of function g is large at all points on all branches except branch ⑧. A zero-value value point exists on branch ⑧. Thus, branch ⑧ is determined to be the faulty line, and the zero-value point on branch ⑧ is the accurate fault position. Then, all nine lines were set as the fault branch separately, and similar results were obtained. It is deduced that the novel fault location algorithm can locate the fault branch accurately and locate the fault distance effectively. Different types of faults at different positions on line ⑧ were simulated, and the location error was used to measure the location accuracy. It is defined as $h = (d_0 - \alpha)/l \times 100$, where d_0 is the calculated fault distance using the proposed algorithm, and l is the total length of the faulty line. Location results are shown in Table 6.5. The maximum location error is −0.09%. Hence, high-precision fault distance location is realized.

To verify the adaptability to a distributed parameter model, the line parameters in Figure 6.8 are changed as follows. The positive and negative sequence parameters are $R_1 = 0.02083\ \Omega/\text{km}$, $X_1 = 0.2811\ \Omega/\text{km}$, $G_1 = 0$, $B_1 = 4.0527\text{E} - 6$(S/km; and the

Table 6.5 Fault distance location simulation results.

Actual distance	Fault type	AG		BC		BCG		ABC	
	Fault resistance (Ω)	0	300	0	300	0	300	0	300
10	Fault distance (km)	9.91	10	9.99	10.01	9.96	10.01	9.94	10
	Location error (%)	−0.09	0	−0.01	0.01	−0.04	0.01	−0.06	0
20	Fault distance (km)	19.93	20	19.99	20.01	19.97	20.01	19.95	20
	Location error (%)	−0.07	0	−0.01	0.01	−0.03	0.01	−0.05	0
30	Fault distance (km)	29.95	29.99	29.99	30	29.97	30	29.97	30
	Location error (%)	−0.05	−0.01	−0.01	0	−0.03	0	−0.03	0
40	Fault distance (km)	39.96	39.98	39.98	39.99	39.97	39.98	39.96	39.98
	Location error (%)	−0.04	−0.02	−0.02	−0.01	−0.03	−0.02	−0.04	−0.02
50	Fault distance (km)	49.96	49.97	49.96	49.97	49.97	49.97	49.97	49.97
	Location error (%)	−0.04	−0.03	−0.04	−0.03	−0.03	−0.03	−0.03	−0.03
60	Fault distance (km)	59.97	59.95	59.96	59.96	59.96	59.96	59.97	59.95
	Location error (%)	−0.03	−0.05	−0.04	−0.04	−0.04	−0.04	−0.03	−0.05
70	Fault distance (km)	69.99	69.95	69.93	69.95	69.93	69.95	69.96	69.95
	Location error (%)	−0.01	−0.05	−0.07	−0.05	−0.07	−0.05	−0.04	−0.05
80	Fault distance (km)	80.02	79.95	79.94	79.95	79.93	79.95	79.96	79.95
	Location error (%)	0.02	−0.05	−0.06	−0.05	−0.07	−0.05	−0.04	−0.05
90	Fault distance (km)	90.06	89.96	89.96	89.97	89.96	89.97	89.94	89.97
	Location error (%)	0.06	−0.04	−0.04	−0.03	−0.04	−0.03	−0.06	−0.03

zero-sequence parameters are $R_0 = 0.1148\,\Omega/\text{km}$, $X_0 = 0.7190\,\Omega/\text{km}$, $G_0 = 0$, $B_0 = 1.6431\text{E} - 6\,\text{S/km}$. The transmission line uses a distributed parameter model. The system voltage is 500 kV, and each line length is 300 km. A single-phase fault is applied at line ⑧, and the fault position is 40% from bus 7 of the total line length. The calculation results of the location function are shown in Figure 6.10.

Similar results to the lumped parameter system are obtained by the proposed algorithm. The function value is large on fault-free branches, and close to 0 at the fault point. Different types of fault at different positions on line ⑧ were simulated. The fault distance location results are shown in Table 6.6. Considering the results in Figure 6.10 and Table 6.6, the proposed fault location algorithm can locate the faulty branch accurately and locate the fault distance with precision, although the fault distance location error increases.

6.2.2.2 Wide Area Protection Based on a Distributed Parameter Model

A. Introduction to the Protection Scheme

1. Theory of Distribution of the Fault Addition Current As can be found from the analysis in Section 6.2.1, after a faulty branch current fault, the steady-state component of the branch current vector can be expressed as

$$\vec{I} = \vec{I}_g - \vec{I}_w - \vec{I}_z \tag{6.35}$$

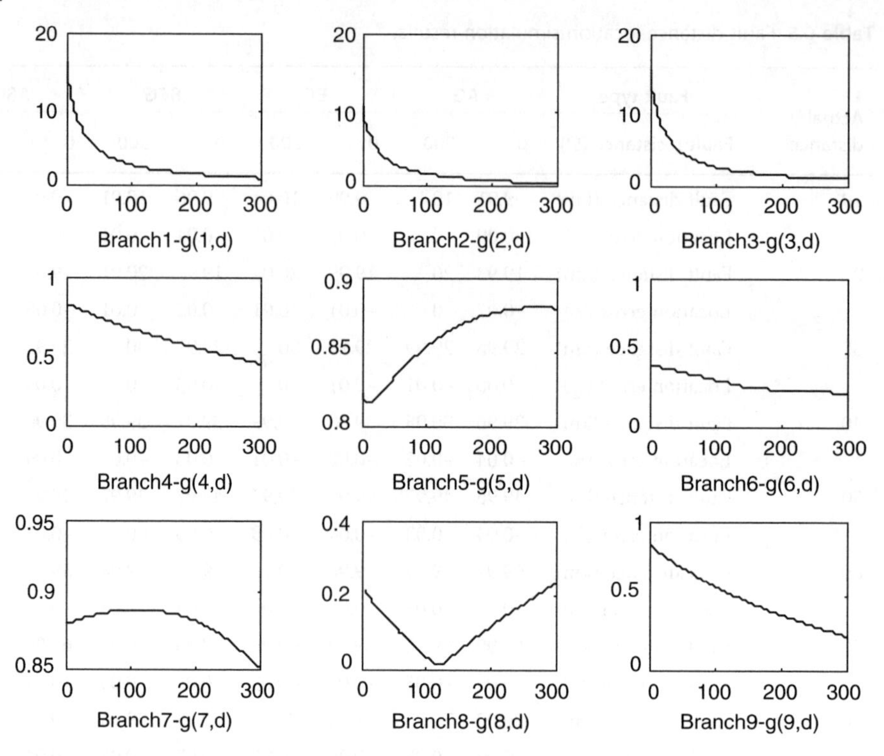

Figure 6.10 Location function value on different branches.

Table 6.6 Fault location simulation results.

	Fault type	AG		BC		BCG		ABC	
AD	Fault resistance (Ω)	0	300	0	300	0	300	0	300
60	Fault distance (km)	61.59	59.73	62.82	59.91	62.67	58.56	63	60.84
	Location error (%)	0.53	−0.09	0.94	−0.03	0.89	−0.48	1	0.28
120	Fault distance (km)	120.45	119.91	120.87	120.21	120.9	119.94	120.9	120.36
	Location error (%)	0.15	−0.03	0.29	0.07	0.3	−0.02	0.3	0.12
180	Fault distance (km)	178.89	179.58	178.02	179.94	178.08	179.88	177.9	179.97
	Location error (%)	−0.37	−0.14	−0.66	−0.02	−0.64	−0.04	−0.7	−0.01
240	Fault distance (km)	238.65	239.46	238.35	239.73	238.29	239.76	237.36	239.67
	Location error (%)	−0.45	−0.18	−0.55	−0.09	−0.57	−0.08	−0.88	−0.11

Note: AD is an abbreviation of actual distance.

Figure 6.11 presents a branch b_0 in the distributed parameter system with buses K_1 and K_2, where the equivalent impedance Z' and equivalent admittance Y' can be calculated as

$$\begin{cases} Z' = Z_C \sinh[\gamma l] \\ Y' = \dfrac{\cosh[\gamma l] - 1}{Z_C \sinh[\gamma l]} \end{cases} \tag{6.36}$$

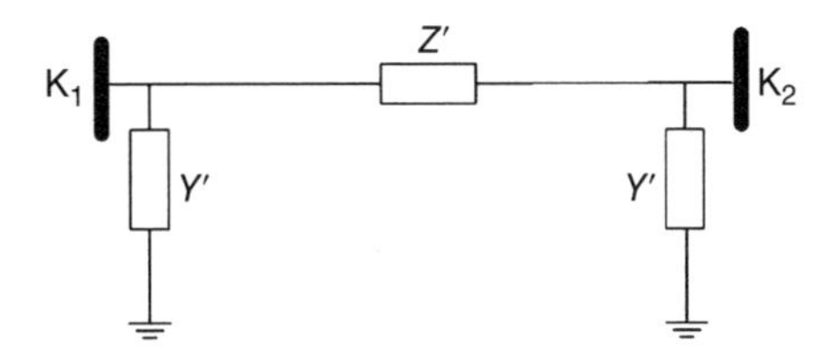

Figure 6.11 Branch model in normal operation.

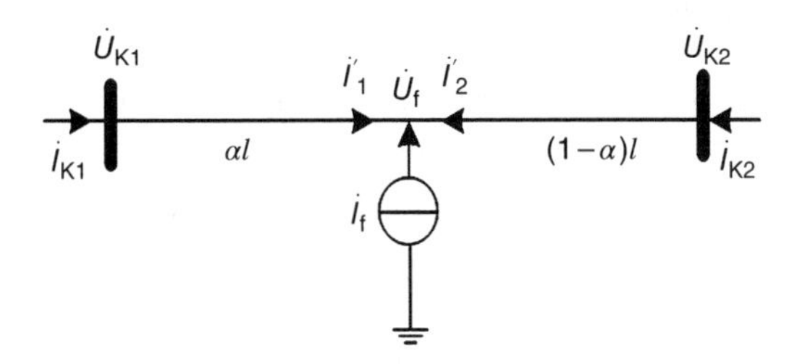

Figure 6.12 Fault steady-state model of branch.

The bus admittance matrix of this branch is given by

$$\mathbf{Y_K} = \begin{bmatrix} \dfrac{\cosh[\gamma l]}{Z_C \sinh[\gamma l]} & -\dfrac{1}{Z_C \sinh[\gamma l]} \\ -\dfrac{1}{Z_C \sinh[\gamma l]} & \dfrac{\cosh[\gamma l]}{Z_C \sinh[\gamma l]} \end{bmatrix} \tag{6.37}$$

When a fault occurs, the fault steady-state model of the branch b_0 is as shown in Figure 6.12.

The nodal equation of the branch is given by

$$\begin{cases} u_f = u_{K1} \cosh[\gamma \alpha l] - i_{K1} Z_C \sinh[\gamma \alpha l] \\ u_f = u_{K2} \cosh[\gamma (1-\alpha) l] - i_{K2} Z_C \sinh[\gamma (1-\alpha) l] \\ i_1' = i_{K1} \cosh[\gamma \alpha l] - \dfrac{u_{K1}}{Z_C} \sinh[\gamma \alpha l] \\ i_2' = i_{K2} \cosh[\gamma (1-\alpha) l] - \dfrac{u_{K2}}{Z_C} \sinh[\gamma (1-\alpha) l] \\ 0 = i_1' + i_2' + i_f \end{cases} \tag{6.38}$$

By eliminating u_f, i_1' and i_2', the nodal equation can be written as

$$\begin{bmatrix} i_{K1} + \dfrac{\sinh[\gamma(1-\alpha)l]}{\sinh[\gamma l]} i_f \\ i_{K2} + \dfrac{\sinh[\gamma \alpha l]}{\sinh[\gamma l]} i_f \end{bmatrix} = \begin{bmatrix} \dfrac{\cosh[\gamma l]}{Z_C \sinh[\gamma l]} & -\dfrac{1}{Z_C \sinh[\gamma l]} \\ -\dfrac{1}{Z_C \sinh[\gamma l]} & \dfrac{\cosh[\gamma l]}{Z_C \sinh[\gamma l]} \end{bmatrix} \begin{bmatrix} u_{K1} \\ u_{K2} \end{bmatrix} \tag{6.39}$$

where the bus admittance matrix does not need to be modified when faults occur. In addition, the injected current in the fault point is assigned to the bus injected currents, as shown in Figure 6.13.

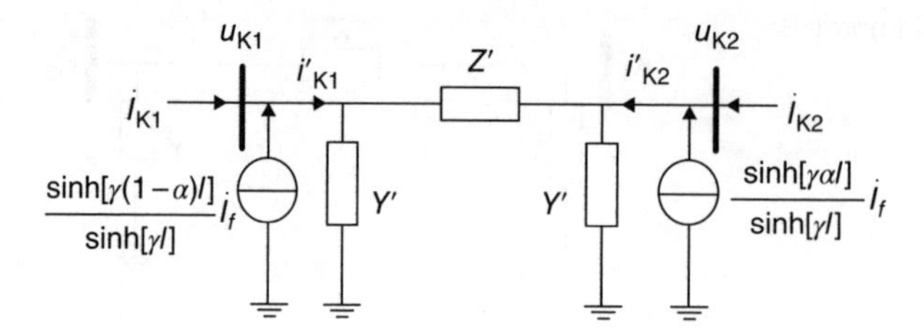

Figure 6.13 Fault steady-state equivalent model of branch.

2. Theory of Fault Matching Degree In the fault steady-state network, the line current vector of the branch with PMU is given by

$$I = CJ - \frac{\sinh[\gamma(1-\alpha)l]}{\sinh[\gamma l]} i_f e_{K1} - \frac{\sinh[\gamma\alpha l]}{\sinh[\gamma l]} i_f e_{K2} \tag{6.40}$$

where $\mathbf{C} = (\mathbf{Y_B A_A^T} + \mathbf{Y_{B1} A_1^T})\mathbf{Y}^{-1}$ is defined as the branch contribution factor matrix. It is clear that the matrix **C** is formed in the normal operating condition and remains unchanged in the progress of a fault location. e_{K1} is a zero vector except for a '1' in the row of directional branch K_1 to K_2. e_{K2} is a zero vector except a '1' in the row of directional branch K_2 to K_1.

Based on the fault addition current distribution theory, there are only two injected currents located at the bus K_1 and bus K_2, respectively. The bus injected current vector can be written as

$$J = \left[0 \cdots \frac{\sinh[\gamma(1-\alpha)l]}{\sinh[\gamma l]} i_f \cdots \frac{\sinh[\gamma\alpha l]}{\sinh[\gamma l]} i_f \cdots 0\right]^{\mathrm{T}} \tag{6.41}$$

From Equations (6.40) and (6.41), the line current vector of the branch with PMU is given by

$$I = [\mathbf{C} - e \quad \mathbf{C} - e] \begin{bmatrix} \frac{\sinh[\gamma(1-\alpha)l]}{\sinh[\gamma l]} i_f \\ \frac{\sinh[\gamma\alpha l]}{\sinh[\gamma l]} i_f \end{bmatrix} = D \cdot G \tag{6.42}$$

where $G = \left[\frac{\sinh[g(1-a)l]}{\sinh[gl]} i_f \quad \frac{\sinh[\gamma\alpha l]}{\sinh[\gamma l]} i_f\right]^{\mathrm{T}}$, $D = [\mathbf{C_{K1}} - e_{K1} \quad \mathbf{C_{K2}} - e_{K2}]$, $\mathbf{C_{K1}}$ and $\mathbf{C_{K2}}$ are the K_1th and K_2th columns of the matrix **C**, respectively.

Then, $\mathbf{C_{K1}}$ and $\mathbf{C_{K2}}$ of every branch in each PCR are used in Equation (6.42) to locate the fault. By using least squares algorithms, vector G can be calculated as

$$G = (D^{\mathrm{T}} D)^{-1} D^{\mathrm{T}} I \tag{6.43}$$

Assuming that g_1 and g_2 are the phasors in G, they are given by

$$\begin{cases} g_1 = \frac{\sinh[\gamma(1-\alpha')l]}{\sinh[\gamma l]} i_f \\ g_2 = \frac{\sinh[\gamma\alpha' l]}{\sinh[\gamma l]} i_f \end{cases} \tag{6.44}$$

From Equations (6.44), virtual fault locations of each branch can be calculated as

$$\alpha' = \frac{1}{2\gamma l}\ln\frac{g_1 + g_2\cosh(\gamma l) + g_2\sinh(\gamma l)}{g_1 + g_2\cosh(\gamma l) - g_2\sinh(\gamma l)} \tag{6.45}$$

Only if the virtual fault location of a branch meets the requirement of $a' \in [0,1]$ is the branch called a suspicious branch. Then, taking vector $\boldsymbol{G}$ of each suspicious branch into Equation (6.40) we can calculate the fault steady state of line current vector $\boldsymbol{I}_{\text{cal}}$. The fault matching degree (FMD) is defined as the distance between $\boldsymbol{I}_{\text{cal}}$ and the fault steady state of line current $\boldsymbol{I}$ measured by PMU. The definition is given by

$$FMD = \frac{\|\boldsymbol{I}_{\text{cal}} - \boldsymbol{I}\|_2}{\|\boldsymbol{I}\|_2} \times 100 \tag{6.46}$$

In the faulted branch, the FMD is near to zero. Although the influence of the calculating error and measurement error in actual working conditions are taken into consideration, the value of the FMD could still be quite small. However, in the non-faulted branch, the FMD is a rather large number. Therefore, the branch with the minimum value of the FMD is considered as the faulted branch.

B. Scheme Verification

The simulation model and PMU configuration are set with reference to Section 6.2.1. On this basis, the protection correlation region and the wide area backup protection algorithm are comprehensively analysed and verified as follows.

1. Formation of Protection Correlation Region Firstly, $\mathbf{A}_{11}$ is formed in the order of nodes with PMU. Then a specialized PCR is established, as shown in Table 6.7. Also, $\mathbf{A}_{22}$ is formed in the order of nodes without PMU. Finally, a generalized PCR is formed, as shown in Table 6.8.

Table 6.7 The specialized protection correlation region.

PCR	Node1	Node2	PCR	Node1	Node2
1	3	4	9	23	36
2	3	18	10	25	26
3	7	8	11	25	37
4	15	16	12	26	27
5	16	21	13	26	28
6	16	24	14	26	29
7	20	34	15	28	29
8	23	24	16	29	38

Table 6.8 The generalized protection associated area.

PCR	Nodes with PMU	Nodes without PMU
17	3,25,30,39	1,2
18	4,7,8,12,15,31,32	5,6,10,11,13,14
19	8,39	9
20	16,18,27	17
21	16,20,33	19
22	21,23,35	22

Table 6.9 Virtual fault locations.

Fault type	AG		BC		BCG		ABC	
Fault resistance (Ω)	0	300	0	300	0	300	0	300
4–5	2.033	2.029	2.030	2.030	2.030	2.030	2.034	2.030
4–14	1.139	1.141	1.140	1.140	1.140	1.140	1.138	1.140
8–5	1.166	1.166	1.166	1.166	1.166	1.166	1.165	1.166
7–6	1.088	1.087	1.088	1.088	1.088	1.088	1.088	1.088
31–6	0.315	0.322	0.318	0.318	0.318	0.319	0.313	0.319
32–10	2.526	2.531	2.531	2.530	2.530	2.531	2.525	2.531
12–11	0.200	0.200	0.200	0.200	0.200	0.200	0.200	0.200
12–13	2.311	2.311	2.311	2.311	2.311	2.311	2.311	2.311
15–14	1.062	1.062	1.062	1.062	1.062	1.062	1.062	1.062
5–6	1.523	1.521	1.522	1.522	1.522	1.521	1.524	1.521
6–11	1.597	1.591	1.588	1.588	1.589	1.589	1.596	1.590
10–11	2.434	2.440	2.438	2.438	2.438	2.439	2.433	2.439
10–13	−2.436	−2.443	−2.441	−2.441	−2.440	−2.441	−2.435	−2.441
13–14	0.417	0.410	0.413	0.413	0.413	0.412	0.419	0.411

2. Determination of Fault Association Area In tests, different faults were applied on the 20% of branch 12–11 in the 18th PCR. The minimum fault steady-state component of the differential current of the 18th PCR is 119.2 A. However, the maximum fault steady-state component of the differential current of the other PCR is less than 1 A. Therefore, the 18th PCR is determined as the fault association area.

3. Determination of Faulted Branch There are 14 branches in the 18th PCR. The virtual fault locations of each branch are presented in Table 6.9, where the branches 31–6, 12–11 and 13–14 are determined as suspicious branches.

Figures 6.14–6.17 show the FMD of three suspicious branches under different fault types and fault resistances. It is clear that the FMD of branch 12–11 is near to zero and much smaller than that of the other two branches.

To verify the accuracy of the proposed method, many tests have been carried out in the EPDL. In these tests, different faults have been applied in branch 12–11 with different fault points. The results of the fault location are presented in Table 6.10. It can be seen that the error of the proposed method is no more than 0.11%. In addition, it is suitable for any PCRs. Hence, the method proposed can not only determine the faulted branch, but also determine the fault point.

6.2.3 Wide Area Protection Suitable for Multiple Fault Identification

6.2.3.1 Introduction to Multi-Fault Identification Algorithm

The lumped-parameter model in Section 6.2.2 is used in this section, shown in Figure 6.6. It can be seen in Figure 6.7 that the equivalent model distributes

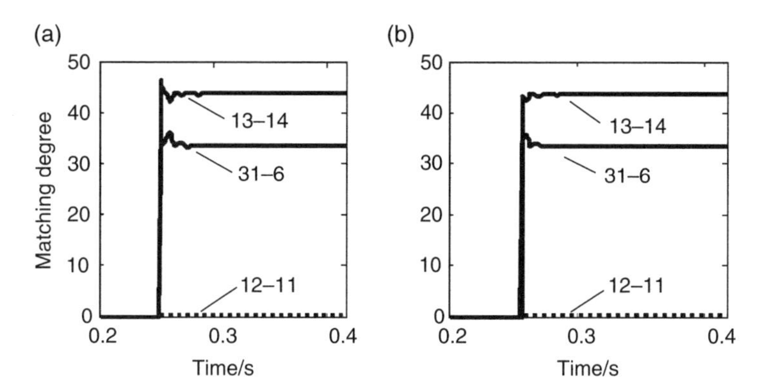

Figure 6.14 Fault matching degree of suspicious branches during single-phase grounding fault. (a) Metal fault. (b) Fault through 300 Ω resistance.

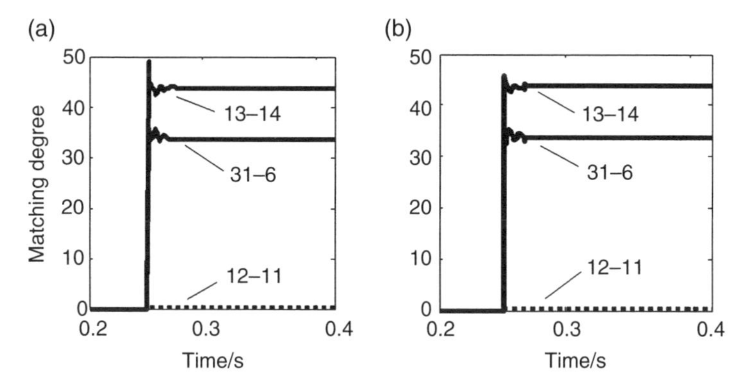

Figure 6.15 Fault matching degree of suspicious branches during two-phase fault. (a) Metal fault. (b) Fault through 300 Ω resistance.

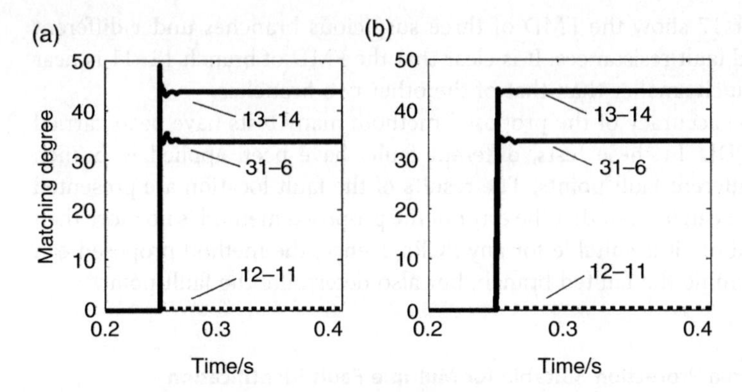

Figure 6.16 Fault matching degree of suspicious branches during two-phase grounding fault. (a) Metal fault. (b) Fault through 300 Ω resistance.

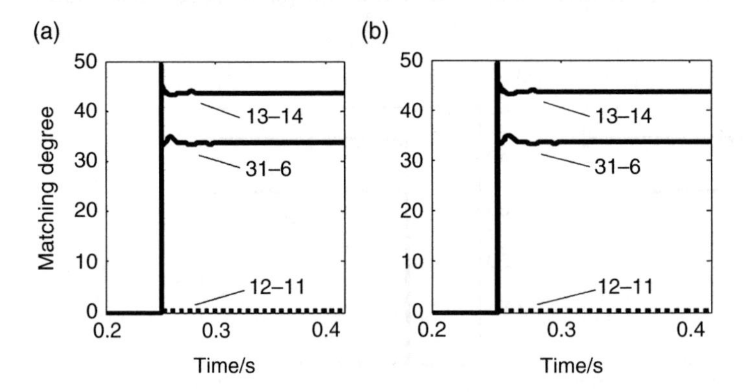

Figure 6.17 Fault matching degree of suspicious branches during three-phase fault. (a) Metal fault. (b) Fault through 300 Ω resistance.

Table 6.10 Fault location results.

	Fault type	**AG**		**BC**		**BCG**		**ABC**	
Fault location (km)	**Fault resistance (Ω)**	**0**	**300**	**0**	**300**	**0**	**300**	**0**	**300**
20	distance	20.02	20.00	20.00	19.99	19.99	19.99	19.98	20.02
	error	0.02	0.00	0.00	−0.01	−0.01	−0.01	−0.02	0.02
40	distance	40.04	40.01	40.00	39.99	40.00	40.00	40.01	40.02
	error	0.04	0.01	0.00	−0.01	0.00	0.00	0.01	0.02
60	distance	60.06	60.01	60.00	60.00	60.01	60.00	60.05	60.02
	error	0.06	0.01	0.00	0.00	0.01	0.00	0.05	0.02
80	distance	80.11	80.01	80.01	80.00	80.02	80.00	80.10	80.02
	error	0.11	0.01	0.01	0.00	0.02	0.00	0.10	0.02

the additional current to the terminals of the branch. The structure and parameters of the branch remain unchanged, so the admittance matrix does not change and thus the computation is reduced.

Here, the situation where two faults occur in different lines simultaneously is taken as an example. According to the superposition theorem, the node of the branch current configuration PMU column vector can be expressed as

$$\vec{I} = \mathbf{P}\vec{R} - (1-a)\dot{I}_{f1}\vec{g}_{K1} - a\dot{I}_{f1}\vec{g}_{K2} - (1-b)\dot{I}_{f2}\vec{g}_{K3} - b\dot{I}_{f2}\vec{g}_{K4} \tag{6.47}$$

$$\mathbf{P} = \left(\mathbf{Y_B}\mathbf{A}^{\mathrm{T}} + \mathbf{Y_{B1}}\mathbf{A}_1^{\mathrm{T}}\right)\mathbf{Y}^{-1} \tag{6.48}$$

where $\mathbf{P}$ is the correlation matrix, $\mathbf{Y_B}$ is the branch admittance matrix, $\mathbf{A}$ is the node-to-branch incidence matrix, $\mathbf{Y_{B1}}$ is the branch-to-ground admittance matrix, $\mathbf{A_1}$ is the node-to-branch incidence matrix and T represents the matrix transposition.

From the previous analysis, the matrix can be formed in the system during normal operation. The fault location process remains the same. $\vec{g}_{K1}$ is a column vector and its element corresponding to K_1–K_2 (from K_1 to K_2) is 1, and the remaining elements are zeros. $\vec{g}_{K2}$ is a column vector and its element corresponding to K_2–K_1 (from K_2 to K_1) is 1, and the remaining elements are zeros. $\mathbf{Y}$ is the admittance matrix of the network, $\vec{R}$ is the current column vector which injects into the network nodes.

Based on the preceding additional fault current distribution principle, steady-state fault slip nodes at both ends of the additional fault network only in K_1 and K_2 have the injection currents, which can get into the current node column vector:

$$\vec{R} = \left[0\cdots(1-a)\dot{I}_{f1}\cdots a\dot{I}_{f1}\cdots(1-b)\dot{I}_{f2}\cdots b\dot{I}_{f2}\cdots 0\right] \tag{6.49}$$

$\vec{g}_{K1}$ and $\vec{g}_{K2}$ represent the K_1th and K_2th column of matrix $\mathbf{P}$, respectively.

$$\vec{I} = \left[\vec{P_{k1}} - \vec{g_{k1}}\ \ \vec{P_{k2}} - \vec{g_{k2}}\ \ \vec{P_{k3}} - \vec{g_{k3}}\ \ \vec{P_{k4}} - \vec{g_{k4}}\right] \bullet \left[(1-a)\dot{I}_{f1}\ \ a\dot{I}_{f1}\ \ (1-b)\dot{I}_{f2}\ \ b\dot{I}_{f2}\right]^{\mathrm{T}} \tag{6.50}$$

The fault line and fault position are unknown. But the fault line and fault location can be found by solving Equation (6.50) for each branch. For any two branches K_1–K_2 and K_3–K_4, assume that the fault occurs where the distance between the fault point 1 and K_1 is a per cent of the total line K_1–K_2, point 2 and K_3 is b per cent of the total line K_3–K_4. We then need to solve

$$\vec{I} = \left[\vec{Q_{K1}}\ \ \vec{Q_{K2}}\ \ \vec{Q_{K3}}\ \ \vec{Q_{K4}}\right]\left[(1-a)\dot{I}_{f1}\ \ a\dot{I}_{f1}\ \ (1-b)\dot{I}_{f2}\ \ b\dot{I}_{f2}\right]^{\mathrm{T}} \tag{6.51}$$

$$\begin{cases} \vec{Q_{K1}} = \vec{P_{K1}} - \vec{g_{K1}} \\ \vec{Q_{K2}} = \vec{P_{K2}} - \vec{g_{K2}} \\ \vec{Q_{K3}} = \vec{P_{K3}} - \vec{g_{K3}} \\ \vec{Q_{K4}} = \vec{P_{K4}} - \vec{g_{K4}} \end{cases} \tag{6.52}$$

Equivalent transformations are made as follows:

$$\mathbf{M} = \left[\vec{Q}_{K1}\ \vec{Q}_{K3}\ \vec{Q}_{K2} - \vec{Q}_{K1}\ \vec{Q}_{K4} - \vec{Q}_{K3}\right] \tag{6.53}$$

$$\vec{N} = [h_1\, h_2\, h_3\, h_4]^{\mathrm{T}} \tag{6.54}$$

$$\begin{cases} h_1 = \dot{I}_{f1} \quad h_2 = \dot{I}_{f2} \\ h_3 = a\dot{I}_{f1}\ h_4 = b\dot{I}_{f2} \end{cases} \tag{6.55}$$

$$\vec{I} = \mathbf{M}\vec{N} \tag{6.56}$$

If the number of branch currents is more than the number of variables, the equation can be solved by the least squares method:

$$\vec{N} = \left(\mathbf{M}^{\mathrm{T}}\mathbf{M}\right)^{-1}\mathbf{M}^{\mathrm{T}}\vec{I} \tag{6.57}$$

$$\begin{cases} a = h_3/h_1 \\ b = h_4/h_2 \end{cases} \tag{6.58}$$

If $a \notin [0,1]$ or $b \notin [0,1]$, this indicates that the relevant lines are not the right combination of the two fault lines, but if $a \in [0,1]$ and $b \in [0,1]$, then the combination is considered as a suspected combination. Substitute all the suspect lines into Equation (6.51) and calculate $\vec{I}_{\mathbf{m}}$ of those lines that are equipped with PMU. The relative distance between $\vec{I}_{\mathbf{m}}$ and the measured current $\vec{I}$ by the PMU is defined as the fault fitting degree T:

$$T = \frac{\left\|\vec{I}_{\mathbf{m}} - \vec{I}\right\|_2}{\left\|\vec{I}\right\|_2} \tag{6.59}$$

Theoretically, the fault fitting degree of the fault combination is zero. However, because of the existence of errors, the actual fault fitting degree is not zero. Based on a large number of simulations, it is reliable to set $T = 1$ as the boundary value. If the fitting degree of a combination is smaller than 1, it can be considered suspicious. For the normal combination, the fault fitting degree should be bigger than 1. So the fault lines and the fault location can be defined.

6.2.3.2 Scheme Verification

This algorithm was tested on a 10-generator, 39-bus New England system, as shown in Figure 6.3. A comprehensive analysis of the algorithm and its validation are given.

The faults are set in branch 8(4–14) and branch 35(11–12) at 0.25 s. Two fault types are analysed.

Case 1 – Three-phase short circuit in branches 8(4–14) and 35(11–12) and the fault points are set at 40% of each line length. Three of the combinations are analysed and the results calculated are shown in Table 6.11. The simulations of T for these combinations are shown in Figures 6.18–6.20.

Table 6.11 Case 1 bank.

8ABC (40%) – 35ABC (40%)	Mode 1		Mode 2		Mode 3	
	8	35	7	36	10	35
Fault fitting degree (T)	0.2663		0.3453		2.3268	
Virtual fault location a	0.4010		1.5309		0.1752	
Virtual fault location b	0.4023		2.6472		0.1479	

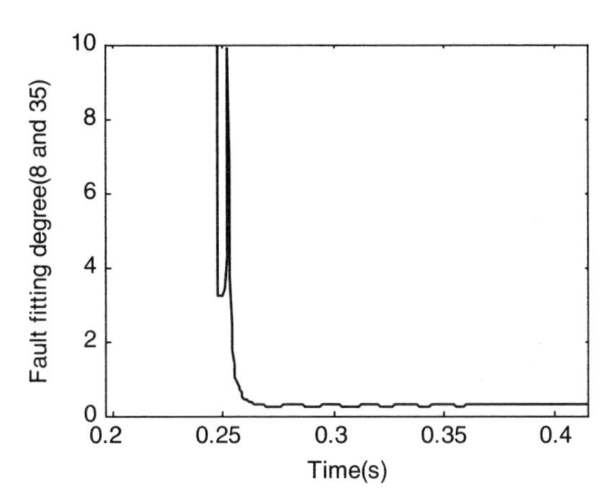

Figure 6.18 Fault fitting degree T (8 and 35).

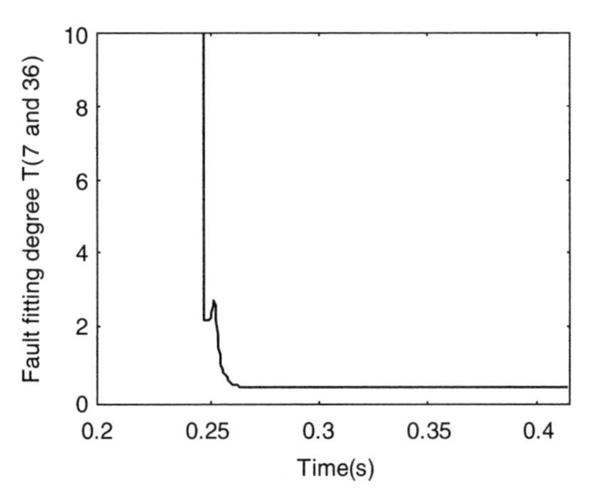

Figure 6.19 Fault fitting degree T (7 and 36).

Firstly, both a and b are judged as to whether they are in the interval [0, 1]. In mode 2, both a and b are bigger than 1, so mode 2 is not a suspect combination. Modes 1 and 3 are suspect combinations because they satisfy the conditions that both a and b are in the interval [0, 1]. Then the fault fitting degree T of each mode is compared to the boundary value. The fault fitting degree of mode 1 is 0.2667, which is smaller than 1 and the fault fitting degree of mode 3 is 2.368 which is bigger than 1. So mode 1 is judged as the fault combination. In the fault combination, one of the fault locations (a) is 0.4010 and the other (b) is 0.4023. They are close to the theoretical value 0.4, which proves the accuracy of this method.

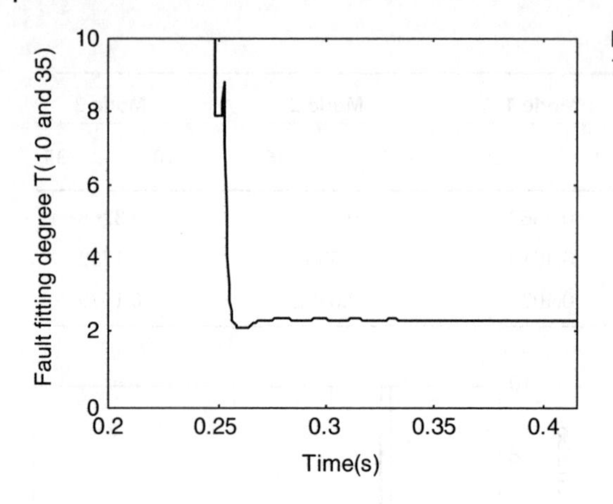

Figure 6.20 Fault fitting degree *T* (10 and 35).

Table 6.12 Case 2 bank.

	Mode 1		Mode 2		Mode 3	
8ABG (60%) – 35ABC (60%)	**8**	**35**	**12**	**36**	**10**	**35**
Fault fitting degree (T)	0.2889		0.3433		2.0318	
Virtual fault location a	0.5678		1.1531		−0.5020	
Virtual fault location b	0.6134		−0.927		1.0173	

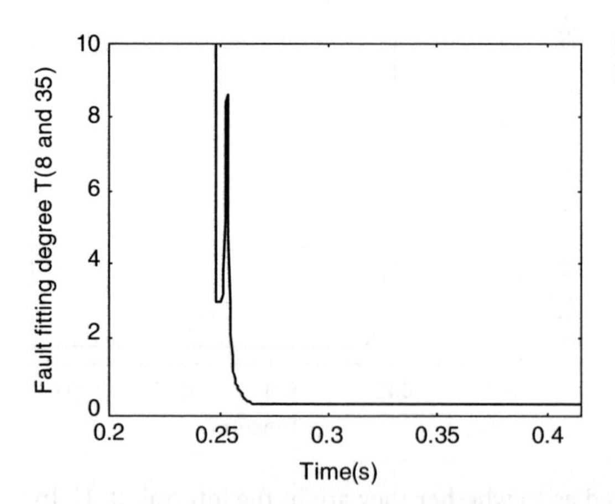

Figure 6.21 Fault fitting degree *T* (8 and 35).

Case 2 – AB two-phase short circuit in branch 8 and three-phase short circuit in branch 35. The failure points are set to 60% of each line. The analysis process is the same as for Case 1; the results are shown in Table 6.12 and the simulations of T in these combinations are shown in Figures 6.21–6.23. So mode 1 is judged as the fault combination. In the fault combination, one of the fault locations (a) is 0.5678; the other (b) is 0.6134. They are close to the theoretical value 0.6, which proves the effectiveness of this method.

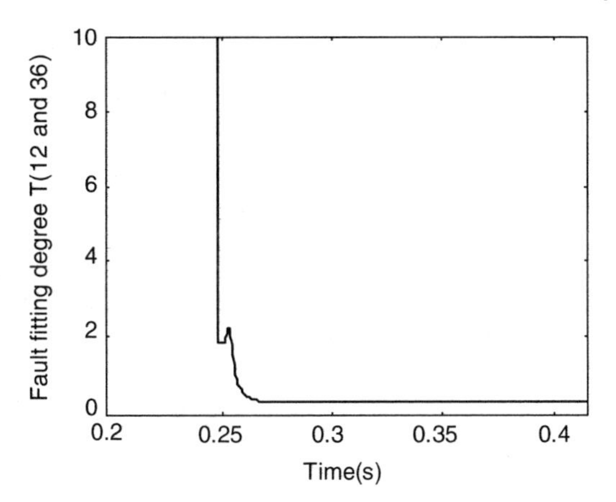

Figure 6.22 Fault fitting degree *T* (12 and 36).

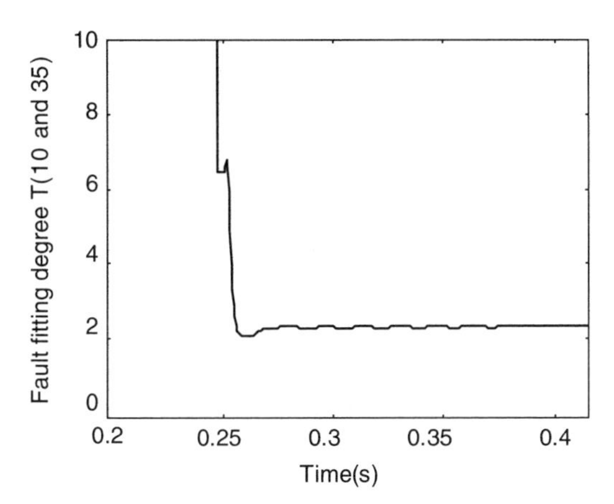

Figure 6.23 Fault fitting degree *T* (10 and 35).

Both the results and simulations indicate that the algorithm is effective and satisfies both situations.

6.3 Wide Area Protection Using Operating Signals

Due to the influence of communication delay, currently, synchronization of wide area information cannot be guaranteed. In order to reduce information synchronization and communication requirements of the wide area protection scheme, this section introduces mainly wide area protection schemes based on logic. The fault feature contained in logical information (distance protection signal, current protection signal and virtual impedance signal) is used to realize fast and reliable faulty component identification in non-synchronized conditions.

The structure of the backup protection system in this section is shown in Figure 6.24. IEDs (intelligent electronic devices) are installed at the circuit breaker (relaying point). The system protection central (SPC) area in the regional network is responsible for realizing the protective algorithm according to the data it gets from the IEDs and for sending trip commands.

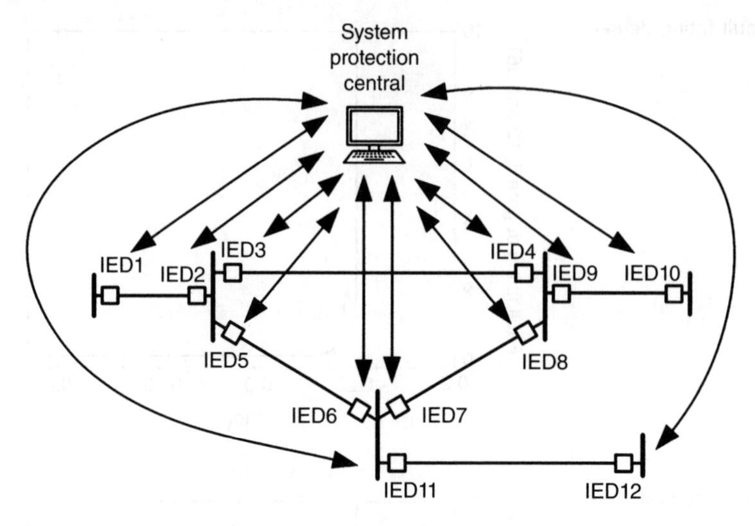

Figure 6.24 Wide area protection system structure.

6.3.1 Wide Area Protection Based on the Distance Protection Operational Signal

In view of the time delay in PMU data acquisition, the wide area backup protection scheme should not rely on the synchronization of data [11]. Furthermore, for different protection schemes, the protection operating information and sensitivity also differ. In order to distinguish between different protection information, the weight setting should be conducted so that the uncertainty problem in traditional weight setting due to reliance on subjective experience can be avoided. Also, for the potential risks in the communication system, such as information loss and error, the wide area backup protection scheme should be fault-tolerant.

A regional centralized protection system parallel to and independent of the existing protection system is introduced in this section, which combines the information of existing distance protection zone-I, zone-II and zone-III to identify the fault. In the normal operating state, this system could accelerate the operation of backup protection and prevent the fault area from expanding. Even under extreme circumstances – for example, when the wide area communication channel is completely inactive – this system could still maintain the function of traditional distance protection to isolate the fault.

6.3.1.1 Protection Range Equivalent Impedance

The impedance corresponding to the protection range on line L_i of a certain distance element is defined as the protection range equivalent impedance that the distance element has on line L_i. Taking Figure 6.25, for example, the protection ranges on line L_1 and L_2 of the zone-II element in distance relay 1 are pq and qr, respectively, thus the corresponding protection range equivalent impedances are

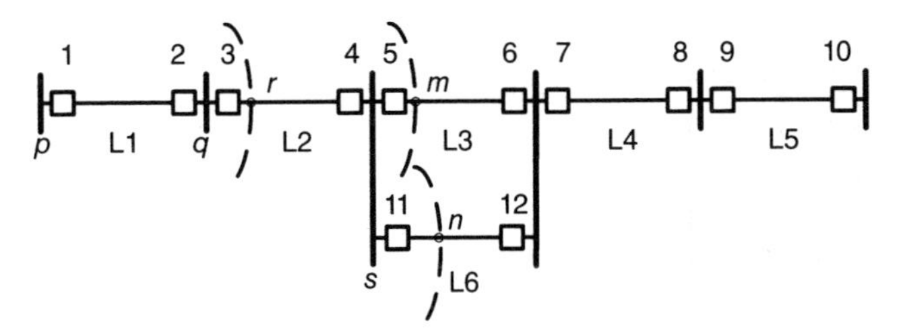

Figure 6.25 Framework of wide area power network.

$$\begin{cases} Z_{pq} = Z_{L1} \\ Z_{qr} = Z^{\mathrm{II}}_{\mathrm{set}-1} - Z_{L1} \end{cases} \tag{6.60}$$

where $Z^{\mathrm{II}}_{\mathrm{set}-1}$ is the distance protection zone-II setting value of relay 1.

The protection ranges of distance protection zone-III in relay 1 on lines L_1, L_2, L_3 and L_6 are *pq*, *qs*, *sm* and *sn*, respectively. Considering the influence of branch lines L_1 and L_6, the protection range equivalent impedances of distance protection zone-III in relay 1 on the lines are

$$\begin{cases} Z_{pq} = Z_{L1} \\ Z_{qs} = Z_{L2} \\ Z_{sm} = \dfrac{Z^{\mathrm{III}}_{\mathrm{set}-1} - Z_{L1} - Z_{L2}}{K_{\mathrm{f}L3-L1}} \\ Z_{sn} = \dfrac{Z^{\mathrm{III}}_{\mathrm{set}-1} - Z_{L1} - Z_{L2}}{K_{\mathrm{f}L6-L1}} \end{cases} \tag{6.61}$$

where $Z^{\mathrm{III}}_{\mathrm{set}-1}$ is the distance protection zone-III setting value of relay 1. $K_{\mathrm{f}L3-L1}$ is the branch coefficient between L_3 and L_1, $K_{\mathrm{f}L6-L1}$ is the branch coefficient between L_6 and L_1.

The branch coefficient is defined as the ratio of the short circuit current on the fault line to the short circuit current on the line where the relay is located.

$$K_{\mathrm{f}L\mathrm{f}-L\mathrm{r}} = \frac{\text{short circuit current on the fault line } L_{\mathrm{f}}}{\text{short circuit current on the line } L_{\mathrm{r}} \text{ where the relay is located}} \tag{6.62}$$

In Figure 6.26, $K_{\mathrm{f}L2-L1} = \dfrac{I_2}{I_1}$ is the branch coefficient between L_2 and L_1, $K_{\mathrm{f}L3-L1} = \dfrac{I_3}{I_1}$ is the branch coefficient between L_3 and L_1, $K_{\mathrm{f}L6-L1} = \dfrac{I_6}{I_1}$ is the branch coefficient between L_6 and L_1.

Considering the branch coefficients, the protection range equivalent impedance of distance protection zone-II in relay *j* on line L_i can be calculated as

$$Z^{\mathrm{II}}_{\mathrm{e}Li-j} = \begin{cases} Z_{Li} & \text{relay } j \text{ is on line } L_i \\ \dfrac{Z^{\mathrm{II}}_{\mathrm{set}-j} - Z_{Li1}}{K_{\mathrm{f}Li-Li1}} & \text{relay } j \text{ is on line } L_{i1} \end{cases} \tag{6.63}$$

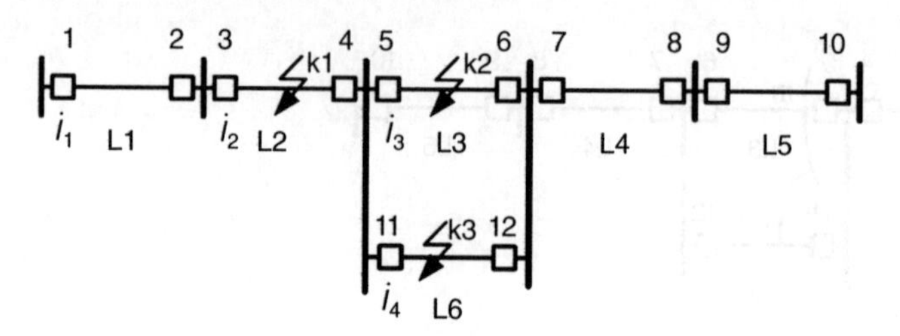

Figure 6.26 Branch coefficient calculation.

where $Z^{\mathrm{II}}_{\mathrm{e}Li-j}$ is the distance protection zone-II setting value of relay j. $K_{\mathrm{f}Li-Li1}$ is the branch coefficient between L_i and L_{i1}. When distance relay j is on line L_i, the protection range equivalent impedance of distance protection zone-II in relay j on line L_i is the line impedance of L_i: Z_{Li}. When distance relay j is on line L_{i1} which is adjacent to line L_i, the protection range equivalent impedance of distance protection zone-II in relay j on line L_i is the zone-II setting impedance of relay j converted to the impedance on line L_i via the branch coefficient $K_{\mathrm{f}Li-Li1}$.

Similarly, considering the branch coefficients, the protection range equivalent impedance of distance protection zone-III in relay j on line L_i can be calculated as follows:

$$Z^{\mathrm{III}}_{\mathrm{e}Li-j} = \begin{cases} Z_{Li} & \text{relay } j \text{ is on line } L_i \text{ or } L_{i1} \\ \dfrac{Z^{\mathrm{III}}_{\mathrm{set}-j} - Z_{Li2} - K_{\mathrm{f}Li1-Li2} Z_{Li1}}{K_{\mathrm{f}Li-Li2}} & \text{relay } j \text{ is on line } L_{i2} \end{cases} \tag{6.64}$$

where $Z^{\mathrm{III}}_{\mathrm{set}-j}$ is the distance protection zone-III setting value of relay j. $K_{\mathrm{f}Li-Li1}$ is the branch coefficient between L_i and L_{i1}. $K_{\mathrm{f}Li-Li2}$ is the branch coefficient between L_i and L_{i2}. When distance relay j is on line L_i, or on line L_{i1} which is adjacent to L_i, the protection range equivalent impedance of distance protection zone-III in relay j on line L_i is the line impedance of L_i: Z_{Li}. When distance relay j is on line L_{i2} which is secondarily adjacent to line L_i, the protection range equivalent impedance of distance protection zone-III in relay j on line L_i is the zone-III setting impedance of relay j converted to the impedance on line L_i via branch coefficient $K_{\mathrm{f}Li-Li1}$ and branch coefficient $K_{\mathrm{f}Li-Li2}$.

6.3.1.2 Distance Protection Contribution Degree

The proposed algorithm is based on the assumptions that a fault occurs randomly at any location on the line, and the probabilities of a fault occurring at different points on the line are evenly distributed [12]. Thus, the distance protection setting principle is a reflection of the fault location in the forward direction of the distance relay. As shown in Figure 6.25, for distance relay 1, if zone-II operates, the fault could be on line pq or qr. The probability of the fault being on line pq is $\dfrac{L_{pq}}{L_{pr}} = \dfrac{Z_{pq}}{Z_{pr}}$. The probability of the fault being on line qr is $\dfrac{L_{qr}}{L_{pr}} = \dfrac{Z_{qr}}{Z_{pr}}$.

Define the distance protection contribution degree ω_{ij}, which represents the probability of the fault being on line L_i when relay j operates. The value of ω_{ij} reflects the contribution of the operating relay j to the identification of whether line L_i is the fault line or not. According to the relationship between the distance protection zone-I, zone-II, zone-III and the fault line, the contribution degrees of the three zones ω_{ij}^{I}, $\omega_{ij}^{\mathrm{II}}$, $\omega_{ij}^{\mathrm{III}}$ are derived as follows.

A. Distance protection zone-I contribution degree

Distance protection zone-I only reflects faults on a certain line. When the relay j is on the fault line L_i, the contribution degree is 1. When the relay j is not on the fault line L_i, the contribution degree is 0. Therefore, the distance protection zone-I contribution degree ω_{ij}^{I} is

$$\omega_{ij}^{\mathrm{I}} = \begin{cases} 1 & \text{component } j \text{ is on line } L_i \\ 0 & \text{otherwise} \end{cases} \tag{6.65}$$

B. Distance protection Zone-II contribution degree

As shown in Figure 6.26, when zone-II in relay 1 operates, the probability of the fault being on L_1 or L_2 is the percentage of the protection range that zone-II has on each line in the total protection range of zone-II. Thus, the distance protection zone-II contribution degree $\omega_{ij}^{\mathrm{II}}$ is

$$\omega_{ij}^{\mathrm{II}} = \frac{Z_{\mathrm{e}Li-j}^{\mathrm{II}}}{\sum\limits_{n}^{N} Z_{\mathrm{e}Ln-j}^{\mathrm{II}}} \tag{6.66}$$

C. Distance protection Zone-III contribution degree

When zone-III in relay 1 operates, the probability of the fault being on L_1, L_2 or L_3 is the percentage of the protection range zone-III has on each line in the total protection range of zone-III. Thus, the distance protection zone-III contribution degree $\omega_{ij}^{\mathrm{III}}$ is

$$\omega_{ij}^{\mathrm{III}} = \frac{Z_{\mathrm{e}Li-j}^{\mathrm{III}}}{\sum\limits_{n}^{N} Z_{\mathrm{e}Ln-j}^{\mathrm{III}}} \tag{6.67}$$

6.3.1.3 Fault Identification Method Based on Distance Protection Fitting Factor

1. Protection Fitness Function The distance protection contribution degrees ω_{ij}^{I}, $\omega_{ij}^{\mathrm{II}}$ and $\omega_{ij}^{\mathrm{III}}$ represent the probability of the fault being on line L_i when the distance element in relay j operates. With the distance protection contribution degrees as the weights, the protection operating information is integrated, thus the protection fitness function of line L_i is

$$E_F(L_i) = \sum_{j}^{B_I} \omega_{ij}^{I} D_j^{I} + \sum_{j}^{B_{II}} \omega_{ij}^{II} D_j^{II} + \sum_{j}^{B_{III}} \omega_{ij}^{III} D_j^{III} \tag{6.68}$$

The protection fitness function $E_F(L_i)$ is calculated with the actual operating status of the distance elements after the fault occurs. In Equation (6.68), D_j^I, D_j^{II} and D_j^{III} represent the operating status of zone-I, zone-II and zone-III in relay j after the fault, respectively. If a certain distance element operates, then the value of the corresponding D is 1; otherwise, the value of the corresponding D is 0. B_I, B_{II} and B_{III} are the number of relays with zone-I, zone-II and zone-III protection.

2. Protection Fitness Expectation Function Define the theoretical operating status. After a fault occurs, theoretically if a distance element should operate, then its theoretical operating status is 1; if it should not operate, then its theoretical operating status is 0. Supposing the fault occurs on line L_i, the protection fitness expectation function is calculated with the theoretical operating status of the distance elements. For different fault locations on L_i, the theoretical operating status of the distance elements also differs, as does the protection fitness expectation function.

Take the system in Figure 6.25, for example. The zone-I protection ranges of relays 5 and 6 are the two ovals in Figure 6.27. The two ovals separate line L_3 into three sections – A, B and C. The theoretical zone-I operating status of the distance relays in the case of different fault locations on L_3 is shown in Table 6.13.

The probabilities of the fault being in sections A, B or C are λ_A^I, λ_B^I and λ_C^I, which can be calculated as

$$\begin{cases} \lambda_A^I = \dfrac{Z_{L3} - Z_{set-6}^I}{Z_{L3}} \\ \lambda_B^I = \dfrac{Z_{set-5}^I + Z_{set-6}^I - Z_{L3}}{Z_{L3}} \\ \lambda_C^I = \dfrac{Z_{L3} - Z_{set-5}^I}{Z_{L3}} \end{cases} \tag{6.69}$$

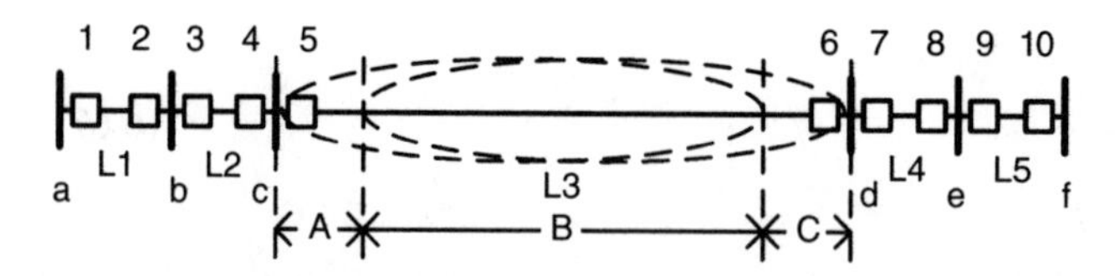

Figure 6.27 Distance protection zone-I operating theoretical value calculation.

Table 6.13 Distance protection zone-I operating theoretical value in the case of L_3 fault.

Distance relay number	1	2	3	4	5	6	7	8	9	10
Fault in section A	0	0	0	0	1	0	0	0	0	0
Fault in section B	0	0	0	0	1	1	0	0	0	0
Fault in section C	0	0	0	0	0	1	0	0	0	0

Similarly, the zone-II protection ranges of relays 3 and 8 also separate line L_3 into three sections – A, B and C, shown in Figure 6.28.

The theoretical zone-II operating status of the distance relays in the case of different fault locations on L_3 is shown in Table 6.14.

The protection range equivalent impedance that zone-II in relays 3 and 8 has on line L_3 could reflect the length of sections A and C in Figure 6.28, respectively. The probabilities of the fault being in sections A, B or C are λ_A^{II}, λ_B^{II} and λ_C^{II}, which can be calculated as

$$\begin{cases} \lambda_A^{II} = \dfrac{Z_{eL3-3}^{II}}{Z_{L3}} \\ \lambda_B^{II} = \dfrac{Z_{L3} - Z_{eL3-3}^{II} - Z_{eL3-8}^{II}}{Z_{L3}} \\ \lambda_C^{II} = \dfrac{Z_{eL3-8}^{II}}{Z_{L3}} \end{cases} \tag{6.70}$$

Similarly, the zone-III protection ranges of relay 1 and relay 10 also separate line L_3 into three sections – A, B and C, shown in Figure 6.29.

The theoretical zone-III operating status of the distance relays in the case of different fault locations on L_3 is shown in Table 6.15.

The protection range equivalent impedance that zone-III in relays 1 and 10 has on line L_3 could reflect the length of sections A and C in Figure 6.29, respectively. The probabilities of the fault being in sections A, B or C are λ_A^{III}, λ_B^{III} and λ_C^{III}, which can be calculated as

Table 6.14 Distance protection zone-II operating theoretical value in the case of L_3 fault.

Distance relay number	1	2	3	4	5	6	7	8	9	10
Fault in section A	0	0	1	0	1	1	0	0	0	0
Fault in section B	0	0	0	0	1	1	0	0	0	0
Fault in section C	0	0	0	0	1	1	0	1	0	0

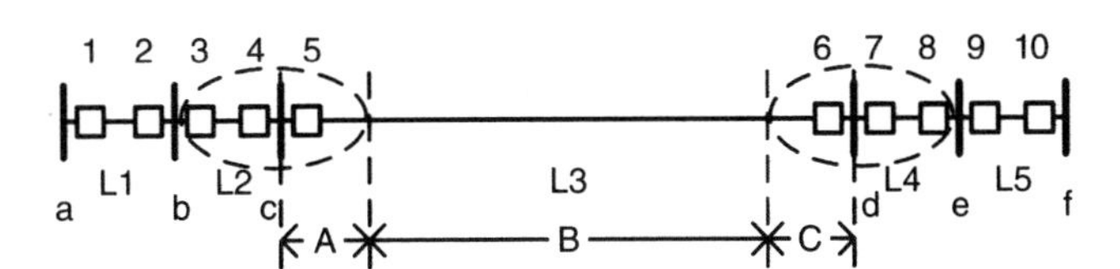

Figure 6.28 Distance protection zone-II operating theoretical value calculation.

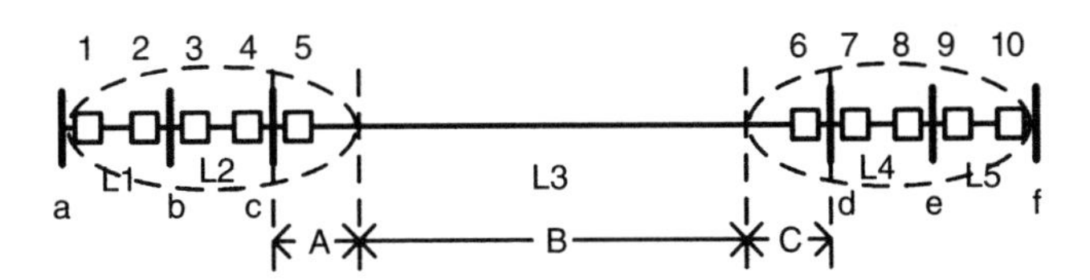

Figure 6.29 Distance protection zone-III operating theoretical value calculation.

Table 6.15 Distance protection zone-III operating theoretical value in the case of L_3 fault.

Distance relay number	1	2	3	4	5	6	7	8	9	10
Fault in section A	1	0	1	0	1	1	0	1	0	0
Fault in section B	0	0	1	0	1	1	0	1	0	0
Fault in section C	0	0	1	0	1	1	0	1	0	1

$$\begin{cases} \lambda_{\mathrm{A}}^{\mathrm{III}} = \dfrac{Z_{eL3-1}^{\mathrm{III}}}{Z_{L3}} \\ \lambda_{\mathrm{B}}^{\mathrm{III}} = \dfrac{Z_{L3} - Z_{eL3-1}^{\mathrm{III}} - Z_{eL3-10}^{\mathrm{III}}}{Z_{L3}} \\ \lambda_{\mathrm{C}}^{\mathrm{III}} = \dfrac{Z_{eL3-10}^{\mathrm{III}}}{Z_{L3}} \end{cases} \tag{6.71}$$

With λ_{A}, λ_{B} and λ_{C} as the weights of different fault locations, the protection fitness expectation function of line L_i can be calculated as

$$\begin{aligned} E^*(L_i) = &\sum_{j}^{B_{\mathrm{I}}} \omega_{ij}^{\mathrm{I}} \left(\lambda_{\mathrm{A}}^{\mathrm{I}} D_{\mathrm{A}ij}^{*I} + \lambda_{\mathrm{B}}^{\mathrm{I}} D_{\mathrm{B}ij}^{*\mathrm{I}} + \lambda_{\mathrm{C}}^{\mathrm{I}} D_{\mathrm{C}ij}^{*\mathrm{I}} \right) \\ &+ \sum_{j}^{B_{\mathrm{II}}} \omega_{ij}^{\mathrm{II}} \left(\lambda_{\mathrm{A}}^{\mathrm{II}} D_{\mathrm{A}ij}^{*\mathrm{II}} + \lambda_{\mathrm{B}}^{\mathrm{II}} D_{\mathrm{B}ij}^{*\mathrm{II}} + \lambda_{\mathrm{C}}^{\mathrm{II}} D_{\mathrm{C}ij}^{*\mathrm{II}} \right) \\ &+ \sum_{j}^{B_{\mathrm{III}}} \omega_{ij}^{\mathrm{III}} \left(\lambda_{\mathrm{A}}^{\mathrm{III}} D_{\mathrm{A}ij}^{*\mathrm{III}} + \lambda_{\mathrm{B}}^{\mathrm{III}} D_{\mathrm{B}ij}^{*\mathrm{III}} + \lambda_{\mathrm{C}}^{\mathrm{III}} D_{\mathrm{C}ij}^{*\mathrm{III}} \right) \end{aligned} \tag{6.72}$$

In Equation (6.72), $D_{\mathrm{A}ij}^{*\mathrm{I}}$, $D_{\mathrm{A}ij}^{*\mathrm{II}}$ and $D_{\mathrm{A}ij}^{*\mathrm{III}}$ represent the theoretical operating status of the protection elements when the fault occurs in section A on line L_i. $D_{\mathrm{B}ij}^{*\mathrm{I}}$, $D_{\mathrm{B}ij}^{*\mathrm{II}}$ and $D_{\mathrm{B}ij}^{*\mathrm{III}}$ are the theoretical operating status of the protection elements when the fault occurs in section B on line L_i. $D_{\mathrm{C}ij}^{*\mathrm{I}}$, $D_{\mathrm{C}ij}^{*\mathrm{II}}$ and $D_{\mathrm{C}ij}^{*\mathrm{III}}$ are the theoretical operating status of the protection elements when the fault occurs in section C on line L_i.

3. Fault Identification Based on Distance Protection Fitting Factor Define the distance protection fitting factor of line L_i as the protection fitness function of line L_i divided by its protection fitness expectation function:

$$P_{\mathrm{e}}(L_i) = \frac{E_{\mathrm{F}}(L_i)}{E^*(L_i)} \tag{6.73}$$

The bigger $P_{\mathrm{e}}(L_i)$ is, the closer the protection fitness function of line L_i is to its protection fitness expectation function, and the more probable that line L_i is the fault line. Therefore, the distance protection fitting factor of the fault line should be close to 1, and the maximum among the distance protection fitting factors of all the lines in the network. Thus, the protection criterion is

$$P_{\mathrm{emax}} = \max_{1 \le i \le n} P_{\mathrm{e}}(L_i) \tag{6.74}$$

where n is the number of lines in the region and the line corresponding to P_{emax} is the fault line.

4. Scheme Under Stressed Conditions The proposed method is based on the startup information of distance protection. Thus, problems that conventional protection would encounter are unavoidable with the proposed method. In stressed conditions, such as power swings, voltage drops, load encroachment or generator outage, some protection may malfunction, even when there is no fault, thus causing wide area protection to malfunction. Accordingly, the schemes of these stressed conditions are discussed here.

Voltage drop, load encroachment and generator outage, etc. may cause distance protection zone-III to malfunction. Considering that the contribution degrees of different protection zones are different, the contribution degree of distance protection zone-III is relatively small. Thus, in these stress conditions the line fitness factor is relatively small. If the fitness factor of the fault line L_i identified by Equation (6.74) satisfies Equation (6.75), then it is taken that the transmitted protection information might be malfunctional information, and the protection operation signals of line L_i need to be checked. For the zone-III relays on the two line ends of line L_i, if only one starts, or neither of them starts, then malfunction is identified, and wide area backup protection will be locked. Otherwise, line L_i is identified as the fault line and a tripping command will be issued. The identification result function $g(L_i)$ is shown in Equation (6.76).

$$P_{\mathrm{e}}(L_i) < P_{\mathrm{eset}}(L_i) \tag{6.75}$$

where $P_{\mathrm{eset}}(L_i)$ is the setting value of the fitness factor of line L_i. The setting principle is that for a certain line, if there exists a fitness factor value that renders both the zone-III of this line and the zone-III of the neighbouring line malfunctional, then this value is the setting value of the fitness factor of this line.

$$g(L_i) = \left(\left(D_j^{\mathrm{III}} + D_{j1}^{\mathrm{III}}\right) \le 1\right) \cap \left(\sum_{k=\mathrm{I}}^{\mathrm{III}} \left(D_j^k + D_{j1}^k\right) \le 1\right) \tag{6.76}$$

where $g(L_i)$ is the malfunction identification result function of line L_i. If $g(L_i) = 1$, then the protection has malfunctioned; otherwise, the protection did not malfunction. D represents the protection operational signal value. Subscripts j and $j1$ represent the protection relays on the two ends of line L_i. Superscripts I, II and III represent distance protection zone-I, zone-II and zone-III.

When there is a power swing, the relays of multi-zone protection may malfunction. In this case, it is difficult to distinguish a fault from a power swing according to the protection startup information transmitted to the protection centre. Consider that the proposed protection system serves as backup protection, and is slow in operation compared with the primary protection. Swing blocking and opening components could be installed in the IEDs, so that by blocking the protection in swing conditions, regional backup protection malfunction could be avoided, and by opening the protection again, distance protection would not refuse to operate when a fault occurred after the power swing [13].

A. Realization of Wide Area Backup Protection Algorithm

1. Distance Protection Fitting Factor Based Backup Protection Process The distance protection contribution degree and distance protection fitness expectation function are calculated off-line and stored in SPC. When the system structure changes, the weight update system in the protection centre is triggered. According to the new system typology, the protection centre will recalculate both the contribution degree of each protection and the protection fitness expectation function of each line in the region, and store these data in the regional host computer.

When a fault is detected by a certain protection in the region, the 'event' flag is activated. This flag will trigger the protection centre processing program, and the IEDs will begin transmitting the protection operation signals to the SPC via the IED at the relaying point. In view of the communication delay, the protection centre needs to wait until all the protection startup signals are acquired. Fault calculation could start after this waiting time. If none of the protection components in the region starts, then the protection centre will maintain the 'waiting' state.

According to the uploaded operating status of the distance elements, the SPC will calculate the distance protection fitness function and distance protection fitting factor of each line. Then, by searching for the maximum value of the distance protection fitting factors, the fault line can be identified. Finally, if the stressed condition criterion is satisfied, the WABP system refuses to issue a tripping command, or the SPC will issue a tripping command to the corresponding IEDs of the fault line. Once the IEDs receive the command, they will break the corresponding circuit breakers and clear the fault.

The proposed wide area backup protection scheme is illustrated in detail in Figure 6.30.

2. Data Requirement and Computational Analysis For the proposed method, there is no need to transmit a large amount of synchronized electrical variable information. After a fault occurs, only the operational signals of distance protection are transmitted to the protection centre.

The distance protection contribution degree and distance protection fitness expectation function are calculated off-line. Thus, their computation is not taken into account. Only the computation of the distance protection fitness function and distance protection fitting factor are considered.

The distance protection fitness function is calculated as in Equation (6.68). After a fault occurs, if there are m items of protection operating status information uploaded to the SPC, then each line requires m multiplications, and n lines require $m \times n$ multiplications. For n lines, there will be at most $6n$ items of protection operating status information uploaded. Thus, at most $6n^2$ multiplications are required.

The distance protection fitting factor is calculated as in Equation (6.75). For n lines, n divisions are required. Therefore, for the proposed algorithm, at most $6n^2$ multiplications and n divisions are required, which is not a high demand on the computing ability of current computer chips.

3. Time Issue Analysis The time delay in a wide area backup protection scheme is mainly caused by the following five links: protection component detecting the fault, protection startup signal acquisition, data transmission, wide area backup protection system processing and command transmission. The time delay of each

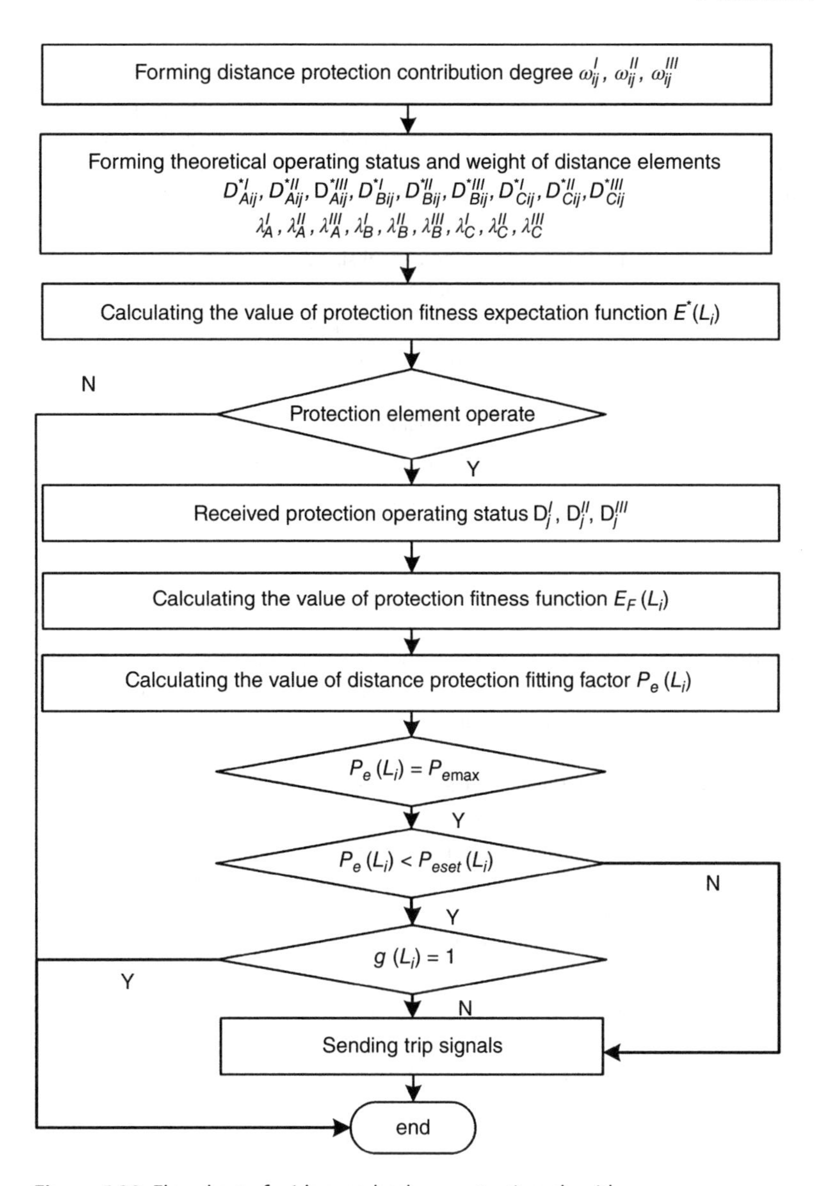

Figure 6.30 Flowchart of wide area backup protection algorithm.

link is shown in Table 6.16. After a fault occurs, local protection could detect the fault and issue the operational signal within 26 ms. The acquisition of protection operational signals takes no more than 10 ms. According to reference [14], the time delay of operational signal transmission is less than 3 ms. Thus, the protection centre could start calculation after a waiting time of 3.5 ms, and it takes no more than 19.5 ms for the protection centre to finish the fault calculation and issue the tripping command. Therefore, the overall time delay of the proposed scheme is within 59 ms.

Table 6.16 Typical delay of the elements.

Type of delay	Element of delay	Typical value of delay (ms)
Local protection fault detection	Protection elements operate	≤26
Protection component operational signal acquisition	Operational signal acquisition	≤10
Data transmission/tripping command transmission	Serializing output	≤3
	communication I/O	
	propagation	
Protection centre processing	Waiting time	3.5
	Executing algorithm	≤16.5

The operational time of distance protection zone-II is usually set to 300–500 ms. Thus, the proposed scheme operates faster than distance protection zone-II to prevent the fault range from expanding.

4. Comparison with Differential and Pilot Protection In existing protection strategies, differential protection is based on real-time electrical variable calculation. The differential protection is simple in principle, but not easy to implement, since its requirement on the synchronization of electrical variables on different sides is very strict. In addition, its requirement on the continuity of data is relatively high. If, for some reason, the communication data are incorrect or the communication is interrupted, differential protection needs to be locked for a considerable length of time before it can be put into operation again. During this period, if an in-zone fault occurs, differential protection will refuse to operate.

As for pilot protection, take directional pilot protection as an example. Its requirement on the synchronization of communication data is not as strict as differential protection. However, if the data from either side of the pilot protection are incorrect, the protection will not operate correctly, which may cause the power outage range to expand. In addition, once a fault occurs in the protection channel and the communication is interrupted, pilot protection may malfunction or refuse to operate.

In the proposed algorithm, although a communication network is needed for the acquisition of input data such as the operating status of protection elements, synchronization is not required. Because these data consist only of 0–1 logical information which does not vary with time, the amount of data is small and a short time delay between the data is allowed. Even if some data are overly delayed, or some data are missing, the fault tolerance ability of the proposed algorithm can still guarantee correct identification of the fault line, which is discussed in Section 6.3.1.3.

6.3.1.4 Scheme verification

A. Simulation Tests on an IEEE 10-Machine, 39-Bus System

This algorithm was tested on a 10-generator, 39-bus New England system, as shown in Figure 6.31. As different regionalization methods do not affect the result of the proposed algorithm, there is no specific requirement on the classification of regions. In Figure 6.31, the protection area in the curve is chosen for simulation. Suppose that the fault occurs on line L_8 close to bus 4.

First, the contribution degree of the protection elements to each line ω_{ij} is formed. The contribution degree ω_{8j} of the protection elements to line L_8 is shown in Table 6.17.

With the distance protection contribution degrees as weights, as well as the theoretical protection operating status, the values of the protection fitness expectation function can be calculated, as shown in Table 6.18.

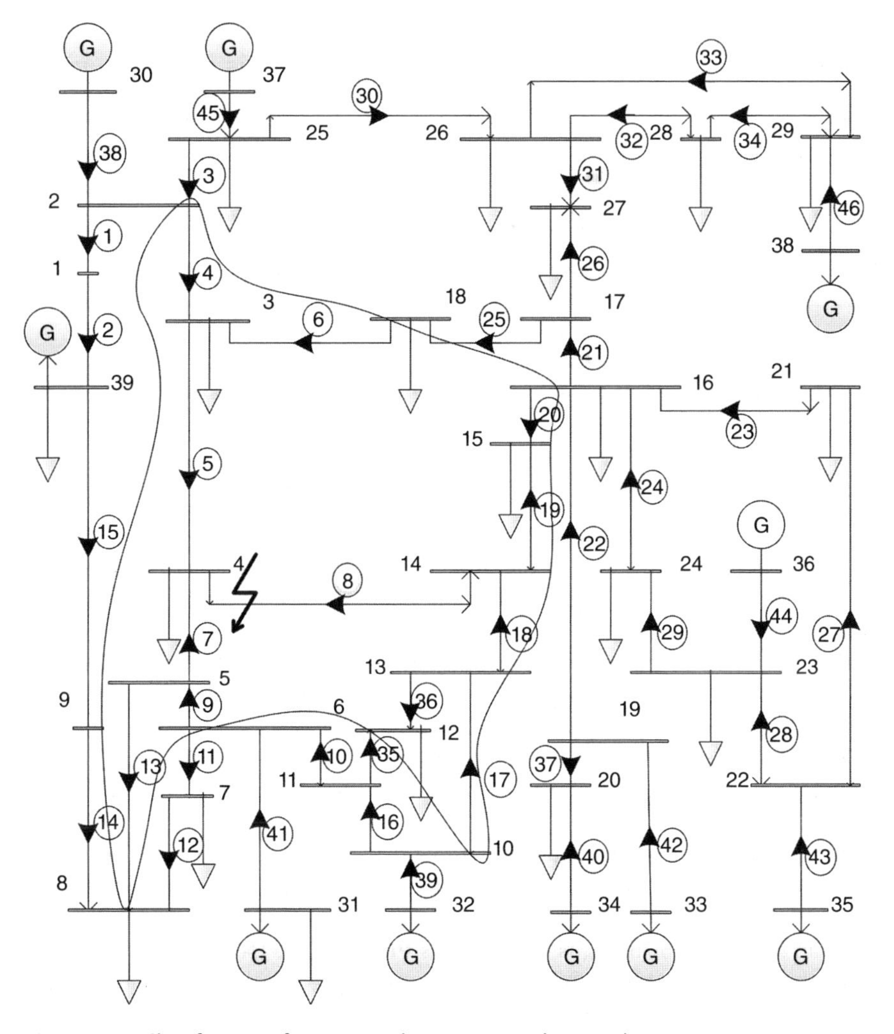

Figure 6.31 Classification of regions in the IEEE 10-machine, 39-bus system.

Table 6.17 Contribution degree of each distance element to fault line L_8 identification.

Relay number	2–3	3–2	3–4	4–3	18–3	3–18	5–4	4–5
Zone-I	0	0	0	0	0	0	0	0
Zone-II	0	0	0.1480	0	0	0	0.1842	0
Zone-III	0.0560	0	0.2745	0	0.0419	0	0.2745	0
Relay number	**4–14**	**14–4**	**6–5**	**5–6**	**8–5**	**5–8**	**13–10**	**10–13**
Zone-I	1	1	0	0	0	0	0	0
Zone-II	1	1	0	0	0	0	0	0
Zone-III	0.2882	0.2745	0	0	0	0	0	0.0102
Relay number	**14–13**	**13–14**	**14–15**	**15–14**	**15–16**	**16–15**	**13–12**	**12–13**
Zone-I	0	0	0	0	0	0	0	0
Zone-II	0	0.1884	0	0.1432	0	0	0	0
Zone-III	0	0.2882	0	0.2882	0	0	0	0

Table 6.18 Protection fitness expectation function value of each line.

Line number	L_4	L_5	L_6	L_7	L_8	L_9
Expected value	5.2311	6.1246	5.2756	5.7398	5.4209	4.8350
Line number	L_{13}	L_{17}	L_{18}	L_{19}	L_{20}	L_{36}
Expected value	4.8635	2.8846	4.6178	6.3087	5.2050	7.0270

The actual operating status of the protection elements is acquired as shown in Table 6.19. Only the 15-digit protection operation information needs to be transmitted.

With the distance protection contribution degrees as weights, as well as the actual operating status of the protection elements, the values of the protection fitness function can be calculated, as shown in Table 6.20.

The distance protection fitting factor can then be obtained by dividing the protection fitness function by the protection fitness expectation function, as shown in Table 6.21.

The simulation results in Table 6.21 show that, when the protection operating information is correct, the calculated distance protection fitting factor of L_8 is close to 1 and the maximum in the chosen area. According to the protection criterion in Equation (6.75), L_8 is identified as the fault line (see Figure 6.32).

To test the fault tolerance ability of the algorithm proposed here, four unfavourable cases are considered below, i.e. when distance protection zone-I or longitudinal distance protection refuses to operate or malfunctions.

Table 6.19 Operating status of each distance element in the case of L_8 fault.

Relay number	2–3	3–2	3–4	4–3	18–3	3–18	5–4	4–5
Zone-I	0	0	0	0	0	0	0	0
Zone-II	0	0	1	0	0	0	1	0
Zone-III	1	0	1	0	1	0	1	0
Relay number	**4–14**	**14–4**	**6–5**	**5–6**	**8–5**	**5–8**	**13–10**	**10–13**
Zone-I	1	0	0	0	0	0	0	0
Zone-II	1	1	0	0	0	0	0	0
Zone-III	1	1	1	0	1	0	0	0
Relay number	**14–13**	**13–14**	**14–15**	**15–14**	**15–16**	**16–15**	**13–12**	**12–13**
Zone-I	0	0	0	0	0	0	0	0
Zone-II	0	0	0	0	0	0	0	0
Zone-III	0	1	0	1	0	0	0	0

Table 6.20 Protection fitness function value of each line.

Line number	L_4	L_5	L_6	L_7	L_8	L_9
Value	0.5784	3.0259	0.5093	2.5949	5.1183	0.1956
Line number	L_{13}	L_{17}	L_{18}	L_{19}	L_{20}	L_{36}
Value	0.8423	0	0.6784	1.4568	0	0

Table 6.21 Distance protection fitting factor of each line.

Line number	L_4	L_5	L_6	L_7	L_8	L_9
Value	0.1106	0.5078	0.0965	0.4617	0.9412	0.0405
Line number	L_{13}	L_{17}	L_{18}	L_{19}	L_{20}	L_{36}
Value	0.1732	0	0.1469	0.2309	0	0

Case 1 – If the distance protection zone-I on the two line ends of line L_8 both refuse to operate, i.e. the zone-I operating value of protection 4–14 in Table 6.18 is 0 (which means that distance protection zone-I fails to identify line L_8 as the fault line).

Case 2 – If the distance protection zone-I on the two line ends of line L_8 both refuse to operate, and distance protection zone-II on one line end of line L_8 refuses to operate, i.e. the zone-I operating value of protection 4–14 in

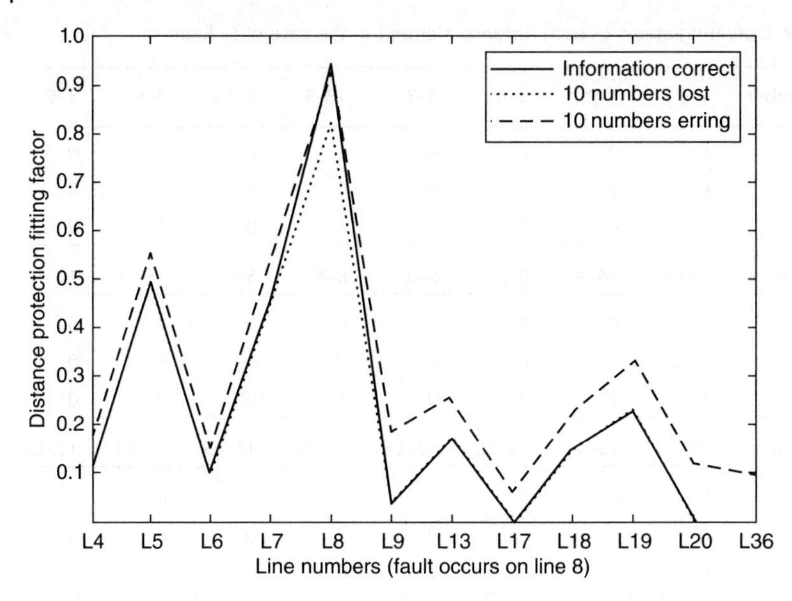

Figure 6.32 Distance protection fitting factor of each line in the case of L_8 fault close to bus 4.

Table 6.18 is 0, and the zone-II operating value of protection 14–4 is 0 (which means that the longitudinal distance protection refuses to operate).

Case 3 – If the distance protection zone-I on one line end of line L_7 malfunctions, i.e. the zone-I operating value of protection 5–4 in Table 6.18 is 1 (which means that distance protection zone-I identifies line L_7 as the fault line by mistake).

Case 4 – If the distance protection zone-II on the two line ends of line L_7 both malfunction, i.e. the zone-II operating value of protection 4–5 in Table 6.18 is 1 (which means the longitudinal distance protection malfunctions).

The distance protection fitting factor of each line in the above four cases is shown in Table 6.22. It can be seen from Table 6.22 that the distance protection fitting factor of line L_8 is the maximum in the region in all four cases. Thus, according to the criterion in Equation (6.74), line L_8 is identified as the fault line. Therefore, even when distance protection zone-I refuses to operate or malfunctions, or the longitudinal distance protection refuses to operate or malfunctions, the proposed algorithm is still able to identify the fault line correctly.

After the fault line is identified, the protection centre will issue a tripping command to protection 4–14 and protection 14–4. Protection 4–14 and protection 14–4 then immediately break the corresponding circuit breakers, thus the fault is cleared.

To test the performance of the algorithm proposed here under stressed conditions, suppose zone-III of protection 5–4 and zone-III of protection 14–4 both malfunction due to load increasing abruptly and significantly at bus 4 (or for other reasons), and the other protection does not operate. The fitness factor value of each line is shown in Table 6.23. The fitness factor setting value of each line in the protection region is shown in Table 6.24. According to Equation (6.75) L_5 is identified as the fault line. However, since the fitness factor value of L_5 is smaller than the setting value, malfunction identification needs to be carried out according to

Table 6.22 Distance protection fitting factor of each line.

Line number	Case 1	Case 2	Case 3	Case 4
L_4	0.1106	0.1106	0.1106	0.1106
L_5	0.4941	0.4341	0.4941	0.4941
L_6	0.0965	0.0965	0.0965	0.0965
L_7	0.4521	0.4521	0.6263	0.6263
L_8	0.7464	0.5825	0.9411	0.9411
L_9	0.0405	0.0405	0.0405	0.0405
L_{13}	0.1732	0.1732	0.1732	0.1732
L_{17}	0	0	0	0
L_{18}	0.1469	0.1469	0.1469	0.1469
L_{19}	0.2309	0.2309	0.2309	0.2309
L_{20}	0	0	0	0
L_{36}	0	0	0	0

Table 6.23 Distance protection fitting factor of each line.

Line number	L_4	L_5	L_6	L_7	L_8	L_9
Value	0	0.1480	0	0.0949	0.1069	0
Line number	L_{13}	L_{17}	L_{18}	L_{19}	L_{20}	L_{36}
Value	0	0	0	0	0	0

Table 6.24 Set fitting factor of each line.

Line number	L_4	L_5	L_6	L_7	L_8	L_9
Value	0.2493	0.3549	0.2403	0.3100	0.3497	0.2271
Line number	L_{13}	L_{17}	L_{18}	L_{19}	L_{20}	L_{36}
Value	0.2922	0.3983	0.0757	0.1844	0.3415	0.2137

Equation (6.76) based on the protection operation status. Since none of the protection relays on line L_5 starts, the malfunction identification function $g(L_5) = 1$, i.e. it is identified as malfunction. Thus, the tripping command will not be issued.

B. Simulation Tests on Guizhou DuYun RTDS

The Guizhou DuYun RTDS (real time digital simulator) shown in Figure 6.33 is taken as an example to verify the reliability of the proposed method in engineering practice.

Suppose the fault occurs on L_6 at the midpoint. When the protection operating information is correct, the simulation result is shown in Figure 6.33. At the same

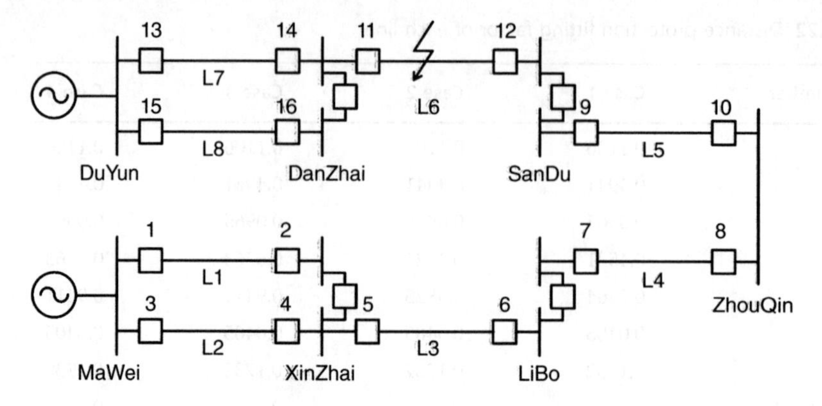

Figure 6.33 Guizhou DuYun System.

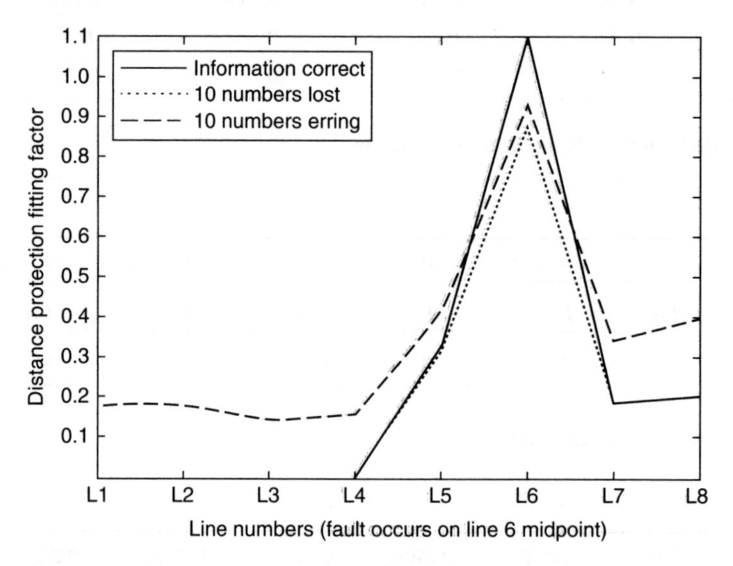

Figure 6.34 Distance protection fitting factor of each line in the case of L_6 fault at the midpoint.

time, when the protection information is incorrect – losing or erring in 10 random numbers (the system in Figure 6.33 contains 8 lines and 16 protection relays, the total number of pieces of protection operating information being 48; thus, 10 random numbers lost or in error means 20% random loss or error in the protection information) – the statistical average of 100 simulation tests is shown in Figure 6.34. It can be seen that, when the protection operating information is correct, the distance protection fitting factor of L_6 is bigger than 1 and is the maximum in the system. According to the protection criterion in Equation (6.75), L_6 is identified as the fault line. On the other hand, when there are 10 random numbers lost or in error, the simulation result shows that line L_6 still meets the protection criterion in Equation (6.75). Therefore, the proposed method is proved reliable and fault-tolerant in engineering practice.

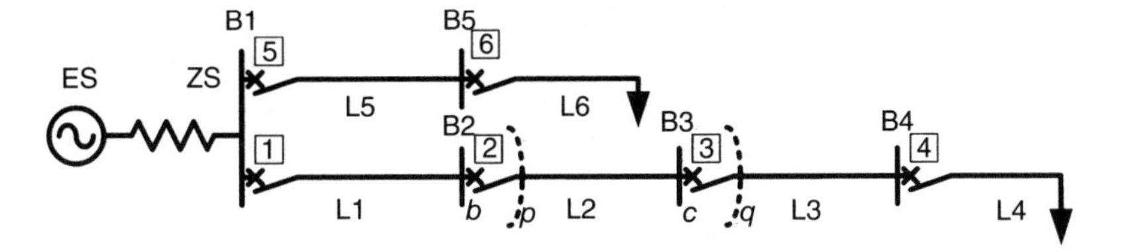

Figure 6.35 Regional power grid structure.

6.3.2 Wide Area Protection Based on the Current Protection Operational Signal

6.3.2.1 Adaptive Current Protection Effectiveness Degree

Adaptive current protection is simple in value setting and reliable in protection performance, thus it could be applied to lines with a voltage level of 110 kV or lower. According to the setting principle of adaptive current protection, the operational value could reflect the fault within a certain range in the forward direction of the protection. As shown in Figure 6.35, if the adaptive current protection zone-I of component 1 operates, the fault can only be on line L_1. If the adaptive current protection zone-II of component 1 operates, the fault can be on line L_1, or on line L_2 close to the bus, i.e. on the left of the dashed line p in Figure 6.35. If the adaptive current protection zone-III of component 1 operates, the fault can be on line L_1, line L_2 or on line L_3 close to the bus, i.e. on the left of the dashed line q in Figure 6.35. It should be noted that the adaptive current protection zone-I, zone-II and zone-III refer to quick break protection, quick break protection with time delay and over-current protection, respectively.

Take component 1 in Figure 6.35, for example. Define adaptive current protection effectiveness degrees α_{11}^{I}, α_{12}^{I} and α_{13}^{I} as the effectiveness of the operation of component 1 zone-I in identifying faults on line L_1, L_2 and L_3. Similarly, adaptive current protection effectiveness degrees $\alpha_{11}^{\mathrm{II}}$, $\alpha_{12}^{\mathrm{II}}$ and $\alpha_{13}^{\mathrm{II}}$ are defined as the effectiveness of the operation of component 1 zone-II in identifying faults on line L_1, L_2 and L_3. Adaptive current protection effectiveness degrees $\alpha_{11}^{\mathrm{III}}$, $\alpha_{12}^{\mathrm{III}}$ and $\alpha_{13}^{\mathrm{III}}$ are defined as the effectiveness of the operation of component 1 zone-II in identifying faults on line L_1, L_2 and L_3.

A. *Effectiveness Degree Calculation for Adaptive Current Protection Zone-I*

Adaptive current protection zone-I only reflects faults on the line where the protection component is situated. Take component 1 in Figure 6.35, for example. Since component 1 is on line L_1 and not on line L_2 or L_3, $\alpha_{11}^{\mathrm{I}} = 1$, $\alpha_{12}^{\mathrm{I}} = 0$, $\alpha_{13}^{\mathrm{I}} = 0$.

B. *Effectiveness Degree Calculation for Adaptive Current Protection Zone-II*

Adaptive current protection zone-II reflects faults on the line where the protection component is situated, and faults on the downstream line close to the bus. Take component 1 in Figure 6.35, for example. The zone-II protection range of component 1 is to the left of the dashed line p. In Figure 6.35, b is the terminal point of line L_2. Define the covered length on each line as the effective distance of zone-II (of component 1) on each line. Since the line impedance is in proportion to the line length in a power system, the effective distance can be expressed with its equivalent impedance. The equivalent impedance of the effective distance that zone-II of component 1 has on L_1 is Z_{L1}, that on L_2 is Z_{bp} and that on the other lines is 0.

When a fault occurs at p, the fault current at component 1 is

$$I_{F1-p} = \frac{K_d E_S}{Z_S + Z_{L1} + Z_{bp}} \tag{6.77}$$

where E_S is the equivalent phase emf of the system and Z_S is the composite impedance on the system source side. Z_{L1} is the line impedance of line L_1, Z_{bp} is the impedance of line length bp on line L_2 and K_d is the fault type coefficient.

The zone-II protection range of component 1 extends to point p on line L_2, thus

$$I_{F1-p} = I_{\text{set}-1}^{\text{II}} \tag{6.78}$$

where $I_{\text{set}-1}^{\text{II}}$ is the current protection zone-II setting value of component 1, which can be calculated according to the setting principle in [15].

Applying Equation (6.78) to Equation (6.77) gives

$$Z_{bp} = \frac{K_d E_S}{I_{\text{set}-1}^{\text{II}}} - Z_S - Z_{L1} \tag{6.79}$$

When zone-II of component 1 operates, the probability of the fault being on L_1 or L_2 is the percentage of the effective distance zone-II has on each line in the total effective distance of zone-II. Therefore, for component 1, the effectiveness degree of the adaptive current protection zone-II is

$$\alpha_{1j}^{\text{II}} = \begin{cases} \dfrac{Z_{L1}}{Z_{L1} + Z_{bp}} & j = 1 \\ \dfrac{Z_{bp}}{Z_{L1} + Z_{bp}} & j = 2 \\ 0 & j \neq 1,2 \end{cases} \tag{6.80}$$

C. Effectiveness Degree Calculation for Adaptive Current Protection Zone-III

Adaptive current protection zone-III reflects faults on the line where the protection component is situated, faults on the downstream line and the second downstream line. Take component 1 in Figure 6.35, for example. The zone-III protection range of component 1 is to the left of the dashed line q. In Figure 6.35, c is the terminal point of line L_3. According to the definition of the equivalent impedance of the effective distance in section B, the equivalent impedance of the effective distance that zone-III of component 1 has on L_1 and L_2 is Z_{L1} and Z_{L2}, respectively; that on L_3 is Z_{cq} and that on the other lines is 0.

When a fault occurs at q, the fault current at component 1 is

$$I_{F1-q} = \frac{K_d E_S}{Z_S + Z_{L1} + Z_{L2} + Z_{cq}} \tag{6.81}$$

where Z_{L2} is the line impedance of line L_2 and Z_{cq} is the impedance of line length cq on line L_3.

The zone-III protection range of component 1 extends to point q on line L_3, thus

$$I_{F1-q} = I_{\text{set}-1}^{\text{III}} \tag{6.82}$$

where $I_{\text{set}-1}^{\text{III}}$ is the current protection zone-III setting value of component 1.

Applying Equation (6.82) to Equation (6.81) gives

$$Z_{cq} = \frac{K_d E_S}{I_{\text{set}-1}^{\text{III}}} - Z_S - Z_{L1} - Z_{L2} \tag{6.83}$$

When zone-III of component 1 operates, the probability of the fault being on L_1, L_2 or L_3 is the percentage of the effective distance zone-III has on each line in the total effective distance of zone-III. Therefore, for component 1, the effectiveness degree of adaptive current protection zone-III is

$$\alpha_{1j}^{\text{III}} = \begin{cases} \dfrac{Z_{L1}}{Z_{L1} + Z_{L2} + Z_{cq}} & j = 1 \\ \dfrac{Z_{L2}}{Z_{L1} + Z_{L2} + Z_{cq}} & j = 2 \\ \dfrac{Z_{cq}}{Z_{L1} + Z_{L2} + Z_{cq}} & j = 3 \\ 0 & j \neq 1,2,3 \end{cases} \tag{6.84}$$

6.3.2.2 Fault Identification Algorithm Based on the Adaptive Current Protection Suiting Factor

A. The Suiting Degree Function Based on Adaptive Current Protection

The adaptive current protection suiting degree function is an integration of the protection operational status information with the effectiveness degree as the weight. By introducing the effectiveness degree as the integration weight, the effect of the more important operational information in fault identification is strengthened, and that of the less important operational information is weakened. In this way, the information integration efficiency is improved, as well as the fault tolerance ability. Take line L_3 in Figure 6.35, for example. The suiting degree function of line L_3 can be expressed as

$$E_{\text{F}}(L_3) = \left(\alpha_{13}^{\text{I}} C_1^{\text{I}} + \alpha_{23}^{\text{I}} C_2^{\text{I}} + \ldots\right) + \left(\alpha_{13}^{\text{II}} C_1^{\text{II}} + \alpha_{23}^{\text{II}} C_2^{\text{II}} + \ldots\right) + \left(\alpha_{13}^{\text{III}} C_1^{\text{III}} + \alpha_{23}^{\text{III}} C_2^{\text{III}} + \ldots\right) \tag{6.85}$$

where C_1^{I}, C_1^{II}, C_1^{III}; C_2^{I}, C_2^{II}, C_2^{III}... represent the zone-I, zone-II and zone-III operational status of component 1, 2, ... after the fault. α_{13}^{I}, α_{13}^{II}, α_{13}^{III}; α_{23}^{I}, α_{23}^{II}, α_{23}^{III}... are the effectiveness degrees of the zone-I, zone-II and zone-III operation of component 1, 2, ... in identifying the fault on line L_3.

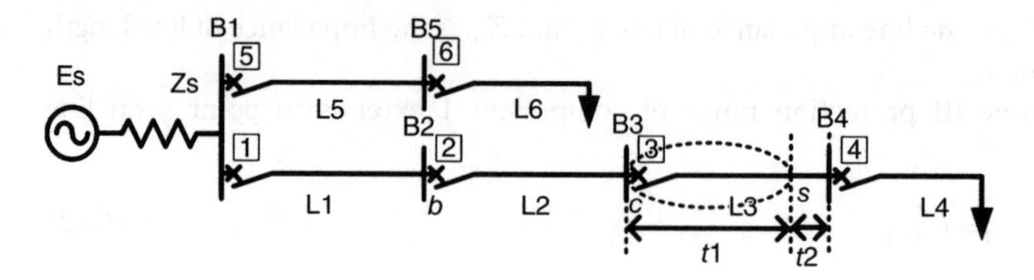

Figure 6.36 Calculation example of the expected zone-I operational status.

Table 6.25 Expected zone-I operational status of the protection components when L_3 is in fault.

Protection component number	1	2	3	4	5	6	7	8
Fault location in section t_1	0	0	1	0	0	0	0	0
Fault location in section t_2	0	0	0	0	0	0	0	0

B. The Suiting Degree Expectation Function of the Adaptive Current Protection

The suiting degree expectation function of the adaptive current protection is calculated with the expected operational status of the protection components when a fault occurs on a certain line. When faults occur at different locations on a particular line, the expected operational status of the components also differs, as does the value of the suiting degree expectation function. Taking line L_3 in Figure 6.35, for example, the suiting degree expectation function is analysed.

The zone-I protection range of component 3 is the oval in Figure 6.36. The oval separates line L_3 into two sections – t_1 and t_2. When faults occur at different locations on L_3, the expected zone-I operational status of the protection components is shown in Table 6.25.

When a fault occurs at point s (the terminal of section $t1$) in Figure 6.36, the fault current at component 3 is

$$I_{F3-s} = \frac{K_d E_S}{Z_S + Z_{L1} + Z_{L2} + Z_{L3t1}} \tag{6.86}$$

where Z_{L3t1} is the impedance of section t_1 on line L_3.

The fault occurs at the end of the zone-I protection range of component 3, thus

$$I_{F3-s} = I^{\mathrm{I}}_{\mathrm{set}-3} \tag{6.87}$$

where $I^{\mathrm{I}}_{\mathrm{set}-3}$ is the current protection zone-I setting value of component 3.

Applying Equation (6.87) to Equation (6.86) gives

$$Z_{L3t1} = \frac{K_d E_S}{\mathrm{I}^{\mathrm{I}}_{\mathrm{set}-3}} - Z_S - Z_{L1} - Z_{L2} \tag{6.88}$$

Also, according to Figure 6.36,

$$Z_{L3t2} = Z_{L3} - Z_{L3t1} \tag{6.89}$$

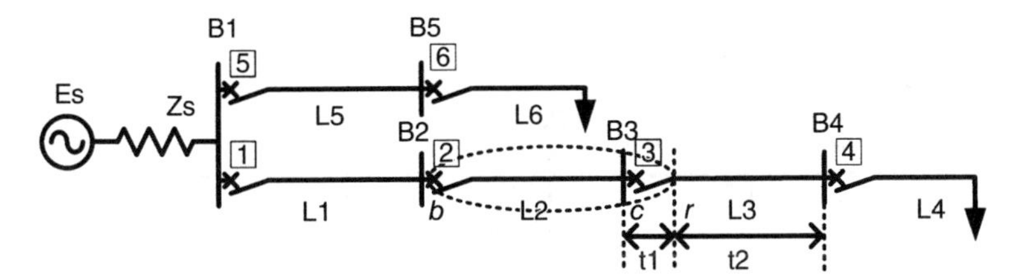

Figure 6.37 Calculation example of the expected zone-II operational status.

Table 6.26 Expected zone-II operational status of the protection components when L_3 is in fault.

Protection component number	1	2	3	4	5	6	7	8
Fault location in section t_1	0	1	1	0	0	0	0	0
Fault location in section t_2	0	0	1	0	0	0	0	0

where Z_{L3t2} is the impedance of section t_2 on line L_3.

Therefore, the probabilities of the fault being in sections t_1 and t_2: β_{t1}^{I} and β_{t2}^{I} can be calculated as

$$\begin{cases} \beta_{t1}^{\mathrm{I}} = \dfrac{Z_{L3t1}}{Z_{L3}} \\ \beta_{t2}^{\mathrm{I}} = \dfrac{Z_{L3t2}}{Z_{L3}} \end{cases} \tag{6.90}$$

The zone-II protection range of component 2 is the oval in Figure 6.37, which separates line L_3 into two sections – t_1 and t_2. When faults occur at different locations on L_3, the expected zone-II operational status of the protection components is shown in Table 6.26.

The equivalent impedance of the effective distance that zone-II of component 2 has on line L_3 (Z_{cr}) could reflect the length of section t_1 in Figure 6.37. Thus, the probabilities of the fault being in sections t_1 and t_2: β_{t1}^{II} and β_{t2}^{II} can be calculated as

$$\begin{cases} \beta_{t1}^{\mathrm{II}} = \dfrac{Z_{cr}}{Z_{L2}} \\ \beta_{t2}^{\mathrm{II}} = \dfrac{Z_{L2} - Z_{cr}}{Z_{L2}} \end{cases} \tag{6.91}$$

The zone-III protection range of component 1 is the oval in Figure 6.38, which separates line L_3 into two sections – t_1 and t_2. When faults occur at different locations on L_3, the expected zone-III operational status of the protection components is as shown in Table 6.27.

The equivalent impedance of the effective distance that zone-III of component 1 has on line L_3 (Z_{cq}) could reflect the length of section t_1 in Figure 6.38. Thus, the probabilities of the fault being in sections t_1 and t_2: $\beta_{t1}^{\mathrm{III}}$ and $\beta_{t2}^{\mathrm{III}}$ can be calculated as

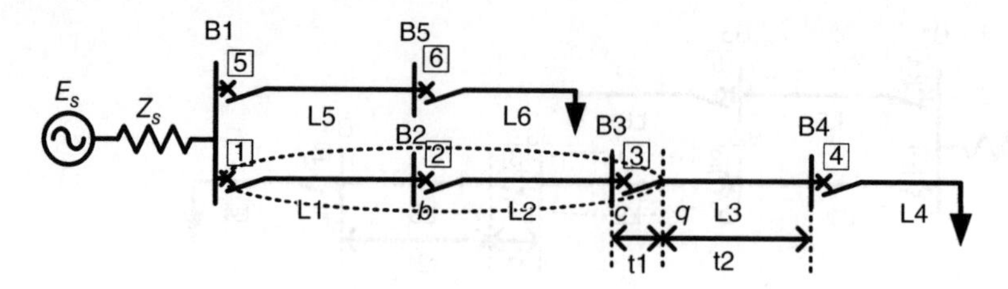

Figure 6.38 Calculation example of the expected zone-III operational status.

Table 6.27 Expected zone-III operational status of the protection components when L_3 is in fault.

Protection component number	1	2	3	4	5	6	7	8
Fault location in section t_1	1	1	0	0	0	0	0	1
Fault location in section t_2	1	1	0	0	0	0	0	1

$$\begin{cases} \beta_{t1}^{\mathrm{III}} = \dfrac{Z_{cq}}{Z_{L3}} \\ \beta_{t2}^{\mathrm{III}} = \dfrac{Z_{L3} - Z_{cq}}{Z_{L3}} \end{cases} \tag{6.92}$$

With β_t^{I}, β_t^{II}, β_t^{III} $(t = t_1, t_2 \ldots)$ as the weights of the expected protection operational status at different fault locations, the suiting degree expectation function of a certain line can be obtained. Taking line L_3 in Figure 6.35, for example, the suiting degree expectation function of line L_3 is

$$\begin{aligned} E^*(L_3) = & \left(\alpha_{13}^{\mathrm{I}} \left(\sum_{t=t_1,t_2..}^{D} \beta_t^{\mathrm{I}} C_{13-t}^{*\mathrm{I}} \right) + \alpha_{23}^{\mathrm{I}} \left(\sum_{t=t_1,t_2..}^{D} \beta_t^{\mathrm{I}} C_{23-t}^{*\mathrm{I}} \right) + \ldots \right) \\ & + \left(\alpha_{13}^{\mathrm{II}} \left(\sum_{t=t_1,t_2..}^{D} \beta_t^{\mathrm{II}} C_{13-t}^{*\mathrm{II}} \right) + \alpha_{23}^{\mathrm{II}} \left(\sum_{t=t_1,t_2..}^{D} \beta_t^{\mathrm{II}} C_{23-t}^{*\mathrm{II}} \right) + \ldots \right) \\ & + \left(\alpha_{13}^{\mathrm{III}} \left(\sum_{t=t_1,t_2..}^{D} \beta_t^{\mathrm{III}} C_{13-t}^{*\mathrm{III}} \right) + \alpha_{23}^{\mathrm{III}} \left(\sum_{t=t_1,t_2..}^{D} \beta_t^{\mathrm{III}} C_{23-t}^{*\mathrm{III}} \right) + \ldots \right) \end{aligned} \tag{6.93}$$

where D means that line L_3 is separated into D sections by the protection components. $t_1, t_2 \ldots$ is the numbering of the sections. $C_{13-t}^{*\mathrm{I}}$, $C_{13-t}^{*\mathrm{II}}$, $C_{13-t}^{*\mathrm{III}}$ represent the expected zone-I, zone-II and zone-III operational status of component 1 when faults occur in section t $(t = t_1, t_2 \ldots)$ on line L_3.

6.3.2.3 Fault Identification Based on the Adaptive Current Protection Suiting Factor

Define the adaptive current protection suiting factor of a certain line as the ratio of its suiting degree function to its suiting degree expectation function. Taking line L_3 in Figure 6.35, for example, the suiting factor of line L_3 is calculated as

$$P_C(L_3) = \frac{E_F(L_3)}{E^*(L_3)} \tag{6.94}$$

Calculate the adaptive current protection suiting factors of all the lines in the region $P_C(L_j)$. The bigger $P_C(L_j)$ is, the closer the suiting degree function of line L_j is to its suiting degree expectation function and the more probable that line L_j is to be the fault line. Therefore, the adaptive current protection suiting factor of the fault line should be the maximum ($P_{C\,max}$) among the suiting factors of all the lines in the network. With this item as the fault identification criterion, the regional backup protection scheme can be mapped, as shown in Figure 6.39.

6.3.2.4 Simulation Verification

A. RTDS Simulation on a 10.5 kV Power Grid in the Tianjin City Distribution Network

A 10.5 kV distribution system in Tianjin city is used for RTDS simulation, and is shown in Figure 6.40. In the system, lines L_1, L_2 and L_5 are overhead lines, lines L_3, L_4 and L_6 are cable lines. The line parameters are shown in Table 6.28. This is an isolated neutral system with regional centralized backup protection.

Suppose a phase-to-phase metallic short circuit fault occurs in the middle section of line L_3; the setting values and operational status of the adaptive current protection components are shown in Table 6.29.

According to the setting values, the adaptive current protection effectiveness degrees α_{Ij}^{I}, α_{ij}^{II} and α_{ij}^{III} can be calculated. The effectiveness degree of the protection components in identifying the fault on L_3: α_{i3} is shown in Table 6.30.

With the adaptive current protection effectiveness degrees as the weights, according to the actual operational status of the protection components in Table 6.29, the values of the suiting degree function can be calculated. Then, according to the expected operational status of the protection components and the corresponding weights given in Section 6.3.2.2, the values of the suiting degree expectation function can be obtained. Finally, the adaptive current protection suiting factor can be obtained according to Equation (6.94). The adaptive current protection suiting degree, suiting degree expectation and suiting factor of the lines are shown in Table 6.31.

The system in Figure 6.40 contains six lines and eight protection components; the adaptive current protection operational status information obtained contains 24 digits in total. Suppose the protection operational information is incorrect – losing or erring in five random digits (five random digits lost or in error means 20% random loss or error), the statistical average of 100 simulation tests is shown in Figure 6.41. It can be seen that, even in the case of information loss or error, the suiting factor of line L_3 is still bigger than that of the other lines, i.e. line L_3 still meets the fault identification criterion and could be correctly identified as the fault line. Thus, the proposed method is proven highly fault-tolerant.

B. Simulation Tests on the IEEE 33-Bus Distribution System

In order to further verify the applicability and feasibility of the proposed method, the IEEE 33-bus distribution system was modelled in RTDS for simulation

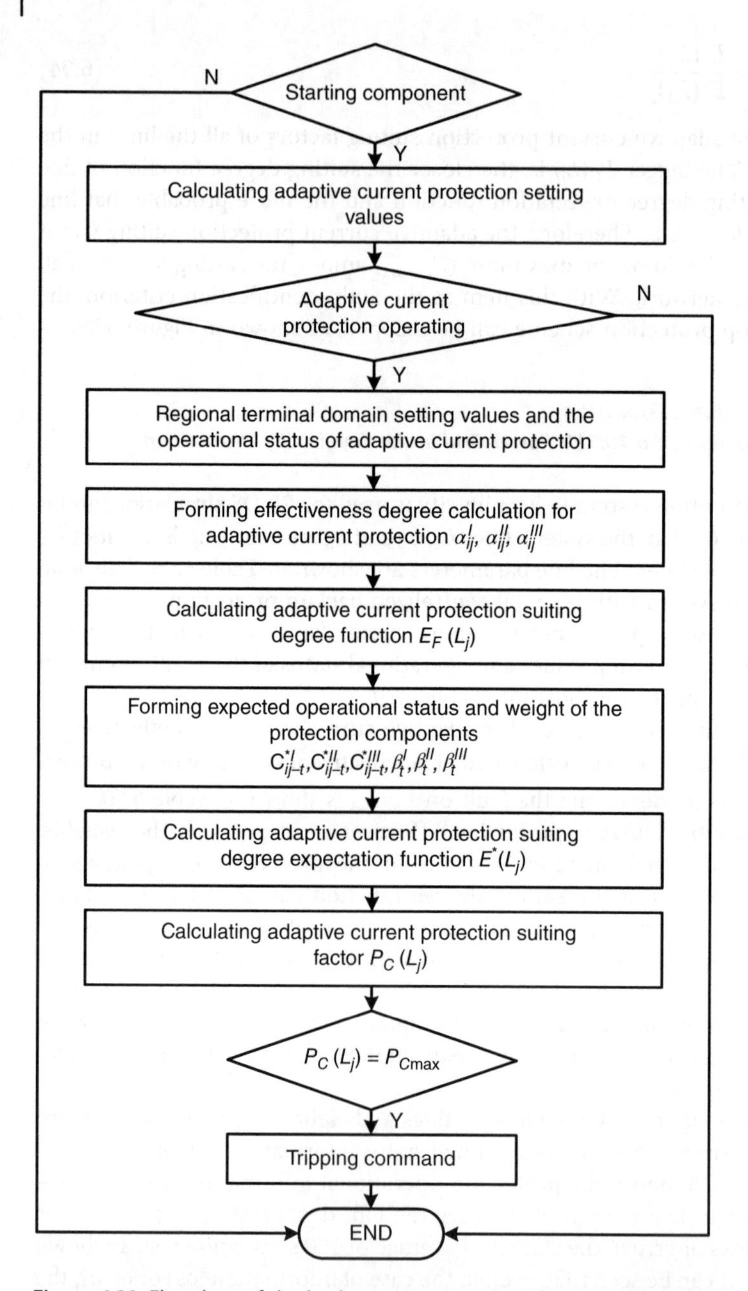

Figure 6.39 Flowchart of the backup protection algorithm based on adaptive current protection suiting factor.

analysis. This is a single supply radial system, the component numbering shown in Figure 6.42.

The backup protection is formed in the regional centralized mode. Substation 33 is set as the main station in the region, which is responsible for receiving the uploaded data from the substations and making a comprehensive decision.

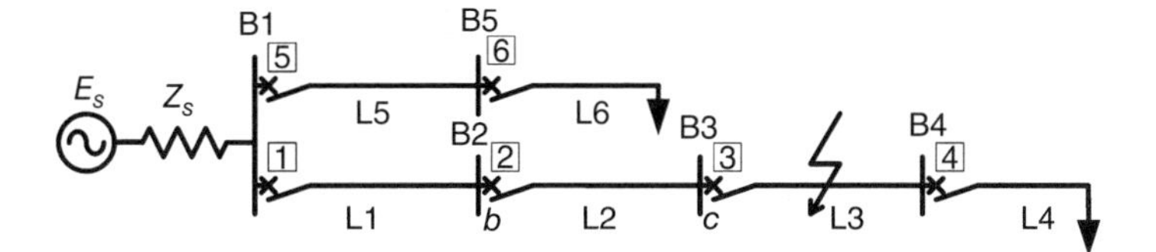

Figure 6.40 A 10.5 kV power grid in the Tianjin city distribution network.

Table 6.28 Line parameters of the 10.5 kV power grid in the Tianjin city distribution network.

Line number	L_1	L_2	L_3	L_4	L_5	L_6
R (Ω/km)	0.027	0.027	0.259	0.259	0.027	0.259
X (Ω/km)	0.347	0.347	0.093	0.093	0.347	0.093
Line length (km)	5	5	7	14	4	6

Table 6.29 Adaptive current protection setting value and operational status of the protection components.

Component number	IsetI	IsetII	IsetIII	Zone-I	Zone-II	Zone-III	Zone-III
1	2.5293	1.7385	1.1590	0	0	0	0
2	1.9603	1.3474	0.6408	0	0	1	0
3	1.0082	0.6930	0.4620	1	1	1	0
4	0.2276	0.1821	0.1365	0	0	0	0
5	0.5114	0.4091	0.3069	0	0	0	0
6	0.2520	0.2016	0.1512	0	0	0	0
7	2.2607	1.5539	0.8217	0	0	0	0
8	1.1594	0.7969	0.5313	0	0	0	0

Table 6.30 Effectiveness degree of each adaptive current protection component to the identification of fault line L_3.

Component number	1	2	3	4	5	6	7	8
Zone-I	0	0	1	0	0	0	0	0
Zone-II	0	0.5651	0.4197	0	0	0	0	0
Zone-III	0.0174	0.4670	0.3333	0	0	0	0	0

Table 6.31 Adaptive current protection suiting degree, suiting degree expectation and suiting factor of the lines.

Line number	L_1	L_2	L_3	L_4	L_5	L_6
Suiting degree	1.5000	1.0330	2.2200	1.2470	0	0
Suiting degree expectation	5.1438	4.0757	2.1330	3.2693	1.3749	2.6724
Suiting factor	0.2916	0.2534	1.0408	0.3814	0	0

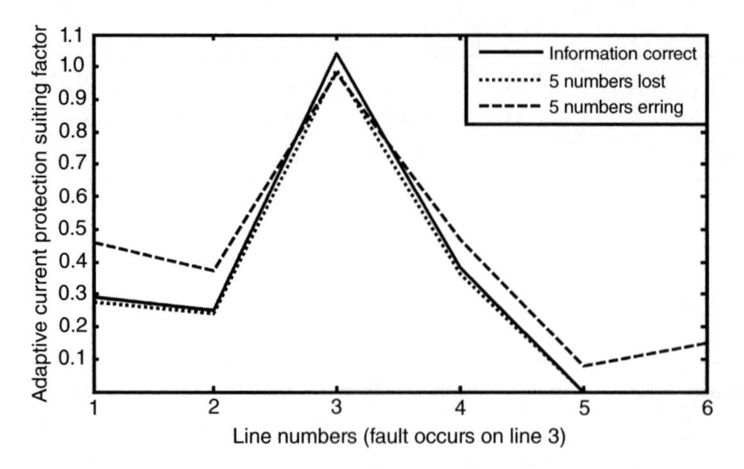

Figure 6.41 Adaptive current protection suiting factor of each line when L_3 is in fault.

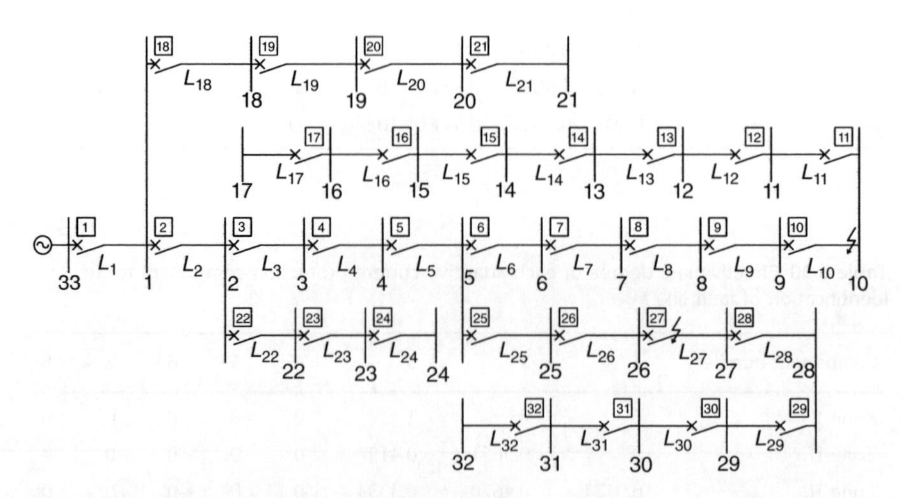

Figure 6.42 IEEE 33-bus distribution system.

Suppose a three-phase metallic grounding fault occurs at the end of line L_{10}; the setting values and operational status of the adaptive current protection components are shown in Table 6.32.

According to the setting values, the adaptive current protection effectiveness degrees α_{ij}^{I}, $\alpha_{ij}^{\mathrm{II}}$ and $\alpha_{ij}^{\mathrm{III}}$ can be formed. The effectiveness degree of the protection components in identifying the fault on L_{10}: α_{i10} is shown in Table 6.33.

With the adaptive current protection effectiveness degrees as the weights, according to the actual operational status of the protection components in Table 6.33, the values of the suiting degree function can be calculated. Then, according to the expected operational status of the protection components and the corresponding weights given in Section 6.3.3.2, the values of the suiting degree expectation function can be obtained. Finally, the adaptive current protection suiting factor can be obtained according to Equation (6.94). The adaptive current protection suiting degree, suiting degree expectation and suiting factor of the lines are shown in Table 6.34.

Table 6.32 Adaptive current protection setting value and operational status of the protection components.

Component number	IsetI	IsetII	IsetIII	Zone-I	Zone-II	Zone-III
1	8.4402	6.1992	4.6964	0	0	0
2	5.6244	4.2609	3.2776	0	0	0
3	4.4746	3.3899	2.6076	0	0	0
4	3.6808	2.7885	2.1450	0	0	0
5	2.5727	1.9490	1.4993	0	0	0
6	2.2679	1.7181	1.3216	0	0	0
7	1.9017	1.4407	1.1082	0	0	0
8	1.4883	1.1275	0.8673	0	0	1
9	1.2200	0.9243	0.7110	0	0	1
10	1.1861	0.8986	0.6912	0	1	1
18	0.0044	0.0036	0.0027	0	0	0
22	0.0116	0.0093	0.0070	0	0	0
25	0.0205	0.0164	0.0123	0	0	0
8	1.4883	1.1275	0.8673	0	0	1

Table 6.33 Effectiveness degree of each adaptive current protection component to the identification of fault line L_{10}.

Component number	1~7	8	9	10	11~32
Zone-I	0	0	0	1	0
Zone-II	0	0	0.3558	0.9865	0
Zone-III	0	0.0133	0.1393	0.3443	0

Table 6.34 Adaptive current protection suiting degree, suiting degree expectation and suiting factor of the lines.

Line number	L_1	L_2	L_3	L_4	L_5	L_6	L_7
Suiting degree	0	0	0	0	0	0	0
Suiting degree expectation	1.242	2.715	2.288	2.266	2.889	2.354	2.399
Suiting factor	0	0	0	0	0	0	0
Line number	L_8	L_9	L_{10}	L_{11}	L_{12}	L_{13}	L_{14}
Suiting degree	0.498	1.363	1.484	0.656	0	0	0
Suiting degree expectation	2.688	2.870	2.355	2.623	3.301	2.651	2.724
Suiting factor	0.185	0.475	0.630	0.250	0	0	0
Line number	L_{15}	L_{16}	L_{17}	L_{18}	L_{19}	L_{20}	L_{21}
Suiting degree	0	0	0	0	0	0	0
Suiting degree expectation	2.634	3.254	0.243	2.140	3.463	2.374	0.654
Suiting factor	0	0	0	0	0	0	0
Line number	L_{22}	L_{23}	L_{24}	L_{25}	L_{26}	L_{27}	L_{28}
Suiting degree	0	0	0	0	0	0	0
Suiting degree expectation	2.482	2.977	0.499	2.344	2.566	3.185	2.882
Suiting factor	0	0	0	0	0	0	0
Line number	L_{29}	L_{30}	L_{31}	L_{32}			
Suiting degree	0	0	0	0			
Suiting degree expectation	2.439	3.246	2.477	0.574			
Suiting factor	0	0	0	0			

For a single power supply system where current protection is only deployed at the power source end of the lines, one of the worst cases is when a fault occurs at the other line end and current protection zone-I does not operate. In this case, the calculated suiting factor of the fault line will decrease. When a fault occurs at the end of line L_{10}, the adaptive current protection suiting factors of the lines are shown in Table 6.34. It can be seen that the suiting factor of line L_{10} is 0.630, which is still the maximum among all the suiting factors in the region. Thus, line L_{10} still meets the fault identification criterion and could be correctly identified as the fault line. In this way, the proposed method is proven highly reliable.

The system in Figure 6.42 contains 32 lines and 32 protection components; the adaptive current protection operational status information obtained contains 96 digits in total. Suppose the protection operational information is incorrect – losing or erring in 20 random digits (20 random digits lost or in error means 21% random loss or error), the statistical average of 100 simulation tests is shown in Figure 6.43.

It can be seen that, even in the case of information loss or error, the suiting factor of line L_{10} is still bigger than that of the other lines, i.e. line L_{10} still meets the fault identification criterion and could be correctly identified as the fault line. Thus, the proposed method is proven highly fault-tolerant.

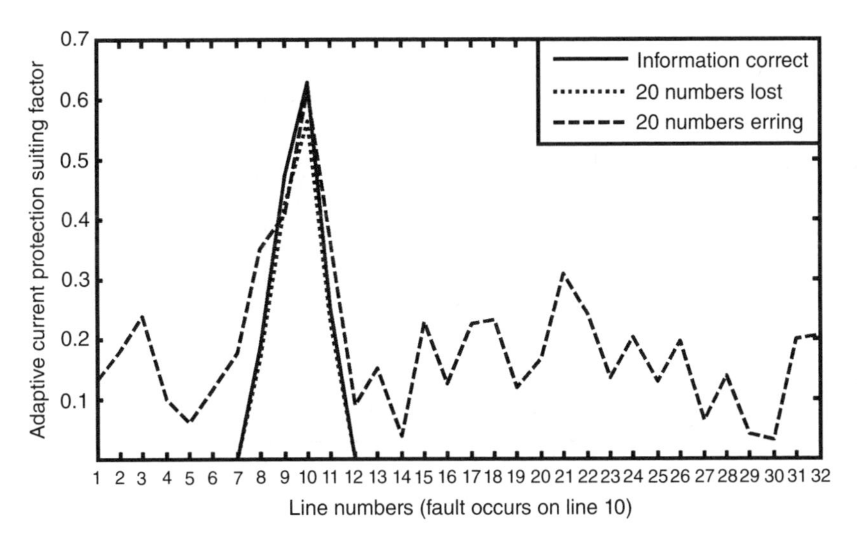

Figure 6.43 Adaptive current protection suiting factor of each line when L_{10} is in fault.

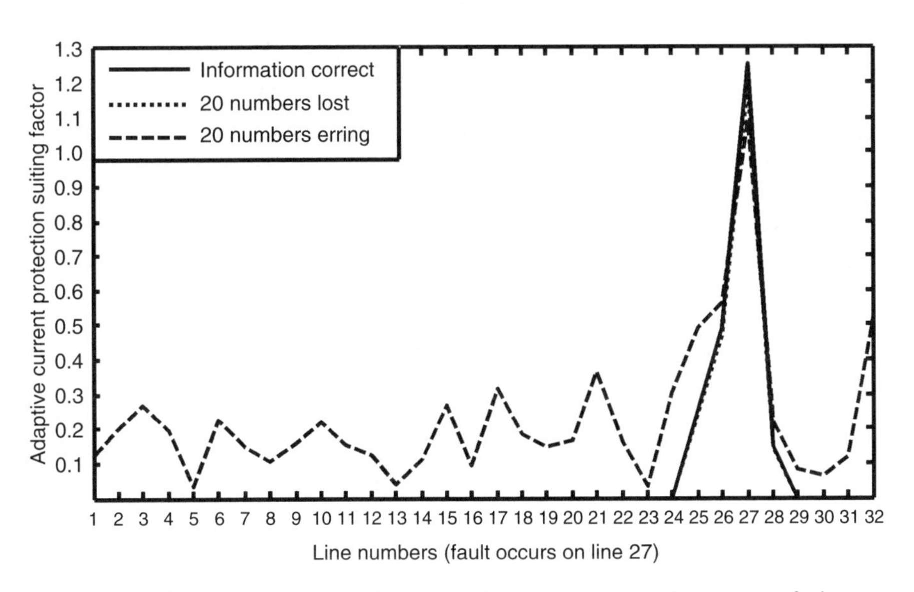

Figure 6.44 Adaptive current protection suiting factor of each line when L_{27} is in fault.

When a single-phase grounding fault occurs at the power source end of line L_{27}, the adaptive current protection suiting factors of the lines are shown in Figure 6.44. It can be seen that, when the protection operational information is correct, the suiting factor of line L_{27} is the maximum among all the suiting factors in the region, which meets the fault identification criterion. When the protection operational information is incorrect – losing or erring in 20 random digits – the statistical result of 100 simulation tests shows that the proposed method is still able to identify the fault line correctly. Therefore, the proposed method is not affected by the fault type.

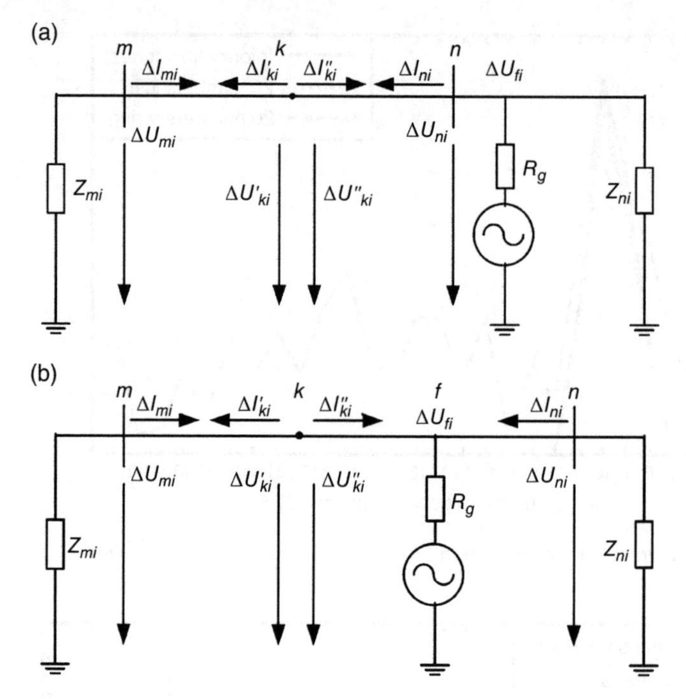

Figure 6.45 Sequence network of the fault system. (a) Sequence network in the case of external fault. (b) Sequence network in the case of internal fault.

6.3.3 Wide Area Protection Based on the Virtual Impedance of the Fault Component

6.3.3.1 Algorithm Principle [15]

Consider a double-ended transmission line with distributed parameters. The sequence networks of the line in the cases of external fault and internal fault are shown in Figure 6.45(a) and (b), respectively. The parameters are all sequence parameters in Figure 6.45. Z_{mi} is the system sequence impedance on the m side, and Z_{ni} is the system sequence impedance on the n side. ΔI_{mi} and ΔI_{ni} are the sequence fault components of the currents at terminal m and terminal n, where $i = 1,2,0$, representing the positive sequence, negative sequence and zero sequence, respectively. It should be noted that, in the following passage, i has the same meaning. ΔU_{mi} and ΔU_{ni} are the sequence fault components of the voltages at terminal m and terminal n. ΔU_{fi} is the equivalent sequence voltage at the fault location in the fault supplementary network. $\Delta I'_{ki}$ and $\Delta U'_{ki}$ are the sequence fault components of current and voltage at point k calculated from terminal m. $\Delta I''_{ki}$ and $\Delta U''_{ki}$ are the sequence fault components of current and voltage at point k calculated from terminal n. R_g is the fault resistance.

Suppose the line length is l, and the distance between k (k is any point on line mn) and terminal m takes up p (in percentage) of the whole line length. The sequence characteristic impedance of the line is Z_{ci}, and the sequence spreading coefficient is r_i. The sequence fault components of current and voltage at point k calculated from terminal m and terminal n are

$$\begin{cases} \Delta U'_{ki} = \Delta U_{mi} \cdot \cosh(r_i \cdot pl) - \Delta I_{mi} \cdot Z_{ci} \sinh(r_i \cdot pl) \\ \Delta I'_{ki} = \dfrac{\Delta U_{mi} \cdot \sinh(r_i \cdot pl)}{Z_{ci}} - \Delta I_{mi} \cdot \cosh(r_i \cdot pl) \\ \Delta U''_{ki} = \Delta U_{ni} \cdot \cosh(r_i \cdot (1-p)l) - \Delta I_{ni} \cdot Z_{ci} \sinh(r_i \cdot (1-p)l) \\ \Delta I''_{ki} = \dfrac{\Delta U_{ni} \cdot \sinh(r_i \cdot (1-p)l)}{Z_{ci}} - \Delta I_{ni} \cdot \cosh(r_i \cdot (1-p)l) \end{cases} \tag{6.95}$$

Suppose that an external fault occurs, as shown in Figure 6.45(a), and the fault location is on the backside of terminal n (or terminal m). Since the line structure is not damaged, the sequence fault components of current and voltage at point k calculated from terminal m and terminal n are both the actual sequence fault components of current and voltage at point k, thus

$$\begin{cases} \Delta I'_{ki} = -\Delta I''_{ki} \\ \Delta U'_{ki} = \Delta U''_{ki} \end{cases} \tag{6.96}$$

Define the virtual impedance as the sequence fault component of voltage divided by the sequence fault component of current

$$\begin{cases} \dfrac{\Delta U'_{ki}}{\Delta I'_{ki}} = Z_{ci} \cdot \dfrac{\Delta U_{mi} \cosh(r_i \cdot pl) + \Delta I_{mi} \sinh(r_i \cdot pl)}{\Delta U_{mi} \sinh(r_i \cdot pl) + \Delta I_{mi} \cosh(r_i \cdot pl)} \\ \dfrac{\Delta U''_{ki}}{\Delta I''_{ki}} = -Z_{ci} \cdot \dfrac{\Delta U_{mi} \cosh(r_i \cdot pl) + \Delta I_{mi} \sinh(r_i \cdot pl)}{\Delta U_{mi} \sinh(r_i \cdot pl) + \Delta I_{mi} \cosh(r_i \cdot pl)} \\ \dfrac{\Delta U'_{ki}}{\Delta I'_{ki}} + \dfrac{\Delta U''_{ki}}{\Delta I''_{ki}} = 0 \end{cases} \tag{6.97}$$

When an internal fault occurs, the line structure is damaged, which is equal to the addition of an equivalent source at the fault location in the fault supplementary network, as shown in Figure 6.45(b). Suppose the fault location f is between terminal n and k, then the sequence fault components of current and voltage at point k calculated from terminal m are the actual values at point k. However, the sequence fault components of point k calculated from terminal n will obviously deviate from the actual values, due to the fault location. In this case, the virtual impedance is

$$\begin{cases} \dfrac{\Delta U'_{ki}}{\Delta I'_{ki}} = Z_{ci} \cdot \dfrac{\Delta U_{mi} \cosh(r_i \cdot pl) + \Delta I_{mi} \sinh(r_i \cdot pl)}{\Delta U_{mi} \sinh(r_i \cdot pl) + \Delta I_{mi} \cosh(r_i \cdot pl)} \\ \dfrac{\Delta U''_{ki}}{\Delta I''_{ki}} = Z_{ci} \cdot \dfrac{\Delta U_{ni} \cosh(r_i \cdot (1-p)l) + \Delta I_{ni} \sinh(r_i \cdot (1-p)l)}{\Delta U_{ni} \sinh(r_i \cdot (1-p)l) + \Delta I_{ni} \cosh(r_i \cdot (1-p)l)} \\ \dfrac{\Delta U'_{ki}}{\Delta I'_{ki}} + \dfrac{\Delta U''_{ki}}{\Delta I''_{ki}} \neq 0 \end{cases} \tag{6.98}$$

Set k to be the midpoint of the line (i.e. $p = 1/2$), so that the following criterion is formed according to different characteristics of the virtual impedance in the cases of out-of-zone and in-zone faults.

$$\begin{cases} \max\left(\Delta I''_{ki}, \Delta I'_{ki}\right) > 0.1 I_N \\ \left|\dfrac{\Delta U''_{ki}}{\Delta I''_{ki}} + \dfrac{\Delta U'_{ki}}{\Delta I'_{ki}}\right| \geq \min\left\{\left|\dfrac{\Delta U''_{ki}}{\Delta I''_{ki}}\right|, \left|\dfrac{\Delta U'_{ki}}{\Delta I'_{ki}}\right|\right\} \end{cases} \tag{6.99}$$

In Inequalities (6.99), the first expression is the fixed threshold of the criterion. When the current calculated from either terminal is detected to be bigger than 0.1 of the rated value, the criterion will start. The second inequality in (6.99) is a ratio restraint criterion. When the absolute value of the sum of two virtual impedances surpasses the absolute value of the smaller virtual impedance, the protection will operate. The protection performance is analysed below.

When an external fault occurs, theoretically the operation quantity should be zero, according to Equations (6.97). The restraint quantity is related to the system impedance of the non-fault terminal and the line impedance. When the system impedance of the non-fault terminal is zero, the restraint quantity is the minimum. Suppose the fault occurs on the backside of terminal m, and the system impedance of terminal n is zero, then $\Delta U_{ni} = 0$, and the restraint quantity is

$$\left|\frac{\Delta U_{ki}}{\Delta I_{ki}}\right| = \left|Z_{ci} \tanh(r_i \cdot l_{nk})\right| > \left|Z_{ci} \sinh(r_i \cdot l_{nk})\right| \tag{6.100}$$

The result in Equation (6.100) is approximately half of the line sequence impedance, thus effective restraint can be realized.

When an internal fault occurs, consider that

$$\begin{cases} Z_{mi} = -\dfrac{\Delta U_{mi}}{\Delta I_{mi}} \\ Z_{ni} = -\dfrac{\Delta U_{ni}}{\Delta I_{ni}} \end{cases} \tag{6.101}$$

Suppose

$$\begin{cases} a = \mathrm{j}\coth\left(r_i \cdot \dfrac{l}{2}\right) = \mathrm{j}\coth\left(r_i \cdot \dfrac{l}{2}\right) \\ b = -\mathrm{j}\dfrac{Z_{mi}}{Z_{ci}} \\ c = -\mathrm{j}\dfrac{Z_{ni}}{Z_{ci}} \end{cases} \tag{6.102}$$

Since the impedance angles of the system impedance and line impedance are both close to 90°, a, b and c in Equations (6.102) are all positive real numbers. a is related to the line length and line parameters, thus a is a constant when line parameters are given. b and c are the ratios of the dorsal system sequence impedance to the line characteristic sequence impedance.

Apply $p = 1/2$ to Equations (6.98), and divide the numerator and denominator by $\Delta I_{\varphi i} Z_{ci} \sinh(r_i \cdot l/2)$ $(\varphi = m, n)$, then Equations (6.98) can be described as

$$\begin{cases} \dfrac{\Delta U'_{ki}}{\Delta I'_{ki}} = Z_{ci} \cdot \dfrac{-Z_{mi}/Z_{ci} \cdot \coth(r_i \cdot l/2) - 1}{-Z_{mi}/Z_{ci} - \coth(r_i \cdot l/2)} \\ \dfrac{\Delta U''_{ki}}{\Delta I''_{ki}} = Z_{ci} \cdot \dfrac{-Z_{ni}/Z_{ci} \cdot \coth(r_i \cdot l/2) - 1}{-Z_{ni}/Z_{ci} - \coth(r_i \cdot l/2)} \end{cases} \tag{6.103}$$

Since $\coth(r_i \cdot l/2)$ is a negative imaginary number, and Z_{mi} and Z_{ni} are positive imaginary numbers, Equations (6.103) can be simplified to

$$\begin{cases} \dfrac{\Delta U'_{ki}}{\Delta I'_{ki}} = \mathrm{j}Z_{ci} \cdot \dfrac{|Z_{mi}|/Z_{ci} \cdot |\coth(r_i \cdot l/2)| + 1}{|Z_{mi}|/Z_{ci} - |\coth(r_i \cdot l/2)|} \\ \dfrac{\Delta U''_{ki}}{\Delta I''_{ki}} = \mathrm{j}Z_{ci} \cdot \dfrac{|Z_{ni}|/Z_{ci} \cdot |\coth(r_i \cdot l/2)| + 1}{|Z_{ni}|/Z_{ci} - |\coth(r_i \cdot l/2)|} \end{cases} \tag{6.104}$$

Apply Equations (6.102) to Equations (6.104), so that

$$\begin{cases} \dfrac{\Delta U'_{ki}}{\Delta I'_{ki}} = \mathrm{j}Z_{ci} \dfrac{ab + 1}{a - b} \\ \dfrac{\Delta U''_{ki}}{\Delta I''_{ki}} = \mathrm{j}Z_{ci} \dfrac{ac + 1}{a - c} \end{cases} \tag{6.105}$$

It can be seen from Equations (6.105) that the value of the virtual sequence impedance is related to the values of b and c. This is discussed in detail below.

Case 1 – If $a > b$ and $a > c$, then, according to Equations (6.105), the virtual sequence impedances calculated from two line terminals both have values in $\left[\mathrm{j}\dfrac{Z_{ci}}{a}, +\infty\right)$, with the same phase. Since the operational quantity is the sum of the negative sequence virtual impedances, and the restraint quantity is the smaller negative sequence virtual impedance, in this case the operational quantity is bigger than the restraint quantity. Therefore, the fault can be correctly identified.

Case 2 – If $a > b$ and $a < c$, or $a < b$ and $a > c$ (i.e. one of the line terminals is weakly fed), suppose terminal m is the weakly fed terminal, i.e. $a < b$ and $a > c$, then, according to Equations (6.105), $\dfrac{\Delta U'_{ki}}{\Delta I'_{ki}}$ has values in $(-\infty, -\mathrm{j}aZ_{ci}]$, and $\dfrac{\Delta U''_{ki}}{\Delta I''_{ki}}$ has values in $\left[\mathrm{j}\dfrac{Z_{ci}}{a}, +\infty\right)$. According to reference [16], when one of the line terminals is weakly fed, the impedance of the other terminal will be much smaller than the line capacitance. Therefore, c is much smaller than a, and $\dfrac{\Delta U''_{ki}}{\Delta I''_{ki}} \approx \mathrm{j}\dfrac{Z_{ci}}{a}$. Although a decreases as the line length and voltage level increase, for a 500 kV line, when the line length is below 400 km, a is not smaller than 3. Therefore, the absolute value of the virtual sequence impedance calculated from terminal m is bigger than the absolute value of the virtual sequence impedance calculated from terminal n. In this case, the restraint quantity is $\left|\mathrm{j}\frac{Z_{ci}}{a}\right|$, and when $\frac{\Delta U'_{ki}}{\Delta I'_{ki}} = -\mathrm{j}aZ_{ci}$ (i.e. $b = \infty$ and there is no load at terminal m) the operational quantity is the minimum. The operational quantity and restraint quantity are

$$\begin{cases} \left|\frac{\Delta U'_{ki}}{\Delta I'_{ki}} + \frac{\Delta U''_{ki}}{\Delta I''_{ki}}\right| = \left|\left(a - \frac{1}{a}\right) Z_{ci}\right| \\ \min\left\{\left|\frac{\Delta U'_{ki}}{\Delta I'_{ki}}\right|, \left|\frac{\Delta U''_{ki}}{\Delta I''_{ki}}\right|\right\} = \left|\frac{1}{a} Z_{ci}\right| \end{cases} \tag{6.106}$$

Consider the value range of a; the operational quantity is bigger than the restraint quantity. In this case of one weakly fed terminal, the proposed method is still able to identify the fault.

Case 3 – If $a < b$ and $a < c$, then, according to the analysis in Case 2, the system sequence impedances of two line terminals are not able to meet all the conditions above. In this case, there is no need to discuss the reliability of the proposed method.

Expanding Inequalities (6.99) to contain the positive, negative, and zero sequence fault component criteria gives

$$\begin{cases} \max\left(\Delta I''_{k1}, \Delta I'_{k1}\right) > 0.1 I_N \\ \left|\frac{\Delta U''_{k1}}{\Delta I''_{k1}} + \frac{\Delta U'_{k1}}{\Delta I'_{k1}}\right| \geq \min\left\{\left|\frac{\Delta U''_{k1}}{\Delta I''_{k1}}\right|, \left|\frac{\Delta U'_{k1}}{\Delta I'_{k1}}\right|\right\} \end{cases} \tag{6.107}$$

$$\begin{cases} \max\left(\Delta I''_{k2}, \Delta I'_{k2}\right) > 0.1 I_N \\ \left|\frac{\Delta U''_{k2}}{\Delta I''_{k2}} + \frac{\Delta U'_{k2}}{\Delta I'_{k2}}\right| \geq \min\left\{\left|\frac{\Delta U''_{k2}}{\Delta I''_{k2}}\right|, \left|\frac{\Delta U'_{k2}}{\Delta I'_{k2}}\right|\right\} \end{cases} \tag{6.108}$$

$$\begin{cases} \max\left(\Delta I''_{k0}, \Delta I'_{k0}\right) > 0.1 I_N \\ \left|\frac{\Delta U''_{k0}}{\Delta I''_{k0}} + \frac{\Delta U'_{k0}}{\Delta I'_{k0}}\right| \geq \min\left\{\left|\frac{\Delta U''_{k0}}{\Delta I''_{k0}}\right|, \left|\frac{\Delta U'_{k0}}{\Delta I'_{k0}}\right|\right\} \end{cases} \tag{6.109}$$

Combining the three sequence criteria to form the protection of the transmission line, the cooperation logic among them is as shown in Figure 6.46. The

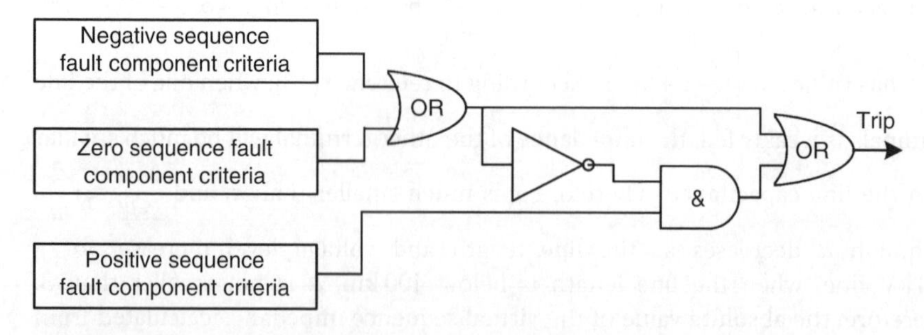

Figure 6.46 Cooperation logic of the sequence criteria.

negative and zero sequence fault component criteria are able to recognize all kinds of asymmetrical faults, and these criteria are effective as long as the fault continues. The positive sequence fault component criterion is able to recognize all kinds of transverse fault. In summary, when the fixed threshold of the negative or zero sequence criterion is detected to start, the negative or zero sequence criterion will be used to identify the fault. When only the fixed threshold of the positive sequence criterion is detected to start, the positive sequence criterion will be used to identify the fault.

6.3.3.2 Performance Analysis

The differential relay modelled is shown in Figure 6.47, where TA and TV are the current transformer and voltage transformer, respectively. First, Relay 1/2 acquires the voltage and current at bus M/N and calculates the virtual impedance. Second, via the communication network, Relay 1 and Relay 2 will exchange the calculated virtual impedances. Then, according to the criteria in Inequalities (6.107)–(6.109), it can be determined whether a certain side meets the operational condition. If a certain side meets the operational condition, then the circuit breaker at this side will trip and tripping information will be sent to the opposite side. If neither side meets the operational condition, then the protection will not operate.

The proposed method has the following characteristics:

1) The proposed method has no requirement for information synchronization. When an internal fault occurs, the virtual impedances calculated from two system ends are shown in Equations (6.103). It can be seen that the values of the virtual impedance calculated from two system ends are related to the backside line impedance, line length, spreading coefficient and the characteristic impedance. Since the four coefficients are all constants, the virtual impedance will also be a constant, i.e. the values of the virtual impedance at different instants are approximately the same. Thus, strict data synchronization of two system ends is not required.
2) The proposed method is immune to transition resistance. For traditional current differential protection, the restraint current has to do with the current

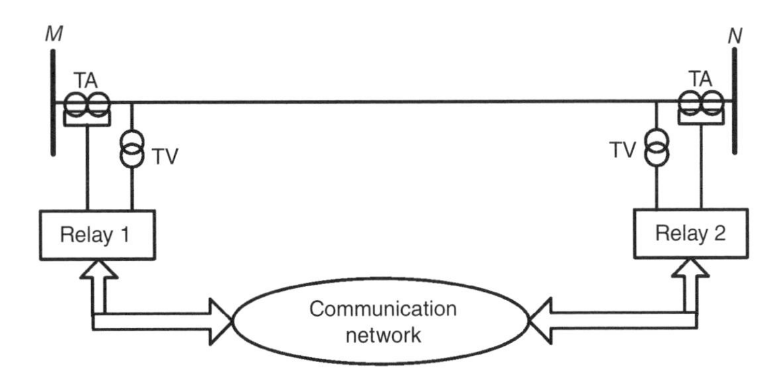

Figure 6.47 Diagram of the differential relay.

distribution coefficients at two sides of the fault point and the load current. When a grounding fault occurs on a heavily loaded line via high transition resistance, the protection sensitivity may not be sufficient. For the proposed method, as can be seen in Equations (6.103), the virtual impedance has nothing to do with the transition resistance. Therefore, no matter how big the transition resistance, the sensitivity of the proposed method remains the same.

3) The proposed method is unaffected by the capacitance current. For traditional current differential protection, the widely applied steady-state capacitance current half compensation method is affected by the fault location, and may lack sensitivity and compensation accuracy. While the proposed method is based on the distributed-parameter line model, which could fully compensate for the capacitance current. Therefore, the reliability of the proposed method is not affected by the capacitance current.
4) The proposed method is not affected by the system parameters. For traditional current differential protection, low sensitivity may result if a fault occurs when the system is in a no-load operating state. However, the proposed method is not affected by no-load operation. According to the analysis of Equations (6.106), the proposed method is still applicable in the case of weak feed end fault.
5) The proposed method is not affected by the fault type. Forming the protection criterion according to whether the line structure is disturbed, the proposed method could operate reliably when an internal fault occurs, and could refuse to operate when an external fault occurs. Even in the cases of transforming fault and developing fault, the proposed method proves to be highly sensitive.

6.3.3.3 Simulation Verification

Referring to line parameters of the Jingjintang 500 kV extra-high-voltage line, a 400 km long line is modelled here using PSCAD/EMTDC, as shown in Figure 6.48. The system rated power $S_N = 500$ MVA, rated voltage $U_N = 500$ kV, fundamental frequency $f = 50$ Hz. The full-cycle Fourier correlation has been used as the filter algorithm with a sampling frequency of 2500 Hz. During the simulation tests, the transition resistance of high-resistance faults was set to 300 Ω, and the fault occurred at $t = 0.5$ s.

The equivalent system parameters are $Z_{M1} = 0.625 + j3.545\ \Omega$, $Z_{M0} = 0.990 + j5.613\ \Omega$, $Z_{N1} = 2.726 + j15.461\ \Omega$, $Z_{N0} = 3.595 + j20.386\ \Omega$. The transmission line parameters are $r_1 = 0.02083\ \Omega/\text{km}$, $l_1 = 0.8948$ mH/km, $C_1 = 0.0129\ \mu\text{F/km}$;

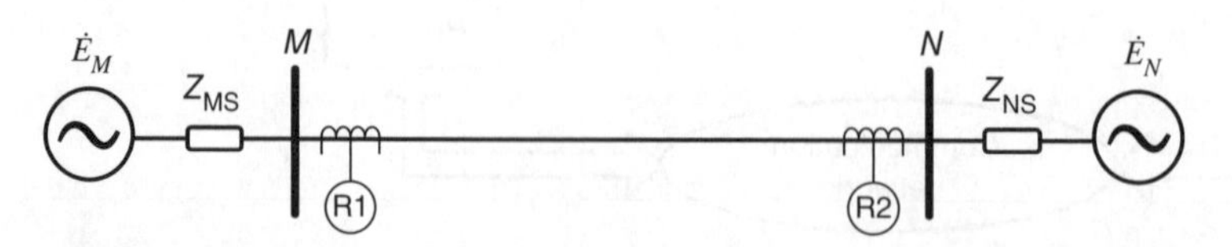

Figure 6.48 Simulation model.

$r_0 = 0.1148\ \Omega/\text{km}$, $l_0 = 2.2886\ \text{mH/km}$, $C_0 = 0.00523\ \mu\text{F/km}$. Define the ratios of the operational quantity to the restraint quantity as K_i:

$$K_i = \frac{\left|\frac{\Delta U''_{ki}}{\Delta I''_{ki}} + \frac{\Delta U'_{ki}}{\Delta I'_{ki}}\right|}{\min\left\{\left|\frac{\Delta U''_{ki}}{\Delta I''_{ki}}\right|, \left|\frac{\Delta U'_{ki}}{\Delta I'_{ki}}\right|\right\}} \tag{6.110}$$

where $i = 1,2,0$. The operational threshold value is set to be 1, i.e. if $K_i \geq 1$ after the sequence criterion starts, the trip command is issued.

A. Internal Fault

When different faults occur at a 40% line length from terminal M, the results of the sequence criterion are shown in Figure 6.49. It can be seen that, for either symmetrical fault or asymmetrical fault, the fault can be identified reliably and with a relatively high sensitivity using the proposed selection method.

To examine the impact of fault location on the proposed method, a series of faults of different types were simulated at several locations along the transmission line. The identification results of the sequence criteria are shown in Table 6.35. It can be seen that the line structure is damaged in the case of internal fault, and the ratios of the operational quantity to the restraint quantity are all bigger than the operational threshold 1, with a considerable margin. Also, the criteria are able to operate reliably in different cases of internal faults, and are unaffected by the fault location.

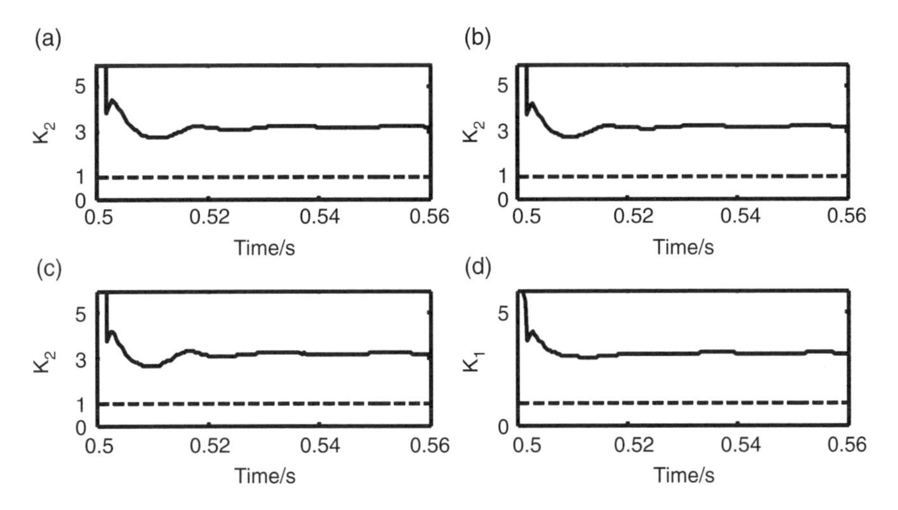

Figure 6.49 Result of criterion when different faults occur at a 40% line length from terminal M. (a) Negative sequence criterion when a single-phase grounding fault occurs. (b) Negative sequence criterion when a phase-to-phase fault occurs. (c) Negative sequence criterion when a phase-to-phase grounding fault occurs. (d) Positive sequence criterion when a three-phase symmetrical fault occurs.

Table 6.35 Results of sequence criterion for an internal metallic fault.

Fault type	Fault location (from terminal M)	K_1	K_2	K_0
Single-phase grounding fault	0%	3.19	3.20	2.53
	30%	3.19	3.22	2.52
	50%	3.19	3.21	2.53
	70%	3.20	3.21	2.53
	100%	3.22	3.23	2.53
Phase-to-phase fault	0%	3.17	3.19	—
	30%	3.21	3.20	—
	50%	3.18	3.20	—
	70%	3.19	3.19	—
	100%	3.18	3.21	—
Phase-to-phase grounding fault	0%	3.19	3.18	2.51
	30%	3.21	3.20	2.53
	50%	3.18	3.20	2.52
	70%	3.20	3.21	2.53
	100%	3.18	3.22	2.55
Three-phase symmetrical fault	0%	3.17	—	—
	30%	3.21	—	—
	50%	3.20	—	—
	70%	3.22	—	—
	100%	3.25	—	—

When faults occur via high transition resistance at several locations along the transmission line, the results of the sequence criteria are as shown in Table 6.36. It can been seen that the proposed method is able to operate reliably for an internal fault via high transition resistance, and is immune to the fault location and fault type.

B. External Fault

When different types of fault occur at the backside of terminal M and terminal N, the identification results of the sequence criteria are as shown in Table 6.37. In the case of an external fault, the line structure is not damaged. It can be seen from Table 6.37 that, for those sequence criteria that start, the ratios of the operational quantity to the restraint quantity are all near to zero, much smaller than the operational threshold value. Therefore, the protection will not operate.

Figure 6.50 shows the identification results of asymmetrical fault at the backside of terminal N and symmetrical fault at the backside of terminal M. It can be seen that the criteria results gradually become stable within a circle after the fault, and when stable they are much smaller than the operational threshold.

Table 6.36 Results of sequence criterion for an internal fault via high transition resistance.

Fault type	Fault location (from terminal M)	K_1	K_2	K_0
Single-phase grounding fault	0%	3.12	3.12	2.51
	30%	3.09	3.09	2.48
	50%	3.08	3.09	2.48
	70%	3.08	3.08	2.48
	100%	3.09	3.11	2.48
Phase-to-phase fault	0%	3.22	3.22	—
	30%	3.22	3.21	—
	50%	3.21	3.22	—
	70%	3.22	3.22	—
	100%	3.20	3.20	—
Phase-to-phase grounding fault	0%	3.27	3.34	2.51
	30%	3.25	3.32	2.52
	50%	3.24	3.31	2.53
	70%	3.24	3.30	2.52
	100%	3.17	3.14	2.51
Three-phase symmetrical fault	0%	3.21	—	—
	30%	3.21	—	—
	50%	3.22	—	—
	70%	3.21	—	—
	100%	3.17	—	—

Table 6.37 Results of sequence criterion for an external fault.

Fault type	Fault location	K_1	K_2	K_0
Single-phase grounding fault	Backside of terminal M	0.08	0.09	0.08
	Backside of terminal N	0.04	0.03	0.04
Phase-to-phase fault	Backside of terminal M	0.08	0.08	—
	Backside of terminal N	0.04	0.05	—
Phase-to-phase grounding fault	Backside of terminal M	0.08	0.09	0.07
	Backside of terminal N	0.07	0.10	0.08
Three-phase symmetrical fault	Backside of terminal M	0.08	—	—
	Backside of terminal N	0.01	—	—

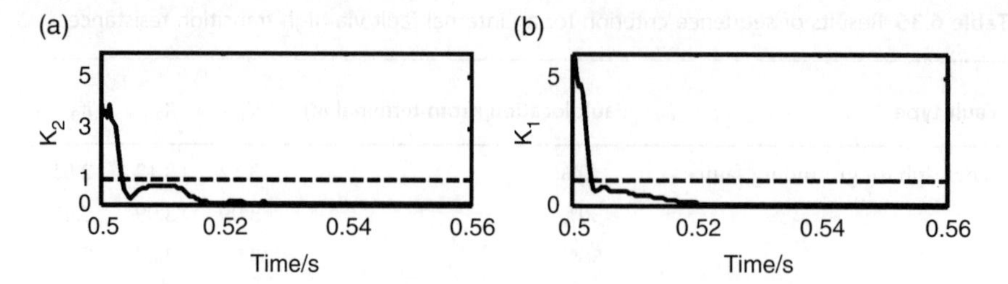

Figure 6.50 Criteria results in the case of an out-of-zone fault. (a) Negative sequence criterion when a single-phase grounding fault occurs at the backside of terminal *N*. (b) Positive sequence criterion when a three-phase symmetrical fault occurs at the backside of terminal *M*.

Table 6.38 Results of sequence criterion for a single-phase grounding fault with no load at terminal *N*.

Fault location (from terminal *M*)	K_1	K_2	K_0
0%	5.59	5.57	9.51
30%	5.62	5.62	9.31
50%	5.55	5.53	9.35
70%	5.53	5.51	9.28
100%	5.55	5.53	9.31

C. *Fault with No Load at One Terminal*

Suppose a single-phase grounding fault occurs with no load at terminal *N*, and the fault location is set at 0%, 30%, 50%, 70% and 100% line length from terminal *M*. In this case, the identification results of the sequence criteria are as shown in Table 6.38. It can be seen that the criteria are able to recognize the fault correctly, unaffected by the system impedance and fault location.

D. *Transitional Fault*

Suppose a single-phase grounding fault at the backside of terminal *N* transfers to different types of internal fault at terminal *N* after 0.03 s. In this case, the identification results of the sequence criteria are as shown in Figure 6.51. It can be seen that the criterion is able to recognize the fault within a short time after the fault transfer, i.e. it is strongly sensitive to the variation in line structure.

E. *Developing Fault*

Suppose a single-phase grounding fault occurs at the forward side and backside of terminal *N*, and after 0.03 s, the single-phase grounding fault develops into a phase-to-phase grounding fault at the same location. In this case, the identification results of the sequence criteria are shown in Figure 6.52. It can be seen that, in the case of internal developing faults, the criteria are able to operate reliably; in the case of external developing faults, the criteria do not operate. This is because the

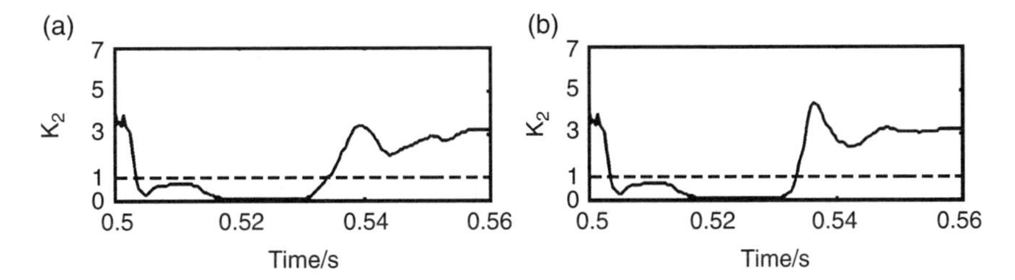

Figure 6.51 Results of negative sequence criterion when a single-phase grounding fault at the backside of terminal *N* transfers to different types of internal fault at terminal *N* after 0.03 s. (a) Single-phase grounding fault transfers to single-phase grounding fault. (b) Single-phase grounding fault transfers to phase-to-phase grounding fault.

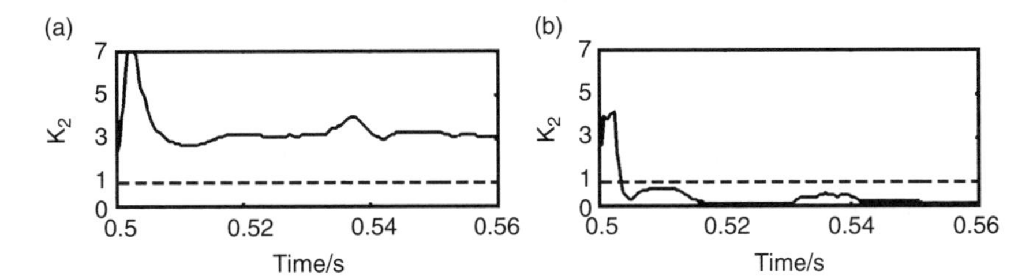

Figure 6.52 Results of criterion when a single-phase grounding fault develops into a phase-to-phase grounding fault at the same location after 0.03 s. (a) Forward side of terminal *N*. (b) Backside of terminal *N*.

damage status of the line structure remains unchanged when one fault develops into another at the same location.

F. Communication Delay

Suppose that a grounding fault via a high transition resistance occurs at the midpoint of line MN, with the communication delay being 5 ms. The reliability of the proposed method is verified. The ratio of the sequence operational quantity to the restraint quantity K_i is shown in Figure 6.53. It can be seen that, even when the information is asynchronous, the proposed method can still identify internal faults quickly and accurately, unaffected by the fault type.

G. Comparison with Traditional Current Differential Protection

Traditional current differential protection is based on the Kirchhoff current law, and the operational equation is

$$\left|\dot{I}_m + \dot{I}_n\right| > K*\left(\left|\dot{I}_m\right| + \left|\dot{I}_n\right|\right) \tag{6.111}$$

where $\dot{I}_m$ and $\dot{I}_n$ are the current phasors at bus m and bus n, respectively. K is the restraint coefficient. When K is fixed, the restraint quantity will increase as the current increases, so that in the case of external faults with high current,

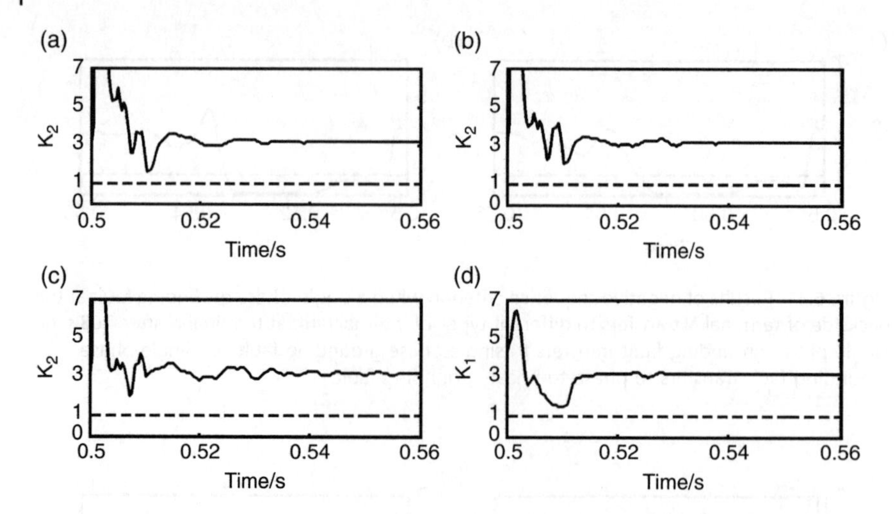

Figure 6.53 Result of criterion in the case of an internal fault for channel delay 5 ms. (a) Negative sequence criterion when a single-phase grounding fault occurs. (b) Negative sequence criterion when a phase-to-phase fault occurs. (c) Negative sequence criterion when a phase-to-phase grounding fault occurs. (d) Positive sequence criterion when a three-phase symmetrical fault occurs.

the restraint ability can be improved. However, in the case of an internal fault, the improved restraint ability will conflict with the requirement for sensitivity.

In order to verify the superiority of the proposed method, comparison is made between the proposed method and traditional current differential protection in the following two aspects – accuracy and speed.

1. **Accuracy** When a single-phase grounding fault via high transition resistance occurs at the midpoint of line MN, the operational curve of traditional current differential protection is as shown in Figure 6.54. It can be seen that the protection refuses to operate. This is because the sensitivity of traditional current differential protection is not high enough in the case of a grounding fault with a high transition resistance. Meanwhile, according to Table 6.36, the proposed method is able to identify grounding faults via high transition resistance accurately, unaffected by the fault location and fault type.

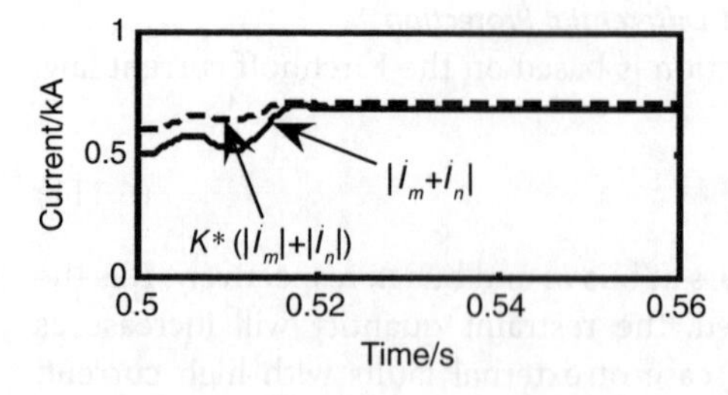

Figure 6.54 Result of traditional current differential protection when a single-phase grounding fault occurs via high fault resistance.

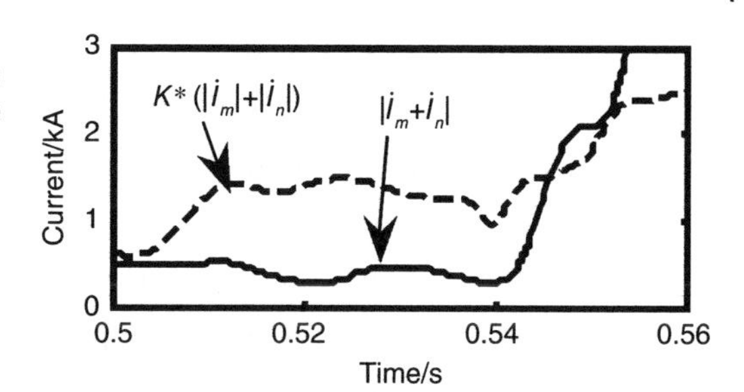

Figure 6.55 Results of traditional current differential protection when a single-phase grounding fault at the backside of terminal *N* transfers to an internal fault at terminal *N* after 0.03 s.

2. **Speed** Suppose that a phase A-to-ground fault occurs at the backside of bus *N* and transforms to the same type of internal fault at the forward side of bus *N* after 0.03 s. In this case, the operational curve of traditional current differential protection is as shown in Figure 6.55. It can be seen that the internal fault is identified at $t = 0.546$ s. Meanwhile, according to Figure 6.51(a), the proposed method can identify the same type of internal fault at $t = 0.534$ s, which is 12 ms faster than traditional current differential protection. In conclusion, compared with traditional current differential protection, the proposed method is faster and more accurate in fault identification.

6.4 Wide Area Tripping Strategy

6.4.1 Tripping Strategy Based on Directional Weighting

6.4.1.1 Node–Branch Correlation Matrix Based on Directional Weighting [17]

According to topology and graph theory, the system wiring could be abstracted to a topological graph $\boldsymbol{G} = \langle \boldsymbol{V}, \boldsymbol{E} \rangle$, where the vertex set $\boldsymbol{V}$ represents the protected nodes in the system, including the generators, buses, transformers and transmission lines. The side set $\boldsymbol{E}$ represents the connection state between different nodes, with '1' meaning 'directly connected' and '0' meaning 'not directly connected'. In a power system, different components are connected via switches. Therefore, the side set $\boldsymbol{E}$ can also be represented by the switch components. Define the node–branch correlation matrix of the power network as $\mathbf{A}$, the element A_{ij} as

$$A_{ij} = \begin{cases} 1 & \text{Node } i \text{ directly connected with switch } j \\ 0 & \text{Node } i \text{ not directly connected with switch } j \end{cases} \tag{6.112}$$

Take a 330 kV local power grid in West China (shown in Figure 6.56) as an example. According to the limited wide area centralized protection area partition scheme in reference [9], substation B_2 is the main station, and its protection area is shown in the dashed line in Figure 6.56, where the bus, line and circuit breaker are represented with *B*, *L* and *CB* respectively. The corresponding node–branch correlation matrix $\mathbf{A}$ is

$$\mathbf{A} = \begin{bmatrix} 0&0&0&1&1&0&1&0&1&0&0&0&0&0&0&0&0&0&0&0 \\ 0&1&0&0&0&1&0&1&0&0&1&0&1&0&1&0&0&0&0&0 \\ 0&0&0&0&0&0&0&0&0&0&0&1&0&1&0&0&1&0&1&0 \\ 1&1&0&0&0&0&0&0&0&0&0&0&0&0&0&0&0&0&0&0 \\ 0&0&1&1&0&0&0&0&0&0&0&0&0&0&0&0&0&0&0&0 \\ 0&0&0&0&1&1&0&0&0&0&0&0&0&0&0&0&0&0&0&0 \\ 0&0&0&0&0&0&1&1&0&0&0&0&0&0&0&0&0&0&0&0 \\ 0&0&0&0&0&0&0&0&1&1&0&0&0&0&0&0&0&0&0&0 \\ 0&0&0&0&0&0&0&0&0&0&1&1&0&0&0&0&0&0&0&0 \\ 0&0&0&0&0&0&0&0&0&0&0&0&1&1&0&0&0&0&0&0 \\ 0&0&0&0&0&0&0&0&0&0&0&0&0&0&1&1&0&0&0&0 \\ 0&0&0&0&0&0&0&0&0&0&0&0&0&0&0&0&1&1&0&0 \\ 0&0&0&0&0&0&0&0&0&0&0&0&0&0&0&0&0&0&1&1 \end{bmatrix} \tag{6.113}$$

where the matrix order is 13 × 20. The rows correspond to substation $B_1 \sim B_3$ and line $L_6 \sim L_{15}$, and the columns correspond to circuit breaker $CB_{11} \sim CB_{30}$. This matrix only reflects the correlation between the circuit breakers and other system components, which depends on the system topology and has nothing to do with the on/off state of the breakers.

Consider that the remote signals of the switch status might be falsely identified, thus affecting the accuracy of topology analysis and tripping sequence searching. Here, the

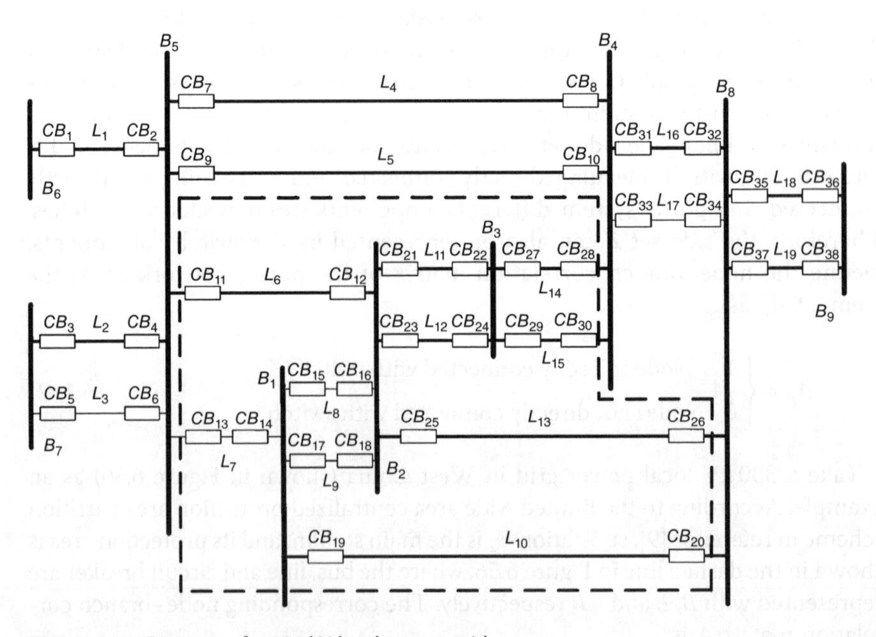

Figure 6.56 Diagram of a 330 kV local power grid.

positive sequence current phasor at the relaying point is used for topology identification (the positive sequence component exists both when the network is in normal operation and when different types of fault occur). First, the positive sequence current phasors at the relaying points in a certain area $\vec{a}_i$ are obtained with the main computer, and then, according to the sequence of breakers in the node–branch correlation matrix, the positive sequence current column vector $\boldsymbol{I}$ is formed.

If the positive sequence current phasor $\vec{a}_i$ at a particular breaker has some information missing, then it is corrected according to Kirchhoff's current law. The correction principle is as follows. When the positive sequence current phasor information at a certain breaker is missing, it is set to be the negative value of the sum of all the current phasors at the outlet line circuit breakers of non-faulty buses in the same substation, or the negative value of the current phasor at the opposite circuit breaker of the non-faulty line. When more than two positive sequence current phasors at the circuit breakers of a node with multiple outlet lines are missing (most faults in the system are single faults), first the missing information of the circuit breaker on the non-faulty line is set to be the negative value of the positive sequence current at the opposite circuit breaker, and then the missing information of the circuit breaker on the faulty line is corrected according to Kirchhoff's current law. Suppose that a fault occurs on line L_9 in the system shown in Figure 6.56, when the information of CB_{15} and CB_{17} in substation B_1 is missing, first the missing information of CB_{15} on the non-faulty line L_8 is set to be the negative value of the positive sequence current at the opposite circuit breaker CB_{16}, and then according to Kirchhoff's current law, the missing information of CB_{17} on the fault line L_9 is the negative value of the sum of the positive sequence currents at circuit breakers CB_{14}, CB_{15} and CB_{19}.

The corrected current column vector is defined as the corrected positive sequence current column vector $\boldsymbol{I}'$, i.e. $\boldsymbol{I}' = \left(\vec{a}_1,\vec{a}_2,\cdots\cdots,\vec{a}_{19},\vec{a}_{20}\right)^{\mathrm{T}}$. Define the current flowing toward the component as positive and the current flowing outward from the component as negative. According to the corrected positive sequence current column vector $\boldsymbol{I}'$, the node–branch correlation matrix could be modified as the directional node–branch correlation matrix $\mathbf{A_f}$. For the system in Figure 6.56, $\mathbf{A_f}$ is

$$\mathbf{A_f} = \begin{bmatrix} 0 & 0 & 0 & -1 & -1 & 0 & -1 & 0 & -1 & 0 & 0 & 0 & 0 & 0 & 0 & 0 & 0 & 0 & 0 & 0 \\ 0 & -1 & 0 & 0 & 0 & -1 & 0 & -1 & 0 & 0 & -1 & 0 & -1 & 0 & -1 & 0 & 0 & 0 & 0 & 0 \\ 0 & 0 & 0 & 0 & 0 & 0 & 0 & 0 & 0 & 0 & 0 & -1 & 0 & -1 & 0 & 0 & -1 & 0 & -1 & 0 \\ 1 & 1 & 0 & 0 & 0 & 0 & 0 & 0 & 0 & 0 & 0 & 0 & 0 & 0 & 0 & 0 & 0 & 0 & 0 & 0 \\ 0 & 0 & 1 & 1 & 0 & 0 & 0 & 0 & 0 & 0 & 0 & 0 & 0 & 0 & 0 & 0 & 0 & 0 & 0 & 0 \\ 0 & 0 & 0 & 0 & 1 & 1 & 0 & 0 & 0 & 0 & 0 & 0 & 0 & 0 & 0 & 0 & 0 & 0 & 0 & 0 \\ 0 & 0 & 0 & 0 & 0 & 0 & 1 & 1 & 0 & 0 & 0 & 0 & 0 & 0 & 0 & 0 & 0 & 0 & 0 & 0 \\ 0 & 0 & 0 & 0 & 0 & 0 & 0 & 0 & 1 & 1 & 0 & 0 & 0 & 0 & 0 & 0 & 0 & 0 & 0 & 0 \\ 0 & 0 & 0 & 0 & 0 & 0 & 0 & 0 & 0 & 0 & 1 & 1 & 0 & 0 & 0 & 0 & 0 & 0 & 0 & 0 \\ 0 & 0 & 0 & 0 & 0 & 0 & 0 & 0 & 0 & 0 & 0 & 0 & 1 & 1 & 0 & 0 & 0 & 0 & 0 & 0 \\ 0 & 0 & 0 & 0 & 0 & 0 & 0 & 0 & 0 & 0 & 0 & 0 & 0 & 0 & 1 & 1 & 0 & 0 & 0 & 0 \\ 0 & 0 & 0 & 0 & 0 & 0 & 0 & 0 & 0 & 0 & 0 & 0 & 0 & 0 & 0 & 0 & 1 & 1 & 0 & 0 \\ 0 & 0 & 0 & 0 & 0 & 0 & 0 & 0 & 0 & 0 & 0 & 0 & 0 & 0 & 0 & 0 & 0 & 0 & 1 & 1 \end{bmatrix} \tag{6.114}$$

Column vector I' could be expressed in the form of a diagonal matrix: $\mathbf{I_D} = \mathrm{diag}(\vec{a}_1, \vec{a}_2, \cdots\cdots, \vec{a}_{19}, \vec{a}_{20})$, thus the node–branch correlation matrix based on directional weighting $\mathbf{A_t}$ is

$$\mathbf{A_t} = \mathbf{A_f}\mathbf{I_D} \tag{6.115}$$

where the direction or symbol represents the direction relationship between the current through the breaker and the component, with positive meaning that the current flows toward the component and negative meaning that the current flows outward from the component. The weight or modulus value represents the electrical correlation between the component and the breaker, with '0' meaning that the breaker has no electrical correlation with the component and 'non-0' meaning that the breaker has electrical correlation with the component. In the tripping strategy, the direction is used to identify whether the breaker is directly connected to the faulty component, and the weight is used to identify the on/off status of the breaker.

6.4.1.2 Wide Area Protection Tripping Strategy

Three basic functions of wide area protection tripping strategy are local backup protection, remote backup protection and breaker failure protection. Take the system in Figure 6.56, for example; suppose all the breakers are switched on. When a fault occurs on line L_{12}, the breakers that are tripped by local backup protection are directly connected breakers, i.e. CB_{23} and CB_{24}. If CB_{23} fails to break, then the other breakers connected to the outlet line of substation B_2, i.e. CB_{12}, CB_{16}, CB_{18}, CB_{21} and CB_{25}, are tripped by the breaker failure protection. When a fault occurs on line L_{12}, and the DC power supply of substation B_3 disappears, the breakers tripped by the remote backup protection are breakers on the opposite side of the outlet line of substation B_3, i.e. CB_{21}, CB_{23}, CB_{28} and CB_{30}. Therefore, in order to realize different protection functions, different tripping strategies need to be constructed.

A. Local Backup Protection Tripping Strategy in the Case of a Line Fault

When a fault occurs within the protection area, first the fault component is identified by the main computer according to the protection algorithm, and the fault column vector $\boldsymbol{D}$ is formed. The number of elements in $\boldsymbol{D}$ represents the number of protection components in the area, and the sequence of elements in $\boldsymbol{D}$ is the same as that in the node–branch correlation matrix. Element D_i is defined as

$$D_i = \begin{cases} 1 & \text{Component } i \text{ is in fault} \\ 0 & \text{Component } i \text{ is not in fault} \end{cases} \tag{6.116}$$

Meanwhile, define the local backup component–breaker correlation vector $\boldsymbol{T}_\mathbf{c}$ as

$$\boldsymbol{T}_\mathbf{c} = \mathbf{A}_\mathbf{t}^\mathrm{T}\boldsymbol{D} \tag{6.117}$$

For the network shown in Figure 6.56, if a fault occurs on line L_{11}, the fault column vector is formed according to Equation (6.116): $\boldsymbol{D} = (0,0,0,0,0,0,0,0,1,0,0,0,0)^\mathrm{T}$, and the local backup component–breaker correlation vector is calculated according to Equation (6.117): $\boldsymbol{T}_{\mathbf{c}} = \left(0,0,0,0,0,0,0,0,0,0,\vec{a}_{21},\vec{a}_{22},0,0,0,0,0,0,0,0\right)^{\mathrm{T}}$. In $\mathbf{T}_{\mathbf{c}}$, the '0' elements represent the fact that the corresponding breakers are not directly connected to the faulty component, and the 'non-0' elements $\vec{a}_{21}$ and $\vec{a}_{22}$ represent the fact that the corresponding breakers are directly connected to the faulty component. For a breaker directly connected to the faulty component, if it is originally switched on, then it is a correlated breaker that needs to be switched off; if it is originally switched off, then it is a correlated breaker that does not need to be switched off. According to reference [10], if the modulus value of the current phasor flowing through CB_i $\left|\vec{a}_i\right|$ exceeds the threshold value δ, then this breaker could be identified as 'switched on'. Otherwise, it is identified as 'switched off'. The threshold value δ is set to $0.08I_N$. Therefore, the on/off status of the corresponding breakers may be identified according to the relationship between the modulus values of $\vec{a}_{21}$ and $\vec{a}_{22}$ and the threshold value δ.

Define mathematical operation $\boldsymbol{X} = [\boldsymbol{Y}]$, where $\boldsymbol{X}$ and $\boldsymbol{Y}$ are both $1 \times n$ column vectors. Element X_i is

$$X_i = \begin{cases} 1 & |Y_i| > \delta \\ 0 & |Y_i| \le \delta \end{cases} \tag{6.118}$$

Thus, the line local backup protection tripping sequence vector $\boldsymbol{T}$ can be expressed as

$$\boldsymbol{T} = \left[\mathbf{A}_{\mathbf{t}}^{\mathrm{T}} \boldsymbol{D}\right] \tag{6.119}$$

When $T_i = 1$, CB_i is a correlated breaker that needs to be switched off; when $T_i = 0$, CB_i is a correlated breaker that does not need to be switched off.

The operating conditions of local backup protection are as follows. When a fault occurs within the local backup protection range, if the main protection or breaker corresponding to the fault component in this substation does not operate, and the local fault electrical variables still exist after the primary protection operational time is over, then local backup protection will operate. The operational time delay is set to be the operational time of the primary protection. The flowchart of the local backup protection tripping strategy is shown in Figure 6.57.

B. *Tripping Strategy in the Case of a Breaker Failure*

When an in-zone fault is detected, the area main computer will search for the local backup protection correlated breakers according to the algorithm in part A in Section 6.4.1.2, and then, tripping commands will be issued to the correlated breakers that need to be switched off. At the same time, the current phasors through the breakers to be tripped are monitored. If, after the arc extinction, the modulus value of the current phasor through a certain breaker $\left|\vec{a}_i\right|$ is still bigger than the threshold value δ, then it is identified as a breaker failure.

In order to reflect the breaker failure information, the breaker failure column vector $\boldsymbol{S}$ is constructed. The number of elements in $\boldsymbol{S}$ represents the number of breakers, and the sequence of elements in $\boldsymbol{S}$ is the same as the sequence of breakers in the node–branch correlation matrix. Element S_i can be expressed as

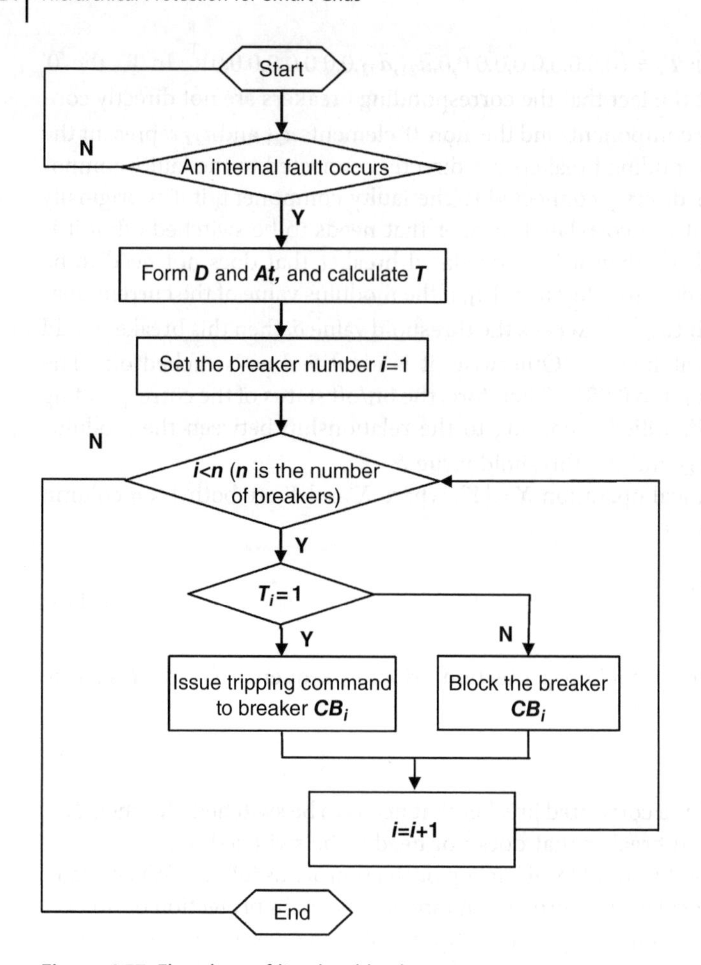

Figure 6.57 Flowchart of line local backup protection tripping strategy.

$$S_i = \begin{cases} 1 & \text{Breaker } i \text{ fails} \\ 0 & \text{Breaker } i \text{ does not fail} \end{cases} \tag{6.120}$$

When breaker failure occurs, if the fault is on the line, then all the branch breakers connected to the same bus as the failing breaker need to be tripped. If the fault is at the bus, then the breaker on the opposite side of the line where the failing breaker is located needs to be tripped. Therefore, when the failing breaker is identified, in order to minimize the fault clearance range, the components connected to the failing breaker need to be searched first, and then the local backup tripping strategy of the connected components can be realized. Taking the system in Figure 6.56, for example, suppose breaker CB_{21} fails when a fault occurs on line L_{11}. According to the location of the failing breaker, the breaker failure column vector is constructed by substation B_2: $\boldsymbol{S} = (0,0,0,0,0,0,0,0,0,0,1,0,0,0,0,0,0,0,0,0,0)^{\mathrm{T}}$. First, the failing component–breaker correlation vector $\boldsymbol{M}_c$ is defined as

$$M_c = A_t^T AS \tag{6.121}$$

where the non-zero elements in AS represent the components connected to the failing breaker. In vector M_c, the '0' elements mean that the corresponding breakers are not directly connected to the failing breaker; the 'non-0' elements mean that the corresponding breakers are directly connected to the failing breaker. For breaker CB_{21}, the failing component–breaker correlation vector is $M_c = \left(0, -\vec{a}_{12}, 0, 0, 0, -\vec{a}_{16}, 0, -\vec{a}_{18}, 0, 0, 0, -\vec{a}_{22}, -\vec{a}_{23}, 0, -\vec{a}_{25}, 0, 0, 0, 0, 0\right)^T$. Then, referring to the analysis in part A in Section 6.4.1.2, the correlated breakers that need to be switched off can be identified by comparing the modulus values of the current phasors with the threshold value δ.

According to the previously defined mathematical operation $X = [Y]$, the breaker failure tripping vector M can be expressed as

$$M = \left[A_t^T AS\right] \tag{6.122}$$

In Equation (6.122), if $M_i = 1$, then CB_i is a correlated breaker that needs to be switched off; if $M_i = 0$, then CB_i does not need to be switched off.

The operating conditions of breaker failure protection are as follows. When a fault occurs within the local backup protection range, if local backup protection has issued the tripping command, and the local fault electrical variables still exist after the maximum time delay of the local backup protection and breaker failure protection, then the breaker failure protection will operate. The operational time delay is set to the primary protection operational time plus the breaker tripping arc extinction time. The flowchart of the breaker failure protection tripping strategy is shown in Figure 6.57.

C. Remote Backup Protection Tripping Strategy in the Case Of Substation and Outlet Line Fault

When the DC power supply of the substation disappears, the breakers will not be able to trip. In this case, if a fault occurs in the substation or on the outlet line, the remote backup protection tripping strategy is needed.

In the protection area with substation B_2 as the main station, if a fault occurs in substation B_1, the area main computer will form a fault vector $D = (1,0,0,0,0,0,0,0,0,0,0,0,0)^T$. However, since the DC power supply of substation B_1 disappears, the fault cannot be cleared by B_1 alone. In addition, the current phasor information of CB_{14}, CB_{15}, CB_{17} and CB_{19} cannot be transmitted to the area main station. Therefore, the corresponding elements in the positive sequence current column vector are set as 0, i.e. $I = \left(\vec{a}_{11}, \vec{a}_{12}, \vec{a}_{13}, 0, 0, \vec{a}_{16}, 0, \vec{a}_{18}, 0, \vec{a}_{20}, \vec{a}_{21}, \vec{a}_{22}, \vec{a}_{23}, \vec{a}_{24}, \vec{a}_{25}, \vec{a}_{26}, \vec{a}_{27}, \vec{a}_{28}, \vec{a}_{29}, \vec{a}_{30}\right)^T$. Now the needed tripping strategy is breaker failure protection for all the outlet lines of the substation. First, the substation correlated breaker vector G is searched:

$$G = A^T D \tag{6.123}$$

In Equation (6.123), if $G_i = 1$, then breaker CB_i is on the outlet line of the substation; if $G_i = 0$, then breaker CB_i is not on the outlet line of the substation. The substation correlated breaker vector G can be viewed as the failing breaker vector.

All other breakers connected to the failing breakers need to be tripped. Define the remote backup component–breaker correlation vector N_c as

$$N_c = A_t^T A G \tag{6.124}$$

where the '0' elements mean that the corresponding breakers are not directly connected to the faulty component; the'non-0' elements mean that the corresponding breakers are directly connected to the faulty component. For the directly connected breakers, if the breaker is originally switched on, then it needs to be switched off; otherwise it does not need to be switched off. For the system in Figure 6.58, the remote backup tripping initial vector is $I = (0,0,\vec{a}_{13},0,0,\vec{a}_{16}, 0,\vec{a}_{18},0,\vec{a}_{20},0,0,0,0,0,0,0,0,0,0)^T$. If $|\vec{a}_i| > \delta$ (i = 13, 16, 18, 20), then breaker CB_i needs to be switched off.

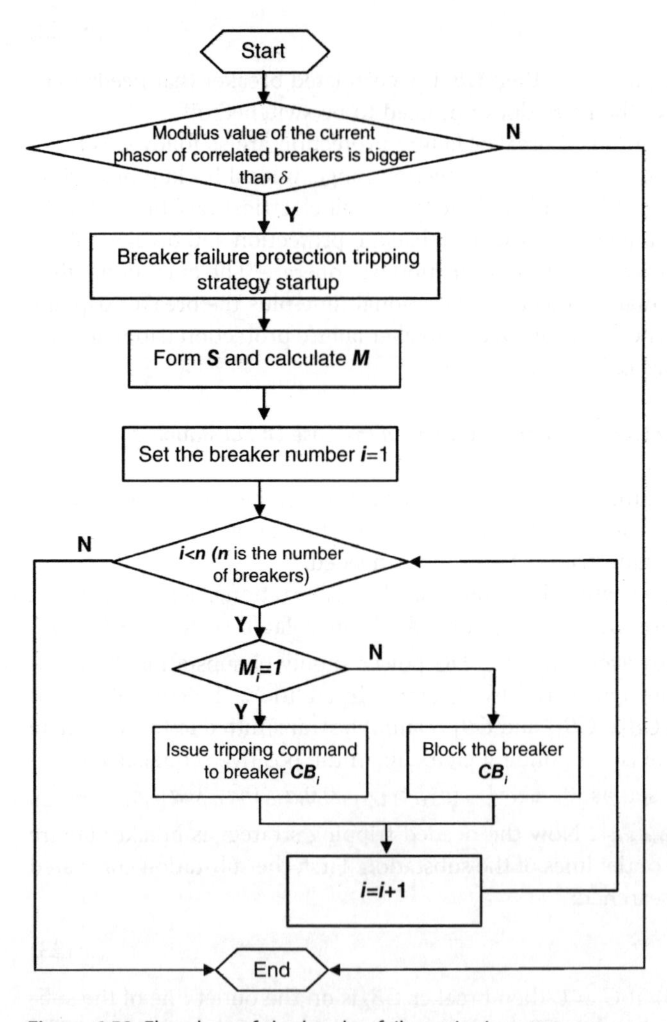

Figure 6.58 Flowchart of the breaker failure tripping strategy.

According to the previously defined mathematical operation $X = [Y]$, the breaker failure tripping vector N can be expressed as

$$N_c = \mathbf{A}_t^T \mathbf{A} G \tag{6.125}$$

In Equation (6.125), if $N_i = 1$, then CB_i is a correlated breaker that needs to be switched off; if $N_i = 0$, then CB_i does not need to be switched off.

Suppose a fault occurs on outlet line L_7 when the DC power supply of substation B_1 disappears; the fault vector formed in the main station is $D = (1,0,0,0,0,0,0,0,0,0,0,0,0)^T$. Combined with the missing positive sequence current column vector, it can be identified that the fault is in substation B_1 where the DC power supply disappears. Thus, the fault vector of the substation where DC power supply disappears could be formed: $Ds = (1,0,0,0,0,0,0,0,0,0,0,0,0)^T$. Then, search for the breakers correlated to the substation where DC power supply disappears, which form vector G:

$$G = \mathbf{A}^T Ds$$

where $G_i = 1$ means that breaker CB_i is on the outlet line of the substation where DC power supply disappears, and $G_i = 0$ means that breaker CB_i is not on the outlet line of the substation where DC power supply disappears. Vector G can be viewed as the failing breaker vector. Define the remote backup component–breaker correlation degree vector N_c as

$$N_c = \mathbf{A}_t^T \mathbf{A} G$$

where a '0' element indicates that the corresponding breaker is not directly related to the faulty component, and a 'non-0' element indicates that the corresponding breaker is directly related to the faulty component. If the breaker directly related to the faulty component is originally switched on, then it needs to be switched off. If it is originally switched off, then it remains unchanged.

According to the defined operation $X = [Y]$, the breaker failure tripping vector N can be expressed as

$$N = \left[\mathbf{A}_t^T \mathbf{A} \mathbf{A}^T Ds\right]$$

where $N_i = 1$ means that breaker CB_i needs to be switched off, and $N_i = 0$ means that breaker CB_i needs not be switched off.

The operating conditions of remote backup protection are as follows. When a fault occurs within the remote backup protection range, if local fault electrical variables still exist after the maximum time delay of local backup protection and breaker failure protection, then the remote backup protection will operate. The operational time delay is set as the primary protection operational time plus twice the breaker tripping arc extinction time. The flowchart for the remote backup protection tripping strategy is shown in Figure 6.59.

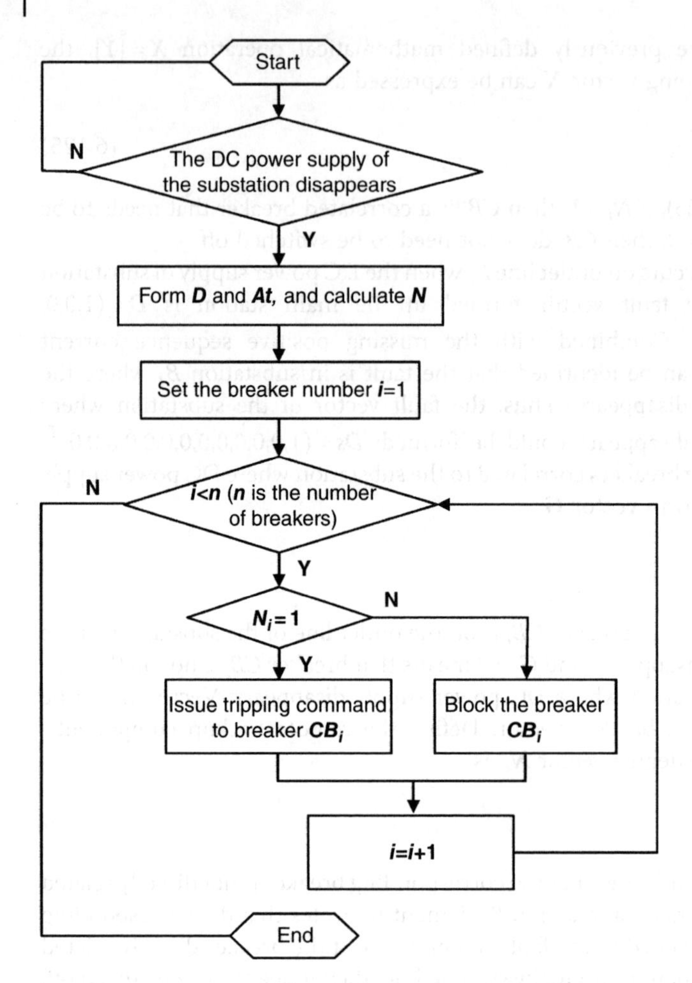

Figure 6.59 Flowchart for the substation and outlet line remote backup protection tripping strategy.

6.4.2 Simulation Verification

Take the New England 10-machine, 39-bus system, for example; the area in the dashed line is a wide area protection area with substation B_2 as the main station, shown in Figure 6.60. In the simulation system, the rated power is $S_N = 100$ MVA, rated voltage $U_N = 100$ kV, rated current $I_N = 0.5774$ kA, the threshold value $\delta = 0.0462$ kA. A fault occurs at $t = 0.3$ s and is cleared at $t = 0.4$ s. In the protection area, the protection components are substations B_1, B_2, B_3, B_4, B_{25}, B_{30}, B_{37} and lines L_{1-2}, L_{2-3}, L_{2-25}, L_{2-30}, L_{3-4}, L_{25-37}. The breakers are CB_{1-2}, CB_{1-39}, CB_{2-1}, CB_{2-3}, CB_{2-25}, CB_{2-30}, CB_{3-2}, CB_{3-4}, CB_{3-18}, CB_{4-3}, CB_{4-5}, CB_{4-14}, CB_{25-2}, CB_{25-26}, CB_{25-37}, CB_{30-2}, CB_{30G}, CB_{37-25}, and CB_{37G}.

According to the sequence of protection components and breakers listed above, the directional node–branch correlation matrix $\mathbf{A_f}$ is formed:

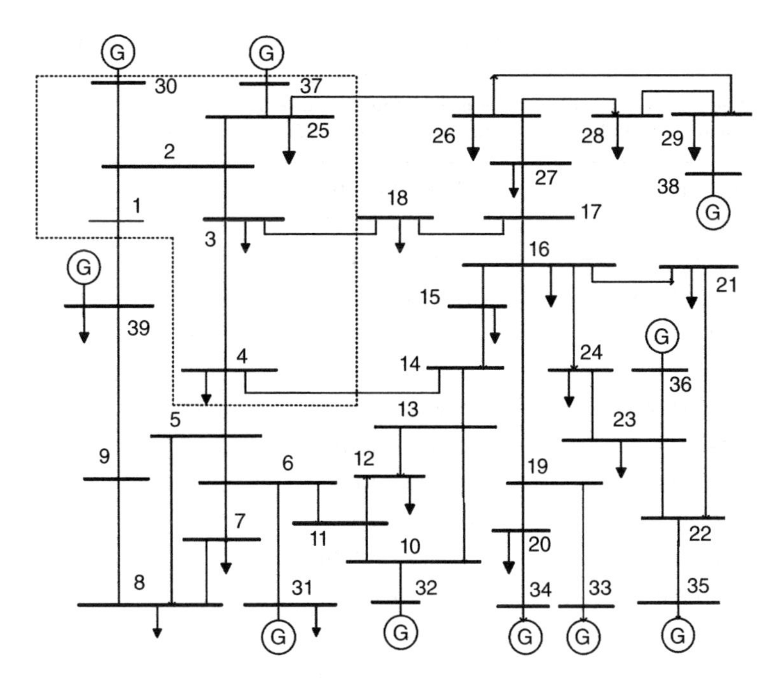

Figure 6.60 Diagram of the New England 10-machine, 39-bus system.

$$\begin{bmatrix}
-1 & -1 & 0 & 0 & 0 & 0 & 0 & 0 & 0 & 0 & 0 & 0 & 0 & 0 & 0 & 0 & 0 & 0 & 0 \\
0 & 0 & -1 & -1 & -1 & -1 & 0 & 0 & 0 & 0 & 0 & 0 & 0 & 0 & 0 & 0 & 0 & 0 & 0 \\
0 & 0 & 0 & 0 & 0 & 0 & -1 & -1 & -1 & 0 & 0 & 0 & 0 & 0 & 0 & 0 & 0 & 0 & 0 \\
0 & 0 & 0 & 0 & 0 & 0 & 0 & 0 & 0 & -1 & -1 & -1 & 0 & 0 & 0 & 0 & 0 & 0 & 0 \\
0 & 0 & 0 & 0 & 0 & 0 & 0 & 0 & 0 & 0 & 0 & 0 & -1 & -1 & -1 & 0 & 0 & 0 & 0 \\
0 & 0 & 0 & 0 & 0 & 0 & 0 & 0 & 0 & 0 & 0 & 0 & 0 & 0 & 0 & -1 & -1 & 0 & 0 \\
0 & 0 & 0 & 0 & 0 & 0 & 0 & 0 & 0 & 0 & 0 & 0 & 0 & 0 & 0 & 0 & 0 & -1 & -1 \\
1 & 0 & 1 & 0 & 0 & 0 & 0 & 0 & 0 & 0 & 0 & 0 & 0 & 0 & 0 & 0 & 0 & 0 & 0 \\
0 & 0 & 0 & 1 & 0 & 0 & 1 & 0 & 0 & 0 & 0 & 0 & 0 & 0 & 0 & 0 & 0 & 0 & 0 \\
0 & 0 & 0 & 0 & 1 & 0 & 0 & 0 & 0 & 0 & 0 & 0 & 1 & 0 & 0 & 0 & 0 & 0 & 0 \\
0 & 0 & 0 & 0 & 0 & 1 & 0 & 0 & 0 & 0 & 0 & 0 & 0 & 0 & 0 & 1 & 0 & 0 & 0 \\
0 & 0 & 0 & 0 & 0 & 0 & 0 & 1 & 0 & 1 & 0 & 0 & 0 & 0 & 0 & 0 & 0 & 0 & 0 \\
0 & 0 & 0 & 0 & 0 & 0 & 0 & 0 & 0 & 0 & 0 & 0 & 0 & 0 & 1 & 0 & 0 & 1 & 0
\end{bmatrix} \tag{6.126}$$

The line local backup tripping strategy, breaker failure tripping strategy and substation and outlet line remote backup tripping strategy are analysed below.

6.4.2.1 Local Backup Protection Tripping Strategy in the Case of a Line Fault

Suppose a fault occurs on line 2–3; the fault column vector is $\boldsymbol{D} = (0,0,0,0,0,0,0,0,1,0,0,0,0)^{\mathrm{T}}$. The positive sequence current phasors of breakers before

Table 6.39 Positive sequence current phasor and T_{ci} before and after the tripping of breakers.

Circuit breaker	Positive sequence current phasor at $t = 0.29$ s/kA	Positive sequence current phasor at $t = 0.39$ s/kA	T_{ci}/kA
CB_{1-2}	−0.3144 − j0.3070	−4.9490 + j0.0047	0
CB_{1-39}	0.3144 + j0.3070	4.9490 − j0.0047	0
CB_{2-1}	0.7188 + j0.2726	5.2062 − j0.0190	0
CB_{2-3}	−3.0184 − j2.7102	−34.5723 − j1.9616	−34.5723 − j1.9616
CB_{2-25}	0.3223 + j0.2345	10.5394 + j1.8436	0
CB_{2-30}	1.9773 + j2.2031	18.8267 + j0.1371	0
CB_{3-2}	3.1582 + j2.6912	−17.3950 + j2.7135	−17.3971 + j2.7138
CB_{3-4}	−0.9178 − j0.4723	8.9445 − j1.0482	0
CB_{3-18}	−0.0420 − j0.5554	9.2349 − j1.1767	0
CB_{4-3}	1.0299 + j0.4512	−8.8947 + j1.0366	0
CB_{4-5}	1.0676 + j1.1921	6.7374 + j0.5450	0
CB_{4-14}	1.0728 + j1.0236	4.7021 + j0.4748	0
CB_{25-2}	−0.2391 − j0.2425	−10.4935 − j1.8493	0
CB_{25-26}	−0.1861 − 1.0234	1.3507 − j0.3845	0
CB_{25-37}	0.8207 + j2.5348	9.6517 + j3.4661	0
CB_{30-2}	−1.9773 − j2.2031	−18.8267 − j0.1371	0
CB_{30G}	1.9773 + j2.2031	18.8267 + j0.1371	0
CB_{37-25}	−0.8207 − j2.5348	−9.6517 − j3.4661	0
CB_{37G}	0.8207 + j2.5348	9.6517 + j3.4661	0

and after the fault, and the line local backup component–breaker correlation vector $\boldsymbol{T}_{\mathbf{c}}$ are shown in Table 6.39.

With the directional node–branch correlation matrix $\mathbf{A_f}$, considering the corrected current column vector $\boldsymbol{I}'$ formed after the fault, the positive sequence current diagonal matrix $\mathbf{I_D}$ can be formed first. Then, applying the data in Table 6.39 and $\delta = 0.0462$ kA to $\mathbf{A_t} = \mathbf{A_f I_D}$, $\boldsymbol{T} = \left[\mathbf{A}_{\mathbf{t}}^{\mathrm{T}} \boldsymbol{D}\right]$, so that $\boldsymbol{T} = (0,0,0,1,0,0,1,0,0,0,0,0,0,0,0,0,0,0,0)^{\mathrm{T}}$, the area main computer will issue a tripping command to breakers CB_{2-3} and CB_{3-2}.

6.4.2.2 Tripping Strategy in the Case of a Breaker Failure

Suppose that a fault occurs on line 3–4; according to the algorithm in Section 6.4.1.1, the correlated breaker sequence is $\boldsymbol{T} = (0,0,0,0,0,0,0,1,0,1,0,0,0,0,0,0,0,0,0)^{\mathrm{T}}$, i.e. breakers CB_{3-4} and CB_{4-3}. After the area main computer issues the tripping command at $t = 0.4$ s, the current through breaker CB_{4-3} is still bigger than δ, thus breaker CB_{4-3} fails and the breaker failure algorithm is needed. The positive sequence current phasors of breakers before and after the tripping, and the failing component–breaker correlation vector $\boldsymbol{M}_{\mathbf{c}}$

Table 6.40 Positive sequence current phasor and M_{ci} before and after the tripping of breakers.

Circuit breaker	Positive sequence current phasor at t = 0.29 s/kA	Positive sequence current phasor at t = 0.45 s/kA	M_{ci}/kA
CB_{1-2}	−2.1494 − j0.1328	−0.4003 − j0.3012	0
CB_{1-39}	2.1494 + j0.1328	0.4003 + j0.3012	0
CB_{2-1}	2.4958 + j0.1078	0.8022 + j0.2670	0
CB_{2-3}	−14.9485 − j1.7377	−3.4236 − j2.5697	0
CB_{2-25}	3.8051 + j0.4427	0.2991 + j0.1454	0
CB_{2-30}	8.6475 + j1.1872	2.3223 + j2.1573	0
CB_{3-2}	15.0343 + j1.7281	3.5613 + j2.5514	0
CB_{3-4}	−20.5615 + j0.3881	−0.0000 − j0.0001	−0.0000 − j0.0001
CB_{3-18}	6.7529 − j1.2046	−1.1377 − j0.7180	0
CB_{4-3}	−18.5541 + j1.7673	−19.2523 + j2.0694	0
CB_{4-5}	11.2159 − j0.1198	11.1376 − j0.1887	−11.1376 + j0.1887
CB_{4-14}	Information missing	Information missing	−9.6697 + j0.5085
CB_{25-2}	−3.7372 − j0.4493	−0.2168 − j0.1533	0
CB_{25-26}	−0.4090 − j0.8365	−0.4877 − j1.0717	0
CB_{25-37}	4.6428 + j2.7667	1.1119 + j2.5291	0
CB_{30-2}	−8.6475 − j1.1872	−2.3223 − j2.1573	0
CB_{30G}	8.6475 + j1.1872	2.3223 + j2.1573	0
CB_{37-25}	−4.6428 − j2.7667	−1.1119 − j2.5291	0
CB_{37G}	4.6428 + j2.7667	1.1119 + j2.5291	0

are shown in Table 6.40. Then the breaker failure column vector is formed: $\boldsymbol{T} = (0,0,0,0,0,0,0,\ 0,0,1,0,0,0,0,0,0,0,0,0)^{\mathrm{T}}$. It can be seen that the information of breaker CB_{4-14} is missing. Since substation 4 is not in fault, according to Kirchhoff's current law, the positive sequence current phasor through breaker CB_{4-14} is corrected as 9.6697 − j0.5085 kA. Applying the positive sequence current phasors after the tripping in Table 6.40 and $\delta = 0.0462$ kA to $\boldsymbol{M} = \left[\mathbf{A}_{\mathrm{t}}^{\mathrm{T}} \mathbf{A} \boldsymbol{S}\right]$, so that $\boldsymbol{M} = (0,0,\ 0,0,0,0,0,0,0,0,1,1,0,0,0,0,0,0,0)^{\mathrm{T}}$, the area main computer will issue a tripping command to breakers CB_{4-5} and CB_{4-14}.

6.4.2.3 Remote Backup Protection Tripping Strategy in the Case of Substation and Outlet Line Fault

Suppose that a fault occurs in substation B_2, and the fault cannot be cleared because the DC power supply of the substation disappears. The fault column vector is $\boldsymbol{D} = (0,1,0,0,0,0,0,0,0,0,0,0,0,0)^{\mathrm{T}}$. The positive sequence current phasors of breakers before and after the fault, and the remote backup component–breaker correlation vector $\boldsymbol{N}_{\mathrm{c}}$ are shown in Table 6.41.

Table 6.41 Positive sequence current phasor and N_{ci} before and after the tripping of breakers.

Circuit breaker	Positive sequence current phasor at $t = 0.29$ s/kA	Positive sequence current phasor at $t = 0.39$ s/kA	N_{ci}/kA
CB_{1-2}	−0.3625 − j0.3422	−8.9121 − j0.1609	−8.9121 − j0.1609
CB_{1-39}	0.3625 + j0.3422		0
CB_{2-1}	0	0	0
CB_{2-3}	0	0	0
CB_{2-25}	0	0	0
CB_{2-30}	0	0	0
CB_{3-2}	3.0635 + j2.6106	−13.6544 + j2.3286	−13.6557 + j2.3288
CB_{3-4}	−0.2060 − j1.6388	−0.9465 − j1.7161	0
CB_{3-18}	0.0198 − j0.5063	7.9410 − j0.8248	0
CB_{4-3}	0.9880 + j0.4212	−6.8927 + j0.7291	0
CB_{4-5}	1.0938 + j1.2079	5.6305 + j0.8213	0
CB_{4-14}	1.0945 + j1.0367	4.1001 + j0.6779	0
CB_{25-2}	0	0	0
CB_{25-26}	−0.2606 − j1.1019	−1.4604 − j1.1437	0
CB_{25-37}	0.6453 + j2.3677	1.8722 + j2.4672	0
CB_{30-2}	−2.1590 − j2.3231	−33.3448 + j0.1517	−33.3448 + j0.1517
CB_{30G}	2.1590 + j2.3231	33.3448 − j0.1517	0
CB_{37-25}	−0.6453 − j2.3677	−1.8722 − j2.4672	0
CB_{37G}	0.6453 + j2.3677	1.8722 + j2.4672	0

It can be seen from Table 6.41 that the positive sequence current phasor of breaker CB_{25-2} is 0 both before and after the fault, thus breaker CB_{25-2} is switched off. Applying the positive sequence current data in Table 6.41 and $\delta = 0.0462$ kA to $\boldsymbol{G} = \mathbf{A}^{\mathrm{T}}\boldsymbol{D}$ and $\boldsymbol{N} = \left[\mathbf{A}_{\mathrm{t}}^{\mathrm{T}}\mathbf{A}\boldsymbol{G}\right]$, so that $\boldsymbol{N} = (1,0,0,0,0,0,1,0,0,0,0,0,0,0,0,1,0,0,0)^{\mathrm{T}}$, the area main computer will issue a tripping command to breakers CB_{1-2}, CB_{3-2} and CB_{30-2}.

6.5 Summary

In this chapter, wide area protection algorithms and tripping strategies are introduced in detail. Wide area protection algorithms are not affected by the system operating mode, fault type or fault resistance, and can identify the fault location quickly, reliably and accurately, making up for the deficiencies of traditional protection. Wide area protection algorithms can be divided into wide area protection based on electrical variables and wide area protection based on logical variables, which are applicable to different cases. When the communication of synchronous information is not available, wide area protection schemes based on logical variables are used. When the communication of synchronous information is available, wide area protection schemes based on electrical variables are used.

Wide area protection schemes based on electrical variables are not affected by the system operating mode, thus problems common in traditional backup protection such as improper cooperation and insufficient sensitivity can be avoided. In addition, only a limited amount of PMU needs to be configured, thus the calculation effort is small, it is easy to implement and is unaffected by the fault type and fault resistance.

For wide area protection schemes based on logical variables, the amount of information to be uploaded is small and there is no strict requirement on information synchronization, thus the reliability is relatively high. Even when some digits of the signal are missing or wrong, the faulty line can still be correctly identified. In addition, the setting calculation of protection and the whole fault identification process are completed online in real time, and are adaptable to the variation of system operating mode and fault type.

Concerning the tripping strategy, the wide area backup protection tripping strategy introduced in this chapter could simplify the cooperation in backup protection and limit fault clearance to the minimum range for all types of faults, without strict requirements on the synchronization of information uploaded from substations, even when part of the uploaded electrical variable information is missing.

References

1 Ma, J., Li, J., Thorp, J. S. *et al.* (2011) A fault steady state component-based wide area backup protection algorithm, *IEEE Transactions on Smart Grids*, **2** (3), 468–475.

2 Ma, J., Li, J. L., Wang, Z. P. *et al.* (2010) A novel wide-area fault location algorithm based on fault model, *Power and Energy Engineering Conference*, pp. 1–4.

3 Ma, J., Xu, D., Wang, T. *et al.* (2012) A novel wide-area multiple-fault location algorithm based on fault confirming degree, *Power System Technology*, **36** (12), 88–93.

4 Chakrabarti, S. and Kyriakides, E. (2008) Optimal placement of phasor measurement units for power system observability, *IEEE Trans. Power Syst.*, **23**, 1433–1440.

5 Abbasy, N. H. and Ismail, H. M. (2009) Unified approach for the optimal PMU location for power system state estimation, *IEEE Trans. Power Syst.*, **24**, 806–813.

6 Terzija, V. V. and Radojevic, Z. M. (2004) Numerical algorithm for adaptive autoreclosure and protection of medium-voltage overhead lines, *IEEE Trans. Power Del.*, **19**, 554–559.

7 Bi, T. S., Qin, X. H. and Yang, Q. X. (2008) A novel hybrid state estimator for including synchronized phasor measurements, *Electric Power Systems Research*, **78**, 1343–1352.

8 Zhang, F., Liang, J., Zhang, L., Yun, Z. H. (2008) A new fault location method avoiding wave speed and based on traveling waves for EHV transmission line, *Third International Conference on Electric Utility Deregulation and Restructuring and Power Technologies*, pp. 1753–1757.
9 Ma, J., Wang, X. and Wang, Z. P. (2012) Partition of protection zone with circular overlapping coverage for wide-area protection system, *Electr. Power Autom. Equip.*, **32** (9), 50–54.
10 Qian, C., Wang, Z. P. and Zhang, J. F. (2010) New algorithm of topology analysis based on PMU information, *Int. Conf. Crit. Infrastruct., CRIS – Proc*, pp. 1–5.
11 Neyestanaki, M. K. and Ranjbar, A. M. (2015) An adaptive pmu-based wide area backup protection scheme for power transmission lines, *IEEE Trans, Smart Grid*, **6** (3), 1550–1559.
12 Jafari, R., Moaddabi, N., Eskandari-Nasab, M. *et al.* (2014) A novel power swing detection scheme independent of the rate of change of power system parameters, *IEEE Trans. Power Del.*, **29** (3), 1192–1202.
13 IEC 61850 Communication networks and systems in substations, *Institute of Electrical and Electronics Engineers, Tech. Rep*, 2002–2005.
14 Ge, Y. Z. (2007) *Novel principles and techniques of protection and fault location*, Xi'an Jiaotong University Press, Xi'an.
15 Ma, J., Pei, X., Ma, W. *et al.* (2015) A new transmission line pilot differential protection principle using virtual impedance of fault component, *Canadian Journal of Electrical & Computer Engineering*, **38** (1), 37–44.
16 Suonan, J., Yang, C., Yang, Z. L., Shen, L. M. and Jiao, Z. B. (2008) New type of transmission line pilot protection based on model identification, *Automation of Electric Power Systems*, **32** (24), 30–34.
17 Ma, J., Chen, Y., Liu, C. *et al.* (2016) Directional weighting based wide area backup protection tripping strategy, *International Transactions on Electrical Energy Systems*, **26** (3), 573–585.

Appendices

Appendix A

Calculation of parameters in Equations (3.81)

$$\begin{cases} D_{01} = 2U_{dc1} + U_{dc2} + 2U_{dc3} + 2U_{dc4} \\ D_{02} = 2U_{dc1} + U_{dc2} + 2U_{dc3} \\ D_{03} = 2U_{dc1} + U_{dc2} + 2U_{dc4} \\ D_{04} = 2U_{dc1} + U_{dc2} \end{cases} \tag{A1}$$

In Equations (A1), k is the terminal voltage drop rate after a short circuit fault occurs in the grid. $U_{dc1} = m_{dc}I_{rq0}$, $U_{dc2} = m_{dc}I_{rqm}$, $U_{dc3} = m_{dc}I_{rq1}$, $U_{dc4} = m_{dc}I_{rq2}$, $\eta_1 = \tau_1 - \alpha_1$, $\eta_2 = \tau_1 - \alpha_2$, $m_{dc} = 6(1-k)(1-s)U_{s0}e^{\tau_1 t_1}/(C_{dc}\omega_1)$, s is the slip, U_{s0} is the steady-state stator voltage amplitude, C_{dc} is the DC-side capacitance, ω_1 is the synchronous angular speed. t_1 is the fault instant, $\tau_1 = R_s/L_s + j\omega_1$ is the attenuation constant of the stator transient flux. α_1 and α_2 are the attenuation constants of transient natural current components $\boldsymbol{\Delta i}_{r1}$ and $\boldsymbol{\Delta i}_{s1}$.

$$\begin{cases} D_{0m} = \tau_1^2 D_{01} - \tau_1\alpha_2 D_{02} - \tau_1\alpha_1 D_{03} + \alpha_1\alpha_2 D_{04} \\ D_0 = K_{i1}D_{0m}/(2\tau_1(\tau_1-\alpha_1)(\tau_1-\alpha_2)) \\ D_1 = (K_{p1}\tau_1 - K_{i1})U_{dc1}/\tau_1 \\ D_2 = (2K_{p1}\tau_1 - K_{i1})U_{dc2}/(2\tau_1) \\ D_3 = (K_{p1}\tau_1 - K_{p1}\alpha_1 - K_{i1})U_{dc3}/(\tau_1-\alpha_1) \\ D_4 = (K_{p1}\tau_1 - K_{p1}\alpha_2 - K_{i1})U_{dc4}/(\tau_1-\alpha_2) \end{cases} \tag{A2}$$

In Equations (A2), K_{p1} and K_{i1} are the proportion parameter and integration parameter of the grid-side converter voltage outer-loop PI controller.

$$\begin{cases} C_{g0m} = \alpha_1\alpha_2 Q_{gm} - \tau_1\alpha_2 Q_{g1} - \tau_1\alpha_1 Q_{g2} \\ C_{g0} = K_{i2}C_{g0m}/(\tau_1\alpha_1\alpha_2) \\ C_{gm} = (K_{p2}\tau_1 - K_{i2})Q_{gm}/\tau_1 \\ C_{g1} = (K_{p2}\alpha_1 + K_{i2})Q_{g1}/\alpha_1 \\ C_{g2} = (K_{p2}\alpha_2 + K_{i2})Q_{g2}/\alpha_2 \end{cases} \tag{A3}$$

Hierarchical Protection for Smart Grids, First Edition. Jing Ma and Zengping Wang.

In Equations (A3), $Q_{g0} = 3kU_{s0}sI_{rq0}$, $Q_{gm} = 3kU_{s0}sI_{rqm}$, $Q_{g1} = 3kU_{s0}sI_{rq1}$, $Q_{g2} = 3kU_{s0}sI_{rq2}$. K_{p2} and K_{i2} are the proportion parameter and integration parameter of the grid-side converter power outer-loop PI controller.

Appendix B

Calculation of parameters in Equations (3.83)

$$\begin{cases} A_{01} = 2D_1 + D_2 + 2D_3 + 2D_4 \\ A_{02} = 2D_1 + D_2 + 2D_3 \\ A_{03} = 2D_1 + D_2 + 2D_4 \\ A_{04} = 2D_1 + D_2 \end{cases} \tag{B1}$$

$$\begin{cases} A_{0m} = \tau_1\alpha_2 A_{02} + \tau_1\alpha_1 A_{03} - \tau_1^2 A_{01} - \alpha_1\alpha_2 A_{04} \\ A_0 = K_{i3}A_{0m}/(2\tau_1(\tau_1 - \alpha_1)(\tau_1 - \alpha_2)) - \omega_1 LC_{g0} - kU_{s0} \\ A_1 = (K_{i3} - K_{p3}\tau_1)D_1/\tau_1 \\ A_2 = (K_{i3} - 2K_{p3}\tau_1)D_2/\tau_1 \\ A_3 = (K_{i3} - K_{p3}\tau_1 + K_{p3}\alpha_1)D_3/(\tau_1 - \alpha_1) \\ A_4 = (K_{i3} - K_{p3}\tau_1 + K_{p3}\alpha_2)D_4/(\tau_1 - \alpha_2) \end{cases} \tag{B2}$$

In Equations (B2), K_{p3} and K_{i3} are the proportion parameter and integration parameter of the grid-side converter d-axis current loop PI controller.

$$\begin{cases} B_{0m} = \tau_1\alpha_2 C_{g1} + \tau_1\alpha_1 C_{g2} - \alpha_1\alpha_2 C_{gm} \\ B_0 = K_{i4}B_{0m}/(\tau_1\alpha_1\alpha_2) + \omega_1 LD_0 \\ B_m = (K_{i4} - K_{p4}\tau_1)C_{gm}/\tau_1 \\ B_1 = (-K_{p4}\alpha_1 - K_{i4})C_{g1}/\alpha_1 \\ B_2 = (-K_{p4}\alpha_2 - K_{i4})C_{g2}/\alpha_2 \end{cases} \tag{B3}$$

In Equations (B3), K_{p4} and K_{i4} are the proportion parameter and integration parameter of the grid-side converter q-axis current loop PI controller.

Appendix C

Parameters and eigenvalues of Equation (3.92)

$$\begin{cases} \lambda_1 = \dfrac{R_{rc}L_s + R_sL_r + j\omega_1\sigma L_rL_s + j\omega\sigma L_rL_s}{\sigma L_rL_s} \\ \lambda_2 = \dfrac{R_sR_{rc} + j\omega_1 R_{rc}L_s + j\omega R_sL_r - \sigma L_rL_s\omega\omega_1}{\sigma L_rL_s} \end{cases} \tag{C1}$$

$$\begin{cases} s_1 = \dfrac{-\lambda_1 + \sqrt{\lambda_1^2 - 4\lambda_2}}{2} \\ s_2 = \dfrac{-\lambda_1 - \sqrt{\lambda_1^2 - 4\lambda_2}}{2} \end{cases} \tag{C2}$$

Appendix D

Parameters and eigenvalues of Equation (3.107)

$$\begin{cases} p_1 = \dfrac{R_s - j\omega_1\sigma L_s}{\sigma L_s} \\ p_2 = -\dfrac{R_s L_m}{\sigma L_r L_s} \\ p_3 = \dfrac{R_{rc} - j(2-s)\omega_1\sigma L_r}{\sigma L_r} \\ p_4 = -\dfrac{R_{rc} L_m}{\sigma L_r L_s} \end{cases} \tag{D1}$$

$$\begin{cases} s_3 = \dfrac{-(p_1 + p_4) + \sqrt{(p_1 + p_4)^2 - 4(p_1 p_4 - p_2 p_3)}}{2} \\ s_4 = \dfrac{-(p_1 + p_4) - \sqrt{(p_1 + p_4)^2 - 4(p_1 p_4 - p_2 p_3)}}{2} \end{cases} \tag{D2}$$

Appendix E

Parameters and eigenvalues of Equation (3.116)

$$\begin{cases} \lambda_1' = \dfrac{R_{rc}' L_s + R_s L_r + j\omega_1\sigma L_r L_s + j\omega\sigma L_r L_s}{\sigma L_r L_s} \\ \lambda_2' = \dfrac{R_s R_{rc}' + j\omega_1 R_{rc}' L_s + j\omega R_s L_r - \sigma L_r L_s \omega\omega_1}{\sigma L_r L_s} \end{cases} \tag{E1}$$

$$\begin{cases} s_1' = \dfrac{-\lambda_1' + \sqrt{\lambda_1^{2\prime} - 4\lambda_2'}}{2} \\ s_2' = \dfrac{-\lambda_1' - \sqrt{\lambda_1^{2\prime} - 4\lambda_2'}}{2} \end{cases} \tag{E2}$$

Index

Hierarchical Protection for Smart Grids, First Edition. Jing Ma and Zengping Wang.

t

W